Bioactive Polymer Materials with Antibacterial Properties

Bioactive Polymer Materials with Antibacterial Properties

Guest Editors

**Md. Amdadul Huq
Shahina Akter**

Basel • Beijing • Wuhan • Barcelona • Belgrade • Novi Sad • Cluj • Manchester

Guest Editors

Md. Amdadul Huq
Department of Life Science
Gachon University
Seongnam
Korea, South

Shahina Akter
Department of Food Science
and Biotechnology
Gachon University
Seongnam
Korea, South

Editorial Office
MDPI AG
Grosspeteranlage 5
4052 Basel, Switzerland

This is a reprint of the Special Issue, published open access by the journal *Polymers* (ISSN 2073-4360), freely accessible at: www.mdpi.com/journal/polymers/special_issues/Bioact_Mater_Antibact_Prop.

For citation purposes, cite each article independently as indicated on the article page online and using the guide below:

Lastname, A.A.; Lastname, B.B. Article Title. *Journal Name* **Year**, *Volume Number*, Page Range.

ISBN 978-3-7258-3694-9 (Hbk)
ISBN 978-3-7258-3693-2 (PDF)
https://doi.org/10.3390/books978-3-7258-3693-2

© 2025 by the authors. Articles in this book are Open Access and distributed under the Creative Commons Attribution (CC BY) license. The book as a whole is distributed by MDPI under the terms and conditions of the Creative Commons Attribution-NonCommercial-NoDerivs (CC BY-NC-ND) license (https://creativecommons.org/licenses/by-nc-nd/4.0/).

Contents

About the Editors vii

Md. Amdadul Huq and Shahina Akter
Bioactive Polymer Materials with Antibacterial Properties: An Editorial
Reprinted from: *Polymers* 2025, *17*, 394, https://doi.org/10.3390/polym17030394 1

Anh Quoc Le, Van Phu Dang, Ngoc Duy Nguyen, Chi Thuan Nguyen and Quoc Hien Nguyen
Antibacterial Activity against *Escherichia coli* and Cytotoxicity of Maillard Reaction Product of Chitosan and Glucosamine Prepared by Gamma Co-60 Ray Irradiation
Reprinted from: *Polymers* 2023, *15*, 4397, https://doi.org/10.3390/polym15224397 5

Yue Zhao, Xiaoyu Wang, Ruilian Qi and Huanxiang Yuan
Recent Advances of Natural-Polymer-Based Hydrogels for Wound Antibacterial Therapeutics
Reprinted from: *Polymers* 2023, *15*, 3305, https://doi.org/10.3390/polym15153305 17

Raja Venkatesan, Krishnapandi Alagumalai and Seong-Cheol Kim
Preparation and Antimicrobial Characterization of Poly(butylene adipate-*co*-terephthalate)/Kaolin Clay Biocomposites
Reprinted from: *Polymers* 2023, *15*, 1710, https://doi.org/10.3390/polym15071710 41

David Alejandro Aguilar-Perez, Cindy Maria Urbina-Mendez, Beatriz Maldonado-Gallegos, Omar de Jesus Castillo-Cruz, Fernando Javier Aguilar-Ayala and Martha Gabriela Chuc-Gamboa et al.
Mechanical Properties of Poly(Alkenoate) Cement Modified with Propolis as an Antiseptic
Reprinted from: *Polymers* 2023, *15*, 1676, https://doi.org/10.3390/polym15071676 54

Clarissa Martins Leal Schrekker, Yuri Clemente Andrade Sokolovicz, Maria Grazia Raucci, Claudio Alberto Martins Leal, Luigi Ambrosio and Mário Lettieri Teixeira et al.
Imidazolium Salts for *Candida* spp. Antibiofilm High-Density Polyethylene-Based Biomaterials
Reprinted from: *Polymers* 2023, *15*, 1259, https://doi.org/10.3390/polym15051259 69

Md. Amdadul Huq, Md. Ashrafudoulla, Md. Anowar Khasru Parvez, Sri Renukadevi Balusamy, Md. Mizanur Rahman and Ji Hyung Kim et al.
Chitosan-Coated Polymeric Silver and Gold Nanoparticles: Biosynthesis, Characterization and Potential Antibacterial Applications: A Review
Reprinted from: *Polymers* 2022, *14*, 5302, https://doi.org/10.3390/polym14235302 83

Pantu Kumar Roy, Sung-Hee Park, Min Gyu Song and Shin Young Park
Antimicrobial Efficacy of Quercetin against *Vibrio parahaemolyticus* Biofilm on Food Surfaces and Downregulation of Virulence Genes
Reprinted from: *Polymers* 2022, *14*, 3847, https://doi.org/10.3390/polym14183847 102

Gunawan Setia Prihandana, Tutik Sriani, Aisyah Dewi Muthi'ah, Siti Nurmaya Musa, Mohd Fadzil Jamaludin and Muslim Mahardika
Antibacterial Activity of Silver Nanoflake (SNF)-Blended Polysulfone Ultrafiltration Membrane
Reprinted from: *Polymers* 2022, *14*, 3600, https://doi.org/10.3390/polym14173600 117

Nobchulee Nuanaon, Sharad Bhatnagar, Tatsuya Motoike and Hideki Aoyagi
Light-Emitting-Diode-Assisted, Fungal-Pigment-Mediated Biosynthesis of Silver Nanoparticles and Their Antibacterial Activity
Reprinted from: *Polymers* 2022, *14*, 3140, https://doi.org/10.3390/polym14153140 133

Paulina Latko-Durałek, Michał Misiak, Monika Staniszewska, Karina Rosłoniec, Marta Grodzik and Robert P. Socha et al.
The Composites of Polyamide 12 and Metal Oxides with High Antimicrobial Activity
Reprinted from: *Polymers* **2022**, *14*, 3025, https://doi.org/10.3390/polym14153025 . . . . **152**

Ngoc-Thang Nguyen and Thi-Lan-Huong Vo
Fabrication of Silver Nanoparticles Using *Cordyline fruticosa* L. Leave Extract Endowing Silk Fibroin Modified Viscose Fabric with Durable Antibacterial Property
Reprinted from: *Polymers* **2022**, *14*, 2409, https://doi.org/10.3390/polym14122409 . . . . **173**

Zehra Edis, Samir Haj Bloukh, Hamed Abu Sara and Nur Izyan Wan Azelee
Antimicrobial Biomaterial on Sutures, Bandages and Face Masks with Potential for Infection Control
Reprinted from: *Polymers* **2022**, *14*, 1932, https://doi.org/10.3390/polym14101932 . . . . **194**

Kuen Hee Eom, Shuwei Li, Eun Gyeong Lee, Jae Ho Kim, Jung Rae Kim and Il Kim
Synthetic Polypeptides with Cationic Arginine Moieties Showing High Antimicrobial Activity in Similar Mineral Environments to Blood Plasma
Reprinted from: *Polymers* **2022**, *14*, 1868, https://doi.org/10.3390/polym14091868 . . . . **223**

Xiaoqing Wang, Sun-Young Lee, Shahina Akter and Md. Amdadul Huq
Probiotic-Mediated Biosynthesis of Silver Nanoparticles and Their Antibacterial Applications against Pathogenic Strains of *Escherichia coli* O157:H7
Reprinted from: *Polymers* **2022**, *14*, 1834, https://doi.org/10.3390/polym14091834 . . . . **236**

Peace Saviour Umoren, Doga Kavaz, Alexis Nzila, Saravanan Sankaran Sankaran and Saviour A. Umoren
Biogenic Synthesis and Characterization of Chitosan-CuO Nanocomposite and Evaluation of Antibacterial Activity against Gram-Positive and -Negative Bacteria
Reprinted from: *Polymers* **2022**, *14*, 1832, https://doi.org/10.3390/polym14091832 . . . . **254**

Fan Wang, Ronghan Wang, Yingjie Pan, Ming Du, Yong Zhao and Haiquan Liu
Gelatin/Chitosan Films Incorporated with Curcumin Based on Photodynamic Inactivation Technology for Antibacterial Food Packaging
Reprinted from: *Polymers* **2022**, *14*, 1600, https://doi.org/10.3390/polym14081600 . . . . **274**

Md. Amdadul Huq, Md. Ashrafudoulla, M. Mizanur Rahman, Sri Renukadevi Balusamy and Shahina Akter
Green Synthesis and Potential Antibacterial Applications of Bioactive Silver Nanoparticles: A Review
Reprinted from: *Polymers* **2022**, *14*, 742, https://doi.org/10.3390/polym14040742 . . . . **290**

About the Editors

Md. Amdadul Huq

Dr. Md. Amdadul Huq is currently working as an Assistant Professor in the Department of Life Sciences, College of BioNano Technology, Gachon University, South Korea. He received his PhD in Biotechnology from Kyung Hee University, South Korea, in 2015. His main interest is the development of novel antimicrobial and anticancer agents to control multidrug-resistant pathogenic microorganisms, as well as cancer cells.

Shahina Akter

Dr. Shahina Akter was an Assistant Professor in the Department of Food Science and Biotechnology, Gachon University, South Korea. She received her PhD in Plant Biotechnology from Hankyong National University, South Korea, in 2019. Her main research interest is the development of transgenic plants as well as novel antimicrobial agents to control multidrug-resistant pathogenic microorganisms.

Editorial

Bioactive Polymer Materials with Antibacterial Properties: An Editorial

Md. Amdadul Huq [1,*] and Shahina Akter [2]

1. Department of Food and Nutrition, College of Biotechnology and Natural Resource, Chung-Ang University, Anseong-si 17546, Republic of Korea
2. Department of Food Science and Biotechnology, Gachon University, Seongnam 13120, Republic of Korea; shahinabristy16@gmail.com
* Correspondence: amdadbge@gmail.com or amdadbge100@cau.ac.kr

Bioactive materials have a wide range of applications, and bioactive materials with antibacterial properties, in particular, have attracted significant medical interest [1]. Antibacterial bioactive materials can be obtained from nature or can be synthesized in different ways. They can be prepared via the simple combination of antibacterial substances with materials such as metals or polymers and applied through delivery strategies to kill pathogenic bacteria. The emergence of multidrug-resistant (MDR) bacteria due to the uncontrolled, immoderate, and multiple use of antibiotics and chemotherapeutics is a serious threat to the world population [2], as antibiotic-resistant microorganisms cause life-threatening diseases in humans. The development of novel, safe, and effective antibacterial agents is the decisive solution to this issue. In this context, bioactive polymer materials with antibacterial properties could be promising agents to control MDR bacteria [3]; therefore, this Special Issue focuses on such materials. It is an open forum where scientists/researchers may share their knowledge, investigations, and findings in this promising field.

This Special Issue contains a total of 17 articles. Among these 17 articles, 3 are comprehensive review articles, and the remaining 14 are research articles. One review article in this Special Issue provides an overview of the green synthesis of AgNPs using different biological resources; the various parameters essential for stable, easy synthesis and high yields; the antibacterial applications and mechanisms of biosynthesized AgNPs; and the prospects for their future development and potential antibacterial applications [4]. Another review article in this Special Issue describes the biosynthesis of chitosan-coated polymeric silver and gold nanoparticles, their characterization, and their potential antibacterial applications. The authors also highlight various mechanisms of the biosynthesized chitosan-coated polymeric silver and gold nanoparticles against pathogenic bacteria [3]. The third review article in this Special Issue focuses on recent advances in natural polymer-based hydrogels for antibacterial wound therapeutics [5].

Fourteen original articles on this research topic are devoted to the antimicrobial applications and mechanisms of various bioactive materials against pathogenic microorganisms. Wang et al. [6] describe the probiotic-mediated biosynthesis of silver nanoparticles (AgNPs) and discover their potent antimicrobial activity against pathogenic strains of *Escherichia coli* O157:H7. Their study suggests that the biosynthesized AgNPs could be used as an excellent new type of antimicrobial agent to control multidrug-resistant strains of *Escherichia coli* O157:H7 [6]. Umoren et al. [7] report the biosynthesis and characterization of chitosan–CuO nanocomposite and evaluate their antibacterial activity against various Gram-positive and Gram-negative bacteria. Their study demonstrates that the biogenic chitosan–CuO nanocomposite exhibits strong antimicrobial activity against the Gram-positive bacteria

Received: 20 December 2024
Accepted: 25 January 2025
Published: 1 February 2025

Citation: Huq, M.A.; Akter, S. Bioactive Polymer Materials with Antibacterial Properties: An Editorial. *Polymers* **2025**, *17*, 394. https://doi.org/10.3390/polym17030394

Copyright: © 2025 by the authors. Licensee MDPI, Basel, Switzerland. This article is an open access article distributed under the terms and conditions of the Creative Commons Attribution (CC BY) license (https://creativecommons.org/licenses/by/4.0/).

Bacillus licheniformis, *Bacillus cereus,* and *Micrococcus luteus* and the Gram-negative bacteria *Pseudomonas aeruginosa*, *Pseudomonas citronellolis*, *E. coli*, *Klebisiella* sp., *Bradyrhizobium japonicum*, and *Ralstonia pickettii*. Eom et al. [8] report that synthetic polypeptides with cationic arginine moieties show high antimicrobial activity against Gram-positive *B. subtilis* and Gram-negative *E. coli*. Wang and coworkers [9] conclude that the inclusion of curcumin in a biopolymer-based film transport system in combination with photodynamic activation represents a promising option for the preparation of food packaging films. Edis et al. [10] conduct research on antimicrobial biomaterials for sutures, bandages, and face masks with the potential for infection control. They conclude that the facile combination of *Aloe Vera Barbadensis* Miller (AV), trans-cinnamic acid (TCA), and iodine (I2) encapsulated in a polyvinylpyrrolidone (PVP) matrix seems a promising alternative to common antimicrobials. Nguyen and Vo [11] report the green synthesis of AgNPs using *Cordyline fruticosa* L. leaf extract as a reducing and capping agent and investigate their bactericidal effect against six pathogenic bacteria: *Escherichia coli*, *Pseudomonas aeruginosa*, *Salmonella enterica*, *Staphylococcus aureus*, *Bacillus cereus,* and *Enterococcus faecalis*. Their study demonstrates that the green synthesized AgNPs exhibit strong antimicrobial activity against the tested pathogenic bacteria. According to Latko-Durałek et al. [12], composites of polyamide 12 and metal oxides show high antimicrobial activity against *Escherichia coli*, *Candida albicans*, and *Herpes simplex* 1.

Nuanaon et al. [13] report the light-emitting-diode-assisted, fungal-pigment-mediated biosynthesis of silver nanoparticles, along with their antibacterial activity. Their study demonstrates that all LED-synthesized AgNPs exhibit antimicrobial potential against pathogenic *Escherichia coli* and *Staphylococcus aureus*. Prihandana et al. [14] investigate the antibacterial activity of a silver nanoflake (SNF)-blended polysulfone ultrafiltration membrane against *Escherichia coli* taken from river water. The SNFs show strong antimicrobial activity against the tested *Escherichia coli*. Roy et al. [15] evaluate the antimicrobial efficacy of quercetin against *Vibrio parahaemolyticus* biofilm on food surfaces and the downregulation of virulence genes. Their findings suggest that plant-derived quercetin should be used as an antimicrobial agent in the food industry to inhibit the establishment of *V. parahaemolyticus* biofilms. Schrekker and coworkers [16] describe the synthesis of high-density polyethylene–imidazolium salt (HDPE-IS) films and discover their potent antibiofilm activity against *Candida albicans*, *C. parapsilosis*, and *C. tropicalis*. Their study suggests that the synthesized HDPE-IS films demonstrate potential as biomaterials for the development of effective medical devices and tools that reduce the risk of fungal infections. Aguilar-Perez et al. [17] assess the effect of propolis on the antibacterial, mechanical, and adhesive properties of commercial poly(alkenoate) cement. They find that the modified cement shows high activity against *Streptococcus mutans*. Le et al. [18] develop antibacterial agents against *Escherichia coli* using the Maillard reaction product of chitosan and glucosamine prepared under gamma co-60 ray irradiation. Venkatesan et al. [19] report the synthesis and antimicrobial characterization of poly(butylene adipate-co-terephthalate)/kaolin clay biocomposites. Their study demonstrates that poly(butylene adipate-co-terephthalate)/kaolin clay biocomposites have great potential as food packaging materials due to their ability to decrease the growth of bacteria and improve the shelf life of packaged foods.

We hope that the reader will find useful references in this Special Issue for the development of novel, safe, and effective bioactive materials with antibacterial properties to control various drug-resistant pathogenic bacteria.

Author Contributions: Conceptualization, M.A.H.; writing—original draft preparation, M.A.H.; writing—review and editing, M.A.H. and S.A. All authors have read and agreed to the published version of the manuscript.

Funding: This research received no external funding.

Institutional Review Board Statement: Not applicable.

Informed Consent Statement: Not applicable.

Data Availability Statement: Not applicable.

Conflicts of Interest: The authors declare no conflicts of interest.

References

1. Huq, M.A.; Akter, S. Biosynthesis, Characterization and Antibacterial Application of Novel Silver Nanoparticles against Drug Resistant Pathogenic *Klebsiella pneumoniae* and *Salmonella* Enteritidis. *Molecules* **2021**, *26*, 5996. [CrossRef] [PubMed]
2. Huq, M.A. Biogenic silver nanoparticles synthesized by *Lysinibacillus xylanilyticus* MAHUQ-40 to control antibiotic-resistant human pathogens *Vibrio parahaemolyticus* and *Salmonella typhimurium*. *Front. Bioeng. Biotechnol.* **2020**, *8*, 1407. [CrossRef] [PubMed]
3. Huq, M.A.; Ashrafudoulla, M.; Parvez, M.A.K.; Balusamy, S.R.; Rahman, M.M.; Kim, J.H.; Akter, S. Chitosan-Coated Polymeric Silver and Gold Nanoparticles: Biosynthesis, Characterization and Potential Antibacterial Applications: A Review. *Polymers* **2022**, *14*, 5302. [CrossRef] [PubMed]
4. Huq, M.A.; Ashrafudoulla, M.; Rahman, M.M.; Balusamy, S.R.; Akter, S. Green Synthesis and Potential Antibacterial Applications of Bioactive Silver Nanoparticles: A Review. *Polymers* **2022**, *14*, 742. [CrossRef] [PubMed]
5. Zhao, Y.; Wang, X.; Qi, R.; Yuan, H. Recent Advances of Natural-Polymer-Based Hydrogels for Wound Antibacterial Therapeutics. *Polymers* **2023**, *15*, 3305. [CrossRef] [PubMed]
6. Wang, X.; Lee, S.-Y.; Akter, S.; Huq, M.A. Probiotic-Mediated Biosynthesis of Silver Nanoparticles and Their Antibacterial Applications against Pathogenic Strains of Escherichia coli O157:H7. *Polymers* **2022**, *14*, 1834. [CrossRef] [PubMed]
7. Umoren, P.S.; Kavaz, D.; Nzila, A.; Sankaran, S.S.; Umoren, S.A. Biogenic Synthesis and Characterization of Chitosan-CuO Nanocomposite and Evaluation of Antibacterial Activity against Gram-Positive and -Negative Bacteria. *Polymers* **2022**, *14*, 1832. [CrossRef] [PubMed]
8. Eom, K.H.; Li, S.; Lee, E.G.; Kim, J.H.; Kim, J.R.; Kim, I. Synthetic Polypeptides with Cationic Arginine Moieties Showing High Antimicrobial Activity in Similar Mineral Environments to Blood Plasma. *Polymers* **2022**, *14*, 1868. [CrossRef] [PubMed]
9. Wang, F.; Wang, R.; Pan, Y.; Du, M.; Zhao, Y.; Liu, H. Gelatin/Chitosan Films Incorporated with Curcumin Based on Photodynamic Inactivation Technology for Antibacterial Food Packaging. *Polymers* **2022**, *14*, 1600. [CrossRef]
10. Edis, Z.; Bloukh, S.H.; Sara, H.A.; Azelee, N.I.W. Antimicrobial Biomaterial on Sutures, Bandages and Face Masks with Potential for Infection Control. *Polymers* **2022**, *14*, 1932. [CrossRef] [PubMed]
11. Nguyen, N.-T.; Vo, T.-L.-H. Fabrication of Silver Nanoparticles Using *Cordyline fruticosa* L. Leave Extract Endowing Silk Fibroin Modified Viscose Fabric with Durable Antibacterial Property. *Polymers* **2022**, *14*, 2409. [CrossRef] [PubMed]
12. Latko-Durałek, P.; Misiak, M.; Staniszewska, M.; Rosłoniec, K.; Grodzik, M.; Socha, R.P.; Krzan, M.; Bażanów, B.; Pogorzelska, A.; Boczkowska, A. The Composites of Polyamide 12 and Metal Oxides with High Antimicrobial Activity. *Polymers* **2022**, *14*, 3025. [CrossRef] [PubMed]
13. Nuanaon, N.; Bhatnagar, S.; Motoike, T.; Aoyagi, H. Light-Emitting-Diode-Assisted, Fungal-Pigment-Mediated Biosynthesis of Silver Nanoparticles and Their Antibacterial Activity. *Polymers* **2022**, *14*, 3140. [CrossRef] [PubMed]
14. Prihandana, G.S.; Sriani, T.; Muthi'ah, A.D.; Musa, S.N.; Jamaludin, M.F.; Mahardika, M. Antibacterial Activity of Silver Nanoflake (SNF)-Blended Polysulfone Ultrafiltration Membrane. *Polymers* **2022**, *14*, 3600. [CrossRef] [PubMed]
15. Roy, P.K.; Park, S.-H.; Song, M.G.; Park, S.Y. Antimicrobial Efficacy of Quercetin against *Vibrio parahaemolyticus* Biofilm on Food Surfaces and Downregulation of Virulence Genes. *Polymers* **2022**, *14*, 3847. [CrossRef] [PubMed]
16. Martins Leal Schrekker, C.; Sokolovicz, Y.C.A.; Raucci, M.G.; Leal, C.A.M.; Ambrosio, L.; Lettieri Teixeira, M.; Meneghello Fuentefria, A.; Schrekker, H.S. Imidazolium Salts for *Candida* spp. Antibiofilm High-Density Polyethylene-Based Biomaterials. *Polymers* **2023**, *15*, 1259. [CrossRef]
17. Aguilar-Perez, D.A.; Urbina-Mendez, C.M.; Maldonado-Gallegos, B.; Castillo-Cruz, O.d.J.; Aguilar-Ayala, F.J.; Chuc-Gamboa, M.G.; Vargas-Coronado, R.F.; Cauich-Rodriguez, J.V. Mechanical Properties of Poly(Alkenoate) Cement Modified with Propolis as an Antiseptic. *Polymers* **2023**, *15*, 1676. [CrossRef] [PubMed]

18. Le, A.Q.; Dang, V.P.; Nguyen, N.D.; Nguyen, C.T.; Nguyen, Q.H. Antibacterial Activity against *Escherichia coli* and Cytotoxicity of Maillard Reaction Product of Chitosan and Glucosamine Prepared by Gamma Co-60 Ray Irradiation. *Polymers* **2023**, *15*, 4397. [CrossRef] [PubMed]
19. Venkatesan, R.; Alagumalai, K.; Kim, S.-C. Preparation and Antimicrobial Characterization of Poly(butylene adipate-co-terephthalate)/Kaolin Clay Biocomposites. *Polymers* **2023**, *15*, 1710. [CrossRef]

Disclaimer/Publisher's Note: The statements, opinions and data contained in all publications are solely those of the individual author(s) and contributor(s) and not of MDPI and/or the editor(s). MDPI and/or the editor(s) disclaim responsibility for any injury to people or property resulting from any ideas, methods, instructions or products referred to in the content.

Article

Antibacterial Activity against *Escherichia coli* and Cytotoxicity of Maillard Reaction Product of Chitosan and Glucosamine Prepared by Gamma Co-60 Ray Irradiation

Anh Quoc Le [1,2,3], Van Phu Dang [3], Ngoc Duy Nguyen [3], Chi Thuan Nguyen [3] and Quoc Hien Nguyen [4,*]

1. Faculty of Biology-Biotechnology, University of Science, Ho Chi Minh City 700000, Vietnam; anhquoc1704@gmail.com
2. Vietnam National University, Ho Chi Minh City 700000, Vietnam
3. Research and Development Center for Radiation Technology, Vietnam Atomic Energy Institute, Ho Chi Minh City 700000, Vietnam
4. Vietnam Atomic Energy Institute, Hanoi 100000, Vietnam
* Correspondence: hien7240238@yahoo.com; Tel.: +84-913-667-966

Citation: Le, A.Q.; Dang, V.P.; Nguyen, N.D.; Nguyen, C.T.; Nguyen, Q.H. Antibacterial Activity against *Escherichia coli* and Cytotoxicity of Maillard Reaction Product of Chitosan and Glucosamine Prepared by Gamma Co-60 Ray Irradiation. *Polymers* **2023**, *15*, 4397. https://doi.org/10.3390/polym15224397

Academic Editors: Shahina Akter and Md. Amdadul Huq

Received: 20 October 2023
Revised: 7 November 2023
Accepted: 10 November 2023
Published: 13 November 2023

Copyright: © 2023 by the authors. Licensee MDPI, Basel, Switzerland. This article is an open access article distributed under the terms and conditions of the Creative Commons Attribution (CC BY) license (https:// creativecommons.org/licenses/by/ 4.0/).

Abstract: In this study, the gamma ray-induced Maillard reaction method was carried out for chitosan (CTS) and glucosamine (GA) to improve the water solubility and antibacterial activity. The mixture solution of CTS and GA was exposed to gamma rays at a dose of 25 kGy and freeze-dried to obtain a Maillard reaction product (MRP) powder. The physicochemical and biological properties of the CTS-GA MRP powder were investigated. The CTS-GA MRP powder expressed good solubility at a concentration of 0.05 g/mL. In addition, the result of the antibacterial activity test against *Escherichia coli* revealed that the CTS-GA MRP powder exhibited highly antibacterial activity at pH 7; in particular, bacterial density was reduced by over 4 logs. Furthermore, the cytotoxicity test of the CTS-GA MRP powder on mouse fibroblast cells (L929) showed non-cytotoxicity with high cell viability (>90%) at concentrations of 0.1–1 mg/mL. Owing to the high antibacterial activity and low cytotoxicity, the water-soluble CTS-GA MRP powder can be used as a favorable natural preservative for food and cosmetics.

Keywords: chitosan; glucosamine; Maillard reaction; antibacterial effect; cytotoxicity

1. Introduction

Food safety is a topic that has never run out of steam. The World Health Organization (WHO) estimated that unsafe food caused 600 million cases of foodborne diseases and 420,000 deaths each year between 2007 and 2015 [1]. Furthermore, food waste and food loss are other challenges threatening food security. Every year, 1.3 billion tons, or one-third of total global food production, is lost or wasted, which is estimated at about USD 1 trillion [2]. Hence, developing a new preservative to prevent foodborne illness, as well as to reduce food waste, could save billions of US dollars every year for the world's economy. Because of the growing awareness and concern regarding food safety and the harm of chemical and synthetic preservatives, alternative strategies for food preservatives are required, and natural additives are arguably the most promising suggestion.

Chitosan (CTS), a polycationic biopolymer derived through the alkaline deacetylation of chitin, is mainly produced from shellfish-processing waste [3,4]. Besides some characteristics such as being biodegradable, non-allergenic, and nontoxic, CTS also exhibits versatile biological activities such as antioxidative, antimicrobial, and anticancer activity. Hence, CTS and its derivatives have gained much interest as a potential food preservative of natural origin [5]. Although the antibacterial mechanism of CTS is still inconclusive, three proposals receive the most consensus: (1) CTS interacts with macromolecules on the bacterial cell wall by electrostatic adsorption and alters its permeability; (2) CTS enters the inside of bacterial cells, binds to DNA, and causes the inhibition of RNA and protein synthesis; and (3) CTS forms a chelate with essential nutrients for cell growth [6]. In terms of safety, CTS derived from shrimp shells achieved GRAS (Generally Recognized as Safe) status from the Food and Agriculture Organization of the United Nations (US-FDA) in 2001 and thus broadened its application objects for food preservation purposes [7], including fruit and vegetables [8], seafood [9], meat, and meat products [10]. However, the application of CTS in many fields is still restricted because of its insolubility in water as well as a reduction in biological activities at neutral or basic pH [3]. CTS derivatives with good solubility in water can be easily applied in many fields. Therefore, there have been many efforts to improve the solubility and biological activities of CTS based on chemical or enzymatic modifications, in which chemical modifications are generally not preferred in food applications [4].

The Maillard reaction (MR) is usually known as a non-enzymatic browning reaction between the carbonyl groups of reducing ends in carbohydrates and the amino groups of amino acids, proteins, or any nitrogenous compounds by heating or irradiating [11]. The MR is a very complex reaction that occurs spontaneously during thermal food processing and produces a wide range of Maillard reaction products (MRPs), which effectively contribute to the flavor formation and antibacterial and antioxidant activities of foods [11,12]. It is generally agreed that there is a substantial amount of MRPs in the average human diet [13]. Therefore, this reaction is considered a friendly green method to improve the properties of CTS for food applications. In the last decade, numerous studies have investigated the effect of CTS–sugar MRPs to increase the nutritional qualities of food as well as to extend its shelf life. The heating-induced MRPs of CTS or its derivatives with saccharides or proteins have been reported to be promising preservative agents for many kinds of food, such as fish, seafood, meats, and noodles, due to their good solubility and bacterial activity. In fact, the MR not only incorporates hydrophilic groups such as the hydroxyl group of monosaccharide onto the CTS chain to improve its solubility but also preserves its global properties, as well as enhances some biological activities. Among the Maillard derivatives of CTS with various saccharides, the one of CTS with GA, which contains both a hydroxyl group and an active amino group, exhibits good solubility in a relatively wide pH range and excellent antibacterial activities [5,14]. Recently, it has been recorded that gamma-ray irradiation could induce Maillard reactions like heating [15,16]. More interestingly, the gamma ray-induced MR can take place rapidly at room temperature without forming any toxic byproducts, such as 5-hydroxymethylfurfural [17]. However, up to now, the number of publications on the preparation of gamma ray-induced MRPs of chitosan–glucosamine, especially for food applications, is still limited. In this study, the CTS-GA MRP solution was prepared using a gamma Co-60 ray irradiation method. The solution was freeze-dried to obtain CTS-GA MRP powder. Its solubility in water was determined. Furthermore, the antibacterial activity against *Escherichia coli* and cytotoxicity on mouse fibroblast cells (L929) of CTS-GA MRP powder were also investigated.

2. Materials and Methods

2.1. Materials

Chitosan from shrimp shell with an average molecular weight (Mw) of ~97 ± 5 kDa and a degree of deacetylation of ~90 ± 3% was supplied by Sun Eco Green Import Export Company Limited, Ho Chi Minh City, Vietnam. Glucosamine was purchased from Merck (Darmstadt, Germany). The *Escherichia coli* ATCC 51813 was provided by the Metabolic Biology Laboratory, University of Science, Ho Chi Minh City. The Mueller–Hinton medium and agar were purchased from Himedia Laboratories Private Limited, Thane, India. Other chemicals such as lactic acid, NaOH of analytical grade, and distilled water were used for all experiments.

2.2. Preparation of CTS-GA MRPs

The preparation of CTS-GA solutions was carried out according to the method described in our previous research [18]. Briefly, a solution of 2% (w/v) CTS in 1% (v/v) lactic acid was prepared. Similarly, a 2% solution of GA in distilled water was also prepared. Four volumes of 2% CTS solution were mixed with one volume of 2% GA solution to obtain the mixture of CTS (1.6%) and GA (0.4%). The CTS-GA mixture solution was exposed to γ-ray from a Co-60 source with a dose of 25 kGy at a dose rate of 1.3 kGy/h with a Gamma-cell 5000 (BRIT, Mumbai, India) to perform a Maillard reaction. Consequently, the irradiated solution was freeze-dried to obtain CTS-GA MRP powder. To confirm the formation of Maillard reaction product, UV-Vis spectra of CTS-GA MRP solution were obtained with a spectrophotometer (Jasco-V630, Tokyo, Japan) at wavelengths of 284 nm and 420 nm [19]. Furthermore, the free GA content in the solution before and after irradiation was determined using the high-performance liquid chromatography (HPLC) method according to AOAC 2012 (2005.01) on an Agilent 1200 series Infinity with a UV-Vis detector (Agilent, Santa Clara, CA, USA). The efficiency of the Maillard reaction was expressed as the ratio of reacted GA to the initial GA by Equation (1) [18]:

$$\text{Maillard reaction efficiency (\%)} = [(M_0 - M_t)/M_0] \times 100 \qquad (1)$$

where M_0 and M_t are the free GA content of the CTS-GA solution before and after irradiation, respectively.

2.3. Determination of Solubility in Water

Firstly, CTS-GA MRP powder was dissolved in 10 mL water for the final concentration from 0.01 to 0.05 g/mL by stirring for 5 h at room temperature. Afterward, the solutions were filtered through 0.45 µm filter paper, which was dried in a forced-air oven at 60 °C until constant weight. The solubility was calculated from the change in the filter paper's weight [14]. The preparation of CTS powder from 25 kGy irradiated CTS solution (iCTS) was performed similarly. The solubility in water of the samples was calculated with Equation (2):

$$\text{Solubility (\%)} = [1 - (M_{Ft} - M_{F0})/M_{sample}] \times 100 \qquad (2)$$

where M_{Ft} and M_{F0} are the weight of the filter after and before filtering; M_{sample} is the weight of sample that was dissolved in water.

2.4. Fourier-Transform Infrared (FTIR) Spectroscopy Analysis

FTIR spectra were recorded with an NIR MIR Frontier (Perkin Elmer, Waltham, MA, USA) by scanning along a spectrum range of 400–4000 cm^{-1}. The CTS or CTS-GA MRP powder was ground with KBr in a ratio of 1/100 (1 mg sample with 100 mg KBr) and then compressed to form discs [12].

2.5. Proton Nuclear Magnetic Resonance (^{1}H-NMR) Analysis

^{1}H-NMR analysis was performed on a Brucker Avance 500 MHz at 70 °C. The CTS or CTS-GA MRP powder was dissolved in D$_2$O solution to obtain a polymer concentration of 5 mg/mL, and measurements were carried out at 70 °C [12].

2.6. Detection of 5-Hydroxylmethylfurfural

The presence of 5-hydroxylmethylfurfural (5-HMF) in the CTS-GA MRP powder was detected using reverse-phase HPLC with the method described by Theobald et al. (1998) [20]. Chromatography analyses were carried out on an Agilent 1260 infinity HPLC system with a UV-Vis detector at 284 nm and a Restek Ultra Aqueous C$_{18}$ column (250 mm × 4.6 mm). The CTS-GA MRP powder was dissolved in water and filtered through a 0.45 µm PTFE membrane. The injection volume was 10 µL, and the run time was 20 min at room temperature. A mixture of acetonitrile/water (15/85, v/v) was used as a mobile phase with a flow rate of 1 mL/min. A 5-HMF solution (20 µg/mL) was used as a standard, and the CTS-GA solution heated at 65 °C for 5 days was used to compare with the MRP prepared with the gamma-ray irradiation method [3].

2.7. Evaluation of Antibacterial Activity

The antibacterial activity of the CTS-GA MRP was investigated against E. coli ATCC 51813 using the disk diffusion method and poisoned food method described by Balouiri et al. (2016), with some modifications [21]. In the disk diffusion test, a Mueller–Hinton agar plate was spread with E. coli (~10^4 CFU/mL). Afterward, filter paper discs (about 6 mm in diameter) coated with the powder of CTS, iCTS, and CTS-GA MRP were placed on the plate. A blank disc was also placed on the plate to serve as a control. Then, the plate was incubated overnight at 37 °C and colony formation was monitored.

In the poisoned food test, 0.04 g the powder of iCTS or CTS-GA MRP was added into a flask containing 99 mL of Mueller–Hinton broth medium at pH 7. Afterward, 1 mL of the bacterial suspension (~10^8 CFU/mL) was aseptically inoculated into each flask. And then, the flasks were shaken at 150 rpm for 4 h at room temperature. A flask containing only broth and bacterial suspension was established in parallel to serve as the control. After shaking, the survival cell density in each mixture was determined using the spread plate technique [21]. The antimicrobial activity of samples was expressed by the reduction in bacterial density (log CFU/mL) in the testing samples compared with the control.

2.8. Evaluation of Cytotoxicity

The cytotoxicity of the CTS-GA MRP was estimated using a 3-(4,5-dimethyl-2-thiazolyl)-2,5-diphenyl-2H-tetrazolium chloride (MTT) viability assay, as described by Zhang et al. (2016), with slight modifications [22]. The L939 cells were cultured in Dulbecco's Modified Eagle's Medium (DMEM) supplemented with 10% BSA (w/v) and 100 U/mL of antibiotic. Cells were grown in 96-well plates at 37 °C in a humidified atmosphere (5% CO_2) for a density of 1×10^4 cells/well. After 24 h, the cells were treated with a medium supplemented with the CTS-GA MRP for the final concentrations of 0.1–1.0 mg/mL. The cells treated with DMEM medium or DMSO 20% served as negative or positive controls, respectively. All samples were incubated for 24 h, and the culture of each well was replaced by a 100 μL aliquot of MTT solution (1 mg/mL) and incubated for 4 h. Then, the supernatant in each well was aspirated and replaced by a 100 μL aliquot of DMSO/ethanol (1/1) solution to solubilize formazan crystals. The absorbance at 540 nm (OD540) of the wells was measured using an ELISA plate reader EZ Read 400 (Biochrom, Cambourne, UK). The relative growth rate (RGR) of cells treated with the CTS-GA MRP in the tested concentration range was determined according to Equation (3) [22]:

$$\text{RGR (\%)} = (OD540_{\text{test sample}} / OD540_{\text{negative control}}) \times 100 \tag{3}$$

2.9. Statistical Analysis

Data of UV-Vis spectra measurements and cytotoxicity evaluation were expressed as the mean ± standard error (SE). One-way ANOVA was performed for each sample, including three replicates. Significant differences between means were determined with the Turkey test at a 0.05 probability level ($p < 0.05$).

3. Results

3.1. Preparation of the MRPs and UV-Vis Spectrophotometric Analyses

The visual colors of the CTS-GA solution before/after irradiation and CTS-GA MRP powder are presented in Figure 1. The change in color of the CTS-GA solution was easily observed after irradiation, namely, the color of the solution became browner. Moreover, this color change was also expressed by the results of the absorbance measurements in Table 1. The absorbance intensities at 284 and 420 nm were increased significantly after irradiation. The same results were also recorded in other studies where the protein–sugar solutions were treated by heating [4] or irradiating [10,16]. In the Maillard reaction, the intermediate stage products can be detected by UV absorbance at 284 nm, while absorbance at 420 nm is preferred for the final stage products [15,19]. Furthermore, upon irradiation, the free GA content in the solution decreased significantly from 4.05 to 0.84 ± 0.09 mg/mL, corresponding to a reaction efficiency of 79.28% (Table 1). Therefore, the results confirmed that MRPs were formed effectively with 25 kGy irradiation.

Figure 1. The visual color of the CTS-GA solution (**A**), 25 kGy irradiated CTS-GA solution (**B**), and CTS-GA MRP powder (**C**).

Table 1. Absorbance at 284 nm and 420 nm of the CTS-GA solution before and after irradiation at 25 kGy and reaction efficiency.

	Absorbance at 284 nm	Absorbance at 420 nm	Reaction Efficiency (%)
Before irradiation	0.5540 ± 0.0353 [a]	0.0868 ± 0.0088 [a]	-
After irradiation	2.9612 ± 0.0278 [b]	0.3695 ± 0.0314 [b]	79.28 ± 1.79

The mean values in the same column with different letters are significantly different ($p < 0.05$) according to the Tukey test.

3.2. Solubility of CTS-GA MRP Powder in Water

The solubility of the iCTS and CTS-GA MRP powders at various concentrations in water is shown in Figure 2. At a concentration of 0.01 g/mL, both powders were completely dissolved. Along with the increase in the concentration, the solubility of the iCTS powder gradually decreased, whereas the solubility of the CTS-GA MRP powder still remained at 100% up to 0.05 g/mL. This result indicated that the MR effectively improved the solubility of CTS. This can be attributed to the increase in the hydroxyl groups in the CTS chains and the decrease in hydrophobic intermolecular interactions that favor polymer aggregation. The same phenomena were also obtained in previous studies [3,23].

3.3. FTIR Spectroscopy

The FTIR spectra of CTS and CTS-GA MRP in the range from 4000 to 400 cm^{-1} are presented in Figure 3. The characteristic bands can be observed at 3473 cm^{-1} (O–H and N–H stretching), 2879 cm^{-1} (C–H stretching vibration), 1619 cm^{-1} (C=O stretching vibrations, amide I), 1600 cm^{-1} (amide II), 1414 cm^{-1} (C–H$_2$ stretching), and 1086 and 1040 cm^{-1} (C–O stretching) on the CTS pattern [24,25]. After irradiation, these bands of CTS were changed. In the spectrum of the CTS-GA MRP, the band at 1619 cm^{-1} shifted to 1602 cm^{-1}, suggesting that Schiff base (C=N double bond) was formed between the carbonyl groups of GA and the amino groups of CTS. In addition, the band at 1414 cm^{-1} shifted to 1410 cm^{-1} in the spectrum of the CTS-GA MRP, indicating an increase in the number of –CH$_2$ groups due to the introduction of GA into the CTS molecule. Similar results were also recorded in other studies [12,24]. Furthermore, the spectrum of the CTS-GA MRP also showed the bands at 1089 cm^{-1} and 1038 cm^{-1} assigned to the C–O stretching vibration. In the region of 1200–400 cm^{-1}, the variation in the bands in the CTS-GA MRP spectrum was more pronounced than that in CTS [12]. These changes demonstrated that the structure of CTS was modified by gamma ray-induced MR.

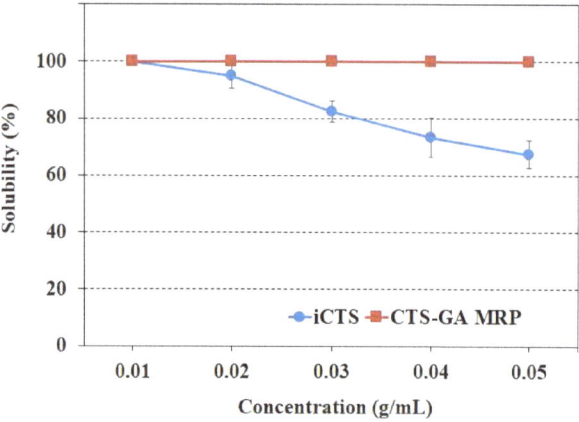

Figure 2. The solubility of iCTS and CTS-GA MRP powder in water.

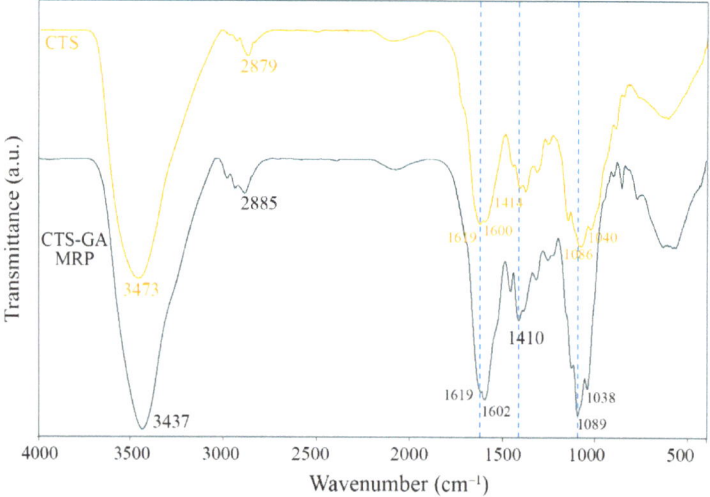

Figure 3. FTIR spectra of CTS and CTS-GA MRP.

3.4. ^{1}H-NMR Spectroscopy

^{1}H-NMR analysis was used to confirm the introduction of the GA unit into the CTS molecule. The ^{1}H-NMR spectra of CTS and CTS-GA MRP are displayed in Figure 4. The characteristic signals in the CTS spectrum were observed. The peak at δ 2.0 ppm is assigned to the methyl protons of N-acetyl glucosamine (H-Ac). The peak at δ 3.1 ppm belongs to the H-2 proton of the GA ring. The multiplet at δ 3.5–4.0 ppm is attributed to the H-3–H-6 protons of the pyranose ring. The signal at δ 4.4 ppm represents the H-1 proton of N-acetyl glucosamine (H-1A), and the other one at δ 4.7–4.9 ppm is the H-1 proton of GA (H-1D) that was overlapped by the strong signal at δ 4.8 ppm of the solvent (D$_2$O) [25,26].

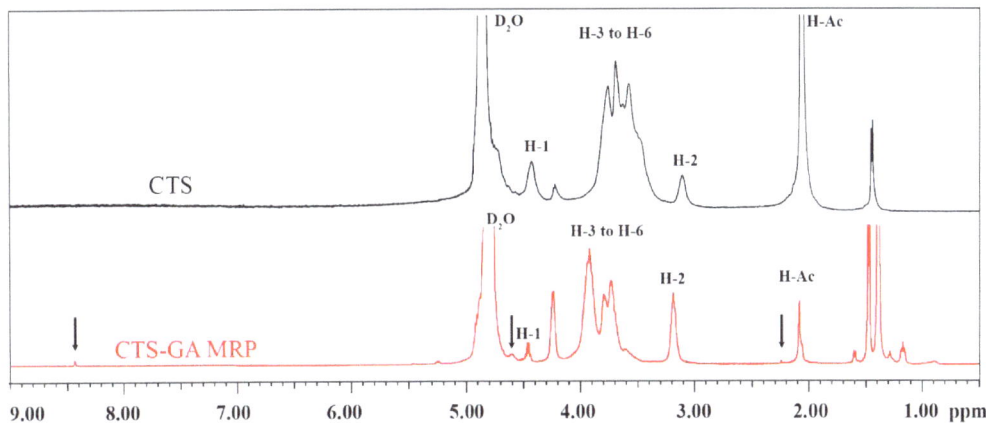

Figure 4. ^{1}H-NMR spectra of CTS and CTS-GA MRP.

By comparing the ^{1}H-NMR spectra, some differences between CTS and CTS-GA MRP were found. In particular, a new signal for the –N=CH– group (Schiff base), an intermediate product of MR, was observed at δ 8.45 ppm in the CTS-GA MRP spectrum. In addition, a new signal appeared at δ 2.25 ppm, corresponding to the –CH$_2$ of GA linked to the –NH$_2$ of CTS, indicating a displacement of the –N=CH– linkage (Schiff base) toward –NH–CH$_2$– [12,23]. A new signal at δ 1.6 ppm may be assigned to the proton of the alkyl group [25]. Furthermore, in the region of δ 4.5–4.6 ppm, some signals were also observed, indicating the N-substitution of the –NH$_2$ groups of CTS [12,25]. The new peaks in the region of δ 1.3–1 ppm in the CTS-GA MRP spectrum in Figure 4 are still unknown. Therefore, the changes in the ^{1}H-NMR spectrum in Figure 4 revealed the alteration in the structure of CTS by the binding of GA to the CTS backbone due to gamma ray-induced MR.

3.5. Detection of 5-Hydroxylmethylfurfural

5-Hydroxylmethylfurfural (5-HMF) is an intermediate product of the Maillard reaction [27]. It is cytotoxic at high concentrations and irritating to the eyes, upper respiratory tract, skin, and mucous membranes; the oral LD$_{50}$ value in rats was determined to be 3.1 g/kg body weight [28]. Under the chromatographic HPLC conditions described above, the retention time of the 5-HMF peak in the standard solution appeared at about 7.0 min (Figure 5). It was also observed in Figure 5 that a characteristic peak of 5-HMF appeared in the heat-induced CTS-GA MRP chromatogram which was not detected in the gamma ray-induced CTS-GA MRP chromatogram. This result indicated that the gamma-ray irradiation method could induce the MR between CTS and GA without forming 5-HMF. The same results were also reported in previous research [16,17]. Oh et al. (2005) [16] reported that no furfurals were detected in irradiated sugar–amino acid solution, whereas these compounds were found in heated solutions.

3.6. Antibacterial Activity

The results in Figure 6 indicated that both iCTS and CTS-GA MRP disks were able to form an inhibition zone on the *E. coli* plate, while the control and CTS disks were not. This result may be attributed to the solubility of the powder on paper disks. On the *E. coli* plate, paper disks absorbed moisture and dissolved the powders. Because of the poor solubility in water, CTS powder did not show an antibacterial activity. On the other hand, the iCTS and CTS-GA MRP powders were well soluble in water (see Figure 2), so they exhibited highly antibacterial activities. In addition, the antibacterial activity of these powders could be evaluated by the diameter of the inhibition zones [21].

Therefore, the antibacterial activity against *E. coli* of the CTS-GA MRP powder was estimated to be higher than that of the iCTS powder.

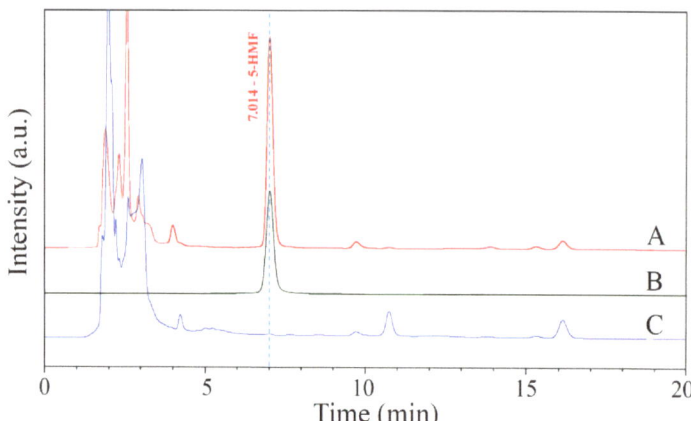

Figure 5. HPLC chromatograms of heat-induced CTS-GA MRP (A), 5-HMF standard (B), and gamma ray-induced CTS-GA MRP (C).

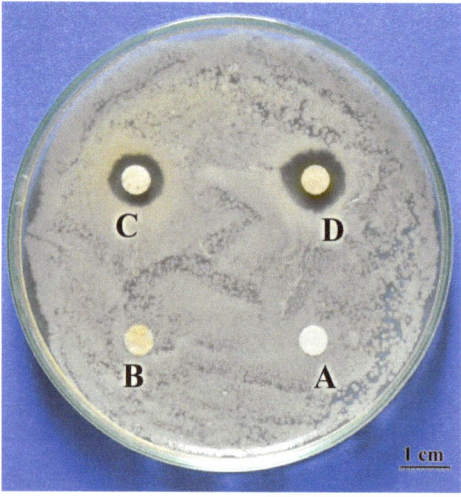

Figure 6. Photograph of the antibacterial activity assessed using the disk diffusion method. (A) Control; (B) CTS powder; (C) iCTS powder; and (D) CTS-GA MRP powder.

The results in Figure 7 indicated that after 4 h of shaking, the bacterial density in both agents of the iCTS and CTS-GA MRP was significantly decreased in comparison with that of the control. It has been shown that the lower the density of the viable bacteria, the higher the antibacterial activity. Furthermore, the results also revealed that the antibacterial activity of the CTS-GA MRP was higher than that of iCTS. Compared to the control, the CTS-GA MRP reduced the bacterial density by up to over 4 log (Figure 7B). This result is almost consistent with the results estimated with the disk diffusion test in Figure 6. The antibacterial ability of MRPs against *E. coli* was also reported by previous authors [3,14,17]. In the study of Rao et al. (2011) [17], the *E. coli* density after 24 h of treatment with a chitosan–glucose MRP was also decreased to 4 log CFU/mL. The obtained results revealed that the gamma ray-induced MR effectively improved the antibacterial activity of the CTS.

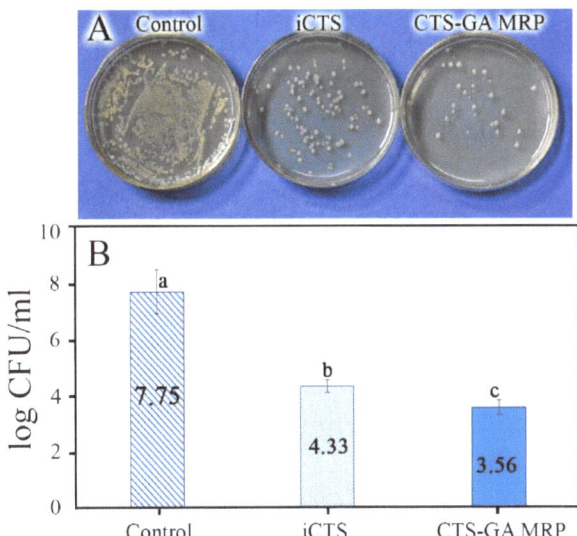

Figure 7. Viable bacterial colonies at the same dilution of 10^{-2} (**A**) and bacterial density (**B**) of the control, iCTS, and CTS-GA MRP samples. The mean values in (**B**) with different letters are significantly different ($p < 0.05$) according to the Tukey test.

3.7. Evaluation of Cytotoxicity

The effect of the CTS-GA MRP concentration on the cell viability of the L929 cell line after incubation for 24 h is presented in Table 2. In the positive control, DMSO significantly reduced the cell viability compared to that of the negative control, in which the RGR reached 36.36%. On the other hand, there was no significant difference in the relative cell viability among different tested concentrations of the CTS-GA MRP. Moreover, the RGR of the CTS-GA MRP sample was also significantly high (>91%). Several authors have evaluated the toxicity of a compound as nontoxic, weakly toxic, or toxic when the relative cell viability is >70%, between 50% and 70%, or <50%, respectively [29]. Therefore, the result in Table 2 indicated that the CTS-GA MRP was non-cytotoxic in the range from 0.1 to 1.0 mg/mL. This result was also further confirmed by the microscope images of L929 cells treated with different concentrations of the CTS-GA MRP in Figure 8, where the number and morphology of treated cells were almost unchanged in comparison with the control. The low cytotoxicity of the MRPs of CTS was also recorded in other studies [22,26]. These results demonstrated that the MRP prepared with the gamma-ray irradiation method did not cause cytotoxicity.

Table 2. Viability of L929 cells treated with CTS-GA MRP at different concentrations for 24 h.

Sample	Cell Viability (RGR %)
Negative control	100 ± 3.61 [a]
Positive control	31.36 ± 4.55 [b]
MRP CTS-GA 0.1 mg/mL	91.29 ± 4.58 [c]
MRP CTS-GA 0.25 mg/mL	92.86 ± 3.02 [c]
MRP CTS-GA 0.5 mg/mL	94.32 ± 2.69 [c]
MRP CTS-GA 1.0 mg/mL	96.36 ± 2.76 [c]

The mean values in the same column with different letters are significantly different ($p < 0.05$) according to the Tukey test.

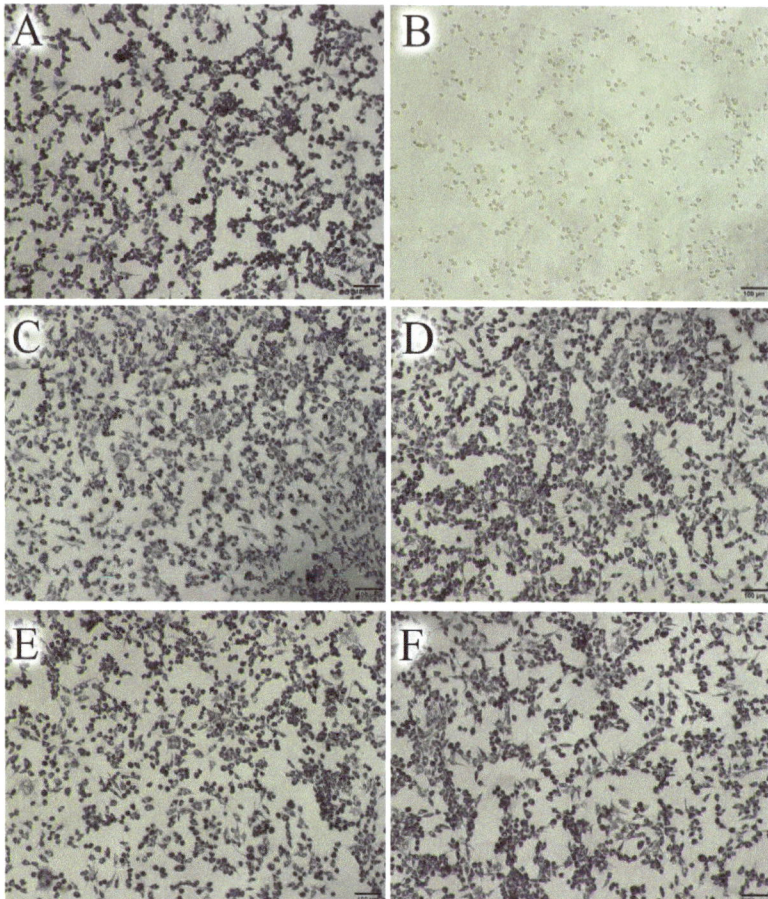

Figure 8. Microscopic images of L929 cells after 24 h of treatment with CTS-GA MRP at different concentrations. (**A**) Control (−); (**B**) positive control; and (**C–F**) CTS-GA MRP at 0.1, 0.25, 0.5, and 1.0 mg/mL, respectively.

4. Conclusions

In this study, a CTS-GA MRP solution was successfully prepared with the gamma ray-induced Maillard reaction at a dose of 25 kGy and freeze-dried to obtain a CTS-GA MRP powder. The as-prepared CTS-GA MRP powder exhibited good solubility in water and highly antibacterial activity against *E. coli* at a neutral pH. In addition, the obtained CTS-GA MRP was free from 5-HMF, an undesirable toxic byproduct. Furthermore, the CTS-GA MRP powder manifested non-cytotoxicity at a concentration of 0.1–1.0 mg/mL. Thus, the CTS-GA MRP can be used as a novel preservative for food and cosmetics.

Author Contributions: Conceptualization: A.Q.L. and Q.H.N.; methodology: V.P.D., C.T.N. and A.Q.L.; formal analysis: C.T.N. and N.D.N.; writing—original draft: A.Q.L. and V.P.D.; writing—review and editing: Q.H.N. and N.D.N. All authors have read and agreed to the published version of the manuscript.

Funding: This research received no external funding.

Institutional Review Board Statement: Not applicable.

Data Availability Statement: The data used to support the findings of this study are included in the article.

Acknowledgments: The authors would like to thank VINAGAMMA Center for supporting favorable conditions for performing this research and the University of Science—Vietnam National University-HCM for providing the cell line for the cytotoxicity evaluation.

Conflicts of Interest: The authors declare no conflict of interest.

References

1. World Health Organization [WHO]. *WHO Estimates of the Global Burden of Foodborne Diseases: Foodborne Disease Burden Epidemiology Reference Group 2007–2015*; World Health Organization: Geneva, Switzerland, 2015; pp. 3–4. ISBN 978-92-4-156516-5.
2. Gustavsson, J.; Cederberg, C.; Sonesson, U.; Van Otterdijk, R.; Meybeck, A. *Global Food Losses and Food Waste—Extent, Causes and Prevention*; FAO: Rome, Italy, 2011; ISBN 978-92-5-107205-9.
3. Chung, Y.C.; Kuo, C.L.; Chen, C.C. Preparation and important functional properties of water-soluble chitosan produced through Maillard reaction. *Bioresour. Technol.* **2005**, *96*, 1473–1482. [CrossRef]
4. Kanatt, S.R.; Chander, R.; Sharma, A. Chitosan glucose complex—A novel food preservative. *Food Chem.* **2008**, *106*, 521–528. [CrossRef]
5. Dutta, P.K.; Dutta, J.; Tripathi, V.S. Chitin and chitosan: Chemistry, properties and applications. *J. Sci. Ind. Res.* **2004**, *63*, 20–31.
6. Hosseinnejada, M.; Jafari, S. Evaluation of different factors affecting antimicrobial properties of chitosan. *Int. J. Biol. Macromol.* **2016**, *85*, 467–475. [CrossRef] [PubMed]
7. Mahae, N.; Chalat, C.; Muhamud, P. Antioxidant and antimicrobial properties of chitosan-sugar complex. *Int. Food Res. J.* **2011**, *18*, 1543–1551.
8. Ng, K.R.; Lyu, X.; Mark, R.; Chen, W.N. Antimicrobial and antioxidant activities of phenolic metabolites from flavonoid-producing yeast: Potential as natural food preservatives. *Food Chem.* **2019**, *270*, 123–129. [CrossRef]
9. Tsai, G.J.; Su, W.H.; Chen, H.C.; Pan, C.L. Antimicrobial activity of shrimp chitin and chitosan from different treatments and applications of fish preservation. *Fish. Sci.* **2002**, *68*, 170–177. [CrossRef]
10. Darmadji, P.; Izumimoto, M. Effect of chitosan in meat preservation. *Meat Sci.* **1994**, *38*, 243–254. [CrossRef]
11. Hafsa, J.; Smach, M.A.; Mrid, R.B.; Sobeh, M.; Majdoub, H.; Yasri, A. Functional properties of chitosan derivatives obtained through Maillard reaction: A novel promising food preservative. *Food Chem.* **2021**, *349*, 129072. [CrossRef] [PubMed]
12. Gullón, B.; Montenegroa, M.I.; Ruiz-Matuteb, A.I.; Cardelle-Cobasa, A.; Corzob, N.; Pintado, M.E. Synthesis, optimization and structural characterization of a chitosan-glucose derivative obtained by the Maillard reaction. *Carbohydr. Polym.* **2016**, *137*, 382–389. [CrossRef]
13. Liu, S.; Sun, H.; Ma, G.; Zhang, T.; Wang, L.; Pei, H.; Li, X.; Gao, L. Insights into flavor and key influencing factors of Maillard reaction products: A recent update. *Front. Nutr.* **2022**, *9*, 973677. [CrossRef] [PubMed]
14. Chung, Y.C.; Yeh, J.Y.; Tsai, C.F. Antibacterial characteristics and activity of water-soluble chitosan derivatives prepared by the Maillard reaction. *Molecules* **2011**, *16*, 8504–8514. [CrossRef]
15. Chawla, S.P.; Chander, R.; Sharma, A. Antioxidant formation by γ-irradiation of glucose-amino acid model systems. *Food Chem.* **2007**, *103*, 1297–1304. [CrossRef]
16. Oh, S.; Lee, Y.; Lee, J.; Kim, M.; Yook, H.; Byun, M. The effect of γ-irradiation on the non-enzymatic browning reaction in the aqueous model solutions. *Food Chem.* **2005**, *92*, 357–363. [CrossRef]
17. Rao, M.S.; Chawla, S.P.; Chander, R.; Sharma, A. Antioxidant potential of Maillard reaction products formed by irradiation of chitosan-glucose solution. *Carbohydr. Polym.* **2011**, *83*, 714–719. [CrossRef]
18. Quoc, L.A.; Phu, D.V.; Duy, N.N.; Thuan, N.C.; Khanh, C.N.; Hien, N.Q.; Nghiep, D.N. Study on the minimum bactericidal concentration (MBC) of Maillard reaction products of chitosan and glucosamine prepared by gamma-irradiation method. *Nucl. Sci. Technol.* **2021**, *11*, 44–51.
19. Chawla, S.P.; Chander, R.; Sharma, A. Antioxidant properties of Maillard reaction products by gamma-irradiation of whey protein. *Food Chem.* **2009**, *116*, 122–128. [CrossRef]
20. Theobald, A.; Müller, A.; Anklam, E. Determination of 5-hydroxymethylfurfural in vinegar samples by HPLC. *J. Agric. Food Chem.* **1998**, *46*, 1850–1854. [CrossRef]
21. Balouiri, M.; Sadiki, M.; Ibnsouda, S.K. Methods for in vitro evaluating antimicrobial activity: A review. *J. Pharm. Anal.* **2016**, *6*, 71–79. [CrossRef]
22. Zhang, H.; Zhang, Y.; Bao, E.; Zhao, Y. Preparation, characterization and toxicology properties of α- and β-chitosan Maillard reaction products nanoparticles. *Int. J. Biol. Macromol.* **2016**, *89*, 287–296. [CrossRef]
23. Braber, N.L.V.; Vergaraa, L.I.D.; Vieyrab, F.E.M.; Borsarellib, C.D.; Yossenc, M.M.; Vegac, J.V.; Corread, S.G.; Montenegroa, M.A. Physicochemical characterization of water-soluble chitosan derivatives with singlet oxygen quenching and antibacterial capabilities. *Int. J. Biol. Macromol.* **2017**, *102*, 200–207. [CrossRef]
24. Tran, T.N.; Doan, C.T.; Nguyen, V.B.; Nguyen, A.D.; Wang, S.L. Anti-oxidant and anti-diabetes potential of water-soluble chitosan-glucose derivatives produced by Maillard reaction. *Polymers* **2019**, *11*, 1714. [CrossRef]

25. Kraskouski, A.; Hileuskaya, K.; Nikalaichuk, V.; Ladutska, A.; Kabanava, V.; Yao, W.; You, L. Chitosan-based Maillard self-reaction products: Formation, characterization, antioxidant and antimicrobial potential. *Carbohydr. Polym. Technol. Appl.* **2022**, *4*, 100257. [CrossRef]
26. Badano, J.A.; Braber, N.V.; Rossi, Y.; Vergara, L.D.; Bohl, L.; Porporatto, C.; Falcone, R.D.; Montenegro, M. Physicochemical, in vitro antioxidant and cytotoxic properties of water-soluble chitosan-lactose derivatives. *Carbohydr. Polym.* **2019**, *224*, 115158. [CrossRef] [PubMed]
27. Ferrer, E.; Algría, A.; Farré, R.; Abellán, P.; Romero, F. High-performance liquid chromatographic determination of furfural compounds in infant formulas: Changes during heat treatment and storage. *J. Chromatogr. A* **2002**, *947*, 85–95. [CrossRef] [PubMed]
28. Ulbritch, R.J.; Northup, S.J.; Thomas, J.A. A review of 5-hydroxymethyl furfural (HMF) in parenteral solutions. *Fundam. Appl. Toxicol.* **1984**, *4*, 843–853. [CrossRef] [PubMed]
29. Abdillahi, H.S.; Verschaeve, L.; Finnie, J.F.; Van Staden, J. Mutagenicity, antimutagenicity and cytotoxicity evaluation of South African Podocarpus species. *J. Ethnopharmacol.* **2012**, *139*, 728–738. [CrossRef] [PubMed]

Disclaimer/Publisher's Note: The statements, opinions and data contained in all publications are solely those of the individual author(s) and contributor(s) and not of MDPI and/or the editor(s). MDPI and/or the editor(s) disclaim responsibility for any injury to people or property resulting from any ideas, methods, instructions or products referred to in the content.

Review

Recent Advances of Natural-Polymer-Based Hydrogels for Wound Antibacterial Therapeutics

Yue Zhao [1], Xiaoyu Wang [2], Ruilian Qi [1] and Huanxiang Yuan [1,*]

1. Department of Chemistry, College of Chemistry and Materials Engineering, Beijing Technology and Business University, Beijing 100048, China
2. School of Materials Science and Engineering, University of Science and Technology Beijing, Beijing 100083, China
* Correspondence: yhx@iccas.ac.cn

Abstract: Hydrogels have a three-dimensional network structure and high-water content, are similar in structure to the extracellular matrix, and are often used as wound dressings. Natural polymers have excellent biocompatibility and biodegradability and are commonly utilized to prepare hydrogels. Natural-polymer-based hydrogels can have excellent antibacterial and bioactive properties by loading antibacterial agents or being combined with therapeutics such as phototherapy, which has great advantages in the field of treatment of microbial infections. In the published reviews of hydrogels used in the treatment of infectious wounds, the common classification criteria of hydrogels include function, source of antibacterial properties, type of antibacterial agent, etc. However, there are few reviews on the classification of hydrogels based on raw materials, and the description of natural-polymer-based hydrogels is not comprehensive and detailed. In this paper, based on the principle of material classification, the characteristics of seven types of natural polymers that can be used to prepare hydrogels are discussed, respectively, and the application of natural-polymer-based hydrogels in the treatment of infectious wounds is described in detail. Finally, the research status, limitations, and prospects of natural-polymer-based hydrogels are briefly discussed.

Citation: Zhao, Y.; Wang, X.; Qi, R.; Yuan, H. Recent Advances of Natural-Polymer-Based Hydrogels for Wound Antibacterial Therapeutics. *Polymers* **2023**, *15*, 3305. https://doi.org/10.3390/polym15153305

Academic Editors: Md. Amdadul Huq and Shahina Akter

Received: 28 June 2023
Revised: 26 July 2023
Accepted: 3 August 2023
Published: 4 August 2023

Copyright: © 2023 by the authors. Licensee MDPI, Basel, Switzerland. This article is an open access article distributed under the terms and conditions of the Creative Commons Attribution (CC BY) license (https:// creativecommons.org/licenses/by/ 4.0/).

Keywords: hydrogel; natural polymer; antibacterial; wound healing

1. Introduction

The skin is the largest organ of the human body, which can protect the organism from external damage, but it is also prone to wounds [1,2]. According to the pathogenesis and effects of wounds, they can be divided into acute wounds and chronic wounds [3]. In particular, chronic wounds that are usually caused by endocrine diseases such as diabetes and neurological diseases have a profound impact on the quality of life as cardiovascular diseases and are particularly difficult to heal [4,5]. According to the Global Wound Care Market report released in 2016, the global wound care industry is expected to be worth 26.24 billion USD by 2023 [6]. Improper wound care is prone to microbial infection, causing long-term inflammation and making the wound difficult to heal [7]. More than 80 years ago, Alexander Fleming discovered penicillin, which made it possible to treat diseases caused by microbial infections [8]. However, the abuse of antibiotics has led to the gradual development of resistance in various pathogens [9], such as vancomycin-resistant *Enterococcus* (VRE) and multidrug-resistant *S. aureus* (MRSA) [10], making the wound more difficult to heal. Therefore, how to effectively treat chronic wounds caused by multidrug-resistant bacteria is a challenge.

As shown in Figure 1 [11], wound healing undergoes the following four processes: hemostasis, inflammation, proliferation, and remodeling, among which a variety of growth factors, enzymes, and cytokines play an important role in the synergistic regulation of cell activity [12]. Generally, acute wounds can heal normally, but chronic wounds are susceptible to various adverse factors such as microbial infection [13] and allergy [14], which makes the wound healing process more difficult to proceed smoothly. For example, during the repair process of a diabetic wound, due to long-term stimulation of hyperglycemia, ROS, and proteases (such as matrix metalloproteinases), the wound tissue will be affected by water loss, microbial infection, fibroblast migration, and reduced proliferation, and it will be difficult to heal [15,16].

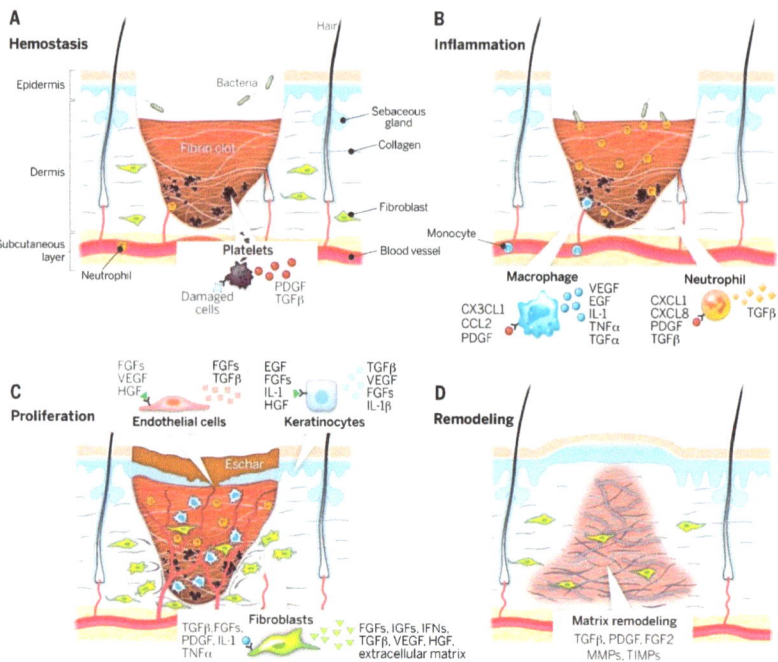

Figure 1. Four stages of wound healing: (**A**) hemostasis, (**B**) inflammation, (**C**) proliferation, (**D**) remodeling. Reprinted with permission from Ref. [11]. Copyright 2018 Elsevier Ltd.

In recent years, strategies based on bioactive materials such as electrospinning and hydrogels combined with phototherapy have been used to treat resistant bacterial infections and have achieved excellent therapeutic effects [17–19]. Particularly, hydrogels have a three-dimensional network structure and high-water content, similar in structure to extracellular matrix (ECM), which is a good wound dressing and can effectively heal wounds [20,21]. However, there are many defects in traditional hydrogels, such as poor mechanical properties, only a wet environment, and the inability to mimic natural tissue microstructure, which limit their application in wound treatment [19,22]. As a wound dressing, hydrogels should meet the following requirements [23–25]: (1) Good biocompatibility and non-cytotoxicity without causing wound allergy and inflammation; (2) Good moisture and moisture absorption which can maintain the wet environment of the wound to promote cell hydration and absorb the tissue exudate of the wound; (3) Good mechanical properties for perfectly attaching to the surface of the wound to form a barrier to prevent secondary infection by microorganisms; (4) Easy to peel without causing wound pain and secondary damage; (5) Good antibacterial, anti-inflammatory, antioxidant and hemostatic effects; (6) Induction of fibroblast adhesion and proliferation to promote wound tissue regeneration.

In recent years, the research on hydrogels has shown an increasing trend, and their functions have become more and more abundant [26–28]. For example, Liu et al. [29] and Fu et al. [30] achieved high antimicrobial efficiency by loading antibacterial agents into hydrogels or by combining them with phototherapy. Han et al. [31] used bionic mussels to achieve high viscosity and antibleeding to form a barrier to prevent microbial infection and achieve painless peel after adding strong oxidants without causing secondary damage to the skin. Shao et al. [32] effectively cleared ROS by introducing boron ester bonds to alleviate oxidative stress. Shen et al. [33] cross-linked hydrogels with dynamic acylhydrazone bonds to engender self-healing and luminescence properties of hydrogels, accompanied by loading growth factors to promote tissue regeneration. However, at present, hydrogels are difficult to achieve multifunctional integration and usually need to load various active substances, resulting in poor biocompatibility and low clinical application potential.

The antibacterial properties of hydrogels are related to the type of loaded antibacterial drugs, the antibacterial therapy, and the mode of drug release. Common loaded drugs include inorganic metals and antibiotics. Among all of the inorganic antibacterial agents, Ag is the most widely used because of its low biological toxicity and good anti-inflammatory and antibacterial properties [34]. Its antibacterial effect is affected by the size of the Ag-loaded hydrogel [35]. Antibiotics that are loaded into hydrogels can reduce drug dosage and improve bioavailability, but there is still a potential risk of drug resistance [10]. Hydrogels can also be used in synergy with photothermal therapy and photodynamic therapy to improve antibacterial activity [30,36]. Photothermal therapy refers to the antibacterial method of photothermal agents by generating thermal effects under the irradiation of near-infrared light [37]. Photodynamic therapy refers to the method of inducing bacterial death via producing ROS by photosensitizer under light of a specific wavelength [38]. The above two therapies combined with hydrogel have the advantages of being non-invasive, high selectivity, and negligible drug resistance, which has been widely used in antimicrobial infection [39–41]. Cross-linking methods of hydrogels, such as π-π stacking and hydrogen bonding, can realize the controlled release of drugs, thereby improving the antibacterial properties [42].

Synthetic polymers, such as polyethylene glycol, polyvinyl alcohol, and polyacrylamide, have good solubility and biocompatibility and are easy to modify, so they are often used in the preparation of hydrogels [43–45].

Natural polymers are macromolecular organic compounds that naturally exist in living organisms, such as polysaccharides and proteins [46]. Due to its excellent biocompatibility and biodegradability, it has been widely used in hydrogel wound dressings, with common materials including chitosan (CS), polydopamine (PDA), gelatin, agarose, sodium hyaluronate (HA), cellulose, and alginate [47–49]. Based on the classification principle of raw materials, the application of natural-polymer-based hydrogels in the treatment of infectious wounds was reviewed in this paper, which is summarized in Table 1. The study of natural macromolecular hydrogels is of great significance for the treatment of wounds, especially those infected by drug-resistant microorganisms. In addition, the research status, limitations, and prospects of natural-polymer-based hydrogels are briefly discussed.

Table 1. Summary of natural-polymer-based hydrogels for antimicrobial therapy.

Matrix Material	Superiority	Hydrogel	Modification	Highlight	Antimicrobial Activity	Mechanism	Reference
Chitosan (CS)	electropositive, effective killing of microorganisms by electrostatic interaction	BP/CS-bFGF hydrogel	carboxymethyl chitosan	Basic fibroblast growth factor was added to promote tissue regeneration.	S. aureus, E. coli > 67%	electrostatic interaction between carboxymethyl chitosan and bacteria	[50]
		CSG-PEG/DMA/Zn hydrogel	amidation reaction of polyethylene glycol monomethyl ether	be prepared by photoinitiated polymerization; multifunctional platform for antibacterial and anti-oxygen adhesion and hemostasis	S. aureus, E. coli, MRSA ≈100%	sustained release of the antimicrobial agent Zn^{2+}	[51]

Table 1. Cont.

Matrix Material	Superiority	Hydrogel	Modification	Highlight	Antimicrobial Activity	Mechanism	Reference
Polydopamine (PDA)	high viscosity; intrinsic photothermal properties	CAC/PDA/Cu(H$_2$O)$_2$ hydrogel	CuSO$_4$ and H$_2$O$_2$ accelerate the deposition rate of PDA	bionic mussels	S. aureus =99.87%; E. coli = 99.14%; MRSA = 99.25%	electrostatic interaction between carboxymethyl chitosan and bacteria; Sustained release of antimicrobial Cu^{2+}	[52]
		CG/PDA@Ag hydrogel	in situ grown Ag; evenly disperse in guar glue hydrogel	Combined with photothermal therapy, the photothermal conversion efficiency of PDA is improved.	S. aureus =99.8%; E. coli = 99.9%	electrostatic interaction between guar gum and bacteria; sustained release of antimicrobial agent Ag$^+$; photothermal effect of Ag@PDA	[53]
Gelatin	beneficial for fibroblast adhesion and growth	Gelatin-based hydrogel loaded with AgPOM	-	responsive to the acidic infectious environment	MRSA > 90%	photothermal effect of AgPOM; ^{1}O$_2$ formed by the reaction with H$_2$O$_2$	[54]
		GelMA-EGF/Gelatin-MPDA-LZM hydrogel	be amidated by methylacrylamide	photothermal and lysozyme synergistically remove biofilm and antibacterial activity	E. coli = 98.08%	photothermal effect of MPDA; lysozyme	[55]
Agarose	Stable thermally reversible hydrogels can be formed by physical cross-linking without the addition of cross-linking agents.	Gel1(Cyan)/Gel2(PCN) hydrogel	-	real-time monitoring; self-oxygenation enhances the photodynamic effect	S. aureus, E. coli, MRSA > 80%	photodynamic effects of oxygen-enhanced PCN-224	[56]
		CMA-Ag hydrogel	be amidated by carboxymethyl group	response to temperature and pH; The onset time of the inflammatory phase was earlier and the duration was shorter.	S. aureus, E. coli ≈ 100%	Sustained release of the antimicrobial agent Ag$^+$	[57]
Hyaluronic acid (HA)	promoting granulation tissue regeneration and re-epithelialization; reducing inflammatory cell infiltration	HA-PEGSB-CMP hydrogel	be amidated by adipyl dihydrazide	good mechanical properties, can be used for motion wounds	EC, MRSA ≈ 100%	photothermal properties of the cuttlefish melanin nanoparticles	[58]
		BSP-U/DAHA hydrogel	be amidated by aldehyde group	Sol-gel transition can be achieved in response to photothermal and pH.	S. aureus =91.68%; E. coli = 94.94%	photothermal properties of hydrogels formed by cross-linking of catechol and Fe^{3+} ligands	[59]
Cellulose	Modifications can be made without compromising the structural and mechanical properties.	RPC/PB hydrogel	-	pH response intelligently releases the drug	S. aureus =84.3%	antimicrobial properties of resveratrol	[60]
		BC/GG-Cu@ZIF/GOx hydrogel	-	response to glucose	S. aureus, E. coli ≈ 100%	nanozymes consume glucose and generate ·OH antimicrobial agents	[61]
Alginate	nonimmunogenic and nonthrombotic	TO/ASP hydrogel	be amidated by diacetone acrylamide	not affected by environmental pH and has good antibacterial activity in the range of pH 4–9	S. aureus =99.92%; E. coli = 99.993%	antimicrobial properties of antimicrobial peptides, thymol, and oligomeric tannic acids	[62]
		ALG-HPR hydrogel	-	NIR responsiveness; long-term release of the drug	S. aureus ≈ 100%	photothermal properties of indocyanine green; antimicrobial properties of rifampicin.	[63]

2. Hydrogels Based on Natural Polymer Matrix for Antibacterial Therapy of Wound Infections

2.1. Chitosan (CS)

Chitosan (CS), a biological polysaccharide extracted from natural chitin, carries a positive charge at physiological pH and is usually present in the shell of arthropods, crustaceans, and fungal cell walls [64,65]. CS has good biocompatibility, biodegradability, bioactivity, and adhesion [66,67]. In addition, CS contains a positive charge and can play a certain killing effect on microorganisms through electrostatic interaction [68]. However, its low solubility and poor mechanical properties limit its clinical application [69]. Hydrogels with good performance can be prepared by modifying the active hydroxyl and amino groups on CS, which can be used for efficient wound treatment [70,71].

Hao et al. [50] designed an injectable BP/CS-bFGF hydrogel to mediate associated cellular responses and promote full-layer diabetic wound healing (Figure 2). In this study, CS was modified to obtain carboxymethyl chitosan (CMCS) to enhance its water solubility and biocompatibility. Compared with straight-chain polyethylene glycol (PEG), branched-chain PEG has stronger mechanical properties and hemostatic properties. Therefore, 4-arm-PEG-CHO and CMCS were selected to form a covalently cross-linked network through Schiff base reaction, and BP/CS hydrogel was obtained. At the same time, the basic fibroblast growth factor (bFGF) is loaded into it to promote fibroblast proliferation, extracellular matrix fibrosis, and blood vessel formation. The swelling rate of BP/CS-bFGF hydrogel is as high as 132% at 37 °C, which can absorb a large amount of tissue exudate and maintain the humidity of the wound environment. The hydrogel has good self-healing ability, and after injection into irregular wounds located in joints, it can be rapidly shaped into defective shapes to enhance the healing of target tissues. Hydrogels have a good hemostatic effect due to their high adhesion and the presence of CMCS. Antibacterial experiments showed that the antibacterial rate of hydrogel for *E. coli* and *S. aureus* was more than 67%. This is because CMCS has a positive charge and can be adsorbed to the surface of bacteria through electrostatic interactions, inhibiting the replication of bacterial DNA. The hydrogel also has good blood compatibility and cell compatibility. Hydrogel was injected into the diabetic wounds of mice. The hydrogel quickly adhered to the wound and stopped bleeding, creating a good healing environment for the wound, while bFGF was released to promote tissue repair. After 14 days of treatment, the expressions of CD31, CD34, and Ki67 increased significantly, and the wound healing rate was 99%. In conclusion, BP/CS-bFGF hydrogel mimics the pathological environment of the extracellular matrix to effectively heal wounds by releasing the bioactive molecule bFGF and antibacterial properties. This work was the first to investigate the effect of hydrogels loaded with basic fibroblast growth factor (bFGF) on diabetic full-layer wound repair. It provides a reference for the treatment of diabetic chronic wounds. However, the antibacterial properties of the hydrogel come from the electrostatic force between CMCS and bacteria, which is far from enough to deal with serious infections, and it is necessary to load antibacterial agents or combine them with other therapies to improve the antibacterial activity.

Although hydrogels can promote wound repair by inducing tissue regeneration, there is still a risk of microbial infection if the desired antibacterial effect is not achieved, which can affect wound healing. Therefore, it is very important to develop a kind of hydrogel with an excellent antibacterial effect. Yang et al. [51] developed a highly antibacterial and antioxidant CSCG-PEG/DMA/Zn hydrogel by adding the antibacterial agent Zn^{2+}, which simultaneously promoted the regeneration of blood vessels and hair follicles and further promoted wound healing. CS was modified with PEG to obtain water-soluble and biocompatible CS-PEG, then the double bond was introduced by grafting glyceryl methacrylate to make it have the ability of photoinitiated polymerization, which was cross-linked with dopamine (DMA) through photoinduced free radical polymerization to form a double-network cross-linked hydrogel. At the same time, the antibacterial agent Zn^{2+} was loaded to improve the antibacterial properties. The double network structure makes the hydrogel have good mechanical properties, such as compressibility and resilience and can withstand large external forces. The degradation rate and swelling rate of the hydrogel can be adjusted by changing the content of DMA, which can absorb tissue exudate on demand while maintaining the humidity of the wound environment. The hydrogel has excellent antioxidant properties, and the DPPH clearance rate can reach more than 95%, which can effectively relieve the oxidative stress of wounds. Most importantly, after the co-culture of CSG-PEG/DMA/Zn hydrogels with bacteria at 37 °C for 2 h, the antibacterial rates of hydrogels containing 0.6% and 0.9% of DMA against *S. aureus*, MRSA, and *E. coli* reached 100%. Moreover, the slow release of Zn^{2+} makes the hydrogel possess long-term antibacterial properties. CSC-PEG/DMA6/Zn hydrogel was used to treat the wounds of MRSA-infected mice. On the 14th day, the wound closure rate was more than 95%, which was much higher than that of other control groups. CSG-PEG/DMA/Zn hydrogel has great potential in the treatment of infected wounds due to its excellent antimicrobial properties, biocompatibility, antibleeding and tissue adhesion. In this study, multifunctional hydrogels were prepared by photoinitiation polymerization, which provided a new idea for wound dressing research.

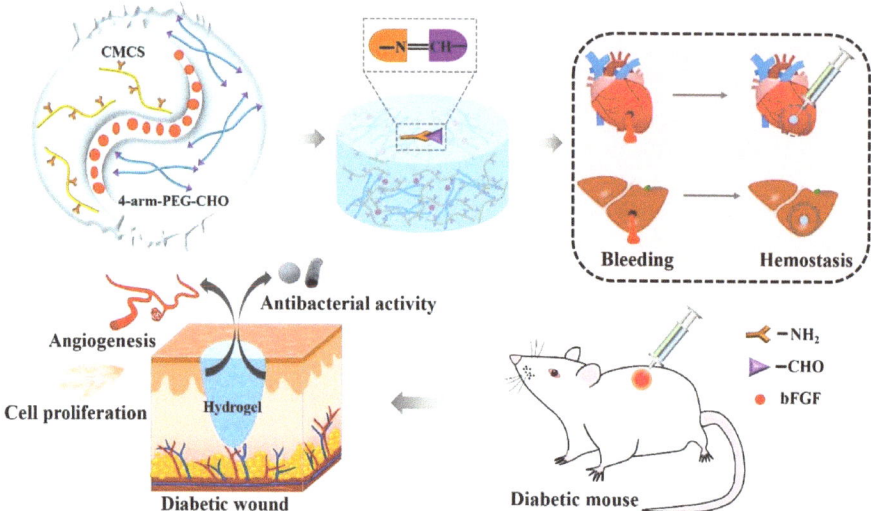

Figure 2. The illustrated preparation process of BP/CS-bFGF hydrogel and the application in promoting diabetic wound healing in mice. Reprinted with permission from Ref. [50]. Copyright 2022 Elsevier Ltd.

2.2. Polydopamine

Polydopamine (PDA) is an important natural melanin analog produced by dopamine autoxidation, which is easy to prepare and has good photostability, biodegradability, and biocompatibility [72–74]. The dynamic bonds based on catechol groups in PDA are self-healing and can be easily broken, making the hydrogel tough and stretchable [75]. PDA can mimic mussels to produce highly viscous hydrogels. Based on this property, Di et al. [76] designed a highly viscous, reusable, and heat-reversible hydrogel for the preparation of wound dressings, electronic skin, and wearable devices. In addition, PDA also has excellent photothermal properties, which can destroy the structure and function of cancer cells and microorganisms through photothermal therapy (PTT) to effectively treat cancer and microbial infection [77,78]. Therefore, PDA-based hydrogels can be loaded with antibacterial agents or combined with PTT to achieve high antibacterial efficiency. However, when PDA is used as a photothermal agent (PTA), there are two defects that limit its application. First, the photothermal conversion efficiency (PCE) is relatively low (about 20%) under the irradiation of an 808 nm near-infrared (NIR) laser, and PCE decreases with the extension of wavelength. Second, there are a large number of active groups on the surface of PDA, such as catechol, primary amine, and secondary amine groups, which make it easy to aggregate, reduce light absorption efficiency, and thus reduce PCE [79]. Therefore, when PDA is used as PTA, how to improve the PCE of PDA and prevent its aggregation is the key to achieving excellent antibacterial properties of photothermal hydrogels.

Liu et al. [52] developed a multifunctional hydrogel CAC/PDA/Cu (H_2O_2) consisting of CMC, alginate, and PDA with antibacterial, anti-inflammatory, and excellent mechanical properties by loading bionic mussel. PDA has strong metal ion trapping properties so that Cu^{2+} can be quickly and evenly distributed on the porous hydrogel surface. The hydrogel has excellent mechanical capacity, and its elongation at break (77.93 ± 4.02%) is greater than that of natural skin (70 ± 5%), which can ensure integrity and mechanical strength when applied to the skin surface. The interconnected high-porosity structure of hydrogels can absorb tissue exudate, maintain wound humidity, and ensure the transport of nutrients and the exchange of biomolecules within cells. PDA/Cu (H_2O_2) coating can induce fibroblasts to migrate to the vicinity of the wound, promote angiogenesis, and thus accelerate tissue repair. The hydrogel can also inhibit inflammation, clear ROS, and improve the wound microenvironment. In vitro antibacterial experiments showed that hydrogels had strong antibacterial activities against *S. aureus* (99.87%), *E. coli* (99.14%), and MRSA (99.25%). In addition, the bactericidal rate of hydrogel against MRSA in vivo was almost 100%, and the wound healing rate was 97.86% on the 14th day. This is mainly because both CMC and Cu^{2+} have antibacterial properties and can enhance the antibacterial effect through synergistic action. In summary, the hydrogel has high anti-inflammatory, antibacterial, and tissue repair-inducing abilities, and its excellent biocompatibility gives it clinical potential for treating resistant bacteria-infected wounds.

Although PDA-based hydrogels containing antibacterial agents can achieve antibacterial purposes, the preparation process is complicated and usually requires a variety of substances to occur cross-linking reactions. The method combined with PTT only requires a simple modification to achieve an excellent antibacterial effect and can prevent temperature during PTT from being too high to cause damage to normal tissue. Qi et al. [53] designed an efficient photothermal treatment nano-platform (CG/PDA@Ag hydrogel) for efficient photothermal antibacterial therapy and promoting wound healing (Figure 3). The charge transfer efficiency and non-radiative transition of PDA@Ag nanoparticles are improved by using Ag for the modification of PDA, and the PCE is increased from 16.6% (PDA) to 36.1%. PDA@Ag nanoparticles were uniformly distributed in CG/PDA@Ag hydrogel without aggregation, and PCE was further increased to 38.2%. The existence of non-covalent bonds in hydrogels, such as hydrogen bonds, π-π stacking, and electrostatic force, make them have good mechanical properties and self-repair capability, which is conducive to injection and prolongation of use time. Cationic guar gum (CG) contains a large number of hydroxyl and quaternary ammonium groups, which are positively charged and can trap negatively charged bacteria through electrostatic interaction and hydrophobic interaction to play a bacteria-killing role. Ag^+ can be released continuously for up to 6 days, giving the hydrogel a long-term bactericidal effect. After 3 min of NIR irradiation (808 nm, 1 W·cm^{-2}), the temperature of CG/PDA@Ag hydrogel could be elevated to 67.3 °C, and the inhibition rates against *E. coli* and *S. aureus* were 99.9% and 99.8%, respectively. Under the combined treatment of CG, Ag^+, and light irradiation, the cell membranes of these two strains of bacteria were severely damaged. Consistent with the results of in vitro experiments, the hydrogel can also play an effective antibacterial role in vivo with good biocompatibility, which can promote rapid wound healing. In addition, organisms can absorb CG/PDA@Ag hydrogels without additional removal. In short, CG/PDA@Ag hydrogel overcomes the shortcomings of PDA and has efficient PCE and strong bacteria trapping/killing ability to effectively promote wound healing in a short time. This research has made a breakthrough in the modification of PDA photothermal agents and realized practical application.

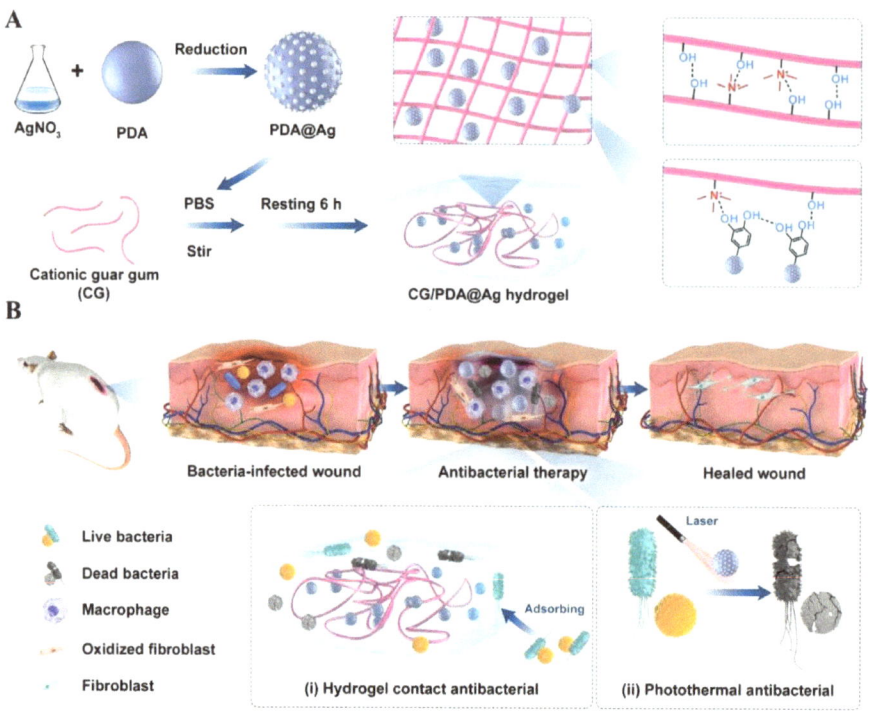

Figure 3. Schematic diagram of (**A**) the hydrogel synthesis mechanism and (**B**) photothermal/chemodynamic synergistic treatment of MRSA-infected wound. Reprinted with permission from Ref. [53]. Copyright 2022 the authors.

2.3. Gelatin

Gelatin is a natural polymer with a collagen sequence, in which the amino acid sequence RGD (Arg/Gly/Asp) can promote cell adhesion and is conducive to cell growth [80]. Due to its good biocompatibility, biodegradability, fibroblast adhesion, and proliferation characteristics, it has great potential in biomedical and tissue engineering applications [81,82]. However, the poor mechanical property, easy fracture, and rapid degradation of gelatin in the dry state limit its application as a hydrogel dressing [83,84]. Recent reports found that hydrogels prepared by using polysaccharide derivatives containing aldehyde group as cross-linking agents and Schiff base reaction between an aldehyde group and amino group of gelatins have both good mechanical properties and little biological toxicity [85,86].

Huang et al. [54] combined hydrogels with PTT and chemodynamic therapy (CDT) to design a bacterial environment-responsive multifunctional injectable hydrogel for treating bacterial infections and promoting wound healing (Figure 4). The natural component tea polyphenols (TPs) were used as cross-linking agents to form a highly cross-linked network with gelatin polymer. At the same time, urea was added to make the hydrogel injectable and have good mechanical properties. Silver polyoxometalate (AgPOM) that is doped in the hydrogel is protonated and aggregated in an acidic environment to exhibit good photothermal performance under NIR-II irradiation. In addition, Mo^{5+} in AgPOM reacts with overexpressed H_2O_2 in the wound to produce $^1O_2$, which achieves a high antibacterial effect in synergy with PTT. The hydrogel containing AgPOM can react with additional H_2O_2 to produce $^1O_2$, and combined with the photothermal effect of AgPOM under NIR-II light irradiation for 10 min (1060 nm, 1 W·cm^{-2}), the antibacterial rate of the hydrogel against

MRSA both in vivo and in vitro can reach more than 90%. Importantly, the good adhesion of the hydrogel also makes it have long-term antibacterial activity. Due to the excellent biocompatibility and antibacterial properties of the hydrogel, a large amount of collagen was deposited on the wound of the mice after 14 days of treatment, and the wound healed almost completely. In conclusion, the hydrogels prepared in this study combined PTT and CDT and opened up new thinking for the development of multifunctional hydrogels.

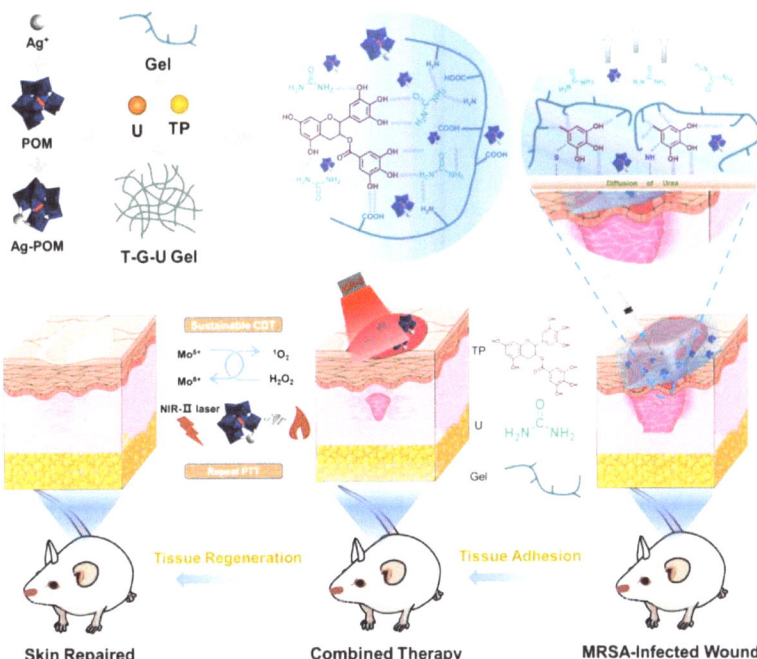

Figure 4. Illustration of the hydrogel synthesis mechanism and photothermal/chemodynamic synergistic treatment of MRSA-infected wound. Reprinted with permission from Ref. [54]. Copyright 2023 American Chemical Society.

When microorganisms infect wounds, they tend to integrate communities and embed themselves in extracellular polymeric substances (EPS), forming a dense structure called biofilm [87]. As a barrier, EPS can resist physical and chemical damage, so ordinary methods such as antibiotics cannot effectively eliminate biofilm, resulting in long-term microbial infection and difficult wound healing [88,89]. Eliminating biofilms is, therefore, extremely important for treating wounds. Wang et al. [55] designed a photothermal bilayer GelMA-EGF/Gelatin-MPDA-LZM hydrogel dressing for efficient biofilm elimination and comprehensive treatment of chronic wounds. The outer layer of hydrogel was photo cross-linked to obtain GelMA with good mechanical properties for the adhesion and proliferation of fibroblasts. Meanwhile, EGF was added to GelMA to promote tissue regeneration and wound re-epithelialization. The mesoporous polydopamine (MPDA) nanoparticles with high PCE were loaded with lysozyme (LZM) to form the inner hydrogel. After 10 min of NIR light irradiation (808 nm, 1 W·cm^{-2}), the photothermal effect of MPDA can destroy and eliminate the biofilm structure, making LZM penetrate into the biofilm and kill bacteria with an antibacterial rate in vivo as high as 98.08%, and the wound temperature only increased to 46 °C which cannot induce damage to the tissue around the wound. SEM observation showed that after photothermal treatment, there was almost no biofilm structure in the wound tissue, and the bacterial morphology was destroyed. After 12 days of treatment, the wound infected with *E. coli* remained at only 7.9% of the area. This bilayer hydrogel can effectively eliminate the biofilm and kill bacteria under the cooperation of photothermal therapy and has a broad application prospect in promoting the repair of infectious chronic wounds.

2.4. Agarose

Agarose is a naturally water-soluble linear polysaccharide found in seaweed that can be dissolved in neutral or alkaline solutions [90]. It has excellent biocompatibility, gelling properties, and physicochemical characteristics, so it is often used as a biomaterial for cell growth and local drug delivery [91,92], but it is not widely used in antibacterial hydrogels. The agarose can form a stable thermo-reversible gel by physical cross-linking without additional cross-linking agents, and the gel temperature is about 25 °C, which is suitable for the preparation of wound dressing [93,94]. However, the low chemical complexity of agarose limits its direct application in fine chemical and biological reactions, and it can be modified by biological, physical, and chemical methods [91,95].

Damaged blood vessels and high metabolic demand can lead to a severe shortage of oxygen supply to diabetic chronic wounds, resulting in more difficult wound healing [96]. Among the existing methods, hyperbaric oxygen therapy and local oxygen therapy have little effect. Although local dissolved oxygen therapy can improve oxygen penetration and absorption, it is dependent on hyperbaric oxygen generators and has a high risk of wound infection [97]. In addition, the high-sugar microenvironment is also conducive to microbial growth, resulting in chronic inflammation and necrosis of capillaries and nerve endings [56]. Therefore, the provision of oxygen to the wound, as well as real-time monitoring and timely sterilization, is essential. Zhu et al. [56] designed a multifunctional double-layer hydrogel that could monitor wound conditions and continuously generate oxygen to enhance the antibacterial effect of PDT (Figure 5). First, sodium alginate/carboxymethyl chitosan is combined with photosensitizer (PCN-224) and pH indicator (bromothymol blue BTB) through Schiff base reaction to obtain the inner hydrogel Gel2. Under white light, PPCN-224 produces ROS, which destroys bacterial cell walls and effectively kills bacteria. Then, the hydrogel obtained by cross-linking agarose and CMCS is loaded with cyanobacteria to obtain the outer hydrogel Gel1. When an infection occurs in the wound, the pH drops, and the color of bromothymol blue (BTB) in the inner Gel2 gradually changes from gray-blue to yellowish-green, indicating the occurrence of a bacterial infection. At the same time, in an acidic environment, the Schiff base bond breaks, and PCN-224 is released, avoiding the loss of bactericide due to premature release. Under natural light conditions, the cyanobacteria in the outer Gel1 can continuously produce oxygen to relieve tissue hypoxia and enhance the antibacterial efficiency of PDT. Adequate oxygen in the environment also has other advantages, such as accelerating cell migration, alleviating inflammation, promoting skin capillary formation, and wound tissue recovery. The self-oxygenated double-layer hydrogel can synergistically treat the refractory wounds infected by anaerobic bacteria and repair the tissue in time through infection monitoring. Chronically infected wounds in mice were largely completely healed after 21 days of treatment with Gel1 (Cyan)/Gel2 (PCN) hydrogel under conditions of white light irradiation. However, it can be seen from the staining results of live/dead bacteria that a certain number of bacteria still survive after 5 min of illumination at 606 nm with the fluence of 200 mW·cm^{-2}. Thus, the antibacterial effect of this method still needs to be strengthened. This study solves the problem of hypoxic infection of chronic wounds and provides insight into real-time monitoring and intelligent treatment of wounds.

Huang et al. [57] designed a macroporous CMA-Ag hydrogel for accelerating wound healing with antibacterial and anti-inflammatory functions (Figure 6). Agarose was modified by introducing carboxymethyl to obtain carboxymethyl agarose (CMA), which improved the surface property of hydrogel and promoted cell proliferation. The carboxymethyl group on CMA deprotonates and interacts with Ag$^+$, and the hydroxyl group also coordinates with Ag$^+$ to obtain CMA-Ag hydrogel. The network formed by CMA and Ag$^+$ coordination makes the porous structure of hydrogel denser with a larger swelling rate, and the tissue exudate can be absorbed more quickly. Low pH will destroy the ionic interaction between hydrogels, and higher temperatures will break the coordination. Therefore, the Ag$^+$ release of the hydrogel is both pH- and temperature-responsive, and its release decreases with the increase in pH during wound healing and increases with the elevation of temperature. A continuous release can be achieved under body temperature conditions. The antibacterial performance of CMA-Ag hydrogel on *E. coli* and *S. aureus* increased with the increase in Ag$^+$ content, and the antibacterial rate could reach as high as 100%. At the same time, in vivo experiments, the hydrogel also has a strong antibacterial activity (100%) against *S. aureus*. The *S. aureus* infected wound healed completely after 14 days of treatment with CMA-Ag hydrogel. CMA-Ag hydrogel can accelerate wound healing by relieving inflammation and promoting tissue regeneration and collagen deposition in the wound. The hydrogel prepared in this study is pH- and temperature-responsive, which can promote the early occurrence and end of inflammation and provide a certain reference for the rapid healing of infected wounds.

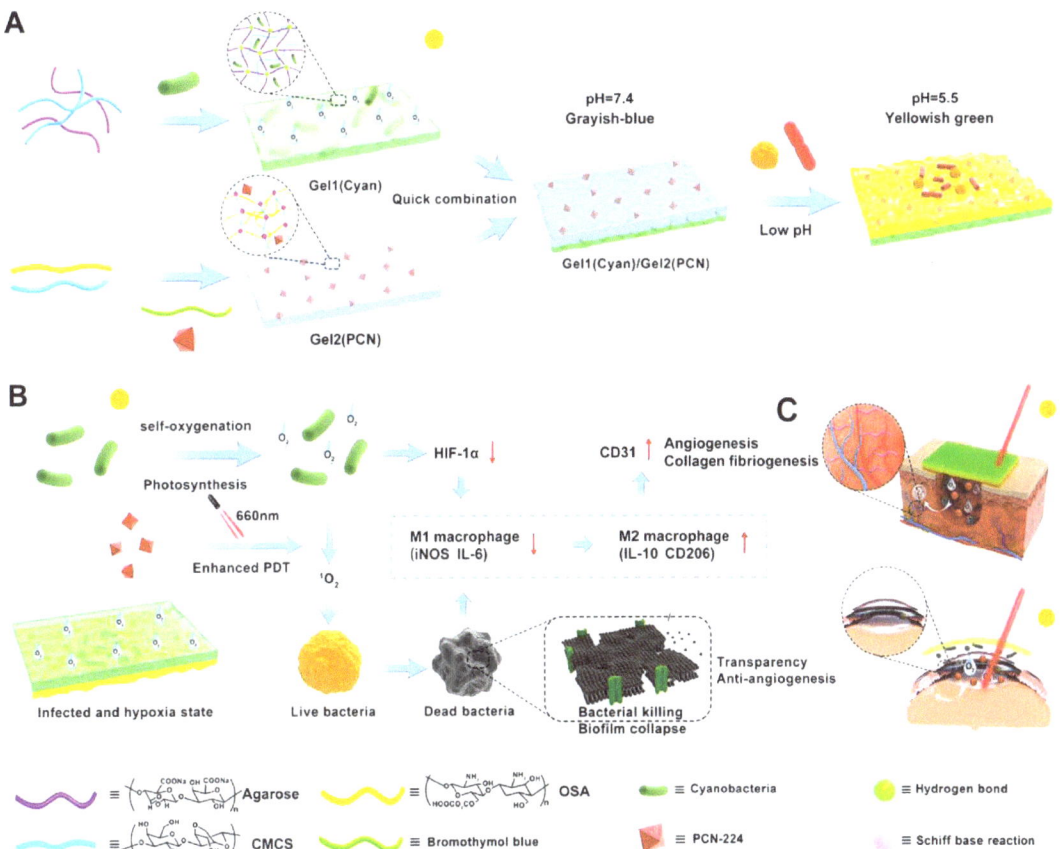

Figure 5. (**A**) Illustration of (**A**) the synthesis of Gel1 (Cyan)/Gel2 (PCN) hydrogel, (**B**) properties of Gel1 (Cyan)/Gel2 (PCN) hydrogel and (**C**) the application of Gel1 (Cyan)/Gel2 (PCN) hydrogel in diabetic wound and keratitis healing. Reprinted with permission from Ref. [56]. Copyright 2022 Wiley-VCH GmbH.

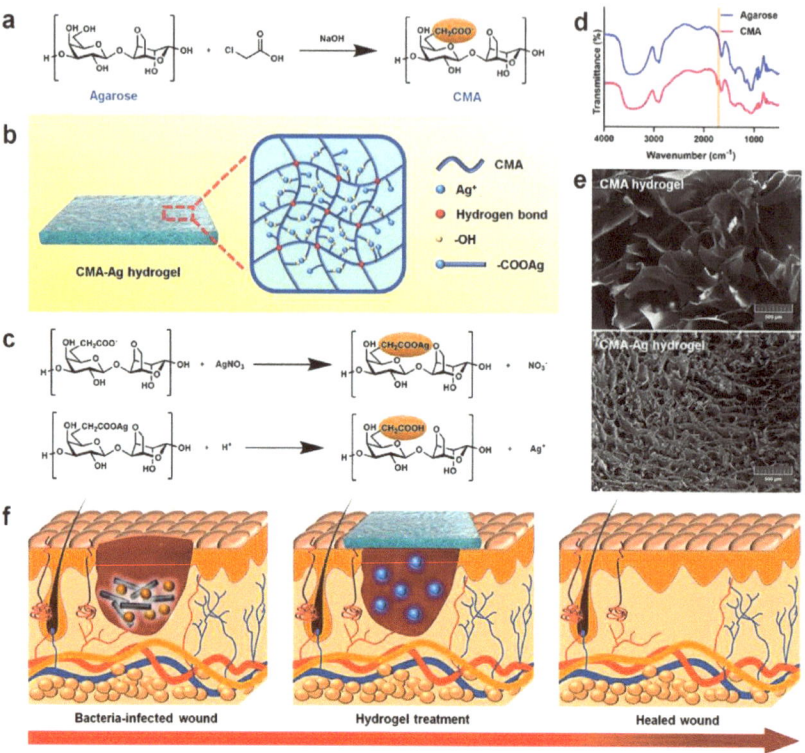

Figure 6. Illustrated preparation of (**a**) CMA molecule and (**b**) CMA–Ag hydrogel. (**c**) The mechanism of Ag$^+$ releasing from CMA–Ag hydrogel. (**d**) FT–IR spectra of the agarose and CMA molecule. (**e**) Morphological characterization of CMA and CMA–Ag hydrogels. (**f**) Schematic application of CMA–Ag hydrogel in the treatment of infected wounds. Reprinted with permission from Ref. [57]. Copyright 2020 WILEY-VCH Verlag GmbH & Co. KGaA, Weinheim.

2.5. Hyaluronic Acid

Hyaluronic acid (HA) is a non-sulfonated glycosaminoglycan and natural anionic polysaccharide consisting of repeated disaccharide units of β-D-glucuronic acid and N-acetyl-D-glucosamine linked by alternating β-1, 3, and β-1, 4-glucosidic bonds, which are produced during fibroblasts proliferation [98–100]. HA can promote wound healing by promoting cell migration and mediating cell signal transduction and has good biocompatibility, biodegradability, and moisture retention [100]. HA can promote the formation of blood clots, the expression of interleukin, the migration and proliferation of keratinocytes and fibroblasts [101], and accelerate the formation and re-epithelialization of granulation tissue. It can also reduce inflammatory cell infiltration and promote wound healing [102].

The skin at the joint is frequently stretched, and traditional non-self-healing wound dressings are easy to damage or fall off, so the wound at this site is not easy to heal and is also susceptible to microbial infection [103]. Therefore, it is extremely important to design a wound dressing that not only has excellent mechanical properties but also can effectively realize antibacterial purpose and promote wound healing. Li et al. [58] designed an HA-PEGSB-CMP hydrogel with adhesion, self-healing, and antibacterial capabilities for the treatment of sports wounds (Figure 7). In this study, HA, as the main component, cross-linked with benzaldehyde functionalized PEG co-polyglycerol caprate (PEGSB) by Schiff base reaction to obtain HA-PEGSB hydrogel that has stretch ability and self-healing properties. The aldehyde group on PEGSB can react with the amino group of wound tissue to introduce tissue adhesion. HA-PEGSB was cross-linked with cuttlefish melanin nanoparticles (CMP) with mussel-like protein structure to obtain HA-PEGSB-CMP hydrogel with enhanced viscosity. A good stretching ability and self-healing capability make the hydrogel adhere firmly to different shapes of wounds. The excellent photothermal performance of CMP endows the hydrogel with antibacterial activity. After 3 min of NIR irradiation (808 nm, 0.8 W·cm^{-2}), *E. coli* and MRSA are killed. The amino group in the dynamic Schiff base and the catechol structure of CMP can promote blood coagulation and give hydrogels good hemostatic properties, which can stop wound bleeding at 88 s. After 14 days of treatment on the hip joint wound of mice, the wound healed almost completely, and the skin surface was smooth. The good biocompatibility of the hydrogel makes it have great potential for clinical application to treat infected sports wounds.

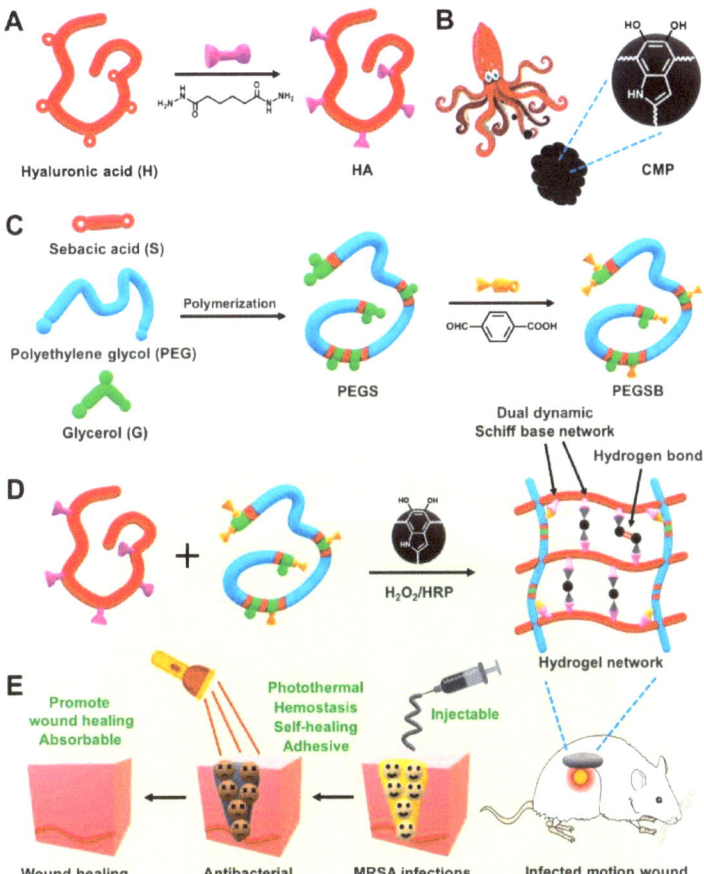

Figure 7. Schematic diagram of the synthesis of HA-PEGSB-CMP hydrogel and its application in promoting wound healing. (**A**) The synthesis of HA. (**B**) The structure of CMP. (**C**) The synthesis of PEGSB. (**D**) The preparation of HA-PEGSB-CMP hydrogel. (**E**) The wound healing application of the hydrogels. Reprinted with permission from Ref. [58]. Copyright 2021 Elsevier B.V.

Yue et al. [59] designed a dual-network multifunctional hydrogel based on polysaccharide materials for the treatment of wounds infected with drug-resistant bacteria (Figure 8). The first hydrogel structure of BSP-U with better physical and chemical properties was obtained by grafting UPy onto the main chain of Bletilla striata polysaccharide (BSP) with the premise of preserving the structure of supramolecular uracymidone (UPy) tetrahydric array. The second hydrogel structure was obtained by grafting dopamine hydrochloride (DP) to aldehyaluronate sodium (AHA) by Schiff base reaction. The quadruple hydrogen bonds of catechols, Fe^{3+}, and UPy dimer form a highly dynamic double cross-linked network structure, which makes the hydrogel have good mechanical properties and can adhere to wound tissue well. The cross-linked structure is sensitive to light, heat, and acid and can achieve sol-gel transition and be easily removed. Hydrogel has excellent swelling, water retention, and air permeability, which can absorb tissue exudate and maintain the humidity and oxygen supply of the environment to create a good microenvironment for the wound. The hydrogels formed by the cross-linking of catechol compound and Fe^{3+} ligand can widely absorb NIR light and have photothermal properties. After 3 min of in vitro irradiation, the antibacterial rate of the hydrogels on *E. coli* and *S. aureus* can reach 91.68% and 94.94%, respectively, showing great potential in protecting wounds from microbial infection. In the wound treatment of *S. aureus*-infected mice wound, the antibacterial efficiency of hydrogel with the aid of NIR light reached 92.4%, and the wound healed almost completely on the 15th day. BSP-U/DAHA Hydrogel has excellent antibacterial ability and bioactivity, which is an ideal candidate for wound healing dressings of drug-resistant bacterial infections.

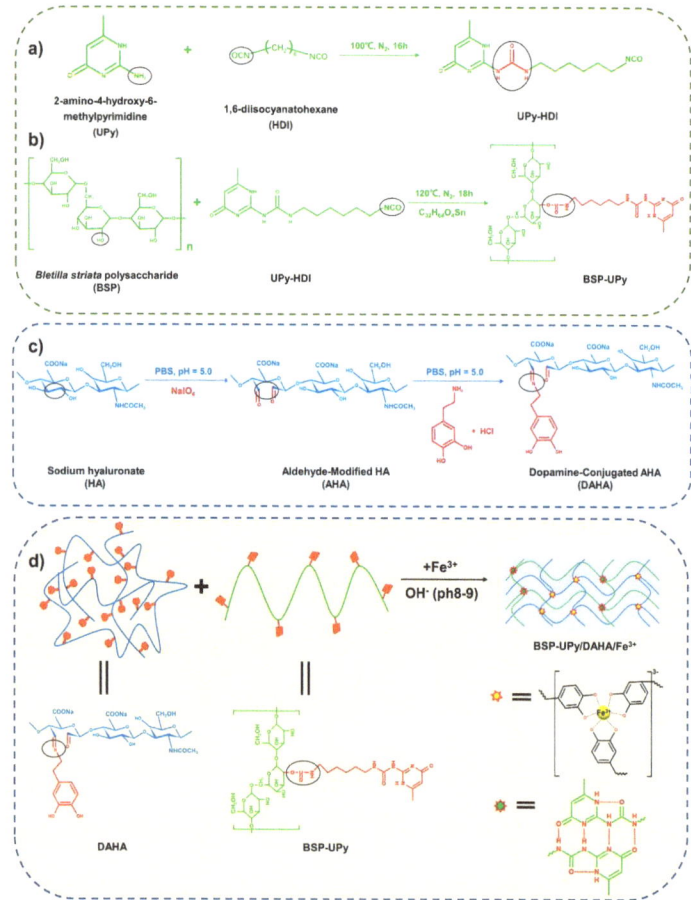

Figure 8. Schematic preparation of (**a**) UPy-HDI, (**b**) BSP-UPy, (**c**) DAHA and (**d**) BSP-U/DAHA hydrogels. Reprinted with permission from Ref. [59]. Copyright 2023 Elsevier Ltd.

2.6. Cellulose

Cellulose is a kind of linear polysaccharide formed from D-glucose via a glycosidic linkage, mainly derived from plant cell walls and bacteria [104,105]. Cellulose has the advantages of good biocompatibility, biodegradability and regeneration, and good mechanical strength [106]. However, the poor solubility of cellulose limits its application [105]. Since cellulose is formed from glucose subunits, it is naturally biocompatible with human tissues and can be modified without affecting structural and mechanical properties [107]. Bacterial cellulose (BC), derived from Gram-negative bacteria, has strong hydrophilicity and a large specific surface area for great application potential [107,108].

The proliferation of microorganisms will lead to the acidity of the wound microenvironment, which can be used to prepare pH-responsive hydrogels that intelligently control drug release [109]. Yang et al. [60] designed a pH-responsive hydrogel for the treatment of infected wounds (Figure 9). First, the natural antibiotic resveratrol (RSV) was esterified by reacting with PEG to obtain RSV-PEG, and then the RPC-conjugated polymer was synthesized by amide reaction with cellulose nanofibers. RPC-conjugated polymers form semi-interpenetrating polymer networks and hydrogen bonds in PVA/Borax networks to enhance the mechanical strength and physicochemical properties of hydrogels. The degradation of RPC/PB hydrogels is pH-dependent, and RSV release can reach 64.2% when pH is 5.4. The dense, porous, and interwoven structure of hydrogel can absorb tissue exudate and increase oxygen permeability to promote wound healing. The physical interactions in the hydrogel network are reversible, allowing it to be reused and stripped without damaging the skin and to have a self-healing capacity. The antibacterial experiment showed that with the increase in RPC content, the antibacterial rate of hydrogel on *S. aureus* was up to 84.3%. The excellent antioxidant and antibacterial properties of RPC, as well as the good mechanical properties and biocompatibility of the hydrogel, allow the wound to heal almost completely after 12 days. This study achieved intelligent controlled drug release, which is expected to replace traditional hydrogels and achieve more extensive clinical application in infectious wounds.

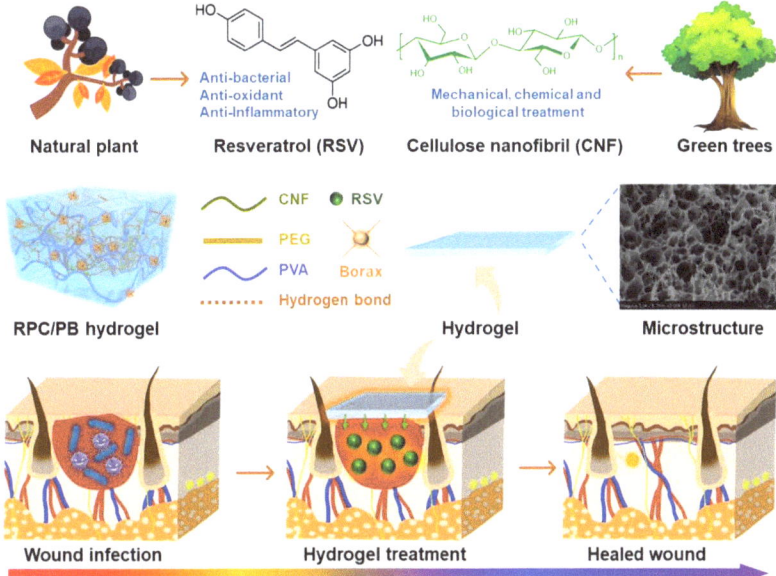

Figure 9. Schematic preparation and application of RPC/PB hydrogel in the treatment of infected wound. Reprinted with permission from Ref. [60]. Copyright 2022 The Authors.

ROS includes H_2O_2, •OH, $^1O_2$, etc. Among them, highly active •OH can oxidize cell membranes and intracellular biomolecules (such as proteins and lipids) and has a stronger killing ability against bacteria than H_2O_2 to avoid microbial resistance [110]. Nanoenzymes can simulate the activity of peroxidase and convert H_2O_2 to •OH for the treatment of diseases caused by microbial infections [111]. Zhang et al. [61] designed a biomimetic hydrogel with the catalytic ability of glucose reaction for synergistic antibacterial and hemostatic use (Figure 10). The prepared Cu@ZIF/GOx was wrapped in a cross-linked network composed of bacterial cellulose (BC) and guar gum (GG) to form a mixed hydrogel. The hydrogel supported by glucose oxidase (GOx) can catalyze glucose at the wound site to form H_2O_2 and gluconic acid. The accumulation of gluconic acid will decrease the pH at the wound site and further activate the activity of Cu@ZIF nano-enzyme, which simulates POD to catalyze the decomposition of H_2O_2 to generate high cytotoxic •OH, which inactivates bacteria. Under the condition of glucose concentration of 10 mM, the inhibition rate of hydrogel on *E. coli* and *S. aureus* can reach 100%, and it can also effectively eliminate biofilms. The recombination of hydrogen bonds at the interface and the reversible recombination of borate diol make the hydrogel have good self-healing and flexibility. The superior water absorption and hydrophilicity of hydrogel can attract red blood cells and platelets to gather in the porous structure and quickly stop bleeding. The hydrogel prepared by this method is glucose-responsive and innovatively realizes the synergistic antibacterial hemostatic treatment of infected wounds through the glucose-catalyzed cascade reaction, providing a good reference for the treatment of diabetic wounds.

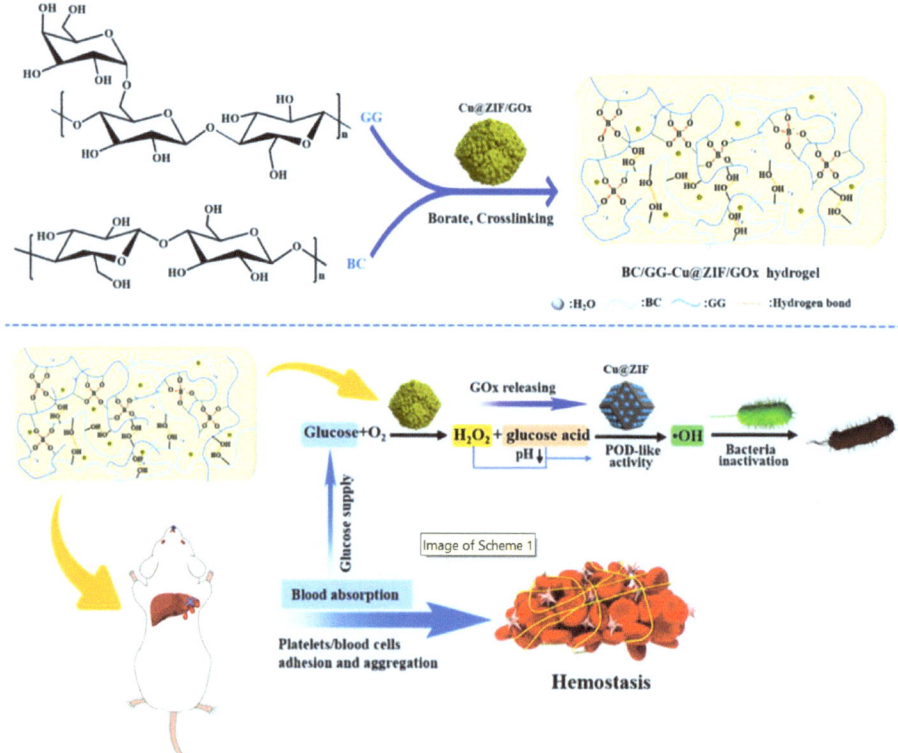

Figure 10. Illustrated preparation of BC/GG-Cu@ZIF/GOx hydrogel, and its application in antibacterial and hemostasis. Reprinted with permission from Ref. [61]. Copyright 2022 Elsevier Ltd.

2.7. Alginate

Alginate is a natural anionic polysaccharide derived from various brown algae and bacteria [112,113]. It is widely used due to its advantages, such as excellent biocompatibility, nonimmunogenic, non-thrombogenic, and high safety, which is a kind of polysaccharide approved by the US Food and Drug Administration [114]. However, due to its high hydrophilicity, it is not conducive to drug-loading and releasing [62]. The hydroxyl and carboxyl groups on its main chain can be modified to make it amphiphilic, which can be used to stabilize the emulsion and load drugs [115,116].

After the wound is infected by microorganisms, the microenvironment becomes acidic due to the bacterial metabolism to produce lactic acid and acetic acid [117]. During the healing process, the pH will gradually increase and fluctuate in the weakly alkaline range. The antibacterial activity of the hydrogel material may be affected by pH, so it is necessary to prepare a wound dressing that is not influenced by pH. Jin et al. [62] designed TO/ASP hydrogel, a highly effective antibacterial platform without influence by pH. Sodium alginate was modified by diacetone acrylamide to obtain amphiphilic sodium alginate (AS), and then formed hydrogel film (ASP) with polylysine (PL) in solution. A TO/ASP film was formed by loading lipophilic thymol (THY), which is a natural antibacterial substance, and hydrophilic oligo-tannic acid (OTA) oxidized by laccase on the ASP (Figure 11). OTA has a polyphenol structure, which is easy to be oxidized to quinone and react with amino groups and can promote antibacterial activity in alkaline environments. THY is a natural lipophilic monoterpene phenol, which can accumulate in the lipids of cell membranes and react with membrane proteins to destroy the structure and function of cell membranes and has excellent antibacterial activity under acidic conditions. The presence of OTA and THY enables TO/ASP hydrogel to maintain consistently high antibacterial activity over a wide pH range (4–9). The antibacterial rates of *E. coli* and *S. aureus* were 99.92% and 99.993%, respectively, in the pH range of 4–9. At the same time, hydrogels can effectively eliminate biofilm and destroy its structural integrity, and the biofilm biomass of *E. coli* and *S. aureus* decreased by 84.97% and 91.01%, respectively, compared with the PBS-treated group. The TO/ASP hydrogel prepared in this study is no longer affected by the change in wound pH and can realize high antibacterial efficacy in the whole process of wound healing. The functional richness of hydrogel was improved.

In the treatment of wounds, drug release decreases, and microorganisms may repopulate over time, so continuous controlled drug release is necessary. Ye et al. [63] designed a NIR-responsive photothermal hydrogel platform (ALG-HPR hydrogel) to control the drug-sustained release and achieve persistent antibacterial effects. First, the antibiotic rifampicin, photothermic agent indocyanine green (ICG), and fatty acids were encapsulated in natural halloysite clay nanotubes (HNTs). When irradiated with NIR, ICG converted light into heat, and phase transition occurred when the temperature rose above the melting point of fatty acids, followed by the release of rifampicin from HNTs, achieving drug release with NIR response. At the same time, the temperature does not exceed 42 °C, which will not burn the surrounding healthy skin tissue. Thus, through the synergistic effect with rifampicin, HNTs can achieve an antibacterial effect at a mild temperature. At room temperature, HNTs with a 25% drug-loading rate were cross-linked with alginate by Ca^{2+}, and hydrogels were obtained by in situ gelation. In vitro antibacterial experiment, rifampicin was continuously released under the control of photo-response of hydrogel after three-time of NIR irradiation, and the antibacterial efficiency against *S. aureus* was nearly 100%. After 21 days of treatment, the granulation tissue became thick, the epidermis regenerated, and finally, the wounds of the mice were almost completely healed. In short, ALG-HPR hydrogel can realize NIR-controlled release for long-term antibacterial performance and has great medical potential. This study proposes a new method of drug-controlled release, which provides insight into the continuous treatment of chronic wounds.

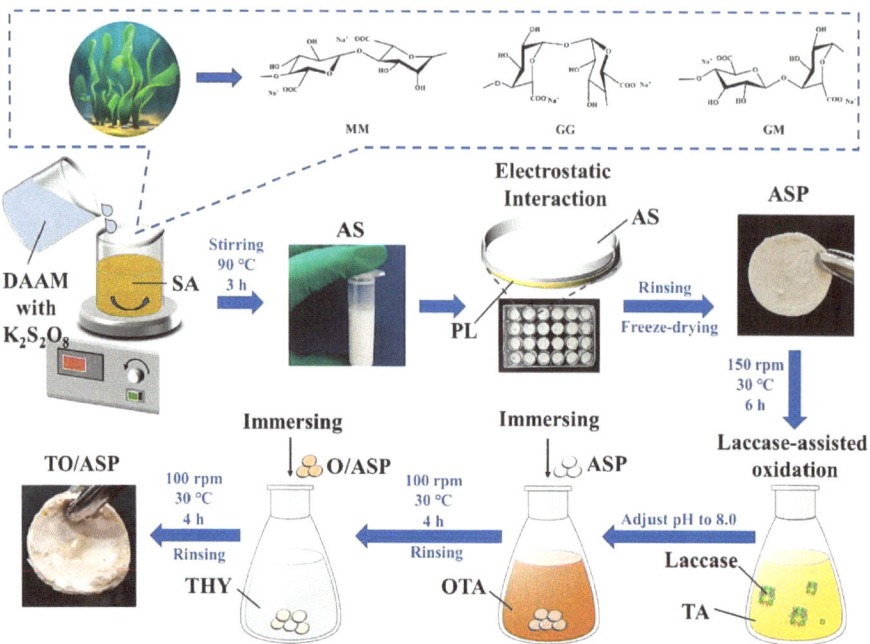

Figure 11. Schematic representation of TO/ASP hydrogel. Reprinted with permission from Ref. [62]. Copyright 2022 Elsevier Ltd.

3. Summary and Prospect

In recent years, the research on natural-polymer-based hydrogels has made great progress. Natural organic polymers come from a wide range of sources whose composition and structure are similar to the extracellular matrix, and they have high biocompatibility and are easy to modify. Hydrogels can achieve a good antibacterial effect and eliminate microbial infection by loading antibacterial agents or combining them with other therapies, such as phototherapy. At the same time, through various modification methods, the hydrogel can have the ability to be anti-inflammatory and antioxidant and promote tissue regeneration and rapid wound healing. Each of the seven natural polymer materials introduced in this review has its own advantages and disadvantages. All of them can be prepared into ideal antibacterial hydrogels with appropriate modifications. However, the positive charge of CS can cause harm to microorganisms through electrostatic interactions. PDA can not only achieve photothermal sterilization but also obtain highly viscous hydrogels by bionic mussels. The inherent antibacterial properties of CS and PDA make them possible to be synergetically used with antibacterial drugs to improve the antibacterial ability of hydrogels. At the same time, these two materials can reduce the dosage of antibacterial drugs and improve biological safety. Therefore, CS and PDA have a broader application prospect in the preparation of antibacterial hydrogels. In short, natural-polymer-based hydrogels have excellent properties in all aspects and are still one of the research hotspots at present.

Although natural-polymer-based hydrogels have made good progress, they still have certain limitations. In the future, the development of natural-polymer-based hydrogels should be closer to the following trends: First, in the treatment process, the wound situation is constantly changing, so the function of hydrogels should also achieve intelligent regulation. For example, effective hemostasis should be provided during the hemostatic phase of the wound, and tissue regeneration should be induced through the release of growth factors or other methods during the proliferation phase. Secondly, hydrogels should have real-time monitoring and continuous release in response to antibacterial drugs because microorganisms have a strong proliferation ability, and secondary infections may occur during the process of wound healing. Third, the precise targeting of drugs is also extremely important because drugs have certain toxicity, which may affect healthy tissues and hinder the repair of wounds. Fourth, the functions of hydrogels should be diversified, such as antibacterial, antioxidant, anti-inflammatory, hemostatic, etc., to meet the needs of the whole process of wound healing. Finally, hydrogels should have excellent biocompatibility and biodegradability so that they can achieve clinical application. In summary, we hope that this review will give readers a systematic and detailed understanding of the current situation of the application of natural-polymer-based hydrogels in the treatment of infectious wounds in the past three years. At the same time, it is expected that this work can provide some guidance for the research of antibacterial hydrogels.

Author Contributions: Conceptualization, H.Y. and R.Q.; writing—original draft preparation, Y.Z. and H.Y.; writing—review and editing, X.W. and R.Q. All authors have read and agreed to the published version of the manuscript.

Funding: This research was funded by the project of Cultivation for young top-notch talents of Beijing Municipal Institutions (BPHR202203045), the Beijing Municipal Education Commission project (KM202210011002) and the Open Research Fund Program of Cultivation Project of Double First-Class Disciplines of Chemistry and Materials Engineering, Beijing Technology and Business University (hc202212). The funders had no role in study design, decision to publish, or preparation of the manuscript.

Institutional Review Board Statement: Not applicable.

Data Availability Statement: Data sharing not applicable.

Acknowledgments: The authors gratefully acknowledge the Beijing Technology and Business University graduate research capability improvement program.

Conflicts of Interest: The authors declare no conflict of interest.

References

1. GBP 2019 Viewpoint Collaborators. Five insights from the Global Burden of Disease Study 2019. *Lancet* **2020**, *396*, 1135–1159. [CrossRef]
2. Dabrowska, A.K.; Spano, F.; Derler, S.; Adlhart, C.; Spencer, N.D.; Rossi, R.M. The relationship between skin function, barrier properties, and body-dependent factors. *Skin Res. Technol.* **2018**, *24*, 165–174. [CrossRef] [PubMed]
3. Liang, Y.; He, J.; Guo, B. Functional Hydrogels as Wound Dressing to Enhance Wound Healing. *ACS Nano* **2021**, *15*, 12687–12722. [CrossRef] [PubMed]
4. Nosrati, H.; Aramideh Khouy, R.; Nosrati, A.; Khodaei, M.; Banitalebi-Dehkordi, M.; Ashrafi-Dehkordi, K.; Sanami, S.; Alizadeh, Z. Nanocomposite scaffolds for accelerating chronic wound healing by enhancing angiogenesis. *J. Nanobiotechnol.* **2021**, *19*, 1. [CrossRef] [PubMed]
5. Tottoli, E.M.; Dorati, R.; Genta, I.; Chiesa, E.; Pisani, S.; Conti, B. Skin Wound Healing Process and New Emerging Technologies for Skin Wound Care and Regeneration. *Pharmaceutics* **2020**, *12*, 735. [CrossRef] [PubMed]
6. Weller, C.D.; Team, V.; Sussman, G. First-Line Interactive Wound Dressing Update: A Comprehensive Review of the Evidence. *Front. Pharmacol.* **2020**, *11*, 155. [CrossRef] [PubMed]
7. Mo, F.; Zhang, M.; Duan, X.; Lin, C.; Sun, D.; You, T. Recent Advances in Nanozymes for Bacteria-Infected Wound Therapy. *Int. J. Nanomed.* **2022**, *17*, 5947–5990. [CrossRef]
8. Yang, K.; Han, Q.; Chen, B.; Zheng, Y.; Zhang, K.; Li, Q.; Wang, J. Antimicrobial hydrogels: Promising materials for medical application. *Int. J. Nanomed.* **2018**, *13*, 2217–2263. [CrossRef]
9. Ng, V.W.; Chan, J.M.; Sardon, H.; Ono, R.J.; Garcia, J.M.; Yang, Y.Y.; Hedrick, J.L. Antimicrobial hydrogels: A new weapon in the arsenal against multidrug-resistant infections. *Adv. Drug Del. Rev.* **2014**, *78*, 46–62. [CrossRef]
10. Li, Z.; Bai, H.; Jia, S.; Yuan, H.; Gao, L.-H.; Liang, H. Design of functional polymer nanomaterials for antimicrobial therapy and combatting resistance. *Mat. Chem. Front.* **2021**, *5*, 1236–1252. [CrossRef]
11. Koehler, J.; Brandl, F.P.; Goepferich, A.M. Hydrogel wound dressings for bioactive treatment of acute and chronic wounds. *Eur. Polym. J.* **2018**, *100*, 1–11. [CrossRef]
12. Yu, R.; Zhang, H.; Guo, B. Conductive Biomaterials as Bioactive Wound Dressing for Wound Healing and Skin Tissue Engineering. *Nano-Micro Lett.* **2021**, *14*, 1. [CrossRef] [PubMed]

13. Wu, Y.K.; Cheng, N.C.; Cheng, C.M. Biofilms in Chronic Wounds: Pathogenesis and Diagnosis. *Trends Biotechnol.* **2019**, *37*, 505–517. [CrossRef] [PubMed]
14. Jiang, Y.; Zhang, X.; Zhang, W.; Wang, M.; Yan, L.; Wang, K.; Han, L.; Lu, X. Infant Skin Friendly Adhesive Hydrogel Patch Activated at Body Temperature for Bioelectronics Securing and Diabetic Wound Healing. *ACS Nano* **2022**, *16*, 8662–8676. [CrossRef] [PubMed]
15. Khan, A.U.R.; Huang, K.; Khalaji, M.S.; Yu, F.; Xie, X.; Zhu, T.; Morsi, Y.; Jinzhong, Z.; Mo, X. Multifunctional bioactive core-shell electrospun membrane capable to terminate inflammatory cycle and promote angiogenesis in diabetic wound. *Bioact. Mater.* **2021**, *6*, 2783–2800. [CrossRef]
16. Jeffcoate, W.J.; Vileikyte, L.; Boyko, E.J.; Armstrong, D.G.; Boulton, A.J.M. Current Challenges and Opportunities in the Prevention and Management of Diabetic Foot Ulcers. *Diabetes Care* **2018**, *41*, 645–652. [CrossRef]
17. Li, B.; Wang, W.; Zhao, L.; Yan, D.; Li, X.; Gao, Q.; Zheng, J.; Zhou, S.; Lai, S.; Feng, Y.; et al. Multifunctional AIE Nanosphere-Based "Nanobomb" for Trimodal Imaging-Guided Photothermal/Photodynamic/Pharmacological Therapy of Drug-Resistant Bacterial Infections. *ACS Nano* **2023**, *17*, 4601–4618. [CrossRef]
18. Xiang, X.; Chen, D.; Li, N.; Xu, Q.; Li, H.; He, J.; Lu, J. PVDF/PLA electrospun fiber membrane impregnated with metal nanoparticles for emulsion separation, surface antimicrobial, and antifouling activities. *Sci. China Technol. Sci.* **2023**, *66*, 1461–1470. [CrossRef]
19. Asadi, N.; Pazoki-Toroudi, H.; Del Bakhshayesh, A.R.; Akbarzadeh, A.; Davaran, S.; Annabi, N. Multifunctional hydrogels for wound healing: Special focus on biomacromolecular based hydrogels. *Int. J. Biol. Macromol.* **2021**, *170*, 728–750. [CrossRef]
20. Op 't Veld, R.C.; Walboomers, X.F.; Jansen, J.A.; Wagener, F. Design Considerations for Hydrogel Wound Dressings: Strategic and Molecular Advances. *Tissue Eng. Part B Rev.* **2020**, *26*, 230–248. [CrossRef]
21. Han, W.; Wang, S. Advances in Hemostatic Hydrogels That Can Adhere to Wet Surfaces. *Gels* **2022**, *9*, 2. [CrossRef] [PubMed]
22. Tonsomboon, K.; Butcher, A.L.; Oyen, M.L. Strong and tough nanofibrous hydrogel composites based on biomimetic principles. *Mater. Sci. Eng. C-Mater. Biol. Appl.* **2017**, *72*, 220–227. [CrossRef] [PubMed]
23. Hu, H.; Xu, F.J. Rational design and latest advances of polysaccharide-based hydrogels for wound healing. *Biomater. Sci.* **2020**, *8*, 2084–2101. [CrossRef] [PubMed]
24. Peng, W.; Li, D.; Dai, K.; Wang, Y.; Song, P.; Li, H.; Tang, P.; Zhang, Z.; Li, Z.; Zhou, Y.; et al. Recent progress of collagen, chitosan, alginate and other hydrogels in skin repair and wound dressing applications. *Int. J. Biol. Macromol.* **2022**, *208*, 400–408. [CrossRef]
25. Lokhande, G.; Carrow, J.K.; Thakur, T.; Xavier, J.R.; Parani, M.; Bayless, K.J.; Gaharwar, A.K. Nanoengineered injectable hydrogels for wound healing application. *Acta Biomater.* **2018**, *70*, 35–47. [CrossRef]
26. Zhong, Y.; Xiao, H.; Seidi, F.; Jin, Y. Natural Polymer-Based Antimicrobial Hydrogels without Synthetic Antibiotics as Wound Dressings. *Biomacromolecules* **2020**, *21*, 2983–3006. [CrossRef]
27. Long, S.; Xie, C.; Lu, X. Natural polymer-based adhesive hydrogel for biomedical applications. *Biosurface Biotribol.* **2022**, *8*, 69–94. [CrossRef]
28. Tiwari, N.; Kumar, D.; Priyadarshani, A.; Jain, G.K.; Mittal, G.; Kesharwani, P.; Aggarwal, G. Recent progress in polymeric biomaterials and their potential applications in skin regeneration and wound care management. *J. Drug Deliv. Sci. Technol.* **2023**, *82*, 104319. [CrossRef]
29. Liu, S.; Yan, Q.; Cao, S.; Wang, L.; Luo, S.-H.; Lv, M. Inhibition of Bacteria In Vitro and In Vivo by Self-Assembled DNA-Silver Nanocluster Structures. *ACS Appl. Mater. Interfaces* **2022**, *14*, 41809–41818. [CrossRef]
30. Fu, H.; Xue, K.; Zhang, Y.; Xiao, M.; Wu, K.; Shi, L.; Zhu, C. Thermoresponsive Hydrogel-Enabled Thermostatic Photothermal Therapy for Enhanced Healing of Bacteria-Infected Wounds. *Adv. Sci.* **2023**, *10*, e2206865. [CrossRef]
31. Han, L.; Lu, X.; Liu, K.; Wang, K.; Fang, L.; Weng, L.T.; Zhang, H.; Tang, Y.; Ren, F.; Zhao, C.; et al. Mussel-Inspired Adhesive and Tough Hydrogel Based on Nanoclay Confined Dopamine Polymerization. *ACS Nano* **2017**, *11*, 2561–2574. [CrossRef] [PubMed]
32. Shao, Z.; Yin, T.; Jiang, J.; He, Y.; Xiang, T.; Zhou, S. Wound microenvironment self-adaptive hydrogel with efficient angiogenesis for promoting diabetic wound healing. *Bioact. Mater.* **2023**, *20*, 561–573. [CrossRef] [PubMed]
33. Shen, J.; Chang, R.; Chang, L.; Wang, Y.; Deng, K.; Wang, D.; Qin, J. Light emitting CMC-CHO based self-healing hydrogel with injectability for in vivo wound repairing applications. *Carbohydr. Polym.* **2022**, *281*, 119052. [CrossRef] [PubMed]
34. Rahimi, M.; Noruzi, E.B.; Sheykhsaran, E.; Ebadi, B.; Kariminezhad, Z.; Molaparast, M.; Mehrabani, M.G.; Mehramouz, B.; Yousefi, M.; Ahmadi, R.; et al. Carbohydrate polymer-based silver nanocomposites: Recent progress in the antimicrobial wound dressings. *Carbohydr. Polym.* **2020**, *231*, 115696. [CrossRef]
35. Dsouza, M.; Jayabalan, S.S. Analysis of the size reduction of AgNPs loaded hydrogel and its effect on the anti-bacterial activity. *IET Nanobiotechnol.* **2021**, *15*, 545–557. [CrossRef]
36. Zhang, Y.K.; Zhang, H.; Zou, Q.L.; Xing, R.R.; Jiao, T.F.; Yan, X.H. An injectable dipeptide-fullerene supramolecular hydrogel for photodynamic antibacterial therapy. *J. Mater. Chem. B* **2018**, *6*, 7335–7342. [CrossRef]
37. Chen, Y.; Gao, Y.J.; Chen, Y.; Liu, L.; Mo, A.C.; Peng, Q. Nanomaterials-based photothermal therapy and its potentials in antibacterial treatment. *J. Control. Release* **2020**, *328*, 251–262. [CrossRef]
38. Zhang, W.S.; Wang, B.J.; Xiang, G.L.; Jiang, T.Z.; Zhao, X. Photodynamic Alginate Zn-MOF Thermosensitive Hydrogel for Accelerated Healing of Infected Wounds. *ACS Appl. Mater. Interfaces* **2023**, *15*, 22830–22842. [CrossRef]
39. Maleki, A.; He, J.; Bochani, S.; Nosrati, V.; Shahbazi, M.-A.; Guo, B. Multifunctional Photoactive Hydrogels for Wound Healing Acceleration. *ACS Nano* **2021**, *15*, 18895–18930. [CrossRef]

40. Badran, Z.; Rahman, B.; De Bonfils, P.; Nun, P.; Coeffard, V.; Verron, E. Antibacterial nanophotosensitizers in photodynamic therapy: An update. *Drug Discov. Today* **2023**, *28*, 103493. [CrossRef]
41. Xu, M.J.; Li, L.; Hu, Q.L. The recent progress in photothermal-triggered bacterial eradication. *Biomater. Sci.* **2021**, *9*, 1995–2008. [CrossRef] [PubMed]
42. Zhang, X.; Tan, B.W.; Wu, Y.T.; Zhang, M.; Liao, J.F. A Review on Hydrogels with Photothermal Effect in Wound Healing and Bone Tissue Engineering. *Polymers* **2021**, *13*, 2100. [CrossRef] [PubMed]
43. Li, Y.Y.; Yang, K.; Wang, Z.F.; Xiao, J.Y.; Tang, Z.F.; Li, H.; Yi, W.H.; Li, Z.L.; Luo, Y.Z.; Li, J.Q.; et al. Rapid In Situ Deposition of Iron-Chelated Polydopamine Coating on the Polyacrylamide Hydrogel Dressings for Combined Photothermal and Chemodynamic Therapy of Skin Wound Infection. *ACS Appl. Bio Mater.* **2022**, *5*, 4541–4553. [CrossRef] [PubMed]
44. Li, Z.H.; Xu, W.L.; Wang, X.H.; Jiang, W.Q.; Ma, X.L.; Wang, F.J.C.; Zhang, C.L.; Ren, C.G. Fabrication of PVA/PAAm IPN hydrogel with high adhesion and enhanced mechanical properties for body sensors and antibacterial activity. *Eur. Polym. J.* **2021**, *146*, 110253. [CrossRef]
45. Liu, S.J.; Jiang, T.; Guo, R.Q.; Li, C.; Lu, C.F.; Yang, G.C.; Nie, J.Q.; Wang, F.Y.; Yang, X.F.; Chen, Z.B. Injectable and Degradable PEG Hydrogel with Antibacterial Performance for Promoting Wound Healing. *ACS Appl. Bio Mater.* **2021**, *4*, 2769–2780. [CrossRef]
46. Jiang, Y.; Wang, Y.; Li, Q.; Yu, C.; Chu, W. Natural Polymer-based Stimuli-responsive Hydrogels. *Curr. Med. Chem.* **2020**, *27*, 2631–2657. [CrossRef]
47. Fan, Z.; Cheng, P.; Zhang, P.; Zhang, G.; Han, J. Rheological insight of polysaccharide/protein based hydrogels in recent food and biomedical fields: A review. *Int. J. Biol. Macromol.* **2022**, *222*, 1642–1664. [CrossRef]
48. Eivazzadeh-Keihan, R.; Ahmadpour, F.; Aliabadi, H.A.M.; Radinekiyan, F.; Maleki, A.; Madanchi, H.; Mahdavi, M.; Shalan, A.E.; Lanceros-Mendez, S. Pectin-cellulose hydrogel, silk fibroin and magnesium hydroxide nanoparticles hybrid nanocomposites for biomedical applications. *Int. J. Biol. Macromol.* **2021**, *192*, 7–15. [CrossRef]
49. Chen, C.; Zhou, P.; Huang, C.; Zeng, R.; Yang, L.; Han, Z.; Qu, Y.; Zhang, C. Photothermal-promoted multi-functional dual network polysaccharide hydrogel adhesive for infected and susceptible wound healing. *Carbohydr. Polym.* **2021**, *273*, 118557. [CrossRef]
50. Hao, Y.; Zhao, W.; Zhang, H.; Zheng, W.; Zhou, Q. Carboxymethyl chitosan-based hydrogels containing fibroblast growth factors for triggering diabetic wound healing. *Carbohydr. Polym.* **2022**, *287*, 119336. [CrossRef]
51. Yang, Y.; Liang, Y.; Chen, J.; Duan, X.; Guo, B. Mussel-inspired adhesive antioxidant antibacterial hemostatic composite hydrogel wound dressing via photo-polymerization for infected skin wound healing. *Bioact. Mater.* **2022**, *8*, 341–354. [CrossRef]
52. Liu, T.; Feng, Z.; Li, Z.; Lin, Z.; Chen, L.; Li, B.; Chen, Z.; Wu, Z.; Zeng, J.; Zhang, J.; et al. Carboxymethyl chitosan/sodium alginate hydrogels with polydopamine coatings as promising dressings for eliminating biofilm and multidrug-resistant bacteria induced wound healing. *Int. J. Biol. Macromol.* **2023**, *225*, 923–937. [CrossRef]
53. Qi, X.; Huang, Y.; You, S.; Xiang, Y.; Cai, E.; Mao, R.; Pan, W.; Tong, X.; Dong, W.; Ye, F.; et al. Engineering Robust Ag-Decorated Polydopamine Nano-Photothermal Platforms to Combat Bacterial Infection and Prompt Wound Healing. *Adv. Sci.* **2022**, *9*, e2106015. [CrossRef]
54. Huang, H.; Su, Y.; Wang, C.; Lei, B.; Song, X.; Wang, W.; Wu, P.; Liu, X.; Dong, X.; Zhong, L. Injectable Tissue-Adhesive Hydrogel for Photothermal/Chemodynamic Synergistic Antibacterial and Wound Healing Promotion. *ACS Appl. Mater. Interfaces* **2023**, *15*, 2714–2724. [CrossRef]
55. Wang, Y.; Lv, Q.; Chen, Y.; Xu, L.; Feng, M.; Xiong, Z.; Li, J.; Ren, J.; Liu, J.; Liu, B. Bilayer hydrogel dressing with lysozyme-enhanced photothermal therapy for biofilm eradication and accelerated chronic wound repair. *Acta Pharm. Sin. B* **2023**, *13*, 284–297. [CrossRef] [PubMed]
56. Zhu, Z.; Wang, L.; Peng, Y.; Chu, X.; Zhou, L.; Jin, Y.; Guo, H.; Gao, Q.; Yang, J.; Wang, X.; et al. Continuous Self-Oxygenated Double-Layered Hydrogel under Natural Light for Real-Time Infection Monitoring, Enhanced Photodynamic Therapy, and Hypoxia Relief in Refractory Diabetic Wounds Healing. *Adv. Funct. Mater.* **2022**, *32*, 1875. [CrossRef]
57. Huang, W.C.; Ying, R.; Wang, W.; Guo, Y.; He, Y.; Mo, X.; Xue, C.; Mao, X. A Macroporous Hydrogel Dressing with Enhanced Antibacterial and Anti-Inflammatory Capabilities for Accelerated Wound Healing. *Adv. Funct. Mater.* **2020**, *30*, 644. [CrossRef]
58. Li, M.; Liang, Y.; Liang, Y.; Pan, G.; Guo, B. Injectable stretchable self-healing dual dynamic network hydrogel as adhesive anti-oxidant wound dressing for photothermal clearance of bacteria and promoting wound healing of MRSA infected motion wounds. *Chem. Eng. J.* **2022**, *427*, 132039. [CrossRef]
59. Yue, X.; Zhao, S.; Qiu, M.; Zhang, J.; Zhong, G.; Huang, C.; Li, X.; Zhang, C.; Qu, Y. Physical dual-network photothermal antibacterial multifunctional hydrogel adhesive for wound healing of drug-resistant bacterial infections synthesized from natural polysaccharides. *Carbohydr. Polym.* **2023**, *312*, 120831. [CrossRef]
60. Yang, G.; Zhang, Z.; Liu, K.; Ji, X.; Fatehi, P.; Chen, J. A cellulose nanofibril-reinforced hydrogel with robust mechanical, self-healing, pH-responsive and antibacterial characteristics for wound dressing applications. *J. Nanobiotechnol.* **2022**, *20*, 312. [CrossRef]
61. Zhang, S.; Ding, F.; Liu, Y.; Ren, X. Glucose-responsive biomimetic nanoreactor in bacterial cellulose hydrogel for antibacterial and hemostatic therapies. *Carbohydr. Polym.* **2022**, *292*, 119615. [CrossRef] [PubMed]
62. Jin, F.; Liao, S.; Li, W.; Jiang, C.; Wei, Q.; Xia, X.; Wang, Q. Amphiphilic sodium alginate-polylysine hydrogel with high antibacterial efficiency in a wide pH range. *Carbohydr. Polym.* **2023**, *299*, 120195. [CrossRef]

63. Ye, J.J.; Li, L.F.; Hao, R.N.; Gong, M.; Wang, T.; Song, J.; Meng, Q.H.; Zhao, N.N.; Xu, F.J.; Lvov, Y.; et al. Phase-change composite filled natural nanotubes in hydrogel promote wound healing under photothermally triggered drug release. *Bioact. Mater.* **2023**, *21*, 284–298. [CrossRef] [PubMed]
64. Motiei, M.; Kashanian, S.; Lucia, L.A.; Khazaei, M. Intrinsic parameters for the synthesis and tuned properties of amphiphilic chitosan drug delivery nanocarriers. *J. Control. Release* **2017**, *260*, 213–225. [CrossRef] [PubMed]
65. Sheng, J.; Han, L.; Qin, J.; Ru, G.; Li, R.; Wu, L.; Cui, D.; Yang, P.; He, Y.; Wang, J. N-trimethyl chitosan chloride-coated PLGA nanoparticles overcoming multiple barriers to oral insulin absorption. *ACS Appl. Mater. Interfaces* **2015**, *7*, 15430–15441. [CrossRef]
66. Zhang, J.; Xia, W.; Liu, P.; Cheng, Q.; Tahirou, T.; Gu, W.; Li, B. Chitosan modification and pharmaceutical/biomedical applications. *Mar. Drugs* **2010**, *8*, 1962–1987. [CrossRef] [PubMed]
67. Chen, W.H.; Chen, Q.W.; Chen, Q.; Cui, C.; Duan, S.; Kang, Y.; Liu, Y.; Liu, Y.; Muhammad, W.; Shao, S.; et al. Biomedical polymers: Synthesis, properties, and applications. *Sci. China Chem.* **2022**, *65*, 1010–1075. [CrossRef]
68. Sahariah, P.; Kontogianni, G.I.; Scoulica, E.; Sigurjonsson, O.E.; Chatzinikolaidou, M. Structure-activity relationship for antibacterial chitosan carrying cationic and hydrophobic moieties. *Carbohydr. Polym.* **2023**, *312*, 120796. [CrossRef]
69. Wang, Y.; Li, B.; Xu, F.; Han, Z.; Wei, D.; Jia, D.; Zhou, Y. Tough Magnetic Chitosan Hydrogel Nanocomposites for Remotely Stimulated Drug Release. *Biomacromolecules* **2018**, *19*, 3351–3360. [CrossRef]
70. Bhattarai, N.; Gunn, J.; Zhang, M. Chitosan-based hydrogels for controlled, localized drug delivery. *Adv. Drug Del. Rev.* **2010**, *62*, 83–99. [CrossRef]
71. Martinez-Martinez, M.; Rodriguez-Berna, G.; Gonzalez-Alvarez, I.; Hernandez, M.J.; Corma, A.; Bermejo, M.; Merino, V.; Gonzalez-Alvarez, M. Ionic Hydrogel Based on Chitosan Cross-Linked with 6-Phosphogluconic Trisodium Salt as a Drug Delivery System. *Biomacromolecules* **2018**, *19*, 1294–1304. [CrossRef] [PubMed]
72. Xue, Y.; Niu, W.; Wang, M.; Chen, M.; Guo, Y.; Lei, B. Engineering a Biodegradable Multifunctional Antibacterial Bioactive Nanosystem for Enhancing Tumor Photothermo-Chemotherapy and Bone Regeneration. *ACS Nano* **2020**, *14*, 442–453. [CrossRef] [PubMed]
73. Li, H.; Marshall, T.; Aulin, Y.V.; Thenuwara, A.C.; Zhao, Y.; Borguet, E.; Strongin, D.R.; Ren, F. Structural evolution and electrical properties of metal ion-containing polydopamine. *J. Mater. Sci.* **2019**, *54*, 6393–6400. [CrossRef]
74. Xi, Y.; Ge, J.; Wang, M.; Chen, M.; Niu, W.; Cheng, W.; Xue, Y.; Lin, C.; Lei, B. Bioactive Anti-inflammatory, Antibacterial, Antioxidative Silicon-Based Nanofibrous Dressing Enables Cutaneous Tumor Photothermo-Chemo Therapy and Infection-Induced Wound Healing. *ACS Nano* **2020**, *14*, 2904–2916. [CrossRef] [PubMed]
75. Balavigneswaran, C.K.; Jaiswal, V.; Venkatesan, R.; Karuppiah, P.S.; Sundaram, M.K.; Vasudha, T.K.; Aadinath, W.; Ravikumar, A.; Saravanan, H.V.; Muthuvijayan, V. Mussel-Inspired Adhesive Hydrogels Based on Laponite-Confined Dopamine Polymerization as a Transdermal Patch. *Biomacromolecules* **2023**, *24*, 724–738. [CrossRef]
76. Di, X.; Hang, C.; Xu, Y.; Ma, Q.; Li, F.; Sun, P.; Wu, G. Bioinspired tough, conductive hydrogels with thermally reversible adhesiveness based on nanoclay confined NIPAM polymerization and a dopamine modified polypeptide. *Mat. Chem. Front.* **2020**, *4*, 189–196. [CrossRef]
77. Fan, S.; Lin, W.; Huang, Y.; Xia, J.; Xu, J.F.; Zhang, J.; Pi, J. Advances and Potentials of Polydopamine Nanosystem in Photothermal-Based Antibacterial Infection Therapies. *Front. Pharmacol.* **2022**, *13*, 829712. [CrossRef]
78. Zou, X.; Ma, G.; Zhu, P.; Cao, Y.; Sun, X.; Wang, H.; Dong, J. A Polydopamine-Coated Platinum Nanoplatform for Tumor-Targeted Photothermal Ablation and Migration Inhibition. *Front. Oncol.* **2022**, *12*, 860718. [CrossRef]
79. Li, X.; Liu, L.; Li, S.; Wan, Y.; Chen, J.X.; Tian, S.; Huang, Z.; Xiao, Y.F.; Cui, X.; Xiang, C.; et al. Biodegradable pi-Conjugated Oligomer Nanoparticles with High Photothermal Conversion Efficiency for Cancer Theranostics. *ACS Nano* **2019**, *13*, 12901–12911. [CrossRef]
80. Mousavi, S.; Khoshfetrat, A.B.; Khatami, N.; Ahmadian, M.; Rahbarghazi, R. Comparative study of collagen and gelatin in chitosan-based hydrogels for effective wound dressing: Physical properties and fibroblastic cell behavior. *Biochem. Biophys. Res. Commun.* **2019**, *518*, 625–631. [CrossRef]
81. Abedinia, A.; Mohammadi Nafchi, A.; Sharifi, M.; Ghalambor, P.; Oladzadabbasabadi, N.; Ariffin, F.; Huda, N. Poultry gelatin: Characteristics, developments, challenges, and future outlooks as a sustainable alternative for mammalian gelatin. *Trends Food Sci. Technol.* **2020**, *104*, 14–26. [CrossRef]
82. Lv, L.-C.; Huang, Q.-Y.; Ding, W.; Xiao, X.-H.; Zhang, H.-Y.; Xiong, L.-X. Fish gelatin: The novel potential applications. *J. Funct. Food* **2019**, *63*, 103581. [CrossRef]
83. Skopinska-Wisniewska, J.; Tuszynska, M.; Olewnik-Kruszkowska, E. Comparative Study of Gelatin Hydrogels Modified by Various Cross-Linking Agents. *Materials* **2021**, *14*, 396. [CrossRef] [PubMed]
84. Dash, R.; Foston, M.; Ragauskas, A.J. Improving the mechanical and thermal properties of gelatin hydrogels cross-linked by cellulose nanowhiskers. *Carbohydr. Polym.* **2013**, *91*, 638–645. [CrossRef]
85. Li, N.; Chen, W.; Chen, G.; Tian, J. Rapid shape memory TEMPO-oxidized cellulose nanofibers/polyacrylamide/gelatin hydrogels with enhanced mechanical strength. *Carbohydr. Polym.* **2017**, *171*, 77–84. [CrossRef]
86. Cui, L.; Jia, J.; Guo, Y.; Liu, Y.; Zhu, P. Preparation and characterization of IPN hydrogels composed of chitosan and gelatin cross-linked by genipin. *Carbohydr. Polym.* **2014**, *99*, 31–38. [CrossRef]
87. Pan, T.; Chen, H.; Gao, X.; Wu, Z.; Ye, Y.; Shen, Y. Engineering efficient artificial nanozyme based on chitosan grafted Fe-doped-carbon dots for bacteria biofilm eradication. *J. Hazard. Mater.* **2022**, *435*, 128996. [CrossRef]

88. Versey, Z.; da Cruz Nizer, W.S.; Russell, E.; Zigic, S.; DeZeeuw, K.G.; Marek, J.E.; Overhage, J.; Cassol, E. Biofilm-Innate Immune Interface: Contribution to Chronic Wound Formation. *Front. Immunol.* **2021**, *12*, 648554. [CrossRef]
89. Kumari, N.; Kumar, S.; Karmacharya, M.; Dubbu, S.; Kwon, T.; Singh, V.; Chae, K.H.; Kumar, A.; Cho, Y.K.; Lee, I.S. Surface-Textured Mixed-Metal-Oxide Nanocrystals as Efficient Catalysts for ROS Production and Biofilm Eradication. *Nano Lett.* **2021**, *21*, 279–287. [CrossRef]
90. Khodadadi Yazdi, M.; Taghizadeh, A.; Taghizadeh, M.; Stadler, F.J.; Farokhi, M.; Mottaghitalab, F.; Zarrintaj, P.; Ramsey, J.D.; Seidi, F.; Saeb, M.R.; et al. Agarose-based biomaterials for advanced drug delivery. *J. Control. Release* **2020**, *326*, 523–543. [CrossRef]
91. Su, Y.; Chu, B.; Gao, Y.; Wu, C.; Zhang, L.; Chen, P.; Wang, X.; Tang, S. Modification of agarose with carboxylation and grafting dopamine for promotion of its cell-adhesiveness. *Carbohydr. Polym.* **2013**, *92*, 2245–2251. [CrossRef]
92. Zarrintaj, P.; Manouchehri, S.; Ahmadi, Z.; Saeb, M.R.; Urbanska, A.M.; Kaplan, D.L.; Mozafari, M. Agarose-based biomaterials for tissue engineering. *Carbohydr. Polym.* **2018**, *187*, 66–84. [CrossRef]
93. Li, M.; Mitra, D.; Kang, E.T.; Lau, T.; Chiong, E.; Neoh, K.G. Thiol-ol Chemistry for Grafting of Natural Polymers to Form Highly Stable and Efficacious Antibacterial Coatings. *ACS Appl. Mater. Interfaces* **2017**, *9*, 1847–1857. [CrossRef]
94. Karimi, T.; Mottaghitalab, F.; Keshvari, H.; Farokhi, M. Carboxymethyl chitosan/sodium carboxymethyl cellulose/agarose hydrogel dressings containing silk fibroin/polydopamine nanoparticles for antibiotic delivery. *J. Drug Deliv. Sci. Technol.* **2023**, *80*, 104134. [CrossRef]
95. Wu, Z.; Li, H.; Zhao, X.; Ye, F.; Zhao, G. Hydrophobically modified polysaccharides and their self-assembled systems: A review on structures and food applications. *Carbohydr. Polym.* **2022**, *284*, 119182. [CrossRef] [PubMed]
96. Yang, Z.; Chen, H.; Yang, P.; Shen, X.; Hu, Y.; Cheng, Y.; Yao, H.; Zhang, Z. Nano-oxygenated hydrogels for locally and permeably hypoxia relieving to heal chronic wounds. *Biomaterials* **2022**, *282*, 121401. [CrossRef]
97. Han, G.; Ceilley, R. Chronic Wound Healing: A Review of Current Management and Treatments. *Adv. Ther.* **2017**, *34*, 599–610. [CrossRef]
98. Fallacara, A.; Baldini, E.; Manfredini, S.; Vertuani, S. Hyaluronic Acid in the Third Millennium. *Polymers* **2018**, *10*, 701. [CrossRef]
99. Yang, J.; Wang, S. Polysaccharide-Based Multifunctional Hydrogel Bio-Adhesives for Wound Healing: A Review. *Gels* **2023**, *9*, 138. [CrossRef]
100. Liang, Y.; Zhao, X.; Hu, T.; Chen, B.; Yin, Z.; Ma, P.X.; Guo, B. Adhesive Hemostatic Conducting Injectable Composite Hydrogels with Sustained Drug Release and Photothermal Antibacterial Activity to Promote Full-Thickness Skin Regeneration During Wound Healing. *Small* **2019**, *15*, e1900046. [CrossRef]
101. Hussain, Z.; Thu, H.E.; Katas, H.; Bukhari, S.N.A. Hyaluronic Acid-Based Biomaterials: A Versatile and Smart Approach to Tissue Regeneration and Treating Traumatic, Surgical, and Chronic Wounds. *Polym. Rev.* **2017**, *57*, 594–630. [CrossRef]
102. Graca, M.F.P.; Miguel, S.P.; Cabral, C.S.D.; Correia, I.J. Hyaluronic acid-Based wound dressings: A review. *Carbohydr. Polym.* **2020**, *241*, 116364. [CrossRef]
103. Li, S.; Wang, L.; Zheng, W.; Yang, G.; Jiang, X. Rapid Fabrication of Self-Healing, Conductive, and Injectable Gel as Dressings for Healing Wounds in Stretchable Parts of the Body. *Adv. Funct. Mater.* **2020**, *30*, 2370. [CrossRef]
104. Kushwaha, J.; Singh, R. Cellulose hydrogel and its derivatives: A review of application in heavy metal adsorption. *Inorg. Chem. Commun.* **2023**, *152*, 721. [CrossRef]
105. Zainal, S.H.; Mohd, N.H.; Suhaili, N.; Anuar, F.H.; Lazim, A.M.; Othaman, R. Preparation of cellulose-based hydrogel: A review. *J. Mater. Res. Technol.* **2021**, *10*, 935–952. [CrossRef]
106. Chen, X.; Chen, J.; You, T.; Wang, K.; Xu, F. Effects of polymorphs on dissolution of cellulose in NaOH/urea aqueous solution. *Carbohydr. Polym.* **2015**, *125*, 85–91. [CrossRef]
107. Nguyen, H.M.; Ngoc Le, T.T.; Nguyen, A.T.; Thien Le, H.N.; Pham, T.T. Biomedical materials for wound dressing: Recent advances and applications. *RSC Adv.* **2023**, *13*, 5509–5528. [CrossRef]
108. Meng, S.; Wu, H.; Xiao, D.; Lan, S.; Dong, A. Recent advances in bacterial cellulose-based antibacterial composites for infected wound therapy. *Carbohydr. Polym.* **2023**, *316*, 121082. [CrossRef]
109. Wu, S.; Yang, Y.; Wang, S.; Dong, C.; Zhang, X.; Zhang, R.; Yang, L. Dextran and peptide-based pH-sensitive hydrogel boosts healing process in multidrug-resistant bacteria-infected wounds. *Carbohydr. Polym.* **2022**, *278*, 118994. [CrossRef]
110. Vatansever, F.; de Melo, W.C.; Avci, P.; Vecchio, D.; Sadasivam, M.; Gupta, A.; Chandran, R.; Karimi, M.; Parizotto, N.A.; Yin, R.; et al. Antimicrobial strategies centered around reactive oxygen species--bactericidal antibiotics, photodynamic therapy, and beyond. *FEMS Microbiol. Rev.* **2013**, *37*, 955–989. [CrossRef]
111. Liu, Y.; Xu, B.; Lu, M.; Li, S.; Guo, J.; Chen, F.; Xiong, X.; Yin, Z.; Liu, H.; Zhou, D. Ultrasmall Fe-doped carbon dots nanozymes for photoenhanced antibacterial therapy and wound healing. *Bioact. Mater.* **2022**, *12*, 246–256. [CrossRef] [PubMed]
112. Lee, K.Y.; Mooney, D.J. Alginate: Properties and biomedical applications. *Prog. Polym. Sci.* **2012**, *37*, 106–126. [CrossRef] [PubMed]
113. Ruvinov, E.; Cohen, S. Alginate biomaterial for the treatment of myocardial infarction: Progress, translational strategies, and clinical outlook: From ocean algae to patient bedside. *Adv. Drug Deliv. Rev.* **2016**, *96*, 54–76. [CrossRef]
114. Ren, J.; Yin, X.; Chen, Y.; Chen, Y.; Su, H.; Wang, K.; Zhang, L.; Zhu, J.; Zhang, C. Alginate hydrogel-coated syringe needles for rapid haemostasis of vessel and viscera puncture. *Biomaterials* **2020**, *249*, 120019. [CrossRef]
115. Zhao, X.; Yu, G.; Li, J.; Feng, Y.; Zhang, L.; Peng, Y.; Tang, Y.; Wang, L. Eco-Friendly Pickering Emulsion Stabilized by Silica Nanoparticles Dispersed with High-Molecular-Weight Amphiphilic Alginate Derivatives. *ACS Sustain. Chem. Eng.* **2018**, *6*, 4105–4114. [CrossRef]

116. Lin, F.; Liu, H.; Zhou, Q.; Zhang, S.; Zhou, Y.; Feng, Y.; Li, J. Amphiphilic alginate-based fluorescent polymer nanoparticles: Fabrication and multifunctional applications. *Int. J. Biol. Macromol.* **2021**, *183*, 2152–2161. [CrossRef]
117. Hu, C.; Long, L.; Cao, J.; Zhang, S.; Wang, Y. Dual-crosslinked mussel-inspired smart hydrogels with enhanced antibacterial and angiogenic properties for chronic infected diabetic wound treatment via pH-responsive quick cargo release. *Chem. Eng. J.* **2021**, *411*, 8564. [CrossRef]

Disclaimer/Publisher's Note: The statements, opinions and data contained in all publications are solely those of the individual author(s) and contributor(s) and not of MDPI and/or the editor(s). MDPI and/or the editor(s) disclaim responsibility for any injury to people or property resulting from any ideas, methods, instructions or products referred to in the content.

Article

Preparation and Antimicrobial Characterization of Poly(butylene adipate-*co*-terephthalate)/Kaolin Clay Biocomposites

Raja Venkatesan *,†, Krishnapandi Alagumalai † and Seong-Cheol Kim *

School of Chemical Engineering, Yeungnam University, Gyeongsan 38541, Republic of Korea
* Correspondence: rajavenki101@gmail.com (R.V.); sckim07@ynu.ac.kr (S.-C.K.)
† These authors contributed equally to this work.

Abstract: The biodegradable polymer poly(butylene adipate-co-terephthalate) (PBAT) starts decomposing at room temperature. Kaolin clay (KO) was dispersed and blended into PBAT composites using a solution-casting method. Fourier-transform infrared spectroscopy (FTIR), X-ray diffraction (XRD), scanning electron microscopy (SEM), and transmission electron microscopy (TEM) were used to evaluate the structure and morphology of the composite materials. PBAT/kaolin clay composites were studied by thermogravimetric analysis (TGA). The PBAT composite loaded with 5.0 wt% kaolin clay shows the best characteristics. The biocomposites of PBAT/kaolin [PBC-5.0 (37.6MPa)] have a good tensile strength when compared to virgin PBAT (18.3MPa). The oxygen transmission rate (OTR), with ranges from 1080.2 to 311.7 (cc/m^2/day), leads the KO content. By including 5.0 wt% kaolin 43.5 (g/m^2/day), the water vapor transmission rate (WVTR) of the PBAT/kaolin composites was decreased. The pure PBAT must have a WVTR of 152.4 (g/m^2/day). Gram-positive (*S. aureus*) and Gram-negative (*E. coli*) food-borne bacteria are significantly more resistant to the antimicrobial property of composites. The results show that PBAT/kaolin composites have great potential as food packaging materials due to their ability to decrease the growth of bacteria and improve the shelf life of packaged foods.

Keywords: poly(butylene adipate-co-terephthalate) (PBAT); kaolin; microstructure; food packaging

Citation: Venkatesan, R.; Alagumalai, K.; Kim, S.-C. Preparation and Antimicrobial Characterization of Poly(butylene adipate-*co*-terephthalate)/Kaolin Clay Biocomposites. *Polymers* **2023**, *15*, 1710. https://doi.org/10.3390/polym15071710

Academic Editor: Md. Amdadul Huq

Received: 8 February 2023
Revised: 15 March 2023
Accepted: 27 March 2023
Published: 29 March 2023

Copyright: © 2023 by the authors. Licensee MDPI, Basel, Switzerland. This article is an open access article distributed under the terms and conditions of the Creative Commons Attribution (CC BY) license (https:// creativecommons.org/licenses/by/ 4.0/).

1. Introduction

Thin-layer materials called biopolymer films and coatings have been used for a long time as food wrappers, carriers for transferring food products, and packing materials to preserve food products [1,2]. They perform as an excellent barrier against the transmission of oxygen and water vapor to prevent the undesired mass transfer and deterioration of food, thereby extending their shelf life and increasing food quality [3–5]. Biodegradable composites based on bioplastics are commonly used to produce food containers, storage carriages, and packing for food products [6,7]. These entirely prevent the circulation of O_2 and H_2O vapor, decreasing unwanted mass transfer and food spoilage to extend shelf life [8].

Biodegradable materials have received a lot of focus in recent times [9]. Because of its multiple benefits, such as being fully biodegradable, thermally stable [10], and flexible [11], for potential uses in food packaging, PBAT is a viable choice. Moreover, with its capacity to be produced through compression molding, it may be used as a material for food packaging [12]. However, to fulfill the specifications for its use in food packaging, PBAT's present antimicrobial activity is poor. As a result, it is still important to enhance the relevant characteristics [13–15]. Biodegradable polymer materials have been investigated as a desirable replacement for petroleum-based polymers to fulfill food packaging demands because of their characteristics of almost unlimited feedstock, biodegradability, and reduced costs [16–18]. PBAT has been chosen over all other biopolymers for its distinctive characteristics of being commonly available and thermally stable [19–21]. However,

PBAT biopolymers are a difficult alternative for food packaging materials due to their poor mechanical properties and limited capacity for forming an efficient oxygen and water barrier. Moreover, PBAT's poor antibacterial characteristics prohibit its application in food packaging [22–24]. Therefore, PBAT's main priority still is to increase mechanical properties and water and oxygen barrier characteristics while maintaining the desired antimicrobial activities.

To enhance the mechanical properties, barrier, and antimicrobial activities of biopolymers, fillers such as metal oxides and montmorillonite have been introduced [25–28]. It is believed that kaolin clays have a high potential for increasing biopolymer composites. Well-known fillers such as clay particles are often used in polymer composites due to their favorable characteristics [29]. They are used in several industrial applications, including biomedical, tissue engineering, and food packaging, as a result of their improved properties [30]. The most common clays are phyllosilicates, namely, layered silicates such as montmorillonite, talc, and kaolin [31,32]. Kaolin is chemically formed by two layers: octahedral $AlO_2(OH)_4$ and tetrahedral SiO_4 [33]. Furthermore, thermoplastic amylose/kaolin composites exhibited mechanical and thermal stability [34]. Recently, the increase in the mechanical and barrier characteristics of semolina film by adding kaolin has led to the use of these composites in packaged foods [35]. The research on PBAT/kaolin composites, therefore, is rarely addressed when considering materials for packaged foods.

In this paper, solution casting of PBAT and kaolin was used to produce several kinds of PBAT/kaolin composites with different concentrations. Characterizations, such as Fourier-transform infrared spectoscopy (FTIR), X-ray diffraction (XRD), transmission electron microscopy (TEM), scanning electron microscopy (SEM), and mechanical tests, were then used to evaluate the physiochemical properties and tensile strength of the composite materials. Kaolin was introduced to PBAT composite materials as an antimicrobial agent and for testing purposes against E. coli and S. aureus, two food-pathogenic bacteria. Finally, we measured the rate of water and oxygen vapor transmission in PBAT and PBAT/kaolin composites in an attempt to further study the barrier properties. The results provided an opportunity for the more extensive use of PBAT/kaolin clay composites in food packaging by demonstrating that kaolin clays increase mechanical, thermal, and antimicrobial activity while lowering barrier characteristics.

2. Materials and Methods

2.1. Materials

PBAT with a molecular weight (Mw) of 14.2×10^4 g mol^{-1} and a melting temperature of 110–120 °C was received from M/s BASF Ltd. in Tokyo, Japan. Samchun Chemicals provided the kaolin (marked as KO) clay for this study, with an average distribution particle size of 4 µm. Table 1 shows the chemical composition of KO, including CaO, K_2O, and Fe_2O_3, acting as the major impurities. Analysis reveals that silica (SiO_2: 47.85%) and aluminum oxide (Al_2O_3: 37.60%) were the major compositions. All chemicals were used from their source without any further purification.

Table 1. Chemical composition of the used kaolin (wt%).

SiO_2	Al_2O_3	Fe_2O_3	MgO	K_2O	CaO	TiO_2	LOI [a]
47.85	37.60	0.83	0.17	0.97	0.57	0.74	11.27

[a] LOI: loss on ignition at 1000 °C.

2.2. Preparation of PBAT/Kaolin Clay Composites

The preparation of PBAT/kaolin composites via solution mixing and drop casting [36] is shown in Figure 1. In 100 mL of chloroform, 2.0 g of PBAT polymer pellets were dissolved over a period of constant stirring to form a clear solution. The (1.0, 2.0, 3.0, and 5.0 wt%) KO materials dispersed immediately in the solution. The solution was placed into a Petri dish and then dried for 48 h at 40 °C in an oven [37,38]. The final stage was to dry the prepared films under a vacuum for 48 h to remove all residual solvents. After the majority of solvents had evaporated, the PBAT/kaolin composites were formed. After the film was dried for 8 h at 60 °C in a vacuum to become clear of residual solvents, the Petri dish was then removed to show the dried composite materials. The ratios are, respectively, PBC-0.0, PBC-1.0, PBC-2.0, PBC-3.0, and PBC-5.0. PBAT/kaolin composite films with wt% of 0.0, 1.0, 2.0, 3.0, and 5.0 of kaolin are summarized in Table 2. Films with a uniform thickness of 0.05 to 0.1 mm were developed. The obtained specimens were separated into 2.0 × 5.0 cm pieces for tensile strength.

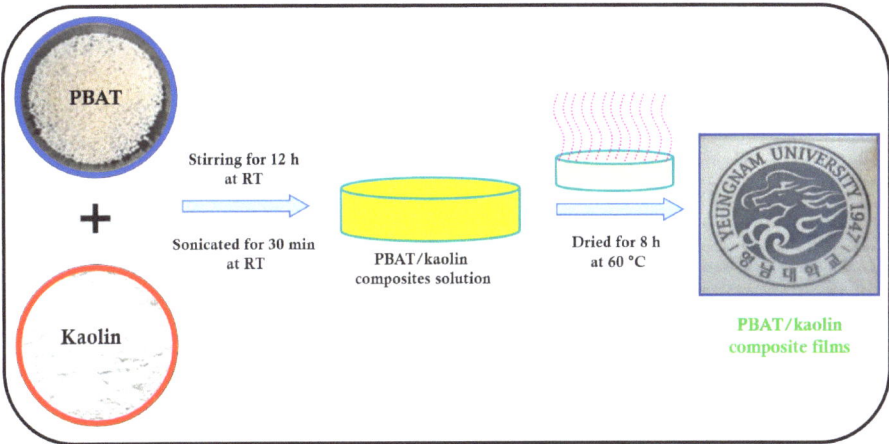

Figure 1. Fabrication diagram for PBAT/kaolin clay biocomposites.

Table 2. Material formulation in blends and composite film preparation.

S. No	Sample Name	PBAT Pellets (wt%)	Kaolin Clay (wt%)
1.	PBC-0.0	100.0	0.0
2.	PBC-1.0	99.0	1.0
3.	PBC-2.0	98.0	2.0
4.	PBC-3.0	97.0	3.0
5.	PBC-5.0	95.0	5.0

2.3. Characterization

2.3.1. Structural Characterization

With the help of a Fourier-transform infrared (FTIR) spectrophotometer (PerkinElmer, Spectrum 100, Waltham, MA, USA) with a resolution of 4 cm^{-1}, the attenuated total reflectance Fourier-transform infrared (ATR-FTIR) spectra of the PBAT composites were studied. The composite film was matched and placed directly on the ray-exposing plate. Around 400 and 4000 cm^{-1} of wavelength was utilized to record the spectra. The characterization of the PBAT composites was performed using an analytical X-ray diffraction meter (Rigaku) at room temperature (PANalytical). CuKα served as the radiation source (λ = 0.15406 nm), and rectangular sections of each film, approximately 2 × 2 cm, were placed on a glass slide to fabricate samples.

2.3.2. Morphological Studies

The morphology and structure of the PBAT composites were studied with scanning electron microscopy (SEM) (Hitachi, S-4800). The 10 nm gold–palladium was sputtered onto dried film samples using a Leica EM ACE200 sputter coater after being attached to a metal stub using double-sided carbon tape. Images were captured at an accelerating voltage of 20 kV. Transmission electron microscope (TEM) images were recorded with a JEM-2100 (JEOL, JEM-2100, Japan) at an acceleration voltage of 300 kV. Direct casting from the solution of the imaging samples onto copper grids followed by room-temperature drying has been used.

2.3.3. Thermogravimetric Analysis (TGA)

The thermal stability evaluation of PBAT composites is performed using TA instruments (QA 50), and the curves are of samples analyzed using the Universal V4.5A program. The weight of each sample is around 5.5 mg, and the temperature range for the measurement is about 40.0–700.0 °C with increments of 10 °C/minute at the rate of 10 °C min^{-1} under nitrogen atmosphere (50 mL min^{-1}).

2.3.4. Mechanical Properties

Instron material testing equipment (Instron 5545, MA) with a 1 kN load cell was used to evaluate the strength of composites in terms of tensile load. According to (ASTM D638-14) ASTM standards [39], the samples were examined at a cross-head speed of 10 mm/min with gauge lengths of 25 mm and dimensions of 150 × 25 × 0.04 mm. The gauge length and thickness values for each sample were noted. The cross-head speed was held constant at 1 mm/s for testing the specimens. At 50 ± 5% RH and 23 ± 2 °C, the tests were conducted. We collected and evaluated strength and distance to look for relationships between stress and strain. Then, average values for each material's tensile strength were calculated for each sample. All items underwent 3 independent tests. The values of the tensile strength were given in MPa.

2.3.5. Barrier Properties

The oxygen transmission rate (OTR) measurements for PBAT/kaolin composites were studied using a Noselab (ATS, Concorezzo, Italy) in accordance with the ASTM D3985-17 standard procedure [40]. Five different testing locations were utilized to evaluate the composite material, and the average result was taken. Whenever the material was at room temperature, preparation occurred. To use a Lyssy L80-5000, the water vapor transmission rate (WVTR) of the PBAT and PBAT/kaolin composites was evaluated in accordance with ASTM F1249-90 [41] at 100% RH and 23 C. The examination was performed 5 times, and an average result was calculated.

2.3.6. Measurements of the Contact Angle

The wettability of PBAT and its composite films was tested using a contact angle analyzer [35]. A small portion of the film was placed on a solid support with a flat base to be evaluated. The sessile drop technique was used to conduct the measurements in ambient conditions. Images were captured every 20 s using a high-resolution CCD camera after a water droplet (~2 µL) was thrown onto the specimens. With a contact angle meter from DataPhysics Instruments, OCA-20 Korea, the surface water contact angles were measured at room temperature and used to determine the hydrophilic nature of PBAT and its composites.

2.3.7. Antimicrobial Property Measurements

The zone inhibition method was employed to investigate the antimicrobial property of PBAT and its composites [42]. Briefly, a specific amount of bacteria culture (0.1 mL) was added to nutrient agar plates. The bacterial zones of inhibition were measured by covering the bacteria colonies with a circle film of 5 mm diameter and incubating them at 37 °C for 24 h.

2.3.8. Statistical Analysis

The statistical significance of every result was evaluated with ANOVA in SPSS 21 (IBM, New York, NY, USA). The data are provided as mean and standard deviation. A one-way analysis of variance was employed to determine statistical differences, and a result of $p < 0.05$ was the value of the maximum.

3. Results and Discussion
3.1. Characterization of Structural and Morphological Analysis

The morphological and structural properties of the PBAT and PBAT/kaolin composites are presented in Figure 2a. The vibration Si-O, which represents the absorption of kaolinite materials, appears in the absorption bands that appear at the wavelength range of 1010 and 1022 cm^{-1}. The peak close to 3000 cm^{-1} is related to C-H stretching for the aliphatic and aromatic regions. With an adjacent CH$_2$ peak at 720 cm^{-1}, C-O in the ester bond showed a strong peak at 1710 cm^{-1}. Figure 2b shows the XRD patterns of PBAT/kaolin composites at different weight percentages. The maximum values of 19.15°, 20.80°, 26.55°, 28.99°, and 50.07° are observed.

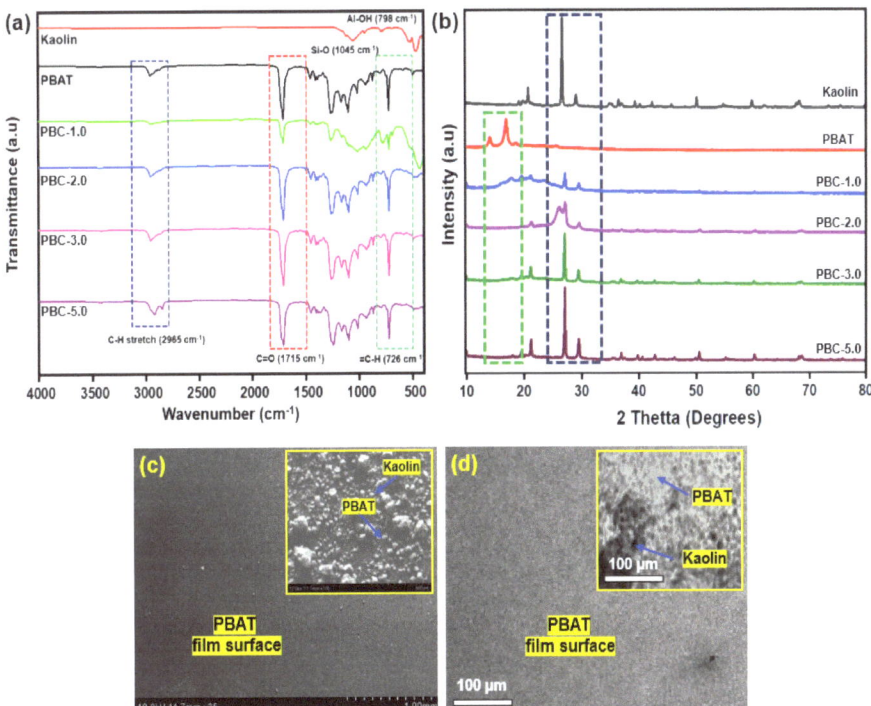

Figure 2. Characterizations of the microstructure and morphology of PBAT/kaolin composites: (**a**) FTIR spectrum; (**b**) XRD; (**c**) SEM image; PBC-5.0 film insets of (**c**); (**d**) TEM image of PBAT and PBAT/kaolin biocomposites (insets (**d**)).

These results, which are in good accordance with the results of additional studies and could be suggestive of the semi-crystalline structure of kaolin [43], are also in line with previous studies. The composite PBAT/kaolin clay film with kaolin clay loadings is shown in SEM images in Figure 2c. The surface of the PBAT/kaolin composite sample in Figure 2c (SEM image) is relatively rough due to the presence of kaolin particles, as compared to the neat PBAT's smooth surface. In the PBAT, the kaolin clay is distributed evenly in this image. The surface of the PBAT/kaolin composites (inset) in Figure 2c (SEM image) is noticeably rougher than the neat PBAT because kaolin powder has been added. The uniform distribution of kaolin in the PBAT/kaolin composite is further confirmed by the TEM image in Figure 2d. The significant shattering of kaolin during the composite preparation process could be a cause of the significant reduction in structured size.

3.2. Thermal Characterization of PBAT/Kaolin Clay Composites

Figure 3a shows TGA thermograms of PBAT and PBAT/kaolin composites. The initial degradation of the PBAT and PBAT/kaolin composite film from 285 to 372 °C can be ascribed to evaporation and the removal of aromatic impurities. With a weight loss of 21.08%, the PBAT matrix (PBAT-0.0) showed the final stage of degradation at 382–460 °C, which was induced by the thermal degradation of -C=O groups in PBAT. The percentage of the Kaolin mineral in the clay samples was interpolated from the TGA peak between 251 °C and 536 °C, usually as a result of dihydroxylation of the kaolinite mineral. Pure kaolin exhibits around 14% weight loss by heating between this temperature range. With a weight loss of 63.2%, PBAT film showed the final stage of degradation from 386 to 441 °C. With the introduction of 1.0, 2.0, 3.0, and 5.0 wt% (PBC-1.0, PBC-2.0, PBC-3.0, and PBC-5.0), kaolin further increased the onset of thermal degradation at 357, 338, 364, and 375 °C, respectively. This could be the result of the C=O groups in PBAT and Si-O-Si groups in kaolin in the composite film producing bonds among themselves and engaging in coordination interaction. When compared to the raw PBAT matrix, the thermal stability of PBC-1.0, PBC-2.0, PBC-3.0, and PBC-5.0 composite film increased by as much as 28–34%. According to our studies, the smaller kaolin clay particles (which have a higher surface area) and more uniform dispersion of these particles in PBAT are responsible for increased thermal degradation temperatures for PBAT/kaolin clay composites. With increasing amounts of particles introduced to the polymer, the residual weight increased, according to the results indicated.

The derivative weight loss curves (DTG), which show the rate at which the materials deteriorate, can be used to study the processes that cause weight loss as a result of heat deterioration more efficiently. For every formulation, these curves appear in Figure 3b. The presence of two peaks in the DTG curves showed that the thermal degradation of the PBAT film occurred in two steps. Kaolin constituents were engaged in the thermal degradation stage, whereas PBAT was the focus of the second. At around 110 °C, a minor peak was observed in the composite formulations. This results from the materials releasing moisture. The low-temperature decomposition (left shoulder of DTG curve) occurred at 300 and 380 °C and could be attributed to the thermal degradation of kaolin, the least thermally stable kaolin component, as well as the decomposition of PBAT. At the 350–450 °C temperature range, the second decomposition step was seen. It was assumed that this area had a connection with the degradation of kaolin. The DTG curve results indicate that the kaolin-added materials followed two-step processes, whereas these samples had a higher decomposition temperature than the control samples exhibited. The amount of char residues that combustion can form on the surface can increase the thermal stability.

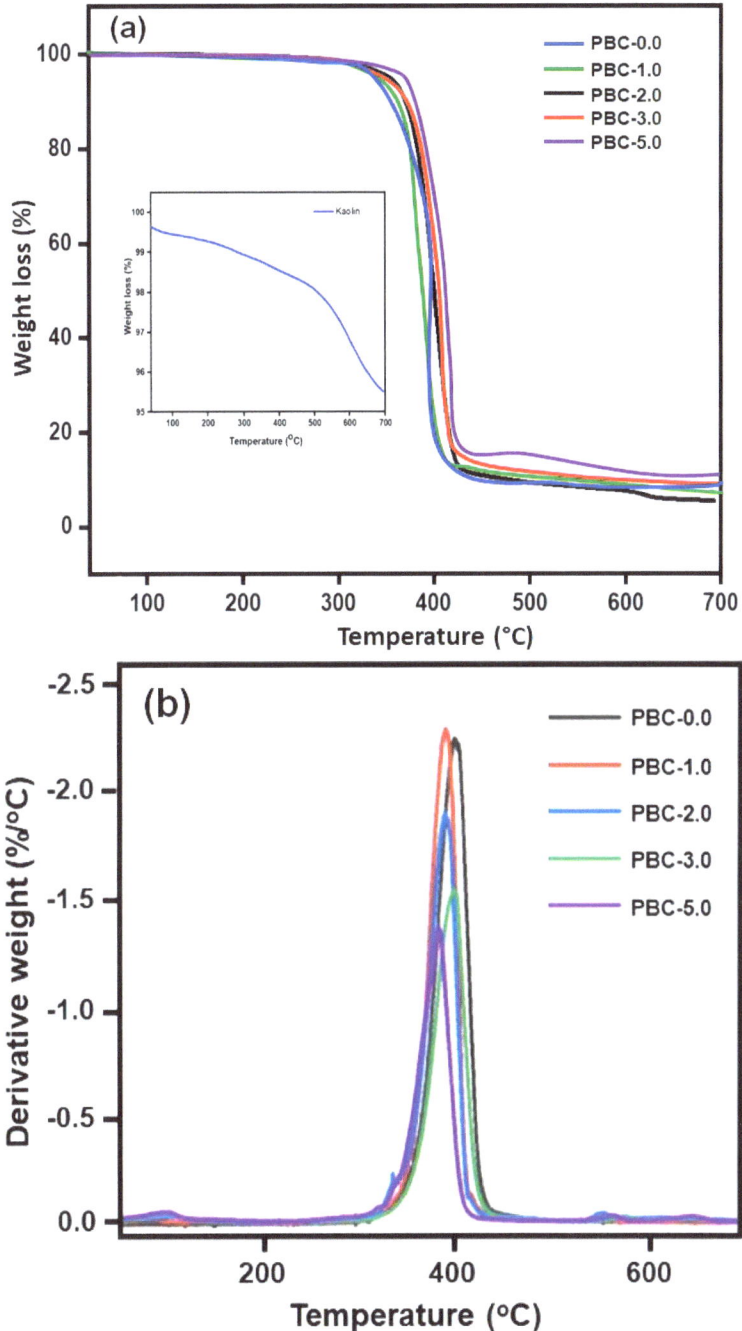

Figure 3. (**a**) TGA curves of PBAT and PBAT/kaolin clay composite samples. (**b**) DTG curves of PBAT and PBAT/kaolin clay composites.

3.3. Characterization of Mechanical and Barrier Properties

Figure 4a shows the properties of the technically described PBAT and PBAT/kaolin clay composite films that have been studied to evaluate their mechanical characteristics. The tensile strength (TS) and elongation at break (EAB) of kaolin-filled PBAT composites are influenced by filler loading, according to Figure 4a. The TS was observed to significantly increase as the kaolin concentration in the PBAT was increased from 1.0% to 5.0%. With respect to pure PBAT film, the PBAT/kaolin (5.0 wt%) composite film has a higher tensile strength (37.60 MPa) (18.34 MPa). While increased filler concentrations decrease the strength and lower EAB due to the agglomeration of fillers and the filler–matrix interface, it might be the reason for this discontinuity. An elastic material, neat PBAT has a high EAB (570.19%) and a low TS (18.34 MPa). When 1.0 wt% of kaolin is added, the tensile strength increases significantly (19.20 MPa), and the EAB decreases to 523.82%. Increased kaolin concentration at 5.0 wt% reduced EAB to 351.22% and increased PBAT (37.60 MPa). It is probably because the kaolin filler and PBAT exhibit strong interfacial bonding. In terms of stress, a successful interfacial bonding between the matrices and the filler is needed. Moreover, the overall altering behavior of TS increased, and EAB reduced as the amount of kaolin in the films decreased.

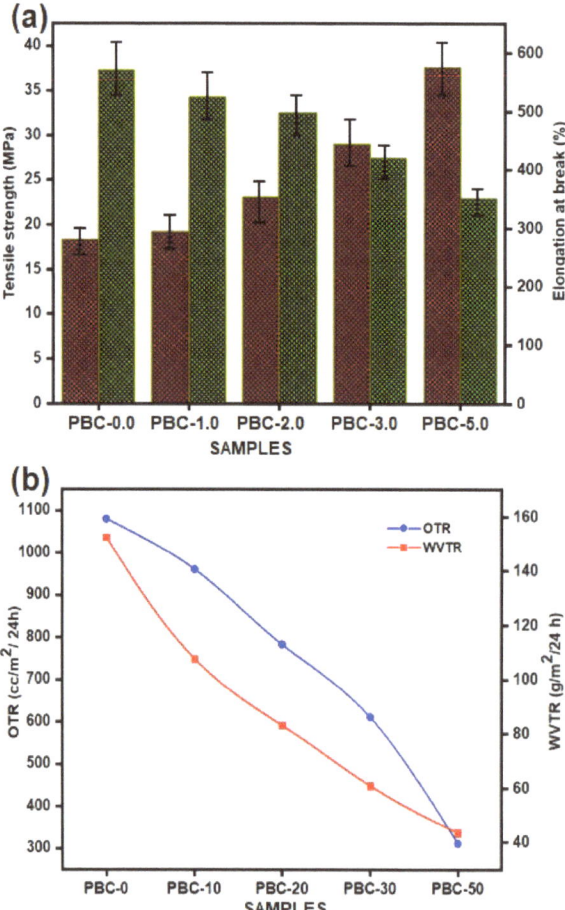

Figure 4. (**a**) Tensile strength and elongation at break of neat PBAT and PBAT/kaolin clay biocomposites; (**b**) OTR and WVTR of PBAT/kaolin clay biocomposites.

In Figure 4b, the WVTR and OTR of PBAT/kaolin composites are presented. H_2O and O_2 molecules can pass directly through PBAT without any barriers [44,45]. The OTR was 1080.21 cc/m^2/day.atm for the PBAT film. After kaolin (5.0 wt%) was mixed on PBAT, it decreased to 311.70 cc/m^2/day.atm. The value is significantly reduced by the addition of kaolin in different weight %. The minimum value of 783.15 cc/m^2/day.atm for kaolin was achieved for 2.0 wt%, irrespective of whether the OTR value was still decreased to 960.67 cc/m2/day.atm for 1.0 wt%. The process leading to the rise in permeability is the development of a tortuous path that poses a challenge for gas molecules to flow through the film. Additionally, the orientation and highest shuck (off) level of kaolin in PBAT led to a decrease in OTR. The addition of kaolin to the PBAT matrix significantly decreased the WVTR of the PBAT. According to the measurements, the WVTR values for PBC-0.0, PBC-1.0, PBC-2.0, PBC-3.0, and PBC-5.0 were 152.45, 107.62, 83.17, 60.92, and 43.50 g/m^2/day, respectively. WVTR decreased with kaolin concentration, and PBAT increased. While the WVTR of pure chitosan is 152.45 g/m^2/day, that of PBAT/kaolin clay composites varies from 152.45 to 43.50 g/m^2/day. The decreases in permeability of composites are attributed to the presence of the uniform dispersion of kaolin with increased percentages in the polymer matrix. However, in the PBAT/kaolin composites, these small molecules have migrated through or around the surfaces of impenetrable kaolin, which results in a long and convoluted pathway.

3.4. Water Contact Angle Measurement

Hydrophobicity experiments, or WCAs, were performed in order to evaluate the surface characteristics of PBAT/kaolin composites and identify how well the filler content impacted these characteristics. The wettability of a surface is evaluated using the contact angle in order to determine whether it is hydrophilic or hydrophobic. Surfaces with a surface contact angle greater than 90° are referred to as hydrophobic. The contact angle values for PBAT/kaolin clay composites with different kaolin wt% are shown in Figure 5. The contact angle of the PBAT film is 70.5°. In the event that PBAT and the PBAT film were mixed, the contact angle value increased to 93.1°.

This exhibits the hydrophobic character of PBAT. With PBAT containing 1.0 wt% kaolin, the contact angle value was increased to 75.1°. The contact angle value increased to 79.4° and 86.9° for kaolin concentrations of 2.0 and 3.0 wt% in the PBAT matrix, respectively. Kaolin was added, and as a result, the contact angle values increased, indicating the hydrophobic character of PBAT/kaolin clay composite films. Kaolin's hydrophobic characteristics were produced by the density-functional theory. In comparison to the absorption coefficient on the surface of kaolin, the bond length of hydrogen atoms is high. Water molecule aggregates therefore form on the surface. Hydrophobicity is strongly affected by a variety of additional factors, such as the degree of loading, compatibility, and polymer matrices. Balaji et al. showed that the water contact angle matched similar results [46].

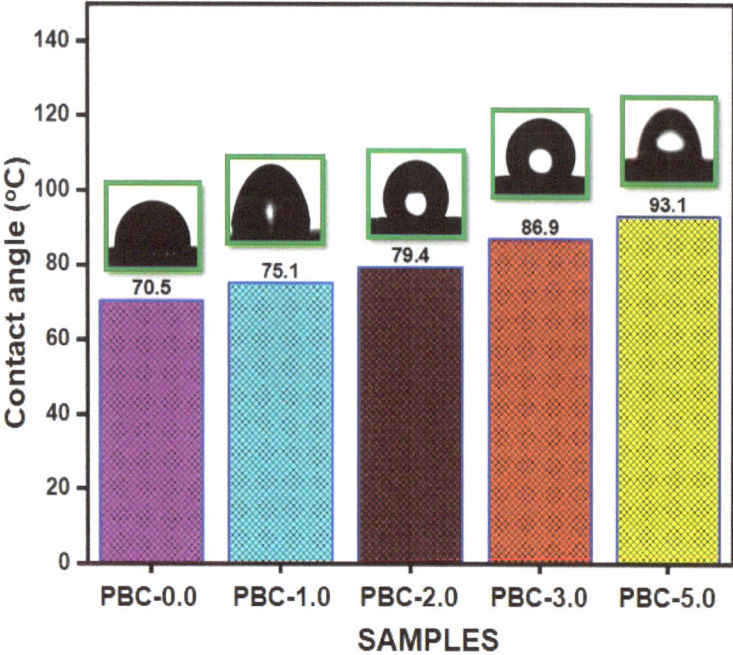

Figure 5. Water contact angle values of PBAT/kaolin biocomposite samples.

3.5. Antimicrobial Activities of PBAT/Kaolin Clay Composites

The inhibition zone against *S. aureus* and *E. coli* shows the results. The PBAT/kaolin composites, in comparison, show a clear antimicrobial zone of inhibition. Figure 6 shows the diameters of the antimicrobial zones in the PBAT/kaolin composites. The (PBC-5.0) composites' powerful antimicrobial activity against *S. aureus* and *E. coli* suggests that kaolin can increase antimicrobial activity. PBAT/kaolin composites revealed dramatically improved antimicrobial properties against food-borne pathogen microorganisms *E. coli* and *S. aureus* compared to neat PBAT. The zones of inhibition diameter of PBAT/kaolin biocomposites are 10.0, 11.3, 12.6, 14.0, and 16.4 mm against *E. coli* and 10.0, 12.4, 15.1, 18.2, and 19.7 mm against *S. aureus* with loadings of 0.0, 1.0, 2.0, 3.0, and 5.0 wt% kaolin, respectively. The slightly lower antimicrobial zone widths for *S. aureus* relative to those for *E. coli* and *S. aureus* provide additional confirmation that PBAT/Kaolin composite effects are real and significant against *S. aureus* microorganisms. With these materials, PBAT-based biodegradable plastics have a maximum effect and efficient antimicrobial activity.

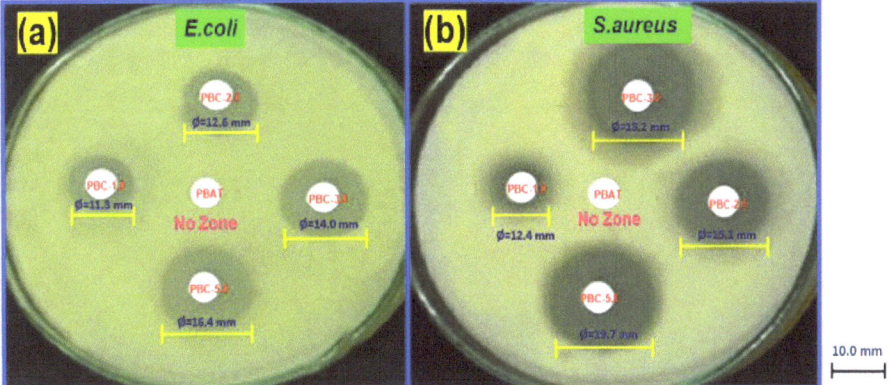

Figure 6. Antimicrobial properties of PBAT/kaolin clay composites with different wt% of kaolin; (**a**) *E. coli*, and (**b**) *S. aureus*.

4. Conclusions

We incorporated kaolin clay into PBAT to fabricate biocomposites. The incorporation of kaolin enhanced the mechanical properties of composites made from PBAT. The good interaction of the materials was shown by the FTIR spectra. According to the compatibility between both PBAT and kaolin, kaolin clay is dispersed uniformly in PBAT based on the morphology of the surface. The addition of kaolin clay was found to increase the thermal stability of PBAT composites, which were directly influenced by kaolin clay dispersion and concentration. Kaolin clay was discovered to improve tensile strength and film thickness when it was introduced to PBAT, but it had an opposite result on the film's elongation at break and barrier properties (OTR and WVTR). The transmission rates to H_2O and O_2 composites significantly decreased. With increasing kaolin, the PBAT's water contact angle improved from 70.5° to 93.1° and improved the hydrophobicity of the composite film. It is interesting to note that the produced composites show significantly improved antimicrobial properties against Gram-positive and Gram-negative bacteria, namely *E. coli* and *S. aureus*. This study suggests that PBAT/kaolin composites offer potential as materials for food packaging to prevent bacterial growth and extend the shelf life of food packages.

Author Contributions: Conceptualization, R.V.; methodology, R.V.; software, K.A.; validation, K.A.; formal analysis, R.V.; investigation, K.A.; resources, R.V.; data curation, K.A.; writing—original draft preparation, R.V.; writing—review and editing, S.-C.K.; visualization, R.V. and K.A.; supervision, S.-C.K.; project administration, S.-C.K.; funding acquisition, S.-C.K. All authors have read and agreed to the published version of the manuscript.

Funding: This research was funded by the Basic Science Research Program through the National Research Foundation of Korea (NRF), funded by the Ministry of Education (2020R1I1A3052258). In addition, the work was also supported by the Technology Development Program (S3060516), funded by the Ministry of SMEs and Startups (MSS, Republic of Korea), in 2021.

Institutional Review Board Statement: Not applicable.

Data Availability Statement: Not applicable.

Conflicts of Interest: The authors declare no conflict of interest.

References

1. Oshani, B.N.; Davachi, S.M.; Hejazi, I.; Seyfi, J.; Khonakdar, H.A.; Abbaspourrad, A. Enhanced compatibility of starch with poly(lactic acid) and poly(ε-caprolactone) by incorporation of POSS nanoparticles: Study on thermal properties. *Int. J. Biol. Macromol.* **2019**, *141*, 578–584. [CrossRef] [PubMed]
2. Rammak, T.; Boonsuk, P.; Kaewtatip, K. Mechanical and barrier properties of starch blend films enhanced with kaolin for ap-plication in food packaging. *Int. J. Biol. Macromol.* **2021**, *192*, 1013–1020. [CrossRef] [PubMed]
3. Castillo, L.; López, O.; López, C.; Zaritzky, N.; García, M.A.; Barbosa, S.; Villar, M. Thermoplastic starch films reinforced with talc nanoparticles. *Carbohydr. Polym.* **2013**, *95*, 664–674. [CrossRef] [PubMed]
4. Romani, V.P.; Martins, V.G.; Goddard, J.M. Radical scavenging polyethylene films as antioxidant active packaging materials. *Food Control* **2020**, *109*, 106946. [CrossRef]
5. Shaikh, S.; Yaqoob, M.; Aggarwal, P. An overview of biodegradable packaging in food industry. *Curr. Res. Food Sci.* **2021**, *4*, 503–520. [CrossRef]
6. Bangar, S.P.; Whiteside, W.S.; Ashogbon, A.O.; Kumar, M. Recent advances in thermoplastic starches for food packaging: A review. *Food Packag. Shelf Life* **2021**, *30*, 100743. [CrossRef]
7. Khalid, M.Y.; Arif, Z.U. Novel biopolymer-based sustainable composites for food packaging applications: A narrative review. *Food Packag. Shelf Life* **2022**, *33*, 100892. [CrossRef]
8. Arrieta, M.P.; López, J.; Hernández, A.; Rayón, E. Ternary PLA–PHB–Limonene blends intended for biodegradable food packaging applications. *Eur. Polym. J.* **2014**, *50*, 255–270. [CrossRef]
9. Yadav, M. Study on thermal and mechanical properties of cellulose/iron oxide bionanocomposites film. *Compos. Commun.* **2018**, *10*, 1–5. [CrossRef]
10. Venkatesan, R.; Rajeswari, N. Nanosilica-reinforced poly(butylene adipate-co-terephthalate) nanocomposites: Preparation, characterization and properties. *Polym. Bull.* **2019**, *76*, 4785–4801. [CrossRef]
11. Guo, G.; Zhang, C.; Du, Z.; Zou, W.; Tian, H.; Xiang, A.; Li, H. Structure and property of biodegradable soy protein isolate/PBAT blends. *Ind. Crop. Prod.* **2015**, *74*, 731–736. [CrossRef]
12. Leporatti, S. Polymer Clay Nano-composites. *Polymers* **2019**, *11*, 1445. [CrossRef]
13. Venkatesan, R.; Zhang, Y.; Chen, G. Preparation of poly(butylene adipate-co-terephthalate)/ZnSnO3 composites with enhanced antimicrobial activity. *Compos. Commun.* **2020**, *22*, 100469. [CrossRef]
14. Arumugam, S.; Kandasamy, J.; Thiyaku, T.; Saxena, P. Effect of Low Concentration of SiO_2 Nanoparticles on Grape Seed Essential Oil/PBAT Composite Films for Sustainable Food Packaging Application. *Sustainability* **2022**, *14*, 8073. [CrossRef]
15. Jaramillo, A.F.; Riquelme, S.A.; Sánchez-Sanhueza, G.; Medina, C.; Solís-Pomar, F.; Rojas, D.; Montalba, C.; Melendrez, M.F.; Pérez-Tijerina, E. Comparative Study of the Antimicrobial Effect of Nanocomposites and Composite Based on Poly(butylene adipate-co-terephthalate) Using Cu and Cu/Cu_2O Nanoparticles and $CuSO_4$. *Nanoscale Res. Lett.* **2019**, *14*, 158. [CrossRef]
16. Ibrahim, I.D.; Hamam, Y.; Sadiku, E.R.; Ndambuki, J.M.; Kupolati, W.K.; Jamiru, T.; Eze, A.A.; Snyman, J. Need for Sustainable Packaging: An Overview. *Polymers* **2022**, *14*, 4430. [CrossRef] [PubMed]
17. Sid, S.; Mor, R.S.; Kishore, A.; Sharanagat, V.S. Bio-sourced polymers as alternatives to conventional food packaging materials: A review. *Trends Food Sci. Technol.* **2021**, *115*, 87–104. [CrossRef]
18. Siracusa, V.; Rocculi, P.; Romani, S.; Rosa, M.D. Biodegradable polymers for food packaging: A review. *Trends Food Sci. Technol.* **2008**, *19*, 634–643. [CrossRef]
19. Dammak, M.; Fourati, Y.; Tarrés, Q.; Delgado-Aguilar, M.; Mutjé, P.; Boufi, S. Blends of PBAT with plasticized starch for packaging applications: Mechanical properties, rheological behaviour and biodegradability. *Ind. Crop. Prod.* **2020**, *144*, 112061. [CrossRef]
20. Lackner, M.; Ivanič, F.; Kováčová, M.; Chodák, I. Mechanical properties and structure of mixtures of poly(butylene-adipate-co-terephthalate) (PBAT) with thermoplastic starch (TPS). *Int. J. Biobased Plast.* **2021**, *3*, 126–138. [CrossRef]
21. Tsou, C.-H.; Chen, Z.-J.; Yuan, S.; Ma, Z.-L.; Wu, C.-S.; Yang, T.; Jia, C.-F.; De Guzman, M.R. The preparation and performance of poly(butylene adipate) terephthalate/corn stalk composites. *Curr. Res. Green Sustain. Chem.* **2022**, *5*, 100329. [CrossRef]
22. Wang, X.; Cui, L.; Fan, S.; Li, X.; Liu, Y. Biodegradable Poly(butylene adipate-*co*-terephthalate) Antibacterial Nanocomposites Reinforced with MgO Nanoparticles. *Polymers* **2021**, *13*, 507. [CrossRef] [PubMed]
23. Zhong, Y.; Godwin, P.; Jin, Y.; Xiao, H. Biodegradable polymers and green-based antimicrobial packaging materials: A mini-review. *Adv. Ind. Eng. Polym. Res.* **2020**, *3*, 27–35. [CrossRef]
24. Díez-Pascual, A.M.; Díez-Vicente, A.L. Antimicrobial and sustainable food packaging based on poly(butylene adipate-co-terephthalate) and electrospun chitosan nanofibers. *RSC Adv.* **2015**, *5*, 93095–93107. [CrossRef]
25. Jamróz, E.; Kulawik, P.; Kopel, P. The Effect of Nanofillers on the Functional Properties of Biopolymer-Based Films: A Review. *Polymers* **2019**, *11*, 675. [CrossRef] [PubMed]
26. Jafarzadeh, S.; Jafari, S.M. Impact of metal nanoparticles on the mechanical, barrier, optical and thermal properties of biodegrad-able food packaging materials. *Crit. Rev. Food Sci. Nutr.* **2021**, *61*, 2640–2658. [CrossRef]
27. Rhim, J.-W.; Wang, L.-F. Preparation and characterization of carrageenan-based nanocomposite films reinforced with clay mineral and silver nanoparticles. *Appl. Clay Sci.* **2014**, *97–98*, 174–181. [CrossRef]
28. Chausali, N.; Saxena, J.; Prasad, R. Recent trends in nanotechnology applications of bio-based packaging. *J. Agric. Food Res.* **2022**, *7*, 100257. [CrossRef]

29. Lee, Y.H.; Kuboki, T.; Park, C.B.; Sain, M.; Kontopoulou, M. The effects of clay dispersion on the mechanical, physical, and flame-retarding properties of wood fiber/polyethylene/clay nanocomposites. *J. Appl. Polym. Sci.* **2020**, *118*, 452–461. [CrossRef]
30. Powell, C.E.; Beall, G.W. Physical Properties of Polymer/Clay Nanocomposites. In *Physical Properties of Polymers Handbook*; Springer: New York, NY, USA, 2007; pp. 561–575. [CrossRef]
31. Hu, L.; Leclair, E.; Poulin, M.; Colas, F.; Baldet, P.; Vuillaume, P.Y. Clay/Polyethylene Composites with Enhanced Barrier Properties for Seed Storage. *Polym. Polym. Compos.* **2016**, *24*, 387–394. [CrossRef]
32. Muthuraj, R.; Misra, M.; Defersha, F.; Mohanty, A.K. Influence of processing parameters on the impact strength of biocomposites: A statistical approach. *Compos. Part A Appl. Sci. Manuf.* **2016**, *83*, 120–129. [CrossRef]
33. Bhattacharyya, K.G.; Gupta, S.S. Adsorption of a few heavy metals on natural and modified kaolinite and montmorillonite: A review. *Adv. Colloid Interface Sci.* **2008**, *140*, 114–131. [CrossRef]
34. Huang, M.; Wang, H.; Yu, J. Studies of biodegradable thermoplastic amylose/kaolin composites: Fabrication, characterization, and properties. *Polym. Compos.* **2006**, *27*, 309–314. [CrossRef]
35. Jafarzadeh, S.; Alias, A.K.; Ariffin, F.; Mahmud, S.; Najafi, A. Preparation and characterization of bionanocomposite films reinforced with nano kaolin. *J. Food Sci. Technol.* **2016**, *53*, 1111–1119. [CrossRef]
36. Venkatesan, R.; Alagumalai, K.; Kim, S.-C. Preparation and Performance of Biodegradable Poly(butylene adipate-*co*-terephthalate) Composites Reinforced with Novel $AgSnO_2$ Microparticles for Application in Food Packaging. *Polymers* **2023**, *15*, 554. [CrossRef]
37. Venkatesan, R.; Alagumalai, K.; Raorane, C.J.; Raj, V.; Shastri, D.; Kim, S.-C. Morphological, Mechanical, and Antimicrobial Properties of PBAT/Poly(methyl methacrylate-*co*-maleic anhydride)–SiO_2 Composite Films for Food Packaging Applications. *Polymers* **2023**, *15*, 101. [CrossRef]
38. Venkatesan, R.; Vanaraj, R.; Alagumalai, K.; Asrafali, S.P.; Raorane, C.J.; Raj, V.; Kim, S.-C. Thermoplastic Starch Composites Reinforced with Functionalized POSS: Fabrication, Characterization, and Evolution of Mechanical, Thermal and Biological Activities. *Antibiotics* **2022**, *11*, 1425. [CrossRef] [PubMed]
39. *ASTM D638-14*; Standard Test Method for Tensile Properties of Plastics. ASTM International: West Conshohocken, PA, USA, 2022.
40. *ASTM D3985-17*; Standard Test Method for Oxygen Gas Transmission Rate through Plastic Film and Sheeting Using a Coulometric Sensor. ASTM International: West Conshohocken, PA, USA, 2022.
41. *ASTM F1249-90*; Standard Test Method for Water Vapor Transmission Rate through Plastic Film and Sheeting Using a Modulated Infrared Sensor. ASTM International: West Conshohocken, PA, USA, 2022.
42. Sundaram, I.M.; Kalimuthu, S.; Ponniah, G. Highly active ZnO modified g-C_3N_4 Nanocomposite for dye degradation under UV and Visible Light with enhanced stability and antimicrobial activity. *Compos. Commun.* **2017**, *5*, 64–71. [CrossRef]
43. Dewi, R.; Agusnar, H.; Alfian, Z. Tamrin Characterization of technical kaolin using XRF, SEM, XRD, FTIR and its potentials as industrial raw materials. *J. Phys. Conf. Ser.* **2018**, *1116*, 042010. [CrossRef]
44. Venkatesan, R.; Rajeswari, N. TiO_2 nanoparticles/poly(butylene adipate-co-terephthalate) bionanocomposite films for packaging applications. *Polym. Adv. Technol.* **2017**, *28*, 1699–1706. [CrossRef]
45. Venkatesan, R.; Rajeswari, N. ZnO/PBAT nanocomposite films: Investigation on the mechanical and biological activity for food packaging. *Polym. Adv. Technol.* **2017**, *28*, 20–27. [CrossRef]
46. Balaji, S.; Venkatesan, R.; Mugeeth, L.; Dhamodharan, R. Hydrophobic nanocomposites of PBAT with Cl-fn-POSS nanofiller as compostable food packaging films. *Polym. Eng. Sci.* **2021**, *61*, 314–326. [CrossRef]

Disclaimer/Publisher's Note: The statements, opinions and data contained in all publications are solely those of the individual author(s) and contributor(s) and not of MDPI and/or the editor(s). MDPI and/or the editor(s) disclaim responsibility for any injury to people or property resulting from any ideas, methods, instructions or products referred to in the content.

Article

Mechanical Properties of Poly(Alkenoate) Cement Modified with Propolis as an Antiseptic

David Alejandro Aguilar-Perez [1], Cindy Maria Urbina-Mendez [1], Beatriz Maldonado-Gallegos [2], Omar de Jesus Castillo-Cruz [2], Fernando Javier Aguilar-Ayala [1], Martha Gabriela Chuc-Gamboa [1], Rossana Faride Vargas-Coronado [2] and Juan Valerio Cauich-Rodriguez [2,*]

[1] Facultad de Odontologia, Universidad Autonoma de Yucatan, Calle 61-A x Av., Itzaes Costado Sur "Parque de la Paz", Col. Centro, Merida 97000, Yucatan, Mexico; david.aguilar@correo.uady.mx (D.A.A.-P.)
[2] Centro de Investigacion Cientifica de Yucatan A.C, Calle 43 # 130 x 32 y 34, Colonia Chuburna de Hidalgo, Merida 97205, Yucatan, Mexico
* Correspondence: jvcr@cicy.mx; Tel.: +52-999-942-83-30 (ext. 424)

Abstract: Background: We assessed the effect of propolis on the antibacterial, mechanical, and adhesive properties of a commercial poly(alkenoate) cement. Methods: The cement was modified with various concentrations of propolis, and antibacterial assays were performed against *S. mutans* by both MTT assays and agar diffusion tests. The compressive, flexural, and adhesive properties were also evaluated. Results: the modified cement showed activity against *S. mutans* in both assays, although reductions in compressive (from 211.21 to 59.3 MPa) and flexural strength (from 11.1 to 6.2 MPa) were noted with the addition of propolis, while adhesive strength (shear bond strength and a novel pull-out method) showed a statistical difference ($p < 0.05$). Conclusion: the antiseptic potential of modified material against *S. mutans* will allow this material to be used in cases in which low mechanical resistance is required (in addition to its anti-inflammatory properties) when using atraumatic restorative techniques, especially in deep cavities.

Keywords: bioactive materials; antibacterial properties; antibiofilm activity

Citation: Aguilar-Perez, D.A.; Urbina-Mendez, C.M.; Maldonado-Gallegos, B.; Castillo-Cruz, O.d.J.; Aguilar-Ayala, F.J.; Chuc-Gamboa, M.G.; Vargas-Coronado, R.F.; Cauich-Rodriguez, J.V. Mechanical Properties of Poly(Alkenoate) Cement Modified with Propolis as an Antiseptic. *Polymers* **2023**, *15*, 1676. https://doi.org/10.3390/polym15071676

Academic Editors: Md. Amdadul Huq and Shahina Akter

Received: 18 February 2023
Revised: 23 March 2023
Accepted: 24 March 2023
Published: 28 March 2023

Copyright: © 2023 by the authors. Licensee MDPI, Basel, Switzerland. This article is an open access article distributed under the terms and conditions of the Creative Commons Attribution (CC BY) license (https://creativecommons.org/licenses/by/4.0/).

1. Introduction

Poly(alkenoate) cement (PAC) is a material that contains calcium fluoroaluminosilicate glass and an aqueous solution of poly(acrylic acid) (PAA) in the original formulation and is commonly used as a base or restorative material in dentistry [1]. High-viscosity PAC is recommended for atraumatic restorative techniques (ARTs) [2], but their applications have been expanded for bone cement replacement when properly formulated [3,4]. Some of their benefits include low pulp inflammation, chemical adhesion to enamel and dentine, low shrinkage, a similar coefficient of thermal expansion to dental tissue, sustained fluoride release, good resistance to marginal filtration, and acceptable mechanical properties [5].

Despite all these advantages, earlier improvements in the original formulation were due to moisture sensitivity and being prone to incomplete crosslinking in the presence of unionized PAA, being especially relevant in PAC as luting cement, as they contain more PAA, which inhibits the formation of apatite interlayers, which are essentials to marginal gap sealing [6]. Therefore, to palliate these undesirable properties and to enhance its properties, the use of additives (i.e., tartaric acid for stronger aluminum complex formation and extending working times) [7], polyelectrolytes based on the copolymers of acrylic acid and unsaturated acids (maleic or itaconic acid for more rigid crosslinking with aluminum) and the inclusion of metals, such as silver and tin, has been proposed [8]. Recently, other types of improvements have considered the use of different fillers and novel compositions to enhance mechanical performance [9].

The mechanical performance of PAC is assessed by both compressive and bending behavior, which are generally defined by international standards despite the fact that the material in oral cavities is subject to complex loads [10,11]. Thus, the strength values measured by using international standards do not represent the mechanical conditions necessary for clinical uses [12]. Therefore, the reported mechanical properties of commercial dental materials vary widely, and generally, when a new formulation of PAC is developed, it is common to consider that higher mechanical properties are desirable [13].

Another important parameter for dental material performance is the adhesion to either the enamel or dentine in the tooth; conventional PAC bonding to dentine, which contains more water while being less mineralized, is more difficult compared to enamel, but due to its hydrophilic nature, it can wet the surface and provide adequate bonding, especially after removing the smear layer with weak (ascorbic) or strong (phosphoric) acids [14].

Many efforts were made to improve adhesion, and a wide variety of tests have been proposed to determine this property among the restorative material and the tooth surface as the shear [13,15], tension [16], and push-out bond strength [17]. Other mechanical methods were proposed in the ISO/TS 11405 standard (dentistry-testing of adhesion to tooth structure) [18] and ISO 29022 standard (dentistry-adhesion-notched edge shear bond strength test) [19].

Fluoride release is another important issue regarding PAC, not only because of its remineralization potential on recurrent caries, but also because it inhibits plaque formation, with claimed antibacterial action [20]. However, this is not enough to overcome secondary caries, and several additives have been used, including chlorhexidine (CHX), quaternary ammonium salts (QAS), metallic particles, modified polymers, etc. [21–24]. Propolis, a natural product with well-documented antibacterial and antimycotic properties, has also been suggested as an additive to dental materials [25–28]. Other natural extracts that have been used include green tea, *Triphala* (an ayurvedic herbal formulation that contains three medicinal plants: *T. chebula*, *T. belerica*, and *Phyllanthus embelica*), *Salvadora persica*, *Olea europaea*, and *Ficus carcia*, which has been proven to have numerous benefits [29–31].

Propolis-modified dental materials can accelerate wound healing due to anti-inflammatory responses, as tissue repair is mediated by their inflammatory mechanism [32]. Furthermore, controversial results exist on propolis addition to PAC; for example, microhardness has been reported to increase via the use of the ethanolic extracts of propolis [33], whereas a reduction in compression strength has been reported due to adding propolis to PAC type II [34], but with an increase in compressive strength regarding high-viscosity PAC [35]. However, in these studies, little emphasis was put on the changes in terms of tooth adhesion. Therefore, the aim of this study was to assess the effect of propolis on *Streptococcus mutans* viability, identified as the most common micro-organism associated with the initial phase of caries [36]. In addition, the mechanical properties of compression, flexion, and adhesion (newly proposed pull-out test) were studied.

2. Materials and Methods

2.1. Materials

Ethanolic solution of propolis (20 wt%, Brand Yucamiel) with a total flavonoid content of 25.94 ± 2.06 mg of quercitin/g, phenol content of 49.68 ± 0.29 mg garlic acid/g, 2.5 µg/mL of average inhibitory concentration (IC50), and antiradical power (1/IC50) of 0.40 was used for all experiments, as reported in a prior publication [32]. Fuji IX PAC was purchased from GC Corporation (Tokyo, Japan), while *S. mutans* was acquired from ATCC (25175).

2.2. Preparation of PAC and Modified PAC (MPAC)

Both materials, PAC and MPAC (formulations described in Table 1), were prepared according to the manufacturer's specifications by hand mixing the powder and aqueous solution of PAA (3.6:1, respectively) and then adding a specific volume of 20 wt% ethanolic solutions of propolis [27,28,33]. After the final mixing, the paste was placed in molds with the corresponding geometry for each test.

Table 1. Compositions and abbreviations used for experimental samples.

Sample	Preparation
PAC	Unmodified material
25MPAC	PAC + 25 µL of propolis
50MPAC	PAC + 50 µL of propolis
MPAC-NA	PAC + 10 µL without acid tissue conditioning
MPAC-A	PAC + 10 µL with acid tissue conditioning
PAC-NA	PAC without acid tissue conditioning
PAC-A	PAC with acid tissue conditioning
FASP	Fluoroaluminosilicate powder

2.3. Physicochemical Characterization of PAC and MPAC

Fourier-transform infrared (FTIR) spectra were obtained using a Nicolet 8700 spectrometer in the spectral range between 4000 and 400 cm^{-1} using KBr pellets. X-ray diffraction (XRD) patterns were obtained using a Siemens D-5000 Bragg diffractometer in the 2θ range from 10° to 60°, with an interval time lapse of 4 s and a step size of 0.02°.

2.4. Microbiological Test

Both materials, PAC and MPAC (12.5 µL, 25 µL, and 50 µL of propolis), were prepared in a silicon mold with a 6.35 mm diameter and 2 mm in thickness. Then, each disc was immersed in 3 mL of sterile distilled water for 24 h at 37 °C. Finally, 100 µL of the obtained eluates were placed in a 96-well plate along with 100 µL of the *S. mutans* inoculum. As a negative control, amikacin (1 mg/mL) was used, while only bacteria were used as the positive control. Bacteria viability was determined using 100 µL of MTT after incubation for 24 h at 37 °C. Additionally, the antibacterial activity of MPAC against *S. mutans* was also assessed by the agar diffusion method; discs of the same dimension (as mentioned above) were placed over bacterial seeding of brain heart infusion (BHI) agar and then placed in an incubator at 37 °C for 24 h. A paper soaked in amikacin was used as the negative control, and a disc without propolis was used as the positive control.

2.5. Mechanical Characterization

2.5.1. Compressive

Tests were conducted according to the ISO 9917-1 standard, using cylindrical samples 6 mm in height and 4 mm in diameter that were obtained using a Teflon mold.

2.5.2. Three-Point Bending

Tests on PAC and propolis-modified MPAC were conducted according to ISO 9917-2. For this, rectangular specimens (25 mm × 5 mm × 1.75 mm) were obtained after Teflon mold casting. Thus, the samples were stored in distilled water at 37 °C for 24 h according to the standard.

Both mechanical tests were conducted on samples containing 25 µL and 50 µL of propolis and carried out in a Shimadzu AGS–X (Kyoto, Japan) universal testing machine with a 5 kN and 1 kN load cell, respectively, and a crosshead speed of 1 mm/min. Five samples were used by the group, then the mean and standard deviation were reported.

2.5.3. Shear Bond Test

Adhesion to dentine was assessed by shear bond testing, suggested by the ISO 29022 standard and by other authors [37,38]; molars were polished with silicon carbide (number 400) abrasive paper until the dentin was exposed, Figure 1a, with an average area of 5 mm^2; then, conditioning the tissue with poly(acrylic acid) for 15 s was carried out, and then a cylinder was cured over the treated surface (MPAC-A). To understand the effect of the propolis, shear adhesive behavior was assessed after four different treatments. For the first case, dentin was propolis-treated with no acid, and then PAC was cured on the surface, referred to as PAC-NA. For the second case, the exposed dentin was treated with propolis and then 1 mL of poly(acrylic acid), instead of the conventional 32–37% phosphoric acid, for 15 s, and then a PAC cylinder was cured over the treated surface (referred as PAC-A). In addition, two controls were manufactured as follows: (a) the first control was prepared with PAC without propolis, while dentin was treated with poly(acrylic acid), denoted as PAC-A, and (b) PAC was cured on dentine without propolis treatment and without poly(acrylic acid) treatment, referred to as PAC-NA. All samples described above were stored in distilled water, Figure 1b, for 24 h at 37 °C according to the standard. The test was carried out with a Shimadzu AGS-X (Kyoto, Japan) universal testing machine, with a 100 N load cell and crosshead speed of 1 mm/min, Figure 1c; five samples were tested, with the mean and standard deviation reported. After detaching, the samples were gold coated and observed by using a JEOL JMS 6360LV scanning electron microscope with an accelerating voltage of 20 kV.

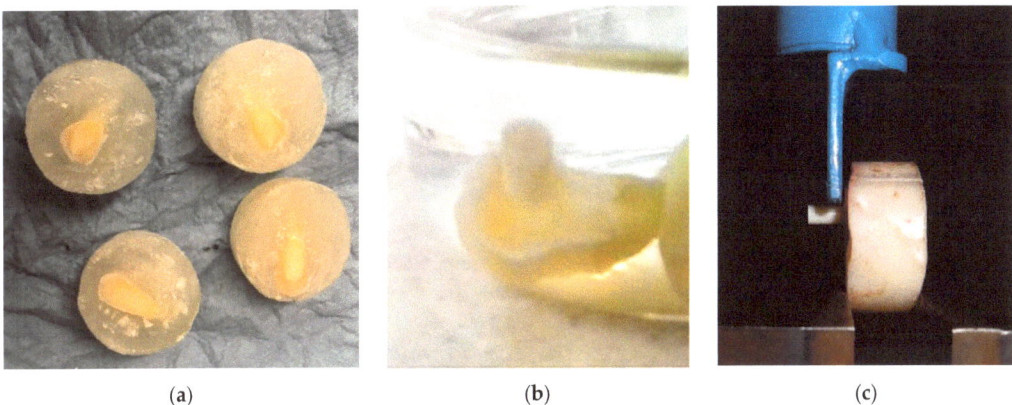

Figure 1. Preparation of samples for shear adhesive test. (**a**) Dentin after polishing. (**b**) MPAC cured on dentin. (**c**) Shear adhesion test.

2.5.4. Pull-Out Test

In addition, a nonstandardized pull-out test was used as a novel method, using the inner surface of the dental tissue treated by different protocols before PAC filling. Groups of 5 premolars per test were embedded vertically in an acrylic resin and then polished until the occlusal portion was removed and a flat surface was obtained (Figure 2a). This flat surface was drilled 6 mm deep while irrigating with distilled water. Strength behavior was assessed prior to the four different treatments; (1) dentin was propolis-treated with no acid (MPAC-NA), then filled with the PAC; (2) the exposed dentin was treated with propolis, then with 1 mL of poly(acrylic acid) for 15 s, and then a PAC cylinder was cured over the treated surface (MPAC-A); (3) PAC without propolis, while the dentin was treated with poly(acrylic acid), denoted as PAC-A, and (4) PAC was cured on dentine without propolis treatment and without poly(acrylic acid) treatment, referred to as PAC-NA. Before the curing process, a metallic root was inserted in the center of the cavity. To verify its alignment, periapical radiographs were obtained, Figure 2b; then, the samples were placed in an incubator in the presence of a saline solution for 24 h at 37 °C. The maximum adhesive force required to remove the complete PAC from the cavity treated was recorded using a Shimadzu AGS-X (Kyoto, Japan) universal testing machine with a 5 kN load cell and a crosshead speed of 1 mm/min, Figure 2c. The mean and standard deviation were reported.

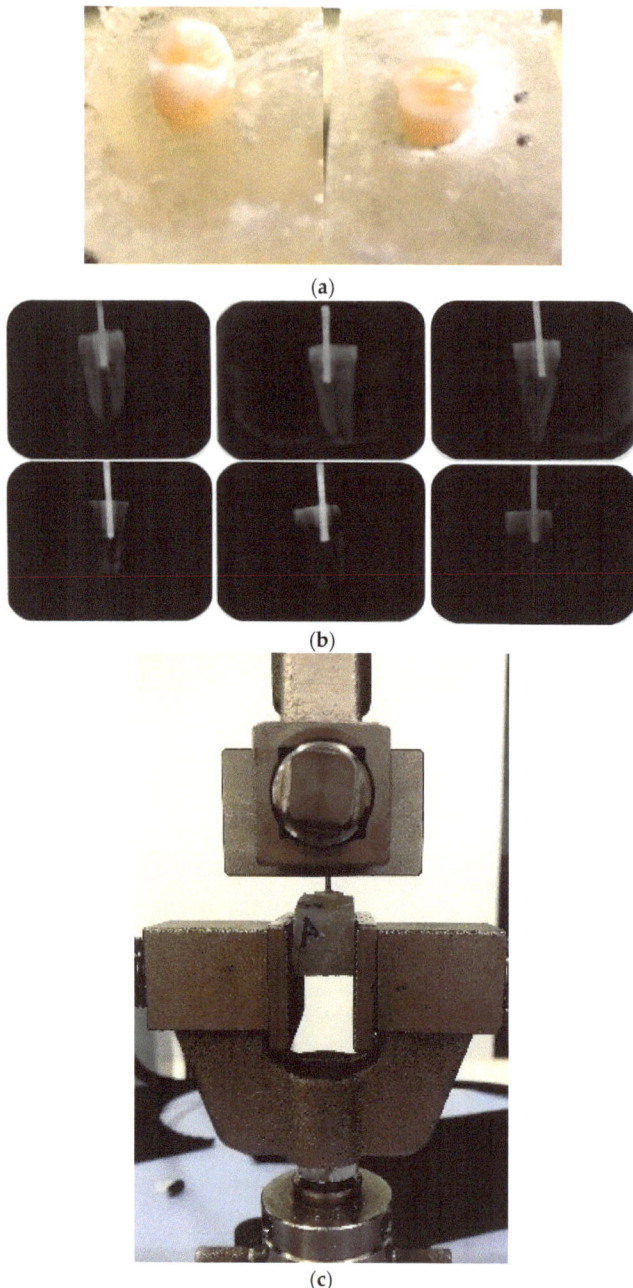

Figure 2. Pull-out test for measuring adhesion strength. (**a**) Premolars embedded in an acrylic resin. (**b**) Radiographs of premolar with metal post. (**c**) Pull-out test.

2.6. Statistical Analysis

The statistical analyses used were a one-way ANOVA with Bonferroni (posthoc), $p < 0.05$ for the compressive, bending, pull-out, and shear bond strength of PAC and MPAC.

3. Results

3.1. Physicochemical Characterization

The unmodified PAC shows absorptions via FTIR (Figure 3a) at 3436 (OH), 3124, 2965 (CH_2), 1720 (COOH), 1635 (C=C vinyl), 1602 (COOasym-Al), 1463 (C-H or COOsym), 1401 (COOsym), 1166 (CO of Poly(acrylic) or tartaric acid), 1076 (C-O), 796, and 644 cm^{-1}. This assignment, however, should consider that, in pure calcium polyacrylates, asymmetric stretch absorptions appear at 1550 cm^{-1}, which correspond to Al at 1599, with calcium and aluminum tartrate at 1595 and 1670 cm^{-1}, respectively. Additionally, we note bands at 1410, 1460, 1385, and 1410 cm^{-1}, corresponding to the symmetric stretching of COO or CH_2 bending. Thus, the absorptions between 1401 and 1463 can be attributed to the polycarboxylates or tartrates of Ca and Al. Finally, the bands between 400 and 800 cm^{-1} could correspond to crystalline structures associated with Al_2O_3, metallic fluorides, etc. As depicted in Figure 3a, little modification to the FTIR spectra was observed after propolis addition.

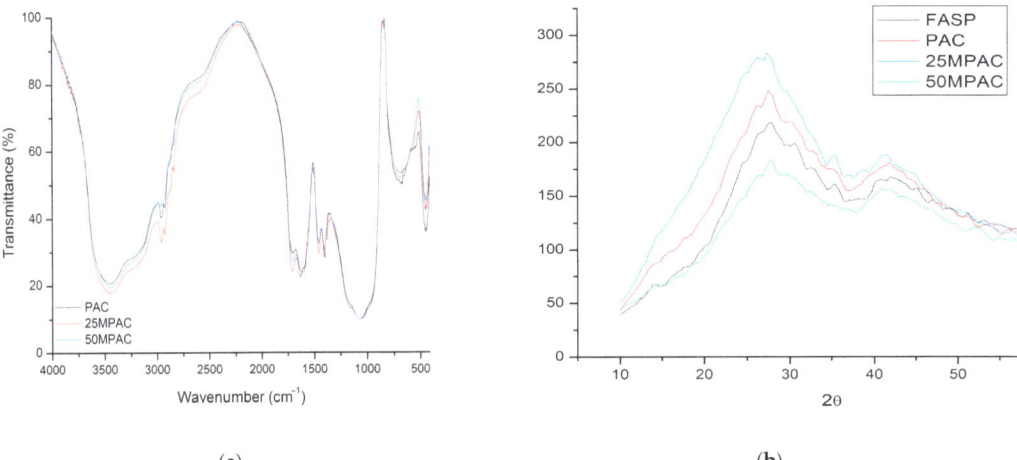

(a)   (b)

Figure 3. Physicochemical characterization of propolis-modified poly(alkenoate) cements; (**a**) FTIR spectra; (**b**) X-ray diffraction pattern.

The X-ray diffraction patterns for PAC and MPAC in Figure 3b show that there were no new reflections after propolis modification, i.e., the fluoroaluminosilicate powder was amorphous in addition to the final poly(alkenoate) cement.

3.2. Antibacterial Properties

Figure 4a shows the results of the MTT viability tests, observing that the MPAC extracts can stop *S. mutans* growth (see yellow color in the second (B) to fourth (D) row in Figure 4a). In contrast, the extract from PAC did not inhibit bacteria growth (see red color in the first row (A) in Figure 4a). The sixth row (F) showed the effectiveness of the amikacin negative control (yellow, F1) and the only bacteria growth, which is stained in deep red (F2).

Figure 4b shows the inhibition halo in the presence of *S. mutans* after contact with the propolis extracts. Amikacin exhibited an inhibition halo of 8 mm (labeled as b.1), while MPAC showed 1.61 mm (labeled as b.3), 1.97 mm (labeled as b.4), and 2.0 mm (labeled as b.5) for 12.5, 25, and 50 µL of propolis in the PAC, respectively, as measured with ImageJ software. No inhibition halo was observed in the PAC-only disc (labeled as b.2).

From these results, it is suggested that MPAC has antiseptic activity against *S. mutans* as the MIC of the ethanol solutions of propolis was 2.5 µg/mL.

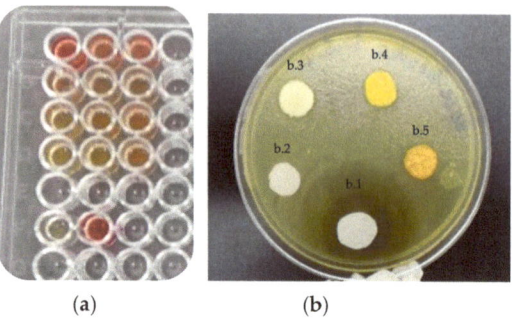

(a)   (b)

Figure 4. Antibacterial activity of propolis-modified PAC against *Streptoccocus mutans*. (**a**) MTT assay. (**b**) Agar diffusion method; b.1 Amikacin, b.2 negative control, b.3 12.5MPAC, b.4 25MPAC, and b.5 50MPAC.

3.3. Mechanical Properties
3.3.1. Compressive Strength

Unmodified PAC exhibited a compressive strength of 211.21 MPa, which was reduced by up to 59.36 MPa with 50 µL of propolis. However, the unmodified PAC value is much higher than the other reported values for the same Fuji IX cement, where 152.4 MPa was reported after 24 h [16,39]. According to ISO 9917-1, the compressive strength should be 100 MPa for restorative polyalkenoates and 50 MPa for base/lining cement. It has been established that the masticatory forces can be 91 N (anterior) and 129 N (posterior), with a maximum of 314 N [40]. Overall, these results suggest that, at the highest concentrations of propolis (50 µL), MPAC was not suitable for restoration, but at the lowest concentration of propolis (25 µL), it is on the limit for its use as a liner. From the fracture surface shown in Figure 5a for compression, it seems that the fracture pattern is not modified, and therefore the poor mechanical properties exhibited are due to curing interference by the propolis compounds, probably with chelating properties, and not due to jvcr@cicy.m the presence of ethanol.

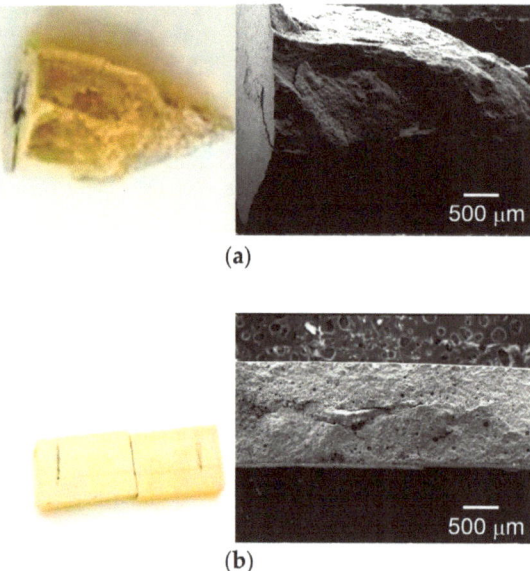

Figure 5. Fracture surface of (**a**) compression and (**b**) bending. Both fracture patterns indicate a rigid material.

3.3.2. Bending Strength

In agreement with this, the bending strength decreased from 11.1 MPa to 6.25 MPa from PAC to MPAC, with the highest concentration of propolis. As mentioned before, the fracture surface pattern was like the unmodified PAC (see Figure 5b). According to the ISO 9917-2 specifications, flexural strength should be no less than 20 MPa for restorative cement and no less than 10 MPa for bases and liners. In this regard, Xie et al. [1] reported Fuji GIC values of 71.1 and 26.1 MPa for Fuji II LC and Fuji II, respectively. In addition, Hu et al. [29] reported 25 MPa of flexural strength for a green tea extract (epigallocatechin-3-gallate)-modified GIC. Therefore, the large amounts of propolis used here (25 µL and 50 µL) do not fulfill the requirements of the standard. The compressive and bending properties are summarized in Table 2.

Table 2. Mechanical properties of PAC and MPAC.

Sample	Compressive Properties			Bending Properties		
	EC (MPa)	σC (MPa) *	εC (%)	EF (GPa)	σF (MPa) *	εF (%)
PAC	89.44 ± 8.88	211.21 ± 8.83 [a]	2.45 ± 0.33	19.72 ± 6.40	11.10 ± 1.71 [a]	0.10 ± 0.02
25MPAC	34.7 ± 7.13	94.40 ± 9.62 [b]	2.43 ± 0.26	7.12 ± 2.51	7.78 ± 1.41 [b]	0.29 ± 0.092
50MPAC	20.4 ± 3.5	59.36 ± 2.45 [b]	3.23 ± 0.15	3.53 ± 1.76	6.25 ± 1.85 [b]	0.42 ± 0.16

MPa = megapascal, GPa = gigapascal, E = elastic modulus, σ = strength, ε_F = maximum deformation; * = one-way ANOVA $p < 0.05$; the groups not sharing letters ([a] or [b]) are statistically different, according to Bonferroni posthoc.

3.3.3. Shear Bond Strength

PAC that was cured in the presence of 25 µL propolis and then adhered to dentin for shear bond testing (25 MPAC) was used additionally for comparison purposes, as the compressive properties suggest their use as a restorative material. This sample showed an adhesion strength of 1.47 ± 0.21 MPa, while PAC only exhibited 1.02 ± 0.01 MPa on nonacid-treated dentin. This increase has been previously reported for nano-MgO-modified PAC, exhibiting a shear bond strength as high as 5 MPa on the enamel of bovine incisors, which increase up to 6 MPa when tested on dentine [41].

Propolis-treated dentine (MPAC-NA) cylinders did not adhere to the dentin surface, probably due to the nonpolar components of propolis. However, in non-propolis-treated dentin and nonacid-treated dentin unmodified glass ionomer cement (PAC-NA), the shear bond strength was 0.62 ± 0.27 MPa. When the dentine surface was acid-treated only with no propolis treatment (PAC-A), the shear bond strength increased up to 1.70 ± 0.53 MPa. For the propolis-treated (followed by acid treatment) dentine surface (MPAC-A), the shear bond strength was 1.71 ± 0.74 MPa, i.e., there was no change in the adhesion. The type of failure (adhesive or cohesive) is shown in Figure 6 only for those samples that adhere to the surface, i.e., MPAC-NA is not shown. In Figure 6a, the PAC-NA samples showed adhesive failure predominantly, but in some cases, mixed types of adhesive failure were observed. In Figure 6b, the PAC-A samples suffered a mixed type of adhesive failure. Finally, in Figure 6c, the MPAC-A samples also suffered a mixed type of adhesive failure.

Pull-out test showed that the force required to remove a PAC cylinder from the tooth cavity is comprehensively higher than in the case of the shear bond test. However, this test allowed us to compare the different treatments on the inner tooth surface, i.e., the force required to remove the PAC after acid treatment was 206.6 N, and that required after propolis treatment was only 57.1 N, being even lower than the force required for PAC removal without propolis and acid etching (151.4 N). Figure 7 shows the type of failure during this test.

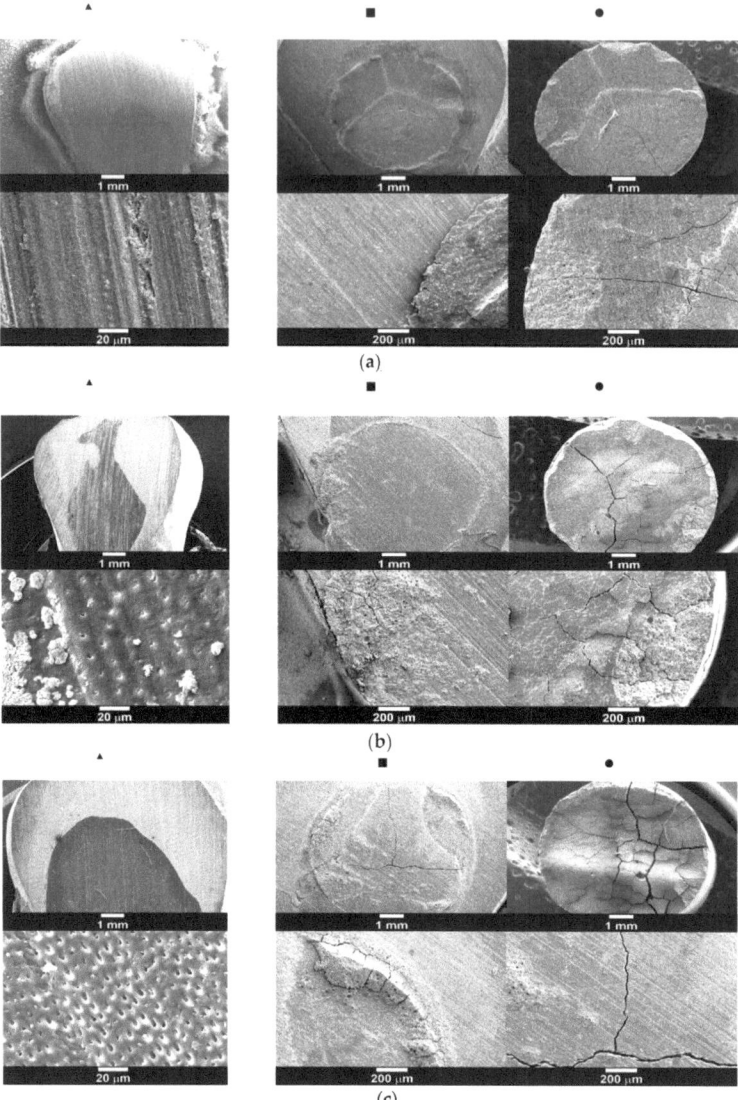

Figure 6. Surface appearance after shear bond testing; PAC-NA (**a**), PAC-A, (**b**), and MPAC-A (**c**). ▲ Tissue appearance after surface treatment (first column); ■ tissue appearance after detachment (second column); ● cylinder surface after detachment (third column). Top row: low magnification (scale bar 1 mm). Bottom row: high magnification (scale bar: 20 μm or 200 μm).

The adhesion strength of MPAC is summarized in Table 3. For comparison purposes, the force reached during the shear adhesion test is reported along with the detaching force during the pull-out test. No experiment was conducted for 25MPAC, as the aim of the experiment was to assess propolis treatment on the dentine surface, not for propolis incorporated into PAC.

Figure 7. Type of failure after the pull-out test. (**a**) No propolis and no acid (PAC-NA), (**b**) No propolis and acid treated (PAC-A), and (**c**) Propolis and acid treated (MPAC-A).

Table 3. Adhesion strength of PAC and MPAC.

Sample	Shear Force (N) *	Shear Strength (MPa) *	Pull-Out Force (N) *
PAC-A	19.3 ± 6.0 [a]	1.70 ± 0.53 [a]	206.6 ± 27.1 [a]
PAC-NA	7.02 ± 3.0 [b]	0.62 ± 0.27 [b]	151.4 ± 92.4 [a]
25MPAC	16.6 ± 2.4 [a]	1.47 ± 0.211 [a]	-
MPAC-A	19.4 ± 8.4 [a]	1.71 ± 0.74 [a]	57.1 ± 12.5 [b]

MPa = megapascal, N = Newtons, * = one-way ANOVA $p < 0.05$; groups not sharing letters ([a] or [b]) are statistically different, according to Bonferroni posthoc.

4. Discussion

Propolis has shown various degrees of antibacterial activity, depending on the source, concentration, solvent, or vehicle used, the bacterial strain tested, and the type of dental material used for modification. In our study, it was shown that 20% ethanolic solutions of propolis added to a PAC exhibited mild antibacterial activity against *S. mutans*, a Gram-positive bacterium, both as an extract and by means of disc-diffusion assay. When comparing this activity to traditional antibiotics, Hatunoglu et al. reported 3.9 μg/mL and 7.8 μg/mL MIC antibiotic values against *S. mutans* for Ampicillin and Gentamicin, respectively, while a value of 15.7 μg/mL for their 10 wt% ethanolic extracts of propolis was reported [27]. Another study reported a MIC of 60 μg/mL and 0.5 μg/mL for Penicillin and Chlorhexidine (0.2%), respectively [42]. This bacterium is involved in secondary caries, but it is also reported that its activity depends on the presence of *S. sanguinis* [43]. In this regard, the bactericidal effect of propolis has been attributed to the presence of flavonoids being more effective against *Streptococcus salivarius* rather than *Streptococcus mutans* [35]. In this line of thought, we have reported that the major components of propolis were pinocembrin, pinobanksin-3-O-acetate, and pinobanksin-3-O-propionate, which can be partly responsible for its mild antibacterial behavior [32].

Mild antibacterial activity, however, can be compensated for by antioxidant activity due to flavonoid and phenol content, in addition to radical scavenging activity. Furthermore, previous results from our group demonstrated the in vitro anti-inflammatory activity of propolis, as the levels of pro-inflammatory IL-1β, IL-6, and TNF-α were low, while the levels of IL-10 and IL-4 were high. Overall, the clinical performance of the propolis MPAC will depend not only on its antibacterial properties, but also on its antioxidant activity and anti-inflammatory properties [32]. In addition, it has been reported that propolis is used for the treatment of candidiasis, acute necrotizing ulcerative gingivitis, gingivitis, periodontitis, and pulpitis in dentistry, and there are reports regarding the antibacterial effects of propolis on methicillin-resistant Staphylococcus aureus. It also exhibits antifungal effects comparable to those of Nystatin, and displays the antimicrobial effects of propolis on anaerobic oral bacteria, such as *S. aureus*, *Actinobacillus*, and oral pathogenic micro-organisms such as *Streptococcus salivarius*, *Streptococcus sanguinis*, *Streptococcus mitis*, *Candida albicans*, *Streptococcus mutans*, and *Shigella* [44].

As propolis exhibited mild antibacterial behavior against *S. mutans*, only high propolis concentrations in PAC were studied, i.e., 25 μL and 50 μL. However, Table 2 clearly shows the deleterious effect on mechanical properties when increasing the concentration of propolis in PAC. The compressive and flexural properties were reduced to levels below the requirements of ISO 9917 for restorative materials. However, those formulations prepared with 25 μL of propolis were close to the required value of 100 MPa for compression in restorative materials but were suitable for base/lining cement and, therefore, are solely used for adhesion tests. The low mechanical properties observed can be explained due to curing kinetics, which is affected by the presence of various components in propolis, such as flavonoids, with the ability to form chelates, which, in turn, can sequester available divalent (calcium) or trivalent (aluminum) ions [45].

The adhesion of PAC to the tooth is the result of two phenomena: micromechanical interlocking and chemical bonding [8]. In this study, two types of adhesion tests were evaluated. In the first case, the traditional shear test was used to assess the propolis effect. The cured cylinders that contained propolis exhibited only 16.6 N of shear force on the unmodified dentine surface, but when dentine was either propolis- or non-propolis-treated and acid etched, the shear force increased up to 19.4 N (equivalent to a shear bond strength of 1.71 ± 0.74 MPa) and 19.3 N (equivalent to a shear bond strength of 1.70 ± 0.53 MPa), respectively, suggesting that the microroughness promoted adhesion [46]. The maximum shear bond force could be achieved via the direct placement of the untreated PAC cylinder onto the acid-treated dentine surface, reaching up to 38 N (3.72 ± 0.65 MPa). In contrast, the minimum shear bond force was 7.02 ± 3.0 N (0.62 ± 0.27 MPa) from the non-propolis-treated dentin and nonacid-treated dentin unmodified glass ionomer cements (PAC-NA), suggesting that the surface macroroughness achieved by polishing alone was not enough to adhere to PAC [47]. During debonding, a mixed type of adhesive failure was noted, i.e., adhesive and cohesive. When no propolis and no acid were used, the responsible factor for the mechanical interlocking is the surface roughness achieved after polishing, Figure 6a. When the dentine surface was acid-treated only, additional changes in surface topography were included, the dentine tubules were exposed, and adhesion was promoted, as shown in Figure 6b. However, the presence of propolis and no acid treatment did not allow PAC adhesion, probably due to the nonpolar compounds found in propolis. This can be alleviated by using acid etching, as shown in Figure 6c. Therefore, it is clear that any PAC modification (not only with propolis) must consider the presence of nonpolar compounds, as they will affect not only their adhesion and the type of failure but also the surface roughness accomplished by polishing (number of abrasive paper or grit size) and acid etching (type of acid, exposure time, etc.). However, other factors, such as dentine/enamel quality and the tubular density used, can alter the outcome [16,48].

In the second variation of the conventional shear test, the effect of propolis was studied by means of the pull-out test. From Table 3, no propolis and no acid treatment resulted in a high detaching force (151.4 N), as more area was exposed to the PAC when the roughness was introduced by drilling. When the sample is propolis-treated but not acid-treated, the detaching force was reduced by up to 57 N, i.e., the nonpolar compounds of propolis are covering the internal surface, limiting the diffusion of the PAC on the dentine [49]. Once again, when the sample is acid-treated only and no propolis is used, the highest detaching force is achieved (206 N). The higher force reached during this type of test in comparison to the conventional shear bond test responds to a higher uncontrolled roughness achieved during drilling and due to the higher exposed area [50]. Even when this type of test is not customary in all laboratories, it provides a more realistic approach to estimating PAC-tooth adhesion, as it involves a higher dentin area. Some of the limitations include the proper alignment of the metallic root, the formation of a true cylindrical cavity in the teeth (different from molars), and the achievement of the same roughness in the internal surface, among others.

When comparing the two sets of results, there is no apparent correlation, and this only confirms that the use of acid etching is enough to achieve the maximum adhesive strength [51]. However, the presence of propolis in PAC can compromise curing and adhesion. When our results are compared with other works, large variations are observed. For example, the tensile bond strength of Fuji IX to dentin was reported to be equal to 3.08 MPa, which is lower than the reported 5.0 MPa for adhesion to enamel [16]. Even when tensile and shear bond strength tend to be very similar, the reported values here are, in general, lower than those reported for the microshear tests that made use of smaller areas (1 mm^2) [38]. Furthermore, Tedesco has reported that the microshear bond strength of Fuji IX depended on the density of the tubules and their location, i.e., from 3.20 MPa in occlusal deep dentin to 4.70 MPa in superficial occlusal dentin [48].

5. Conclusions

The addition of propolis to the PAC mixture had a response against a strain characteristic of the oral cavity. Despite the reduction in both compressive and bending strengths, the adhesion showed no statistical difference at shear bond strength ($p < 0.05$); thus, this material can be used in conditions where the affected property requirements are not critical, i.e., as with the cavity liner to the isolate surfaces near the pulp tissue when the atraumatic restorative technique is used, specifically due to the reported antibacterial effect against *S. mutans*.

Furthermore, the anti-inflammatory properties of propolis incorporated into this modified material could interact with the fluid of the dentinal tubules for the treatment of reversible pulpitis, and these formulations can be explored for pulpar treatments.

Author Contributions: Perform some experiments, write—original draft preparation, review and editing, D.A.A.-P. Perform experiments with different propolis concentrations, C.M.U.-M. and B.M.-G. Data curation and interpretation, drafting of manuscript and critical revision, O.d.J.C.-C. Conceptualization and methodology, F.J.A.-A. and M.G.C.-G. Various materials characterization, R.F.V.-C. Funding, idea generation, writing the manuscript and co-ordinating the team, J.V.C.-R. All authors have read and agreed to the published version of the manuscript.

Funding: This research was funded by CONACYT grant number 1360 (Fronteras de la Ciencia) and 248378 (Atención a Problemas Nacionales).

Institutional Review Board Statement: Not applicable.

Data Availability Statement: The data presented in this study are available on request from the corresponding author.

Acknowledgments: The authors wish to thank Alejandro May Pat for his technical assistance during the pull-out test and Ricardo J. Mis Fernandez for the XRD analysis.

Conflicts of Interest: The authors declare no conflict of interest. The funders had no role in the design of the study; in the collection, analyses, or interpretation of data; in the writing of the manuscript; or in the decision to publish the results.

References

1. Xie, D.; Brantley, W.; Culbertson, B.; Wang, G. Mechanical properties and microstructures of glass-ionomer cements. *Dent. Mater.* **2000**, *16*, 129–138. [CrossRef] [PubMed]
2. Sidhu, S.K.; Nicholson, J.W. A Review of Glass-Ionomer Cements for Clinical Dentistry. *J. Funct. Biomater.* **2016**, *7*, 16. [CrossRef] [PubMed]
3. Hatton, P.V.; Kearns, V.R.; Brook, I.M. Bone-Cement Fixation: Glass-Ionomer Cements. In *Biomaterials*; Woodhead Publishing: Sawston, UK, 2008; pp. 252–263.
4. Zandi Karimi, A.; Rezabeigi, E.; Drew, R.A.L. Glass ionomer cements with enhanced mechanical and remineralizing properties containing 45S5 bioglass-ceramic particles. *J. Mech. Behav. Biomed. Mater.* **2019**, *97*, 396–405. [CrossRef] [PubMed]
5. Calvo, A.F.B.; Kicuti, A.; Tedesco, T.K.; Braga, M.M.; Raggio, D.P. Evaluation of the relationship between the cost and properties of glass ionomer cements indicated for atraumatic restorative treatment. *Braz. Oral Res.* **2016**, *30*, 3–9. [CrossRef]
6. Chen, S.; Mestres, G.; Lan, W.; Xia, W.; Engqvist, H. Cytotoxicity of modified glass ionomer cement on odontoblast cells. *J. Mater. Sci. Mater. Med.* **2016**, *27*, 116. [CrossRef]
7. Nicholson, J.; Brookman, P.; Lacy, O.; Wilson, A. Fourier Transform Infrared Spectroscopic Study of the Role of Tartaric Acid in Glass-ionomer Dental Cements. *J. Dent. Res.* **1988**, *67*, 1451–1454. [CrossRef]
8. Nicholson, J.W. Adhesion of glass-ionomer cements to teeth: A review. *Int. J. Adhes. Adhes.* **2016**, *69*, 33–38. [CrossRef]
9. Fierascu, R.C. Incorporation of Nanomaterials in Glass Ionomer Cements—Recent Developments and Future Perspectives: A Narrative Review. *Nanomaterials* **2022**, *12*, 3827. [CrossRef]
10. *ISO 9917-1*; Dentistry-Water-Based Cements-Part 1: Powder/Liquid Acid-Base Cements. ISO: Geneva, Switzerland, 2007.
11. *ISO 9917-2*; Dental-Water-Based Cements-Part 2: Light-Activated Cements. ISO: Geneva, Switzerland, 2008.
12. Wetzel, R.; Eckardt, O.; Biehl, P.; Brauer, D.; Schacher, F. Effect of poly(acrylic acid) architecture on setting and mechanical properties of glass ionomer cements. *Dent. Mater.* **2020**, *36*, 377–386. [CrossRef]
13. Ilie, N.; Hilton, T.J.; Heintze, S.D.; Hickel, R.; Watts, D.C. Academy of dental materials guidance—Resin composites: Part I—Mechanical properties. *Dent. Mater.* **2017**, *33*, 880–894. [CrossRef]
14. Somani, R.; Jaidka, S.; Jawa, D.; Mishra, S. Comparative evaluation of smear layer removal by various chemomechanical caries removal agents: An in vitro SEM study. *J. Indian Soc. Pedod. Prev. Dent.* **2015**, *33*, 204–207. [CrossRef] [PubMed]
15. Yanıkoglou, N.D.; Sakara, R.E. Test methods used in the evaluation of the structure features of the restorative materials. *J. Mater. Res. Technol.* **2020**, *9*, 9720–9734. [CrossRef]
16. Pereira, L.C.; Nunes, M.C.P.; Dibb, R.G.P.; Powers, J.M.; Roulet, J.-F.; Navarro, M.F.d.L. Mechanical properties and bond strength of glass-ionomer cement. *J. Adhes. Dent.* **2002**, *4*, 73–80. [PubMed]
17. Segreto, D.; Brandt, W.C.; Correr-Sobrinho, L.; Sinhoreti, M.A.; Consani, S. Influence of Irradiance on the Push-out Bond Strength of Composite Restorations Photoactivated by LED. *J. Contemp. Dent. Pract.* **2008**, *9*, 89–96. [CrossRef]
18. *ISO/TS 11405*; Dentistry-Testing of Adhesion to Tooth Structure. ISO: Geneva, Switzerland, 2015.
19. *Standard 11266 I, ISO 29022*; Dentistry-Adhesion-Notched Edge Shear Bond Strength Test. ISO: Geneva, Switzerland, 2014.
20. Ching, H.S.; Luddin, N.; Kannan, T.P.; Ab Rahman, I.; Ghani, N.R.N.A. Modification of glass ionomer cements on their physical-mechanical and antimicrobial properties. *J. Esthet. Restor. Dent.* **2018**, *30*, 557–571. [CrossRef]
21. Farrugia, C.; Camilleri, J. Antimicrobial properties of conventional restorative filling materials and advances in antimicrobial properties of composite resins and glass ionomer cements—A literature review. *Dent. Mater.* **2015**, *31*, e89–e99. [CrossRef]

22. Elsaka, S.E.; Hamouda, I.M.; Swain, M.V. Titanium dioxide nanoparticles addition to a conventional glass-ionomer restorative: Influence on physical and antibacterial properties. *J. Dent.* **2011**, *39*, 589–598. [CrossRef]
23. Du, X.; Huang, X.; Huang, C.; Frencken, J.; Yang, T. Inhibition of early biofilm formation by glass-ionomer incorporated with chlorhexidine in vivo: A pilot study. *Aust. Dent. J.* **2012**, *57*, 58–64. [CrossRef]
24. Xie, D.; Weng, Y.; Guo, X.; Zhao, J.; Gregory, R.L.; Zheng, C. Preparation and evaluation of a novel glass-ionomer cement with antibacterial functions. *Dent. Mater.* **2011**, *27*, 487–496. [CrossRef]
25. Vagner Rodrigues, S. Propolis: Alternative Medicine for the Treatment of Oral Microbial Diseases. In *Alternative Medicine*; Books on Demand: Norderstedt, Germany, 2012.
26. Huang, X.-Y.; Guo, X.-L.; Luo, H.-L.; Fang, X.-W.; Zhu, T.-G.; Zhang, X.-L.; Chen, H.-W.; Luo, L.-P. Fast Differential Analysis of Propolis Using Surface Desorption Atmospheric Pressure Chemical Ionization Mass Spectrometry. *Int. J. Anal. Chem.* **2015**, *2015*, 176475. [CrossRef]
27. Hatunoğlu, E.; Ö Ztü Rkb, F.; Bilenler, T.; Aksakalli, S.; Ş Imşeke, N. Antibacterial and mechanical properties of propolis added to glass ionomer cement. *Angle Orthod.* **2014**, *84*, 368–373. [CrossRef] [PubMed]
28. Topcuoglu, N.; Ozan, F.; Ozyurt, M.; Kulekci, G. In vitro antibacterial effects of glassionomer cement containing ethanolic extract of propolis on *Streptococcus* mutans. *Eur. J. Dent.* **2012**, *6*, 428–433. [CrossRef]
29. Hu, J.; Du, X.; Huang, C.; Fu, D.; Ouyang, X.; Wang, Y. Antibacterial and physical properties of EGCG-containing glass ionomer cements. *J. Dent.* **2013**, *41*, 927–934. [CrossRef] [PubMed]
30. Singer, L.; Bierbaum, G.; Kehl, K.; Bourauel, C. Evaluation of the Flexural Strength, Water Sorption, and Solubility of a Glass Ionomer Dental Cement Modified Using Phytomedicine. *Materials* **2020**, *13*, 5352. [CrossRef] [PubMed]
31. Paulraj, J.; Nagar, P. Antimicrobial efficacy of triphala and propolis-modified glass ionomer cement: An in vitro study. *Int. J. Clin. Pediatr. Dent.* **2020**, *13*, 457–462. [CrossRef] [PubMed]
32. Xool-Tamayo, J.; Chan-Zapata, I.; Arana-Argaez, V.E.; Villa-de la Torre, F.; Torres-Romero, J.C.; Araujo-Leon, J.A.; Aguilar-Ayala, F.A.; Rejón-Peraza, M.E.; Castro-Linares, N.C.; Vargas-Coronado, R.F.; et al. In vitro and in vivo anti-inflammatory properties of Mayan propolis. *Eur. J. Inflamm.* **2020**, *18*, 205873922093528. [CrossRef]
33. Altunsoy, M.; Tanrıver, M.; Türkan, U.; Uslu, M.E.; Silici, S. In Vitro Evaluation of Microleakage and Microhardness of Ethanolic Extracts of Propolis in Different Proportions Added to Glass Ionomer Cement. *J. Clin. Pediatr. Dent.* **2016**, *40*, 136–140. [CrossRef] [PubMed]
34. Subramaniam, P.; Babu, K.G.; Neeraja, G.; Pillai, S. Does Addition of Propolis to Glass Ionomer Cement Alter its Physicomechanical Properties? An In Vitro Study. *J. Clin. Pediatr. Dent.* **2017**, *41*, 62–65. [CrossRef]
35. Andrade, A.L.; Lima, A.M.; Santos, V.R.; Da Costa e Silva, R.M.F.; Barboza, A.P.M.; Neves, B.R.A.; Vasconcellos, W.A.; Domingues, R.Z. Glass-ionomer-propolis composites for caries inhibition: Flavonoids release, physical-chemical, antibacterial and mechanical properties. *Biomed. Phys. Eng. Express* **2019**, *5*, 027006. [CrossRef]
36. Sundeep Hedge, K.; Bhat, S.; Rao, A.; Sain, S. Effect of a propolis extract on Streptococcus mutans counts: An in vivo. *Int. J. Clin. Pediatr. Dent.* **2013**, *6*, 22–25.
37. Braga, R.R.; Meira, J.B.C.; Boaro, L.C.C.; Xavier, T.A. Adhesion to tooth structure: A critical review of "macro" test methods. *Dent. Mater.* **2010**, *26*, e38-49. [CrossRef] [PubMed]
38. Scherrer, S.S.; Cesar, P.F.; Swain, M.V. Direct comparison of the bond strength results of the different test methods: A critical literature review. *Dent. Mater.* **2010**, *26*, e78–e93. [CrossRef] [PubMed]
39. Fareed, M.A.; Stamboulis, A. Effect of Nanoclay Dispersion on the Properties of a Commercial Glass Ionomer Cement. *Int. J. Biomater.* **2014**, *2014*, 1–10. [CrossRef] [PubMed]
40. Morneburg, T.; Pröschel, P. Measurement of masticatory forces and implant loads: A methodologic clinical study. *Int. J. Prosthodont.* **2002**, *15*, 20–27.
41. Noori, A.J.; Kareem, F.A. Setting time, mechanical and adhesive properties of magnesium oxide nanoparticles modified glass-ionomer cement. *J. Mater. Res. Technol.* **2019**, *9*, 1809–1818. [CrossRef]
42. Haghgoo, R.; Mehran, M.; Afshari, E.; Zadeh, H.F.; Ahmadvand, M. Antibacterial Effects of Different Concentrations of Althaea officinalis Root Extract versus 0.2% Chlorhexidine and Penicillin on Streptococcus mutans and Lactobacillus (In vitro). *J. Int. Soc. Prev. Community Dent.* **2017**, *7*, 180–185. [CrossRef]
43. Loesche, W.J. Role of *Streptococcus* mutans in human dental decay. *Microbiol. Rev.* **1986**, *50*, 353–380. [CrossRef]
44. Panahandeh, N.; Adinehlou, F.; Sheikh-Al-Eslamian, S.M.; Torabzadeh, H. Extract of Propolis on Resin-Modified Glass Ionomer Cement: Effect on Mechanical and Antimicrobial Properties and Dentin Bonding Strength. *Int. J. Biomater.* **2021**, *2021*, 1–7. [CrossRef]
45. Geckil, H.; Ates, B.; Durmaz, G.; Erdoğan, S.; Yilmaz, I. Antioxidant, Free Radical Scavenging and Metal Chelating Characteristics of Propolis. *Am. J. Biochem. Biotechnol.* **2005**, *1*, 27–31. [CrossRef]
46. Bala, O.; Arisu, H.D.; Yikilgan, I.; Arslan, S.; Gullu, A. Evaluation of surface roughness and hardness of different glass ionomer cements. *Eur. J. Dent.* **2012**, *6*, 79–86.
47. Ekambaram, M.; Yiu, C. Bonding to hypomineralized enamel—A systematic review. *Int. J. Adhes. Adhes.* **2016**, *69*, 27–32. [CrossRef]
48. Tedesco, T.K.; Calvo, A.F.B.; Domingues, G.G.; Mendes, F.M.; Raggio, D.P. Bond Strength of High-Viscosity Glass Ionomer Cements is Affected by Tubular Density and Location in Dentin? *Microsc. Microanal.* **2015**, *21*, 849–854. [CrossRef] [PubMed]

49. Zheng, B.; Cao, S.; Al-Somairi, M.A.A.; He, J.; Liu, Y. Effect of enamel-surface modifications on shear bond strength using different adhesive materials. *BMC Oral Health* **2022**, *22*, 224. [CrossRef] [PubMed]
50. Kaptan, A.; Oznurhan, F.; Candan, M. In Vitro Comparison of Surface Roughness, Flexural, and Microtensile Strength of Various Glass-Ionomer-Based Materials and a New Alkasite Restorative Material. *Polymers* **2023**, *15*, 650. [CrossRef] [PubMed]
51. Francois, P.; Vennat, E.; Le Goff, S.; Ruscassier, N.; Attal, J.-P.; Dursun, E. Shear bond strength and interface analysis between a resin composite and a recent high-viscous glass ionomer cement bonded with various adhesive systems. *Clin. Oral Investig.* **2018**, *23*, 2599–2608. [CrossRef] [PubMed]

Disclaimer/Publisher's Note: The statements, opinions and data contained in all publications are solely those of the individual author(s) and contributor(s) and not of MDPI and/or the editor(s). MDPI and/or the editor(s) disclaim responsibility for any injury to people or property resulting from any ideas, methods, instructions or products referred to in the content.

Article

Imidazolium Salts for *Candida* spp. Antibiofilm High-Density Polyethylene-Based Biomaterials

Clarissa Martins Leal Schrekker [1], Yuri Clemente Andrade Sokolovicz [2], Maria Grazia Raucci [3], Claudio Alberto Martins Leal [2], Luigi Ambrosio [3], Mário Lettieri Teixeira [4], Alexandre Meneghello Fuentefria [1,5,*] and Henri Stephan Schrekker [2,*]

1. Institute of Basic Health Sciences, Universidade Federal do Rio Grande do Sul (UFRGS), Rua Sarmento Leite 500, Porto Alegre 90050-170, RS, Brazil
2. Laboratory of Technological Processes and Catalysis, Institute of Chemistry, Universidade Federal do Rio Grande do Sul (UFRGS), Avenida Bento Gonçalves 9500, Porto Alegre 91501-970, RS, Brazil
3. Institute of Polymers, Composites and Biomaterials, National Research Council of Italy (IPCB-CNR), Viale John Fitzgerald Kennedy 54, Mostra d'Oltremare Padiglione 20, 80125 Naples, Italy
4. Laboratory of Biochemistry and Toxicology, Instituto Federal Catarinense (IFC), Rodovia SC 283—km 17, Concórdia 89703-720, SC, Brazil
5. Faculty of Pharmacy, Universidade Federal do Rio Grande do Sul (UFRGS), Avenida Ipiranga 2752, Porto Alegre 90610-000, RS, Brazil
* Correspondence: alexandre.fuentefria@ufrgs.br (A.M.F.); henri.schrekker@ufrgs.br (H.S.S.)

Citation: Martins Leal Schrekker, C.; Sokolovicz, Y.C.A.; Raucci, M.G.; Leal, C.A.M.; Ambrosio, L.; Lettieri Teixeira, M.; Meneghello Fuentefria, A.; Schrekker, H.S. Imidazolium Salts for *Candida* spp. Antibiofilm High-Density Polyethylene-Based Biomaterials. *Polymers* **2023**, *15*, 1259. https://doi.org/10.3390/polym15051259

Academic Editors: Md. Amdadul Huq, Shahina Akter and Dimitrios Bikiaris

Received: 24 October 2022
Revised: 7 February 2023
Accepted: 14 February 2023
Published: 1 March 2023

Copyright: © 2023 by the authors. Licensee MDPI, Basel, Switzerland. This article is an open access article distributed under the terms and conditions of the Creative Commons Attribution (CC BY) license (https://creativecommons.org/licenses/by/4.0/).

Abstract: The species of *Candida* present good capability to form fungal biofilms on polymeric surfaces and are related to several human diseases since many of the employed medical devices are designed using polymers, especially high-density polyethylene (HDPE). Herein, HDPE films containing 0; 0.125; 0.250 or 0.500 wt% of 1-hexadecyl-3-methylimidazolium chloride (C_{16}MImCl) or its analog 1-hexadecyl-3-methylimidazolium methanesulfonate (C_{16}MImMeS) were obtained by melt blending and posteriorly mechanically pressurized into films. This approach resulted in more flexible and less brittle films, which impeded the *Candida albicans*, *C. parapsilosis*, and *C. tropicalis* biofilm formation on their surfaces. The employed imidazolium salt (IS) concentrations did not present any significant cytotoxic effect, and the good cell adhesion/proliferation of human mesenchymal stem cells on the HDPE-IS films indicated good biocompatibility. These outcomes combined with the absence of microscopic lesions in pig skin after contact with HDPE-IS films demonstrated their potential as biomaterials for the development of effective medical device tools that reduce the risk of fungal infections.

Keywords: ionic liquid; human mesenchymal stem cells; biocompatibility; melt blending; histopathological evaluation

1. Introduction

Nowadays, polymer-based medical devices such as catheters [1–3], prostheses [1,2], endotracheal tubes [1,2], implants [1,2], tissues for tissue engineering [1], drug delivery systems [3] and heart valves [1], are commonly used in hospitals. *Candida* spp. (e.g., *C. tropicalis*, *C. albicans* and *C. parapsilosis*) have a strong tendency to colonize these polymeric surfaces, forming fungal biofilms [4,5]. By definition, biofilms are complex communities of microorganisms, with a high degree of organization, characterized by cells that are adhered to a surface or interface and embedded in an extracellular matrix of extrapolymeric substances (polysaccharides, proteins, lipids and DNA) of microbial origin, producing a spatially organized three-dimensional structure [6]. Chemical communication between cells, called quorum sensing, allows microorganisms (bacteria and fungi) to coordinate their activity and group together in communities that provide similar benefits as those of multicellular organisms [1,7]. The process of biofilm formation occurs through adhesion to medical devices, which is arbitrated by the proteins of the cell wall. As biofilms are highly adherent, the in vivo destruction of the biofilm requires the removal of the contaminated medical device, and this procedure could result in medical complications [6–8]. Altogether, the *Candida* sp. biofilms elevate the probability of nosocomial infections in immunocompromised patients due to therapeutic failure and the elevated resistance to important antifungal drugs [9], such as amphotericin B and azoles [1,10,11].

Within this context, the development of biomaterials with improved antibiofilm properties is highly desired. As such, the utilization of high-density polyethylene (HDPE) with imidazolium salt (IS) additives is promising for the development of biomaterials for use in medical devices. This polymer has excellent mechanical and biological properties, turning this material extremely suitable for applications in medical devices [12], which has been explored in facial implants as a substitute for the human skeleton in bone regeneration [13,14], tangible bone implants [12], tissue engineering (scaffolds) [15], reconstruction of nasal cartilage [16] and catheters [17,18].

The application of drug additives in HDPE-based medical devices represents an emerging technology. Interestingly, these devices do not have the primary purpose to act as drug reservoirs but may contain the latter, leading to an adjunctive pharmacological action. The incorporation of additives, e.g., antibiotics into temporary or permanent implants is being used in an attempt to reduce infections and to improve the acceptance of organic implanted material, minimizing the possibility of rejection [17,18].

Imidazolium salts (ISs) have an ion that is a cationic version of a neutral imidazole heterocycle and are known for presenting various advanced properties [19–21]. When ISs are in the liquid state at 100 °C, these are classified as ionic liquids. In general, ISs are attractive substances for various chemical and pharmaceutical applications, principally due to their thermal and chemical stability, neglectable volatility, and modifiable physical and chemical properties through structural modifications [22–25]. Various biological activities and applications of ISs have been identified, such as antibacterial [24–26], antifungal [26–28], antitumor [24], antioxidant [24], antifibrous [24], bioengineering [24,26] and anti-inflammatory [29].

Currently, there are few truly effective antifungal drugs against emerging yeasts. A variety of *N*-alkyl-substituted ISs was screened in vitro to verify the antifungal activity against *C. glabrata*, *C. parapsilosis*, *C. tropicalis*, and *Trichosporon asahii*. The best activity against fungal growth was determined for the ISs 1-hexadecyl-3-methylimidazolium chloride ($C_{16}MImCl$) and 1-hexadecyl-3-methylimidazolium methanesulfonate ($C_{16}MImMeS$). This, in combination with the absence of cytotoxicity and damage to human leukocytes, turns these substances into promising drug leads [23]. Interestingly, the pre-treatment of catheter surfaces with $C_{16}MImCl$ impeded the growth of *C. tropicalis* biofilms [30]. Complementing this preventive action and in comparison to chlorhexidine, the gold standard for asepsis in hospitals, much lower concentrations of $C_{16}MImCl$ and $C_{16}MImMeS$ were necessary to effectively remove *C. tropicalis* biofilms on polystyrene microtiter surfaces [31]. In addition to these ISs, *t*-BuOH-functionalized ISs with varying *N*-alkyl chain lengths were studied for their antimicrobial and antibiofilm properties. The one with the longest *N*-alkyl chain, dodecyl, was the most effective to inhibit the biofilm growth of *Staphylococcus epidermidis* and *C. albicans* [32].

Poly(L-lactide) films containing an IS additive ($C_{16}MImCl$ or $C_{16}MImMeS$) have been prepared by solvent casting which presented effective antibiofilm activities against *C. albicans*, *C. parapsilosis* and *C. tropicalis* [33]. The above-mentioned materials are biocompatible, do not cause skin irritation, and retain the original poly(L-lactide)'s mechanical and thermal properties. The incorporation of IS additives in polymers is an encouraging route to obtain biomaterials. Herein, HDPE-based biomaterials were obtained through melt-blending with an IS ($C_{16}MImCl$ or $C_{16}MImMeS$) (Figure 1). The resulting biomaterials were characterized, including their antibiofilm properties against *Candida* spp.

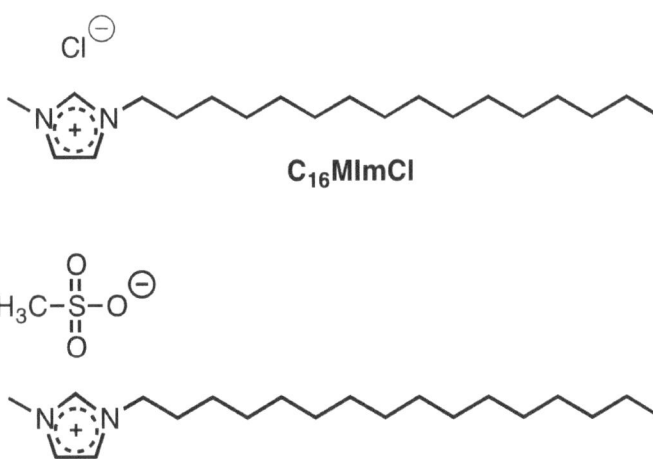

Figure 1. ISs C_{16}MImCl and C_{16}MImMeS applied in this study.

2. Materials and Methods

Materials. HDPE (HA 7260, Braskem S.A., Triunfo, RS, Brazil) and C_{16}MImCl (CJC China Jie Chemical, Shanghai, China) were donated and purchased, respectively. C_{16}MImMeS was prepared using the synthesis reported [20]. For the removal of residual water, HDPE and the ISs were vacuum dried at 60 °C for 5 h.

Yeast Strains. The following phenotypically identified biofilm-forming yeast strains were selected: *C. albicans* CA04; *C. parapsilosis* RL11, RL20 and *C. tropicalis* ATCC750, ATCC950, 17A, 57A, 72A, 102A, 17P, 72P, 94P, RL15, RL16, RL17. These isolates belonged to the mycology collection of the Laboratory of Applied Mycology at UFRGS [33].

Melt-Blended HDPE-IS and Film Formation. HDPE was melt-blended without or with an IS (0, 0.125, 0.250, or 0.500 wt%), using a twin-screw extruder (HAAKE Rheomex PTW 16 OS, Thermo Fisher Scientific, Waltham, MA, USA). The IS was added after 1 min to the molten polymer, and the components were mixed for 6 min, maintaining the screw speed and temperature of 60 rpm and 190 °C, respectively. Processed samples were left cooling to 25 °C, air dried and milled. After vacuum drying at 60 °C for 5 h, the samples were pressed into 0.5 mm thick films using a hydraulic press (Monarch 3710, Carver, Wabash, IN, USA). Initially, the material was molten within 4 min at 190 °C, and then pressed for 30 s at 4 lbf. The obtained films were abbreviated as HDPE.IS.content (e.g., HDPE.MeS.0125 for HDPE containing 0.125 wt% of C_{16}MImMeS).

Scanning Electron Microscopy (SEM). A scanning electron microscope (EVO 50, Carl Zeiss AG, Oberkochen, Germany) was used to study: (A) the morphology of the HDPE and HDPE-IS film surfaces and (B) the biofilm inhibition on these films through reported protocols [33].

Atomic Force Microscopy (AFM). The surfaces of the HDPE and HDPE-IS films were studied with the aid of a scanning probe microscope (5500, Agilent Technologies, Chandler, AZ, USA), using a reported procedure [33].

X-ray Diffraction (XRD). The crystallinity of the HDPE and HDPE-IS films was analyzed with a powder diffractometer (D500, Siemens, Munich, Germany) through a reported procedure [33].

Differential Scanning Calorimetry (DSC). A differential scanning calorimeter (Q20 V24.10 Build 122, TA Instruments, New Castle, DE, USA) was used to study the phase transitions of the HDPE and HDPE-IS films, using a reported protocol [33].

Thermogravimetric Analysis (TGA). The thermal degradation of the HDPE and HDPE-IS films was analyzed in a thermogravimetric analyzer (QA-50, TA Instruments, New Castle, DE, USA), using a reported protocol [33].

Dynamic Mechanical Analysis (DMA). The dynamic mechanical properties of the HDPE and HDPE-IS films were studied using a reported procedure [33], and a dynamic mechanical analyzer (Q800, TA Instruments, New Castle, DE, USA).

Water Contact Angle Measurements. The water contact angles were measured with the aid of a goniometer/drop shape analyzer (DSA100, Krüss, Hamburg, Germany), using a reported procedure [33].

Antibiofilm Assay. Petri dishes with Sabouraud agar containing chloramphenicol (HiMedia Laboratories LLC, Kelton, PA, USA) were employed to grow fresh yeast colonies (36 °C, 24 h). An inoculum (10^6 CFU/mL) of the yeast colonies in tryptone soya broth (6 mL; HiMedia Laboratories LLC, Kelton, PA, USA) was prepared and incubated (36 °C, 24 h). HDPE films (1 × 1 cm) were sterilized (UV), inserted in a composition of peptone water (9'mL; HiMedia Laboratories LLC, Kelton, PA, USA) and tryptone soya broth inoculum (1 mL), and incubated for 96 h. The weakly adherent cells were removed using peptone water, and the films were inserted in flasks with peptone water (50 mL). After treatment under ultrasound (40 KHz, 10 min; Ultrasonic washer, USC-700, Unique Indústria e Comércio de Produtos Eletrônicos Ltda, Jardim Belo Horizonte, SP, Brazil), the solutions with the detached cells were diluted (10^{-1}, 10^{-2}, 10^{-3}). These dilutions (20 µL) were plated in Petri dishes on Sabouraud agar containing chloramphenicol, and incubated (36 °C, 24 h). Finally, the number of CFU/cm^{-2} was determined and given logarithmically (log M, where M is the average value). Pure HDPE (film) was employed as a positive control [34].

Minor Antibiofilm Concentration (MAC) Assay. The protocols CLSI M27-A2 and CLSI M38-A were employed with minor modifications. Initially, the fresh yeast colonies were grown in Petri dishes on Sabouraud agar containing chloramphenicol (36 °C, 24 h; HiMedia Laboratories LLC, Kelton, PA, USA). After preparation of a 10^6 CFU/mL yeast inoculum (100% transmittance for 0.9% saline and 90% transmittance for the 10^6 CFU/mL yeast inoculum) in sterile saline (0.9%), aliquots (20 µL) were pipetted into 96-well microplates and complemented with Roswell Park Memorial Institute culture medium (180 µL; Gibco RPMI 1640, Thermo Fisher Scientific, Waltham, MA, USA). The HDPE films were cut in circles (5 mm diameter), sterilized (UV, 30 min), placed into the 96-well microplates, incubated (36 °C, 24 h), and then washed with sterile saline (0.9%, 1 mL; Sigma-Aldrich, Saint Louis, MO, USA). These films were placed into sterile 96-well microplates and 3-(4,5-dimethylthiazol-2-yl)-2,5-diphenyltetrazolium bromide (160 µL; Sigma-Aldrich, Saint Louis, MO, USA) was added to assess cell viability as a function of redox potential (3 h). The removal of the solution containing 3-(4,5-dimethylthiazol-2-yl)-2,5-diphenyltetrazolium bromide was followed by treatment (15 min) with isopropanol (160 µL; Sigma-Aldrich, Saint Louis, MO, USA). A microplate reader (EZ Read 400, Biochrom, Cambridge, United Kingdom) was used to determine the absorption intensities (570 and 690 nm) using 100 µL of each sample in isopropanol. The pure HDPE film in the mixture of yeast inoculum (20 µL) and RPMI (180 µL) was the positive control. For the negative control, RPMI (200 µL) was used [34]. The percentage of biofilm inhibition was determined through the formula: 100 − [(average assay absorbance)/(average absorbance of the positive control)] × 100.

Biological Analysis on HDPE Films. (A) In vitro cell culture: human Mesenchymal Stem Cells (hMSC, Lonza, Italy) at the fifth passage were employed to perform biological studies, using a reported procedure for the in vitro cell culture [33]. (B) Cell attachment—morphological analysis: A confocal microscope (TCS SP8, Leica Microsystems, Buccinasco, Milan, Italy) was employed to analyze cell-film interactions and spreading, using a fluorescent dye. In particular, HDPE films were cultured with 2×10^4 cells (48 h; 37 °C); later, the non-attached cells were eliminated by careful washing with phosphate buffer solution (PBS; pH = 7.4, 0.01 M, Sigma-Aldrich, Milan, Italy), while the attached cells were treated with cell tracker green 5-chloromethylfluorescein diacetate (Life Technologies, Milan, Italy) in phenol red-free medium (37 °C; 30 min). The last step before the observation by CLSM consisted of washing with PBS and incubation in complete medium (1 h). (C) Biocompatibility test—attachment and proliferation: The biocompatibility test was performed on sterilized HDPE films (ethanol (4 h) and UV light (2 h)) equilibrated in Eagle's alpha minimum essential medium (sterile-filtered, Sigma-Aldrich, Milan, Italy) overnight. Later, HDPE films with and without IS were seeded in triplicate with 1×10^4 hMSCs and cultured (21 days). The effect of HDPE films on cell attachment and proliferation was quantitatively estimated by the Alamar blue assay (Life Technologies, Italy) at different time points. The results were reported as % of Alamar blue reduction (% AB reduction).

Statistical Analysis. One-way Analysis of Variance was employed with the multiple Dunnett comparison test, considering a significant difference for $P < 0.05$. The statistical analysis data were represented as mean ± standard deviation for n = 4 (antibiofilm, and MAC) or n = 3 (biocompatibility assay).

Histopathological Evaluation in Pig Skin with HDPE Films. The pig skin preparation, penetration and histopathological evaluation were performed following the reported procedures [33]. Ethic approval number: 04/2016 of the Animal Use Ethics Committee of the Federal Catarinense Institute - Campus Concórdia, Concórdia, SC, Brazil.

3. Results

HDPE-based biomaterials with IS additives were prepared by melt blending, followed by pressing into films. This resulted in the preparation of HDPE films with 0; 0.125; 0.250 and 0.500 wt% of either $C_{16}MImCl$ or $C_{16}MImMeS$. In comparison to the rigid film of HDPE, increasing the content of IS made the films more flexible and less brittle.

SEM investigations were performed to study the morphology of the HDPE-based films' surfaces. The micrographs of HDPE, HDPE.Cl.0500 and HDPE.MeS.0500 (Figure 2) indicate that the IS incorporation did not have an expressive influence on the surface morphology of these biomaterials. This behavior is different compared to our previous work, in which the addition of $C_{16}MImCl$ and $C_{16}MImMeS$ in the PLLA matrix interfered with the morphology leading to the formation of superficial spheres and increasing the roughness of the final material [33]. This could be related to the difference in the procedures that were applied to obtain the films; solvent casting (PLLA) vs. pressure molding (HDPE), as well as the chemical interactions between the polymer and IS [35]. As such, the eventual effects of the ISs on the surface morphology of HDPE could have been eliminated during the transformation into films under heat and pressure, assuming the flat surface of the hydraulic press plates. This was further supported by the AFM images (Figure 3), where HDPE (roughness = 13.0 nm), HDPE.Cl.0500 (roughness = 17.2 nm) and HDPE.MeS.0500 (roughness = 15.6 nm) presented smooth surfaces. The somewhat higher roughness of HDPE.Cl.0500 was most likely related to its higher crystallinity, which will be presented in Table 1. This was less pronounced for HDPE.MeS.0500.

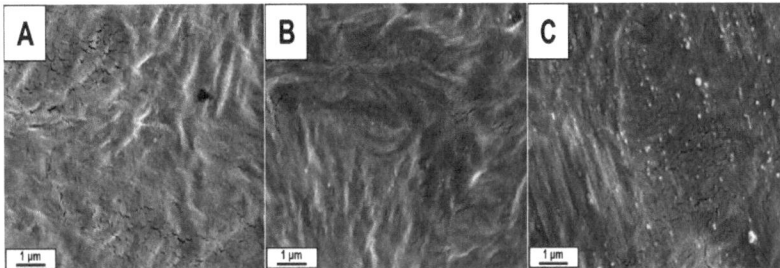

Figure 2. SEM micrographs of (**A**) HDPE, (**B**) HDPE.Cl.0500, and (**C**) HDPE.MeS.0500 (scale bar = 1 μm).

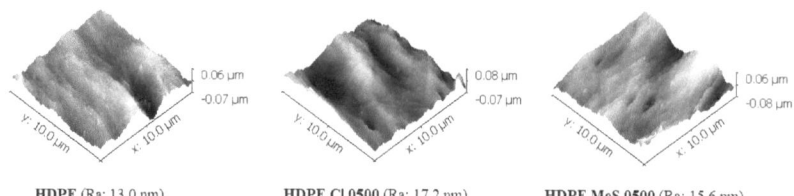

Figure 3. AFM images of HDPE, HDPE.Cl.0500, and HDPE.MeS.0500 (Ra = arithmetic mean roughness).

Table 1. Thermal Properties of HDPE Films.

Sample	T_m [1] [°C]	T_c [2] [°C]	ΔH_m [3] [J/g]	ΔH_c [4] [J/g]	X_c [5] [%]	$T_{5\%}$ [6] [°C]	$T_{10\%}$ [7] [°C]	$T_{50\%}$ [8] [°C]	Residue [9] [%]
HDPE	132.3	116.8	209.8	221.4	71.6	429.6	443.5	483.3	0.1
HDPE.Cl.0125	131.6	117.5	212.9	216.0	72.7	416.7	441.6	482.5	0.2
HDPE.MeS.0125	132.0	117.2	212.8	207.9	72.7	431.0	445.1	483.0	0
HDPE.Cl.0250	131.7	117.6	205.7	223.9	70.4	400.8	429.3	473.1	0.7
HDPE.MeS.0250	131.6	117.4	207.2	211.1	70.9	415.4	434.6	474.1	0
HDPE.Cl.0500	131.5	117.8	220.3	231.6	75.6	414.8	441.9	480.4	0.4
HDPE.MeS.0500	131.8	117.3	206.1	210.1	70.7	412.2	432.6	481.4	0.7

[1] Melting point obtained using DSC. [2] Crystallization temperature obtained using DSC. [3] Melting enthalpy obtained using DSC. [4] Crystallization enthalpy obtained using DSC. [5] Crystallinity obtained using DSC, and Equation (1), where ΔH^0_m = 293 J/g for 100% crystalline HDPE [36], and Fp = polymer fraction. [6] Temperature at decomposition of 5 wt% obtained using TGA. [7] Temperature at decomposition of 10 wt% obtained using TGA. [8] Temperature at decomposition of 50 wt% obtained using TGA. [9] Residual weight at 550 °C obtained using TGA.

Although the surface structure was not affected much by the presence of IS, the crystallinity of the HDPE-based biomaterials was studied by XRD (Figure 4). Independent of the IS (C_{16}MImCl or C_{16}MImMeS) or the IS content (0.125, 0.250 or 0.500 wt%), the type of HDPE crystallinity was not affected by obtaining crystalline HDPE.IS materials. All materials presented the typical HDPE peaks at 21.5° and 23.9°, which correspond to the (110) and (200) planes, respectively [36].

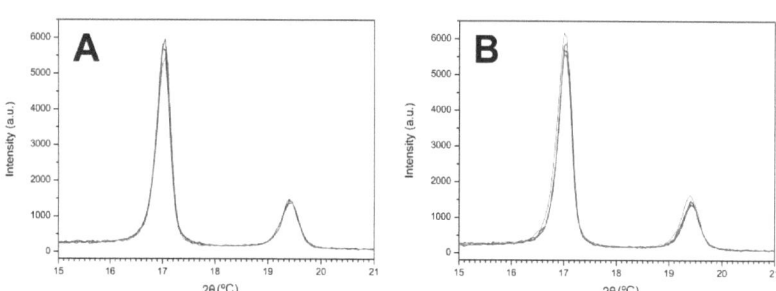

Figure 4. XRD diffractograms (a.u. = arbitrary units) within the 2θ range of (**A**) 15–25°, HDPE (black line), HDPE.Cl.0125 (red line), HDPE.Cl.0250 (blue line), and HDPE.Cl.0500 (green line) and (**B**) HDPE (black line), HDPE.MeS.0125 (red line), HDPE.MeS.0250 (blue line), and HDPE.MeS.0500 (green line).

The thermal properties of the HDPE-based biomaterials were studied by DSC and TGA, and the results are given in Table 1. In general, the incorporation of IS in the contents of 0.125, 0.250 and 0.500 wt% led to subtle modifications in the thermal properties of the

HDPE.IS biomaterials. The melting and crystallization temperatures (HDPE: 132.3 °C and 116.8 °C, respectively) varied within 1 °C. Compared to neat HDPE, HDPE.Cl.0500 showed increases of 5%, 4.4% and 5.6% in the melting enthalpy, the crystallization enthalpy and the crystallinity, respectively. The results also indicate that the ISs can be used as additives in the content range of 0.125–0.500 wt%, without modifying the thermal properties to a large extent. The same properties were studied with PLLA.IS biomaterials which showed an increase in thermal stability of 21 °C, whereas in this study the incorporation of IS into HDPE basically did not affect the thermal stability [33]. This effect may be related to the different intermolecular interactions between the ISs and the polymers. In PLLA.IS relatively strong non-covalent hydrogen bond interactions can take place whereas in HDPE.IS this is dependent on the weaker intermolecular van der Waals forces.

In Table 2 the storage and loss moduli and the stiffness results that were derived by DMA are summarized. Most of the HDPE.IS biomaterials showed similar storage moduli as HDPE, except for HDPE.Cl.0500, which showed lower values. For the loss moduli, the IS-containing HDPE films showed, in general, lower values although this did not follow a clear trend with increasing IS content, and HDPE.Cl.0500 presented a decrease of 24% in the value of HDPE. The lower storage and loss moduli and stiffness for HDPE.Cl.0500 could be related to its higher crystallinity as determined by DSC (Table 1). This could also explain the higher film roughness of HDPE.Cl.0500 (Figure 3). Interestingly, the biomaterials HDPE.Cl.0250, HDPE.MeS.0125 and HDPE.MeS.0500 demonstrated a better stiffness performance than the neat HDPE; the last one showed an increase of 35% in stiffness at 40 °C. No clear trend with the increase in IS load was observed indicating the non-linearity of the results. Generally, the dynamic-mechanical properties balance depended on the IS and its content, which was optimal for HDPE.MeS.0500. These results are in agreement with those obtained for PLLA.IS, which also generally demonstrated better values when C_{16}MImMeS was employed [33].

$$Xc\ (\%) = \frac{\Delta Hm}{\Delta H°m \times Fp} \times 100\% \tag{1}$$

Table 2. Dynamic Mechanical Properties of HDPE Films.

Sample	G'-40 [1] [GPa]	G'40 [2] [GPa]	G''-40 [3] [GPa]	G''40 [4] [GPa]	S-40 [5] [kN/m]	S40 [6] [kN/m]	S90 [7] [kN/m]
HDPE	3.33	1.48	0.07	0.17	315.45	140.50	26.71
HDPE.Cl.0125	3.14	1.44	0.05	0.16	271.63	124.96	22.05
HDPE.MeS.0125	3.17	1.40	0.06	0.16	349.31	154.66	28.11
HDPE.Cl.0250	3.28	1.52	0.07	0.16	346.19	160.34	29.83
HDPE.MeS.0250	3.09	1.38	0.05	0.15	288.79	129.50	23.64
HDPE.Cl.0500	2.44	1.12	0.04	0.12	230.62	106.03	19.45
HDPE.MeS.0500	3.18	1.45	0.05	0.16	416.42	190.24	32.03

[1] StorXage modulus at −40 °C. [2] Storage modulus at 40 °C. [3] Loss modulus at −40 °C. [4] Loss modulus at 40 °C. [5] Stiffness at −40 °C. [6] Stiffness at 40 °C. [7] Stiffness at 90 °C.

To better understand the influence of IS dispersed in the HDPE-IS films regarding wettability properties, the water contact angle technique was applied (Figure 5). Although HDPE.Cl.0250 showed a higher water contact angle than HDPE, the other films with C_{16}MImCl contents of 0.125 and 0.500 wt% only showed minor variations regarding the IS-free film. The increased hydrophobicity of HDPE.Cl.0250 suggests that C_{16}MImCl was present at the surface and that its aliphatic part was preferentially oriented towards the water drop. In contrast, the HDPE films with C_{16}MImMeS showed enhanced hydrophilicity according to the elevation of the IS load. The same effect was observed when 0.5 wt% of C_{16}MImMeS was applied in PLLA [33]. This suggests again that the IS was present at the film surface and that the polar part (imidazolium cation ring and IS anion) was preferentially oriented towards the water drop.

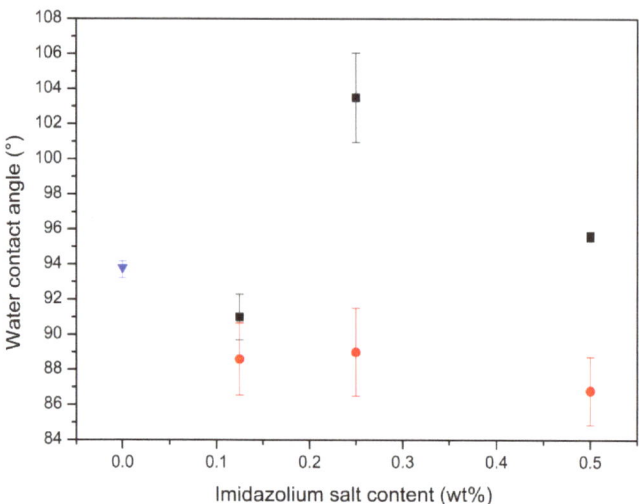

Figure 5. Water contact angles for HDPE films related to the IS content: HDPE (blue ▼), HDPE.Cl (black ■), and HDPE.MeS (red ●).

Initially, the in vitro biofilm assay antibiofilm was performed to verify whether HDPE biomaterials with ISs exhibited an antibiofilm effect compared to HDPE (Figures S1–S3). In this test it was verified that in comparison with HDPE (without the ISs), all tested biomaterials (HDPE.Cl.0125, HDPE.Cl.0250, HDPE.Cl.0500, HDPE.MeS.0125, HDPE.MeS.0250 and HDPE.MeS.0500) showed antibiofilm activity against clinical isolates of *C. tropicalis* 72A, *C. parapsilosis* RL11 and RL20 and *C. albicans* CA04. Furthermore, HDPE films with C_{16}MImCl (excluding HDPE.Cl.0250) also showed antibiofilm activity against *C. tropicalis* RL17.

Subsequently, the in vitro minor antibiofilm assay was performed to verify the percentage of prevention of biofilm formation on films of HDPE containing C_{16}MImCl (HDPE.Cl.0125, HDPE.Cl.0250 and HDPE.Cl.0500) or C_{16}MImMeS (HDPE.MeS.0125, HDPE.MeS.0250 and HDPE.MeS.0500), and the results are represented in Figures 6 and 7, respectively. Those films were differentiated from the neat HDPE against 12 isolates of *C. tropicalis* that are well known to form biofilms [33,37]. The results of the statistical analysis are shown in Figures S4–S9. In general, the obtained results suggest that the presence of IS reduced the growth of biofilms compared to HDPE. The biofilm inhibition varied between 0–75% and 0–64% on the HDPE films containing C_{16}MImCl and C_{16}MImMeS, respectively. The inhibition percentage was dependent on the tested *C. tropicalis* isolate, which was possibly due to genetic mutations that made some isolates more resistant to the HDPE-IS biomaterials [38]. The best percentages of impediment of biofilm formation were obtained using HDPE.Cl.0125 and HDPE.Cl.0250 with 75% for *C. tropicalis* 17P. When HDPE.Cl.0500 was employed, a 65% impediment was obtained for *C. tropicalis* 17P and 47% for *C. tropicalis* 17A (Figure 6). The best percentages of impediment of biofilm formation using HDPE.MeS.0125 were 54% for *C. tropicalis* ATCC 750 and 41% for *C. tropicalis* 17P. HDPE.MeS.0250 demonstrated a 40% impediment for *C. tropicalis* 94P and 37% for *C. tropicalis* ATCC 950. In the case of HDPE.MeS.0500, a 64% impediment was obtained for *C. tropicalis* 17P and 46% for *C. tropicalis* 72P (Figure 7).

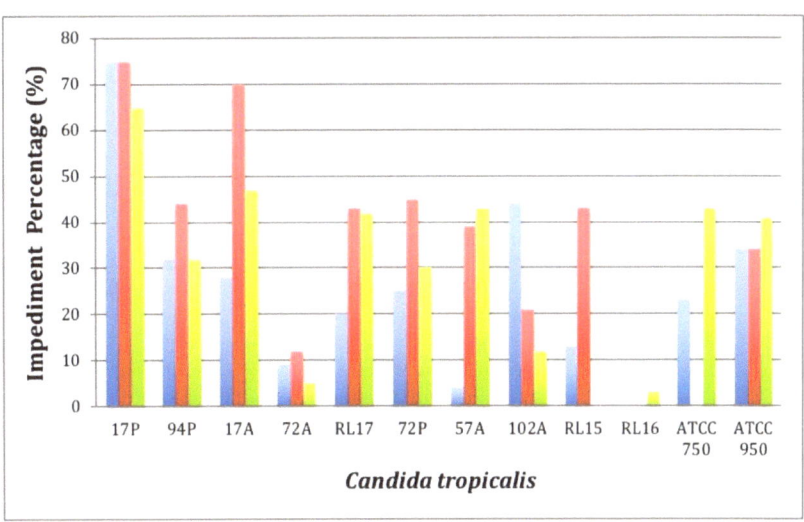

Figure 6. In vitro minor antibiofilm concentration assay: impediment percentage for HDPE.Cl.0125 (blue bars), HDPE.Cl.0250 (red bars), and HDPE.Cl.0500 (green bars).

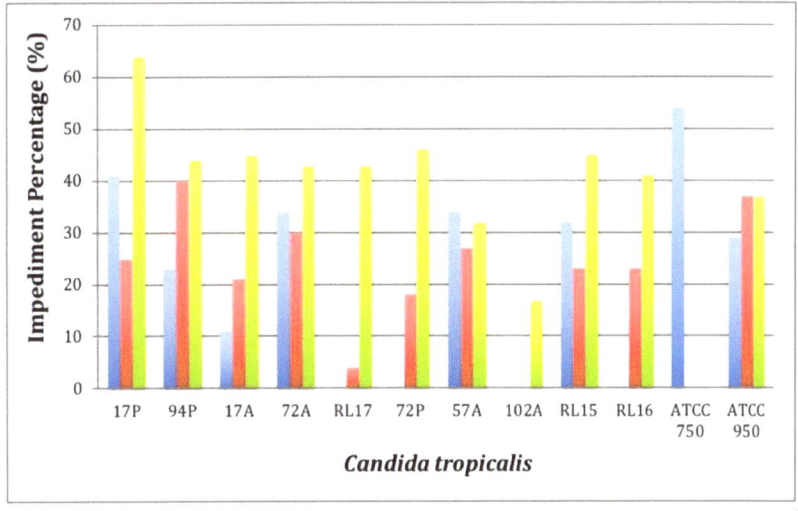

Figure 7. In vitro minor antibiofilm concentration assay: impediment percentage for HDPE.MeS.0125 (blue bars), HDPE.MeS.0250 (red bars), and HDPE.MeS.0500 (green bars).

These results can be ascribed to the intrinsic antibiofilm property of IS, which was previously reported for $C_{16}MImCl$ in the pre-treatment of catheter surfaces [30], $C_{16}MImCl$ and $C_{16}MImMeS$ incorporated in PLA-based biomaterials [33], and imidazolium polymeric materials [27]. Now, this property was effectively transposed after their incorporation in HDPE. As HDPE alone is not an effective antibiofilm material, the ISs must be present on the biomaterial's surface for this antibiofilm property to take place. As such, the prevention of biofilm formation was the result of a surface phenomenon due to the presence of IS at the HDPE surface. Even if IS would leach into the biological medium, its antibiofilm action will only take place when it is present on the surface of the biomaterial.

The results obtained with HDPE films with different contents of C_{16}MImCl and C_{16}MImMeS tested with *C. tropicalis* isolates showed that there is no direct relationship between the IS content used as an additive in HDPE and the percentage of prevention of formation of the biofilm (Figures 6 and 7). Considering the effectiveness of the HDPE biomaterials in relation to the IS content compared to *C. tropicalis* isolates, it was possible to verify that at contents of 0.125 and 0.250 wt%, C_{16}MImCl was more effective in preventing biofilm formation when compared to C_{16}MImMeS. At the content of 0.500 wt%, C_{16}MImMeS was more effective. In the case of PLA.IS, increasing the contents of the IS C_{16}MImCl and C_{16}MImMeS increased the percentage of impediment of biofilm formation [33]. The absence of this trend in the case of HDPE.IS suggests that other biomaterial properties impacted their determined antibiofilm potential including hydrophilicity and roughness [39,40].

Figure 8 shows SEM micrographs of HDPE samples. After 72 h of incubation with the clinical isolate *C. tropicalis* 72A (biofilm builder), the micrographs of HDPE (Figure 8A–D) show the formation of the biofilm with extracellular material and the cells at different stages of growth adhered to the HDPE film surface. In the cases of the films HDPE.Cl.0500 and HDPE.MeS.0500 (Figure 8E–H), the biofilm formation was prevented as no fungal and biofilm growth of *C. tropicalis* 72A was observed on the surfaces of these biomaterials. The results obtained indicate that, in the same way as PLLA.IS, both ISs were effective as anti-biofilm additives for HDPE.IS [33].

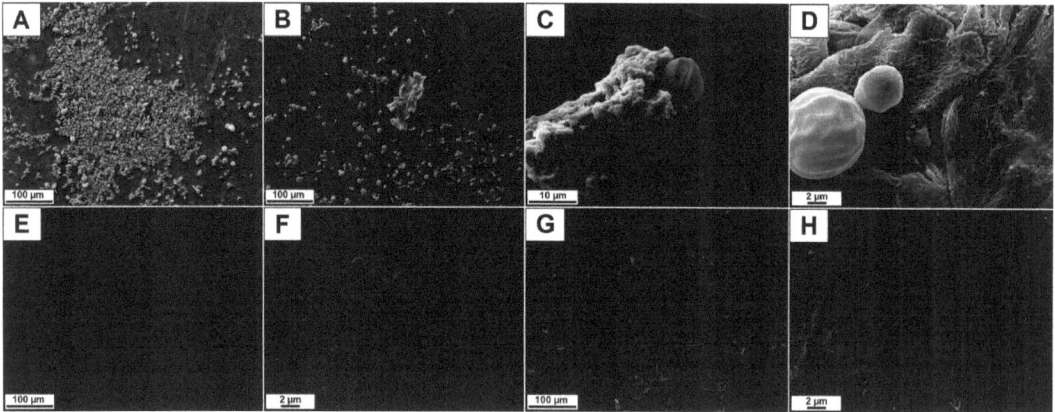

Figure 8. SEM micrographs of (**A**) HDPE (scale bar = 100 μm), (**B**) HDPE (scale bar = 100 μm), (**C**) HDPE (scale bar = 10 μm), (**D**) HDPE (scale bar = 2 μm), (**E**) HDPE.Cl.0500 (scale bar = 100 μm), (**F**) HDPE.Cl.0500 (scale bar = 2 μm), (**G**) HDPE.MeS.0500 (scale bar = 100 μm) and (**H**) HDPE.MeS.0500 (scale bar = 2 μm), after undergoing biofilm growth conditions.

The biocompatibility of a material is the principal parameter that governs the decision about the possibility to apply it in implants for human bodies. Such a biomaterial, when used in tissue engineering, should be non-toxic and biocompatible, without causing an intolerable degree of damage to that body [41]. In general, the in vitro cell-material interaction study is frequently used as an initial preliminary analysis of cell biocompatibility [42]. Herein, human mesenchymal stem cells (hMSC), generally used to evaluate the regeneration of mineralized extracellular matrix (ECM) in bone defects [43–46], were used for the in vitro testing of the biocompatibility of the HDPE-based materials with IS [43]. In particular, the effect of HDPE.IS biomaterials on the hMSC's behavior was evaluated by cell adhesion, which is the first step involved in the biocompatibility process (Figures 9 and 10A,B). Indeed, this cell attachment is the main stage to assess the influence of material surfaces on the hMSC behavior in the first hours of culture time. Both qualitative and quantitative analyses were performed with the aim to obtain information about the cell adhesion process. The morphological analysis (Figure 9) demonstrated a change in morphology with increasing IS contents. hMSC seeded on HDPE (without IS) showed a thin and elongated structure typical of fibroblast cells. Differently, the HDPE.IS biomaterials show a correlation between the presence of IS and the hMSC morphology. HDPE-based biomaterials with the ISs C_{16}MImCl and C_{16}MImMeS induced the stem cells to assume a polygonal structure, typical of osteoblast cells. This behavior was more pronounced for the biomaterials with higher IS contents (Figure 9C,D,F,G), which is highly favorable for bone repair processes. HDPE, **HDPE.Cl** and HDPE.MeS showed excellent values in the quantitative cell adhesion analysis which demonstrated good surface properties, promoting the extension of filopodia from the body cell and ensuring a stable cell attachment in the first 48 h of incubation (Figure 10A). For the C_{16}MImCl-based biomaterials, the cell adhesion increased with an increasing IS content, showing a higher cell adhesion percentage for HDPE.Cl.0500 in comparison to HDPE. All HDPE.MeS samples showed values comparable to those obtained with **HDPE**. The confocal micrographs demonstrate that HDPE.Cl.0500 and HDPE.MeS.0500 improved the spreading of hMSC at the cell-material interface. Indeed, the cells are polygonal in shape, which is different from the elongated morphology observed for the substrates with lower IS amounts. These results were also obtained with PLLA.IS substrates in previous work as reported [33].

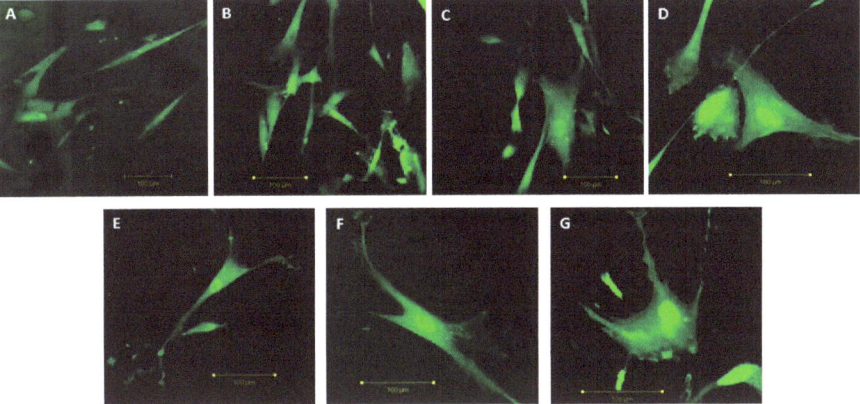

Figure 9. Confocal micrographs of hMSC grown on (**A**) HDPE, (**B**) HDPE.Cl.0125, (**C**) HDPE.Cl.0250, (**D**) HDPE.Cl.0500, (**E**) HDPE.MeS.0125, (**F**) HDPE.MeS.0250, and (**G**) HDPE.MeS.0500 (scale bar = 100 μm).

After the initial cell adhesion (Figure 10A), which is important for the next biocompatibility step, the cell proliferation after longer exposure times was studied (4, 7, 10, 14 and 21 days). This enables evaluating its continued cell development after initial adaptation to the biomaterial. In general, the HDPE materials, without and with IS, showed lower proliferation percentages than the control after 7 days (Figure 10B). Nevertheless, this was compensated for in all materials after 21 days; the cells became acquainted over time with their new environment. HDPE, HDPE.Cl.0250, HDPE.Cl.0500 and HDPE.MeS.0250 exceeded the proliferation percentages of the control after 10 days. The best proliferation results were achieved with the biomaterials containing 0.250 wt% of IS.

Finally, the results of the histopathological evaluation of skin of pig ear incubated with the HDPE films containing the ISs (Figure S10) showed no microscopic lesions.

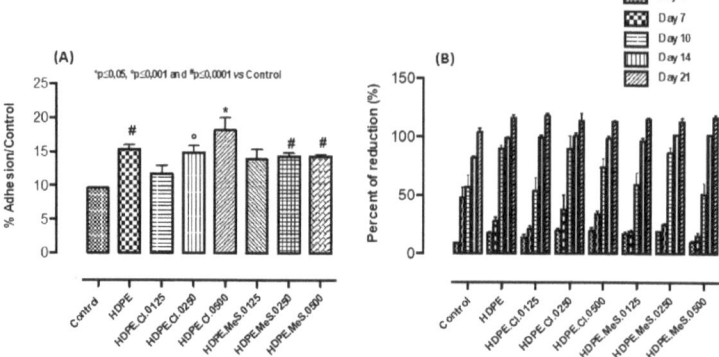

Figure 10. (**A**) hMSC adhesion after 48 h of incubation time and (**B**) hMSC proliferation at 4, 7, 10, 14 and 21 days of culture time.

4. Conclusions

In conclusion, the melt-blending of HDPE with the IS additives C_{16}MImCl and C_{16}MImMeS, and subsequent pressure molding, provided promising biomaterial films. Altogether, the ability of HDPE.IS to act effectively against the biofilm formation of Candida species, being biocompatible with hMSC, affording good cell adhesion and proliferation and being highly favorable for bone repair processes may open alternatives for the development of innovative medical devices.

Supplementary Materials: The following supporting information can be downloaded at: https://www.mdpi.com/article/10.3390/polym15051259/s1.

Author Contributions: Conceptualization, A.M.F. and H.S.S.; methodology, M.G.R., L.A., M.L.T., A.M.F. and H.S.S.; validation, A.M.F. and H.S.S.; formal analysis, C.M.L.S., M.G.R. and C.A.M.L.; investigation, C.M.L.S., Y.C.A.S. and M.L.T.; resources, M.G.R., L.A., M.L.T., A.M.F. and H.S.S.; data curation, C.M.L.S., M.G.R. and M.L.T.; writing—original draft preparation, C.M.L.S., Y.C.A.S. and M.G.R.; writing—review and editing, L.A., A.M.F. and H.S.S.; visualization, C.M.L.S., Y.C.A.S. and M.G.R.; supervision, A.M.F.; project administration, L.A., A.M.F. and H.S.S.; funding acquisition, A.M.F. and H.S.S. All authors have read and agreed to the published version of the manuscript.

Funding: This study was financed in part by the Coordenação de Aperfeiçoamento de Pessoal de Nível Superior—Brasil (CAPES)—Finance Code 001; the Conselho Nacional de Desenvolvimento Científico e Tecnológico—Brasil (CNPq)—Science without Borders Special Visiting Scientist grant number 400531/2013-5; and the Fundação de Amparo à Pesquisa do Estado do Rio Grande do Sul—Brasil (FAPERGS).

Informed Consent Statement: The animal study protocol was approved by the Institutional Ethics Committee of Federal Catarinense Institute - Campus Concórdia (protocol code 04/2016 of April 15, 2016).

Data Availability Statement: The data presented in this study are available on request from the corresponding author.

Acknowledgments: We acknowledge the Center of Microscopy and Microanalysis at UFRGS for the SEM facilities. The Coordenação de Aperfeiçoamento de Pessoal de Nível Superior—Brasil (CAPES), the Conselho Nacional de Desenvolvimento Científico e Tecnológico—Brasil (CNPq) and the the Fundação de Amparo à Pesquisa do Estado do Rio Grande do Sul—Brasil (FAPERGS) are acknowledged for financial support. Alexandre M. Fuentefria and Henri S. Schrekker are grateful to CNPq for the research productivity PQ fellowships.

Conflicts of Interest: The authors declare no conflict of interest.

References

1. Dias, A.S.; Miranda, I.M.; Branco, J.; Soares, M.M.; Vaz, C.P.; Rodrigues, A.G. Adhesion, Biofilm Formation, Cell Surface hydrophobicity, and Antifungal Planktonic Susceptibility: Relationship among *Candida* spp. *Front. Microbiol.* **2015**, *6*, 1–8.
2. Fox, E.P.; Bui, C.K.; Nett, J.E.; Hartooni, N.; Mui, M.C.; Andes, D.R.; Nobile, C.J.; Johnson, A.D. An Expanded Regulatory Network Temporally Controls Candida albicans Biofilm Formation. *Mol. Microbiol.* **2015**, *96*, 1226–1239. [CrossRef] [PubMed]
3. Oncu, S. Optimal Dosage and Dwell Time of Ethanol Lock Therapy on Catheters Infected with *Candida* species. *Clin. Nutr.* **2014**, *33*, 360–362. [CrossRef] [PubMed]
4. Coad, B.R.; Kidd, S.E.; Ellis, D.H.; Griesser, H.J. Biomaterials Surfaces Capable of Resisting Fungal Attachment and Biofilm Formation. *Biotechnol. Adv.* **2014**, *32*, 296–307. [CrossRef]
5. Liu, R.; Chen, X.; Falk, S.P.; Masters, K.S.; Weisblum, B.; Gellman, S.H. Nylon-3 Polymers Active Against Drug-Resistant *Candida albicans* Biofilms. *J. Am. Chem. Soc.* **2015**, *137*, 2183–2186. [CrossRef]
6. Ramage, G.; Robertson, S.N.; Williams, C. Strength in Numbers: Antifungal Strategies Against Fungal Biofilms. *Int. J. Antimicrob. Agents* **2014**, *43*, 114–120. [CrossRef]
7. Pannanusorn, S.; Fernandez, V.; Römling, U. Prevalence of Biofilm Formation in Clinical Isolates of *Candida* species Causing Bloodstream Infection. *Mycoses* **2013**, *56*, 264–272. [CrossRef]
8. Costa, A.C.B.P.; Pereira, C.A.; Freire, F.; Junqueira, J.C.; Jorge, A.O.C. Methods for Obtaining Reliable and Reproducible Results in Studies of *Candida* Biofilms Formed in vitro. *Mycoses* **2013**, *56*, 614–622. [CrossRef]
9. Morace, G.; Perdoni, F.; Borghi, E. Antifungal Drug Resistance in *Candida* species. *J. Glob. Antimicrob. Resist.* **2014**, *2*, 254–259. [CrossRef]
10. Seddiki, S.M.L.; Boucherit-Otmani, Z.; Boucherit, K.; KunKel, D. Infectivités Fongiques des Cathéters Implantes dues à *Candida* sp. Formation des biofilms et Résistance. *J. De Mycol. Médicale* **2015**, *25*, 130–135. [CrossRef]
11. Azizi, M.; Farag, N.; Khardori, N. Antifungal Activity of Amphotericin B and Voriconazole Against the Biofilms and Biofilm-Dispersed Cells of *Candida albicans* Employing a Newly Developed in vitro Pharmacokinetic Model. *Ann. Clin. Microbiol. Antimicrob.* **2015**, *14*, 21. [CrossRef] [PubMed]
12. Pourdanesh, F.; Jebali, A.; Hekmatimoghaddam, S.; Allaveisie, A. In vitro and in vivo Evaluation of a New Nanocomposite, containing High Density Polyethylene, Tricalcium Phosphate, Hydroxyapatite, and Magnesium Oxide Nanoparticles. *Mater. Sci. Eng. C* **2014**, *40*, 382–388. [CrossRef] [PubMed]
13. Niechajev, I. Facial Reconstruction Using Porous High-Density Polyethylene (Medpor): Long-Term Results Aesthetic. *Plast. Surg.* **2012**, *36*, 917–927. [CrossRef] [PubMed]
14. Lim, J.S.; Kook, M.S.; Jung, S.; Park, H.J.; Ohk, S.H.; Oh, H.K. Plasma Treated High-Density Polyethylene (HDPE) Medpor Implant Immobilized With rhBMP-2 for Improving the Bone Regeneration. *J. Nanomater.* **2014**, *2014*, 810404. [CrossRef]
15. Reznickova, A.; Novotna, Z.; Kolska, Z.; Kasalkova, N.S.; Rimpelova, S.; Svorcik, V. Enhanced Adherence of Mouse Fibroblast and Vascular Cells to Plasma Modified Polyethylene. *Mater. Sci. Eng. C* **2015**, *52*, 259–266. [CrossRef]
16. Durbec, M.; Mayer, N.; Ciolino, D.V.; Disant, F.; Gerin, F.M.; Groult, E.P. Reconstruction du Cartilage Nasal par Ingénierie Tissulaire à Base de Polyéthylène de Haute Densité et d'un Hydrogel. *Pathol. Biol.* **2014**, *62*, 137–145. [CrossRef]
17. Sobczac, M.; Debek, C.; Oledzka, E.; Kozlowski, R. Polymeric Systems of Antimicrobial Peptides- Strategies and Potential. *Appl. Mol.* **2013**, *18*, 14122–14137. [CrossRef]
18. Siedenbiedel, F.; Tiller, J. Antimicrobial Polymers in Solution and on Surfaces: Overview and Functional Principles. *Polymers* **2012**, *4*, 46–71. [CrossRef]
19. Borowiecki, P.; Krawczyk, M.M.; Plenkiewicz, J. Chemoenzymatic Synthesis and Biological Evaluation of Enantiomerically Enriched 1-(β-hydroxypropyl)Imidazolium- and Triazolium-Based Ionic Liquids Beilstein. *J. Org. Chem.* **2013**, *9*, 516–525.
20. Pilz-Junior, H.L.; de Lemos, A.B.; de Almeida, K.N.; Corção, G.; Schrekker, H.S.; Silva, C.E.; da Silva, O.S. Microbiota potentialized larvicidal action of imidazolium salts against Aedes aegypti (Diptera: Culicidae). *Sci. Rep.* **2019**, *9*, 16164–16172. [CrossRef]
21. Vekariya, R.L. A review of ionic liquids: Applications towards catalytic organic transformations. *J. Mol. Liq.* **2017**, *227*, 44–60. [CrossRef]
22. Singh, S.K.; Savoy, A.W. Ionic liquids synthesis and applications: An overview. *J. Mol. Liq.* **2020**, *297*, 112038. [CrossRef]
23. Schrekker, H.S.; Donato, R.K.; Fuentefria, A.M.; Bergamo, V.; Oliveira, L.F.; Machado, M.M. Imidazolium salts as antifungal agents: Activity against emerging yeast pathogens, without human leukocyte toxicity. *MedChemComm* **2013**, *4*, 1457–1460. [CrossRef]
24. Riduan, S.T.; Zhang, Y. Imidazolium Salts and Their Polymeric Materials for Biological Applications. *Chem. Soc. Rev.* **2013**, *42*, 9055–9070. [CrossRef]
25. Pendleton, J.N.; Gilmore, B.F. The Antimicrobial Potential of Ionic Liquids: A Source of Chemical Diversity for Infection and Biofilm Control International. *J. Antimicrob. Agents* **2015**, *46*, 131–139. [CrossRef]
26. Smiglak, M.; Pringle, J.M.; Lu, X.; Han, L.; Zhang, S.; Gao, H.; MacFarlane, D.R.; Rogers, R.D. Ionic Liquids for Energy, Materials, and Medicine. *Chem. Commun.* **2014**, *50*, 9228–9250. [CrossRef]
27. Liu, L.; Wu, H.; Riduan, S.N.; Ying, J.Y.; Zhang, Y. Short Imidazolium Chains Effectively Clear Fungal Biofilm in Keratitis Treatment. *Biomaterials* **2013**, *34*, 1018–1023. [CrossRef]
28. McCann, M.; Curran, R.; Shoshan, M.B.; Mckee, V.; Devereux, M.; Kevin, K.; Kellett, A. Synthesis, Sctructure and Biological Activity of Silver(I) Complexes of Substituted Imidazoles. *Polyhedron* **2013**, *56*, 180–188. [CrossRef]

29. Fang, B.; Zhou, C.H.; Rao, X.C. Synthesis and Biological Activities of Novel Amine-Derived bis-Azoles as Potencial Antibacterial and Antifungal Agents European. *J. Med. Chem.* **2010**, *45*, 4388–4398. [CrossRef]
30. Bergamo, V.Z.; Donato, R.K.; Dalla Lana, D.F.; Donato, K.J.Z.; Ortega, G.G.; Schrekker, H.S.; Fuentefria, A.M. Imidazolium salts as antifungal agents: Strong antibiofilm activity against multidrug-resistant *Candida tropicalis* isolates. *Lett. Appl. Microbiol.* **2015**, *60*, 66–71. [CrossRef]
31. Bergamo, V.Z.; Balbueno, E.A.; Hatwig, C.; Pippi, B.; Dalla Lana, D.F.; Donato, R.K.; Schrekker, H.S.; Fuentefria, A.M. 1-n-Hexadecyl-3-methylimidazolium methanesulfonate and chloride salts with effective activities against *Candida tropicalis* biofilms. *Lett. Appl. Microbiol.* **2015**, *61*, 504–510. [CrossRef] [PubMed]
32. Navale, G.R.; Dharne, M.S.; Shinde, S. Antibiofilm Activity of tert-BuOH Functionalized Ionic Liquids With Methylsulfonate Counter Anions. *R. Soc. Chem. Adv.* **2015**, *5*, 68136–68142.
33. Schrekker, C.M.L.; Sokolovicz, Y.C.A.; Raucci, M.G.; Selukar, B.S.; Klitzke, J.S.; Lopes, W.; Leal, C.A.M.; de Souza, I.O.P.; Galland, G.B.; dos Santos, J.H.Z.; et al. Multitask Imidazolium Salt Additives for Innovative Poly(L-lactide) Biomaterials: Morphology Control, *Candida* spp. Biofilm Inhibition, Human Mesenchymal Stem Cell Biocompatibility, and Skin Tolerance. *ACS Appl. Mater. Interfaces* **2016**, *8*, 21163–21176. [CrossRef] [PubMed]
34. Trafny, E.A.; Lewandowski, R.; Marciniak, I.Z.; Stepinska, M. Use of MTT Assay for Determination of the Biofilm Formation Capacity of Microorganisms in Metalworking Fluids Word. *J. Microbiol. Biotechnol.* **2013**, *29*, 1635–1643. [CrossRef] [PubMed]
35. Guo, Q.; Liu, Q.; Zhao, Y. Insights into the Structure and Dynamics of Imidazolium Ionic Liquid and Tetraethylene Glycol Dimethyl Ether Cosolvent Mixtures: A Molecular Dynamics Approach. *Nanomaterials* **2021**, *11*, 2512. [CrossRef] [PubMed]
36. Benabid, F.Z.; Kharchi, N.; Zouai, F.; Mourad, A.-H.; Benachour, D. Impact of co-mixing technique and surface modification of ZnO nanoparticles using stearic acid on their dispersion into HDPE to produce HDPE/ZnO nanocomposites. *Polym. Polym. Compos.* **2019**, *27*, 389–399. [CrossRef]
37. Seddiki, S.M.L.; Boucherit-Otmani, Z.; Boucherit, K.; Kunkel, D. Fungal Infectivities of Implanted Catheters due to *Candida* sp. Biofilms Formation and Resistance. *J. Med. Mycol.* **2015**, *25*, 130–135. [CrossRef]
38. Dühring, S.; Germerodt, S.; Skerka, C.; Zipfel, P.F.; Dandekar, T.; Schuster, S. Host-Pathogen Interactions between the Human Innate Immune System and *Candida albicans*-Understanding and Modeling Defense and Evasion Strategies. *Front. Microbiol.* **2015**, *6*, 625. [CrossRef]
39. Correia, D.M.; Fernandes, L.C.; Martins, P.M.; García-Astrain, C.; Costa, C.M.; Reguera, J.; Lanceros-Méndez, S. Ionic Liquid–Polymer Composites: A New Platform for Multifunctional Applications. *Adv. Funct. Mater.* **2020**, *30*, 1909736. [CrossRef]
40. Le, P.H.; Nguyen, D.H.K.; Aburto-Medina, A.; Linklater, D.P.; Crawford, R.J.; MacLaughlin, S.; Ivanova, E.P. Nanoscale Surface Roughness Influences *Candida albicans* Biofilm Formation. *ACS Appl. Bio Mater.* **2020**, *3*, 8581–8591. [CrossRef]
41. Williams, D.F. On the mechanisms of biocompatibility. *Biomaterials* **2008**, *29*, 2941–2953. [CrossRef] [PubMed]
42. Raucci, M.G.; Guarino, V.; Ambrosio, L. Biomimetic strategies for bone repair and regeneration. *J. Funct. Biomater.* **2012**, *3*, 688–705. [CrossRef] [PubMed]
43. Raucci, M.G.; Alvarez-Perez, M.A.; Demitri, C.; Sannino, A.; Ambrosio, L. Proliferation and osteoblastic differentiation of hMSCs on celulose-based hydrogels. *J. Appl. Biomater. Funct. Mater.* **2011**, *10*, 302–307.
44. Fasolino, I.; Raucci, M.G.; Soriente, A.; Demitri, C.; Madaghiele, M.; Sannino, A.; Ambrosio, L. Osteoinductive and anti-inflammatory properties of chitosan-based scaffolds for bone regeneration. *Mater. Sci. Eng. C* **2019**, *105*, 110046. [CrossRef] [PubMed]
45. Raucci, M.G.; Alvarez-Perez, M.A.; Meikle, S.; Ambrosio, L.; Santin, M. Poly(epsilon-lysine) dendrons tethered with phosphoserine increase mesenchymal stem cell differentiation potential of calcium phosphate gels. *Tissue Eng. Part A* **2014**, *20*, 474–485. [CrossRef] [PubMed]
46. Soriente, A.; Fasolino, I.; Gomez-Sánchez, A.; Prokhorov, E.; Buonocore, G.G.; Luna-Barcenas, G.; Ambrosio, L.; Raucci, M.G. Chitosan/hydroxyapatite nanocomposite scaffolds to modulate osteogenic and inflammatory response. *J. Biomed. Mater. Res. Part A* **2022**, *110*, 266–272. [CrossRef] [PubMed]

Disclaimer/Publisher's Note: The statements, opinions and data contained in all publications are solely those of the individual author(s) and contributor(s) and not of MDPI and/or the editor(s). MDPI and/or the editor(s) disclaim responsibility for any injury to people or property resulting from any ideas, methods, instructions or products referred to in the content.

Review

Chitosan-Coated Polymeric Silver and Gold Nanoparticles: Biosynthesis, Characterization and Potential Antibacterial Applications: A Review

Md. Amdadul Huq [1,*,†], Md. Ashrafudoulla [2], Md. Anowar Khasru Parvez [3], Sri Renukadevi Balusamy [4,*], Md. Mizanur Rahman [5], Ji Hyung Kim [6] and Shahina Akter [6,*,†]

[1] Department of Food and Nutrition, College of Biotechnology and Natural Resource, Chung-Ang University, Anseong-si 17546, Gyeonggi-do, Republic of Korea
[2] Department of Food Science and Technology, Chung-Ang University, Anseong-si 17546, Gyeonggi-do, Republic of Korea
[3] Department of Microbiology, Jahangirnagar University, Savar, Dhaka 1342, Bangladesh
[4] Department of Food Science and Technology, Sejong University, Seoul 143-747, Republic of Korea
[5] Department of Biotechnology and Genetic Engineering, Faculty of Biological Science, Islamic University, Kushtia 7003, Bangladesh
[6] Department of Food Science and Biotechnology, Gachon University, Seongnam 461-701, Republic of Korea
* Correspondence: amdadbge@gmail.com (M.A.H.); renucoimbatore@gmail.com (S.R.B.); shahinabristy16@gmail.com (S.A.)
† These authors contributed equally to this work.

Citation: Huq, M.A.; Ashrafudoulla, M.; Parvez, M.A.K.; Balusamy, S.R.; Rahman, M.M.; Kim, J.H.; Akter, S. Chitosan-Coated Polymeric Silver and Gold Nanoparticles: Biosynthesis, Characterization and Potential Antibacterial Applications: A Review. *Polymers* **2022**, *14*, 5302. https://doi.org/10.3390/polym14235302

Academic Editor: Paolo Ferruti

Received: 4 November 2022
Accepted: 1 December 2022
Published: 4 December 2022

Publisher's Note: MDPI stays neutral with regard to jurisdictional claims in published maps and institutional affiliations.

Copyright: © 2022 by the authors. Licensee MDPI, Basel, Switzerland. This article is an open access article distributed under the terms and conditions of the Creative Commons Attribution (CC BY) license (https:// creativecommons.org/licenses/by/ 4.0/).

Abstract: Biosynthesized metal nanoparticles, especially silver and gold nanoparticles, and their conjugates with biopolymers have immense potential in various fields of science due to their enormous applications, including biomedical applicationS. Polymeric nanoparticles are particles of small sizes from 1 nm to 1000 nm. Among different polymeric nanoparticles, chitosan-coated silver and gold nanoparticles have gained significant interest from researchers due to their various biomedical applications, such as anti-cancer, antibacterial, antiviral, antifungal, anti-inflammatory technologies, as well as targeted drug delivery, etC. Multidrug-resistant pathogenic bacteria have become a serious threat to public health day by day. Novel, effective, and safe antibacterial agents are required to control these multidrug-resistant pathogenic microorganismS. Chitosan-coated silver and gold nanoparticles could be effective and safe agents for controlling these pathogenS. It is proven that both chitosan and silver or gold nanoparticles have strong antibacterial activity. By the conjugation of biopolymer chitosan with silver or gold nanoparticles, the stability and antibacterial efficacy against multidrug-resistant pathogenic bacteria will be increased significantly, as well as their toxicity in humans being decreased. In recent years, chitosan-coated silver and gold nanoparticles have been increasingly investigated due to their potential applications in nanomedicinE. This review discusses the biologically facile, rapid, and ecofriendly synthesis of chitosan-coated silver and gold nanoparticles; their characterization; and potential antibacterial applications against multidrug-resistant pathogenic bacteria.

Keywords: chitosan-coated silver and gold nanoparticles; biosynthesis; antibacterial applications; multidrug-resistant pathogenic bacteria

1. Introduction

In recent years, bio-nanotechnology has attracted remarkable attention from researchers due to its extensive usage in different fields of science, especially for developing new bioactive materialS. Metal nanoparticles are small particles with a large surface area, making them perfect for utilization in various biomedical and industrial sectorS. Among different types of metal nanoparticles, silver and gold nanoparticles have gained significant attention due to their wide range of applications in various fields ofsciencE. In recent years, they have been widely used to develop antibacterial, antifungal, antiviral, and anticancer technologies, as well as gene therapy agents, biosensors, drug delivery, chronic disease diagnostics systems, etc. [1–7]. Recent studies have shown the vigorous antimicrobial activity of silver nanoparticles (AgNPs) against numerous pathogenic microorganisms, including multidrug-resistant bacteria [1,8,9]. AgNPs are often added to topical creams, hand gel, medical catheter coverings, wound dressings, antiseptic sprays, and cosmetics, etc., due to their effective antimicrobial properties [10–12]. Several reports have described the applications of AgNPs as wound-healing agents [13,14]. AgNPs were also effectively used as vehicles for various drugs to treat different diseases [15–18]. Gold nanoparticles (AuNPs) are also extensively used in biomedical science due to their high functionality, ease of detection, biocompatibility, and low toxicity [1,19,20]. Many studies have investigated numerous applications of AuNPs as drug and gene delivery agents to treat different diseases and as antimicrobial agents to control pathogenic microorganisms [1,21–25].

Polymer-coated metallic nanoparticles have gained considerable interest over recent years due to their unique physicochemical properties and wide applicationS. Among various polymeric nanoparticles, chitosan-coated polymeric silver and gold nanoparticles represent an emerging group of bioactive hybrid materials in medical science because of their biodegradability, biocompatibility, high activity, and stability with low toxicity [26,27]. Chitosan (Ch) is a biopolymer and is considered a non-toxic polymer that shows excellent antibacterial and antifungal activities against numerous pathogenic microorganisms, compared to other bioactive polymers [27,28]. There are many reports on the application of Ch inhibiting the growth of pathogenic microorganisms, including both Gram-positive and Gram-negative bacteria [29–31]. Ch is a natural polysaccharide and is widely used in pharmaceutical industries, as well as food industries due to its high biocompatibility and biodegradability with low toxicity [32,33]. Commercially, the bioactive polymer Ch is synthesized through the deacetylation process of chitin, which is collected from the outer skeleton of crab, shrimp, lobster, and crayfish shells [34]. Structurally, Ch is a cationic biopolymer consisting of D-glucosamine and N-acetyl D-glucosamine units attached by β-1,4 glycosidic bondS. Biopolymer Ch has two types of bioactive functional groups, the hydroxyl group and the amino group, and these active groups are responsible for the potential antimicrobial activity of Ch [34,35]. Ch is a positively charged molecule due to the presence of $-NH^{3+}$ groups, and these active amino groups are also responsible for the interaction with the negatively charged cell membranes of bacteria [27,35]. Ch is also used as a stabilizing agent for the synthesis of different metallic nanoparticleS. It can facilitate the modification of the surface physical absorption and electrostatic interaction, thus improving the stability and bioactivity of nanoparticles and making them a perfect candidate as potential therapeutic agents [36–39].

The emergence of multidrug-resistant (MDR) pathogenic bacteria seriously threatens public health worldwide [40]. The MDR bacteria create different health problems, including infectious diseases and threats to decrease the yield of many accomplishments, such as surgical procedures, transplantation, cancer care, etc. [41]. These MDR bacteria include *Staphylococcus aureus, Klebsiella pneumoniae, Streptococcus pneumoniae, Escherichia coli, Acinetobacter baumannii, Pseudomonas aeruginosa, Vibrio parahaemolyticus, Salmonella Typhimurium, Enterococcus faecium, Enterococcus faecalis, Enterobacter* spp., etc. [36,42,43]. Day by day many other bacteria also increasingly becoming resistant to antibioticS. Therefore, the development of new, safe, and effective antibacterial agents is urgently required. Many recent studies showed the efficacy of silver and gold nanoparticles in controlling multidrug-resistant microorganisms [1,2,44,45]. However, the main drawbacks of these nanoparticles are low stability and high toxicity [46,47]. By the conjugation of Ch with silver nanoparticles or gold nanoparticles, the toxicity will be decreased, but the stability and efficacy will be increased significantly [46,47]. Already, several reports showed the high stability and improved efficacy of chitosan-coated silver nanoparticles (Ch-AgNPs) and chitosan-coated gold nanoparticles (Ch-AuNPs) against pathogenic bacteria [46,47]. In this review, we discuss the facile, non-toxic, and eco-friendly method for the synthesis of

Ch-AgNPs and Ch-AuNPs, their characterization, and potential antibacterial applications in controlling multidrug-resistant pathogenic bacteria.

2. Biosynthesis of Ch-Coated Polymeric Silver and Gold Nanoparticles

There are several chemical and biological methods that are commonly applied for the synthesis of silver and gold nanoparticles and their nanocompositeS. Most chemical methods are conducted using different toxic chemicals and produce various hazardous by-products [48,49]. On the other hand, biological methods use eco-friendly and nonhazardous biological agents without using any toxic chemicals [1,12]. Due to the numerous drawbacks of chemical methods, scientists are focusing more on biological techniques for facile, non-toxic, and eco-friendly synthesis of nanoparticles and nanocompositeS. Moreover, biosynthesized nanoparticles and nanocomposites are pharmacologically more active than chemically synthesized nanomaterials [50]. According to Ghetas et al. [50], biologically synthesized nanoparticles showed significantly high antibacterial and antifungal activities against various pathogenic bacteria and fungi, compared to the chemically synthesized nanoparticleS. The biosynthesis of nanoparticles is an eco-friendly process that uses natural compounds from plants or microbes as reducing and stabilizing agents instead of hazardous chemicalS. Various biological resources, including different microbes, such as bacteria, fungi, algae, etc., and different parts of a plant, such as roots, leaves, and fruit, etc., could be utilized for the biosynthesis of nanoparticles, as well as nanocomposites [1,45,51,52].

There are two common biological approaches for the synthesis of Ch-AgNPs and Ch-AuNPs, a one-step process and a two-step procesS. In the one-step process, all materials, including plant or microbial extracts, Ch, and metal salts, such as silver nitrate ($AgNO_3$) or gold (III) chloride trihydrate ($HAuCl_4 \cdot 3H_2O$), are added together in a reaction flask and kept in a magnetic stirrer or in a shaking incubator with optimum conditions until the production of nanoconjugates (Figure 1A). In the two-step process, firstly, silver or gold nanoparticles are synthesized using metal salts and plant extracts or microbial culture supernatant. Then, the Ch is added to the synthesized silver or gold nanoparticles and kept in a magnetic stirrer until the production of Ch-coated polymeric silver or gold nanoparticles (Figure 1B). Paulkumar et al. [51] reported the one-step protocol for the biosynthesis of Ch-AgNPs using the stem extract of *Saccharum officinarum*. Saha et al. [47] also reported the one-step protocol for the biosynthesis of Ch-AuNPs using black pepper (*Piper nigrum*) extract. Rajeshkumar et al. [53] mentioned the two-step protocol for the green synthesis of Ch silver nanocomposites using the leaf extract of *Cissus arnottiana*. Raza et al. [54] also reported the two-step protocol for the biosynthesis of Ch-AgNPs using the cell-free extract of a fungal isolate, *Aspergillus fumigatus* KIBGE-IB33. Figure 1 illustrates the different steps of the biosynthesis of Ch-coated polymeric silver and gold nanoparticles using plants and microbeS.

2.1. Plant-Mediated Biosynthesis

For the biosynthesis of Ch-coated polymeric silver and gold nanoparticles, different plants and their various parts, including leaf, root, shoot, stream, fruit, etc., can be used efficiently. Plant extracts are renewable and nontoxic and are prepared using an environmentally friendly aqueous medium; they require mild reaction conditions without producing any toxic byproducts [55]. Plant extracts contain various bioactive phytochemicals, such as terpenoids, flavonoids, alkaloids, polysaccharides, phenols, organic acids, vitamins, and minerals, as well as various enzymes, amino acids, and proteins [52,56,57]. These bioactive phytochemicals can be used as both reducing and stabilizing agents, as well as capping agents during the synthesis of nanoparticles and nanocomposites [52,56,57]. There are several recent reports about the facile, non-toxic, eco-friendly biosynthesis of Ch-coated polymeric silver and gold nanoparticles using different plantS. For example, the leaf extract of *Cissus arnottiana* was used for the biosynthesis of Ch-AgNPs [53]. This is a two-step process, where the author first synthesized AgNPs using leaf extract of *Cissus arnottiana* as reducing and stabilizing agentS. Then, the biosynthesized AgNPs were added to the Ch solution in a reaction flask and kept in a magnetic stirrer. Then, the synthesized nanocomposite pellet was collected by centrifugation and lyophilized it. Finally, the lyophilized Ch-AgNPs were dissolved in distilled water and used for characterization [53]. In another study, the plant extract of *Cuscuta reflexa* was used for the green synthesis of AgNPs, and then the Ch and synthesized AgNPs were mixed and kept in a stirrer under dark conditionS. Finally, the solution was lyophilized and utilized for characterization and biomedical applications [58]. Shinde et al. [46] synthesized Ch-AgNPs using a one-step procesS. They added the leaf extract of *Prunus cerasus* and Ch solution in a reaction vessel, and then silver nitrate solution was added dropwise to the reaction vessel while it was magnetically stirred. Within 2 h of incubation, the reaction mixture changed from colorless to dark yellowish brown. The color change indicates the formation of Ch silver nanocompositeS. Then, the reaction mixture was lyophilized to obtain the powder of Ch-AgNPS. Paulkumar et al. [51] also used the one-step process for the biosynthesis of Ch-AgNPS. A amount of 1 mM of silver nitrate was mixed with the Ch solution using a magnetic stirrer. Then, the stem extract of *S. officinarum* was added to the Ch silver nitrate suspension. After adding the stem extract, the colorless reaction mixture turned brown which indicates the synthesis of nanocompositeS. The SEM and EDS analysis confirmed the formation of Ch-AgNPs [51]. Saha et al. [47] reported the biosynthesis of Ch-AuNPs using black pepper (*Piper nigrum*) extract. The biosynthesis of Ch-coated polymeric silver and gold nanoparticles and their antimicrobial applications are shown in Table 1.

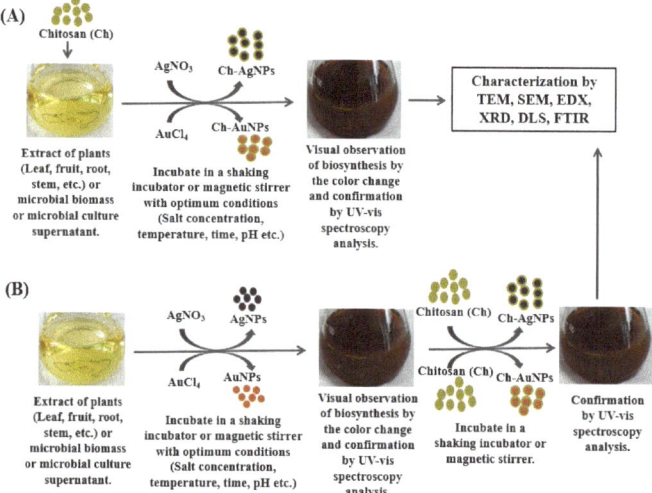

Figure 1. Schematic illustration of the biological synthesis of chitosan-coated polymeric silver and gold nanoparticles using plants and microbeS. (**A**) One-step process, (**B**) two-step process.

Table 1. Biosynthesis of chitosan-coated polymeric silver and gold nanoparticles and their potential antibacterial applications.

Nanoparticles	Synthesis Method	Characterization of Synthesized Nanoparticles	Applications	References
Ch-silver	Green synthesis of Ch silver bionanocomposite using the plant extract of *Saccharum officinarum*.	Characterized by UV-vis spectrophotometer, TEM, and FTIR.	Antibacterial applications against *Bacillus cereus*, *Staphylococcus*, and *Escherichia coli*.	[59]

Table 1. Cont.

Nanoparticles	Synthesis Method	Characterization of Synthesized Nanoparticles	Applications	References
Ch-silver	Synthesis of Ch silver bioconjugates using leaf extract of *Prunus cerasus*.	The bioconjugates were characterized using TEM, DLS, FT-IR, UV–Vis spectroscopy, and a zeta potential analyzer.	Antimicrobial applications *E. coli, Enterococcus faecalis, Klebsiella pneumoniae*, and *S. aureus*.	[46]
Ch-silver	Biosynthesis of Ch silver nanocomposite using Aloe vera extract and *Cuscuta reflexa* extract.	Characterized by UV–vis spectrum, FT-IR, and SEM	Antibacterial activities against *Staphylococcus aureus* ATCC 33592, *K. pneumoniae* ATCC 13884, *Bacillus subtilis* ATCC 55614, and *E. coli* ATCC 11229.	[58]
Ch-silver	Green synthesis of Ch silver nanoparticles using vitamin C as a reducing agent.	SEM, Zeta potential, and XRD.	In vitro antimicrobial activities against *E. coli* and *S.* Typhimurium, and in vivo antibacterial activity against *E. coli* in minced beef meat sampleS.	[60]
Ch-silver	Green synthesis of Ch-AgNPs using Ch as a stabilizer and sodium hydroxide as a reducing agent.	UV–vis spectroscopy, FT-IR spectroscopy, XRD, SEM, EDX, and zeta sizer nano.	Antibacterial activity against *S. aureus, E. coli*, and antifungal activity against *Candida albicans*.	[61]
Ch-silver	Green synthesis by a simple and environmentally friendly in situ chemical reduction process.	UV–Vis, TEM, SEM, XRD, and FTIR.	Antibacterial activity against *S. aureus*, and *E. coli*.	[26]
Ch-silver	Green and rapid synthesis of Ch-AgNPs using economically abundant biopolymer crustacean wastE.	UV–visible spectral, FTIR, XRD, AFM, TEM, and DLS.	Antibacterial activity against *Bacillus* sp., *Staphylococcus* sp., *Pseudomonas* sp., *E. coli, Proteus* sp., *Serratia* sP. and *Klebsiella* sp. Antifungal activity against *Aspergillus niger, A. fumigatus, A. flavus*, and *C. albicans*.	[62]
Ch-silver	Biosynthesis of AgNPs and Ch-AgNPs using the stem extract of *Saccharum officinarum*.	Characterized by UV–vis, TEM, SEM, and FTIR.	Antibacterial activity against *B. subtilis* (MTCC 3053), *K. planticola* (MTCC 2277), *Streptococcus faecalis* (ATCC 8043), *P. aeruginosa* (ATCC 9027), and *E. coli* (ATCC 8739).	[51]
Ch-silver	Biosynthesis of Ch-AgNPs using fungal biomass (*Aspergillus fumigatus* KIBGE-IB33).	UV–vis, SEM, DLS, and FTIR.	Antimicrobial activity against *Enterococcus faecalis* ATCC 29212 *S.* Typhimurium ATCC 3632, *Listeria monocytogenes* ATCC 7644, and *P. aeruginosa* ATCC 27853.	[54]
Ch-silver	Green synthesis of Ch-AgNPs using Ch as a reducing agent, as well as the stabilizing agent.	UV–Vis, FTIR spectroscopy, TEM, XRD, and DLS.	Antibacterial activity against Gram-positive *S. aureus* (KMIEV B161), and Gram-negative *E. coli*.	[63]

Table 1. Cont.

Nanoparticles	Synthesis Method	Characterization of Synthesized Nanoparticles	Applications	References
Ch-silver	The green route was used for the synthesis of Ch-based silver nanoparticles using Ch as a reducing and stabilizing agent.	UV–vis, FTIR, SEM, XRD, and TEM.	Antibacterial activity against *P. aeruginosa*, *E. coli*, and methicillin-resistant *S. aureus*.	[64]
Ch-silver	Ch-AgNPs were synthesized using AgNO$_3$, cysteine, and Ch.	UV–vis, DLS and Zeta potential, TEM, and XRD.	Antifungal activity against *Sporothrix brasiliensis*, and *Sporothrix schenckii*.	[65]
Ch-silver	Biosynthesis of Ch-AgNPs using leaf extract of *Cissus arnottiana*.	UV–Vis, SEM, TEM, AFM (atomic force microscope), XRD, and SAED.	Antibacterial and antifungal activity against *S. aureus*, *Streptococcus mutans*, *E. faecalis*, and *C. albicans*.	[53]
Ch-silver	Ch ascorbic acid-based green synthesis of polymeric silver nanoparticleS.	UV–Vis, TEM, X-ray photoelectron spectroscopy (XPS).	Antibacterial and antifungal activity against *S. aureus*, *P. aeruginosa*, *E. coli*, and *C. albicans*.	[66]
Ch-silver	Green synthesis of Ch-AgNPs using Ch as a reducing agent, as well as the stabilizing agent.	UV–vis, FTIR, XRD, and high-resolution transmission electron microscopy (HRTEM).	Catalytic activity and antibacterial activity against *E. coli*, and *M. luteus*.	[67]
Ch-silver	Synthesis of biogenic Ch-AgNPs using Ch as a reducing agent, as well as the stabilizing agent.	UV–vis, FTIR, EDX, SEM, TEM, and XRD.	Anticancer activity in human hepatocellular carcinoma HepG2 cellS.	[68]
Ch-silver	Biogenic synthesis of Ch functionalized silver nanoparticles using leaf extract of *Carica papaya*.	UV–vis, FTIR, DLS, HRTEM, and zeta potential estimation.	Antibacterial and antibiofilm activities against *E. coli*, and *S. aureus*.	[69]
Ch-silver	Biosynthesis of AgNPs and Ch-AgNPs using seed extract of *Piper nigrum*.	UV–vis, XRD, SEM, TEM, and FTIR.	Antibacterial activity against *E. coli*, and *Bacillus subtilis*.	[70]
Ch-silver and Ch-gold	Biosynthesis of Ch-silver and Ch-gold nanocomposites using *Bacillus Subtilis*.	UV–vis, XRD, SEM, and TEM.	Antibacterial activity against *S. aureus*, and *P. aeruginosa*. Antifungal activity against *A. niger*, and *C. albicans*.	[71]
Ch-silver and Ch-gold	Biosynthesis of Ch-silver and Ch-gold nanoparticles using two endophytic fungi, *Aspergillus* sp., and *Alternaria* sP.	UV–vis, XRD, FTIR, and TEM.	Antibacterial activity against *E. coli*, and *S. aureus*. Antibiofilm activity against *P. aeruginosa*, *B. subtilis*, *E. coli*, and *S. aureus*.	[72]
Ch-gold	The AuNPs and Ch-AuNPs have been biosynthesized using the extract of black pepper (*Piper nigrum*)	UV–vis, DLS, zeta potential, TEM, SAED, and EDX.	Antifilarial activity against *Setaria cervi* causes filarial parasite diseasE.	[47]
Ch-gold	Ch-AuNPs were synthesized using gold (III) chloride trihydrate and Ch.	UV–Vis, FE-TEM, FE-SEM, Zeta potential, and EDX.	Antifungal activity against *C. albicans*.	[73]

Table 1. Cont.

Nanoparticles	Synthesis Method	Characterization of Synthesized Nanoparticles	Applications	References
Ch-gold	Green synthesis of Ch-AuNPs using Ch as a reducing and stabilizing agent.	UV–vis, DLS, and TEM.	Antibacterial activity against *S. aureus* ATCC 29213, *S. aureus* ATCC 43300, and *E. coli* 11046.	[29]
Ch-gold	Green synthesis of Ch-AuNPs using Ch as a reducing and stabilizing agent.	TEM, SEM, FTIR, and XRD.	Antibacterial activity against *P. aeruginosa*, and *S. aureus*. Antifungal activity against *C. albicans*.	[36]

2.2. Microbe-Mediated Biosynthesis

For the biosynthesis of Ch-coated polymeric silver and gold nanoparticles, different microbes, such as bacteria, yeast, fungi, algae, etc., can also be used. These microorganisms are wonderful biological agents for the non-toxic, cost-effective, eco-friendly, and facile synthesis of nanoparticles and nanocomposites [2,45,56]. Microbial cells or cell-free culture supernatants contain many bioactive compounds, including amino acids, proteins, enzymes, flavonoids, organic materials, and many other primary and secondary metabolites [8,56]. These biomolecules of microorganisms serve as the reducing agents and the capping and stabilizing agents during synthesizing nanoparticles and nanocomposites [8,56]. There are some recent reports about the facile, non-toxic, eco-friendly biosynthesis of Ch-coated polymeric silver and gold nanoparticles using microorganismS. For example, the cell-free extract of fungi was utilized for the biosynthesis of silver-based Ch nanocomposites [54]. Initially, they used the cell-free filtrate of the fungal isolate for the biosynthesis of AgNPS. Then, the pre-synthesized AgNPs and Ch solution were mixed and the bioactive Ch-AgNPs were formed under microwave irradiation. The presence of hydroxyl and amino groups on the biopolymer Ch influence the formation of nanocomposites by binding the metallic components of the metal ions [54]. Youssef et al. [71], also used a two-step process for the biosynthesis of Ch-silver and Ch-gold nanocomposites using *Bacillus subtilis* bacterium. In another study, two marine fungi *Aspergillus* sP. Silv2 and *Alternaria* sP. Gol2 were used for the biological synthesis of Ch-silver and Ch-gold nanocomposites [72].

3. Characterization of Synthesized Ch-Coated Polymeric Silver and Gold Nanoparticles

The characterization of nanoparticles and their nanocomposites is necessary for evaluating their physical and chemical properties, such as size, shape, morphology, purity, particle crystallinity, surface chemistry, etC. Different instruments and techniques have been utilized to investigate the physical characteristics of the silver and gold nanoparticles, as well as Ch-coated polymeric silver and gold nanoparticleS. The commonly used instruments are UV-visible spectrophotometry (UV-vis), transmission electron microscope (TEM), scanning electron microscope (SEM), X-ray diffraction (XRD), Fourier transform infrared spectroscopy (FTIR), and dynamic light scattering (DLS), etC. The synthesis of silver and gold nanoparticles and Ch-coated polymeric silver and gold nanoparticles are initially observed by the naked eye due to the color changE. Generally, the brown or dark brown color of the reaction mixture indicates the synthesis of AgNPs and the Ch-coated polymeric silver nanocomposite, and the wine red, pink, violet, or purple color of the reaction mixture indicates the synthesis of AuNPs and the Ch-coated polymeric gold nanocomposite [1,27,46,47]. Then, the formation of AgNPs and AuNPs or Ch-coated polymeric silver or gold nanocomposite is confirmed by UV-visible spectrophotometry. Synthesized AgNPs and Ch-AgNPs showed a strong peak at around 400–500 nm in UV-visible spectrophotometry [1,2,46,58]. Similarly, synthesized AuNPs and Ch-AuNPs showed peaks at around 500–600 nm in UV-visible spectrophotometry [1,47]. The absorption spectra depended on the morphology, size, and shape of the synthesized nanoparticles [2,74]. According to Shinde et al. [46], the biosynthesized AgNPs and Ch-AgNPs using the leaf extract of *Prunus cerasus* showed an absorption peak at 429 and 445 nm, respectively (Figure 2A).

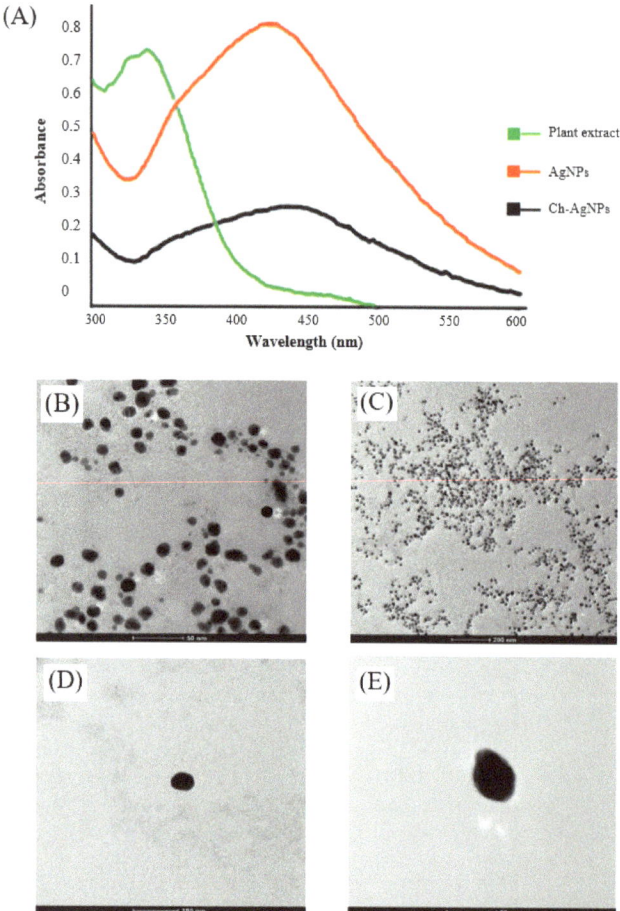

Figure 2. UV–Vis absorption spectra (**A**) and transmission electron microscopy analysis of synthesized AgNPs (**B**,**C**) and Ch-AgNPs (**D**,**E**). This figure has been reprinted with permission from Ref. [46], copyright 2021, MDPI.

The morphology, shape, size, purity, and aggregation of Ch-coated polymeric silver or gold nanoparticles are observed by TEM and SEM analysiS. Shinde et al. [46] utilized TEM to investigate the morphology, purity, and aggregation of biosynthesized Ch-AgNPS. The TEM analysis revealed that the spherical shape and silver nanoparticles were completely coated by Ch and showed a clear layer surrounding their core (Figure 2B–E). In another study by Paulkumar et al. [51], SEM was used to check the morphology of the biosynthesized Ch-AgNPs and found that the synthesized silver nanoparticles were strongly bound on the surface of the biopolymer Ch. X-ray diffraction is used to analyze the structural features of nanoparticles and nanocomposites, such as crystallinity, particle size, etc. [2]. DLS is used for the investigation of particle size distribution and polydispersity index. Shinde et al. [46] reported the average particles of biosynthesized AgNPs and Ch-AgNPs were 32.16 and 50 nm, respectively, with a polydispersity index of 0.2. They also investigated the zeta potential of biosynthesized AgNPs and Ch-AgNPs to check the stability of

AgNPs and Ch-AgNPs in aqueous suspensions [46]. FT-IR analysis of nanoparticles and nanocomposites is performed to identify the available biomolecules responsible for the synthesizing and stabilizing of nanoparticles and nanocomposites [46]. The biosynthesis of Ch-AgNPs using fungal biomass and their characterization by UV–vis, SEM, energy dispersive X-ray analysis, DLS, and FTIR has been reported by Raza et al. [54]. Shinde et al. [46] have also reported the biosynthesis of Ch-coated AgNPs from the leaf extract of *Prunus cerasus* and the synthesized nanocomposites were analyzed by UV-Vis, TEM, FT-IR, DLS, and zeta potential analyzer.

The biosynthesis of AuNPs and Ch-AuNPs using black pepper extract and their characterization by UV–vis, DLS, zeta potential, TEM, SAED, and EDX have been reported by Saha et al. [47]. The ecofriendly synthesis of Ch-AuNPs and their characterization by TEM, SEM, XRD, DLS, and FTIR have been conducted by Hashem et al. [36] (Figure 3). According to Hashem et al. [36], the TEM analysis showed a spherical shape with sizes ranging from 20 to 120 nm (Figure 3A). The DLS analysis showed that the average diameter of synthesized Ch-AuNPs was 218.2 nm (Figure 3B). The DLS analysis revealed the large size of synthesized Ch-AuNPs, compared to TEM analysis because of the presence of water molecules during DLS analysis around the synthesized Ch-AuNPs [36]. The XRD pattern of Ch-AuNPs revealed the crystalline nature of Ch-AuNPs (Figure 3C). The XRD pattern showed five clear diffraction peaks. Among these five diffraction peaks, the peak at 22.8° assured the presence of Ch in crystalline form. Other four peaks at 37.9°, 44.1°, 64.6°, and 77.4° confirmed the presence of AuNPs [36].

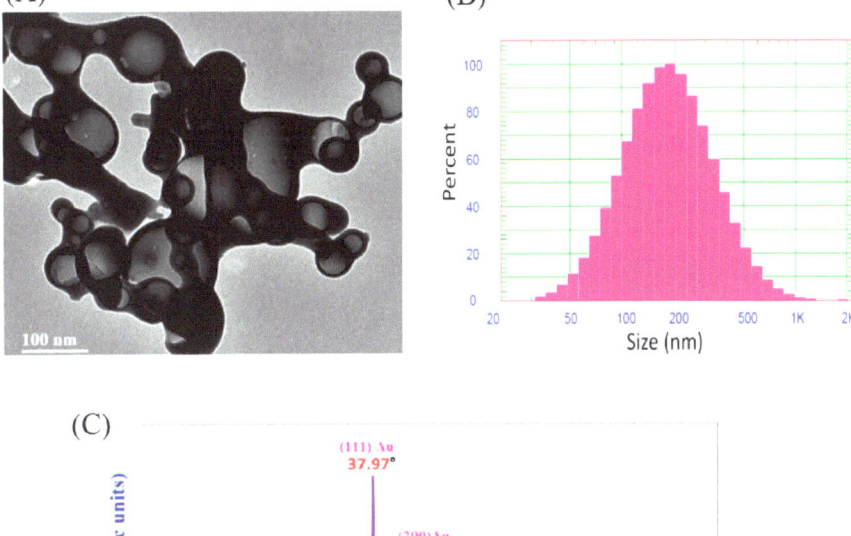

Figure 3. TEM images (**A**), particle size distribution (**B**), and XRD pattern (**C**) of Ch-AuNPS. This figure has been reprinted with permission from Ref. [36], copyright 2022, MDPI.

4. Potential Antibacterial Applications of Ch-Coated Polymeric Silver and Gold Nanoparticles

Bacterial resistance to various antibiotics is a serious problem worldwidE. Numerous infections caused by multidrug-resistant bacteria are sometimes impossible to treat, leading to the death of many people worldwide [75]. According to the World Health Organization, at least 700,000 people are currently dying every year due to drug-resistant diseases and among these 700,000 people, 230,000 people die only from multidrug-resistant tuberculosis caused by a bacterium. If no action is taken, antimicrobial resistance could force up to 24 million people into extreme poverty by 2030 and drug-resistant diseases could cause 10 million deaths every year by 2050 [76]. Therefore, the development of novel, safe, and effective antibacterial agents to control multidrug-resistant bacteria and treat infectious diseases is urgently needed. Ch-coated polymeric silver and gold nanoparticles could be potential and effective antibacterial agents that solve these problemS. The biopolymer Ch shows excellent antibacterial activities against numerous Gram-positive and Gram-negative pathogenic bacteria [29,30]. According to Avadi et al. [77], Ch showed strong antimicrobial activity against pathogenic *E. coli*. Costa et al. [78], investigated the antimicrobial activity of Ch against six oral pathogenic bacterial strains, such as *Prevotella buccae* (CCUG 15,401), *Tannarella forsythensis* (CCUG 51,269), *Aggregatibacter actinomycetemcomitans* (CCUG 13,227), *Streptococcus mutans* (CCUG 45,091), *Porphyromonas gingivalis* (9704 CIP 103,683T), and a clinical isolate of *Prevotella intermedia*, and found that the bioactive Ch effectively inhibits the growth of these pathogens with low MICs (minimum inhibitory concentrations) and shows quick and efficient bactericidal activity [78]. Jiang et al. [79] investigated the antimicrobial activity of two water-soluble chitosans against 31 representative foodborne pathogens and found that the used chitosans effectively controlled most of these foodborne pathogenS. Many other studies also support the antimicrobial efficacy of Ch against different pathogenic Gram-positive and Gram-negative bacteria [29,80,81]. Similarly, many studies showed the efficacy of silver and gold nanoparticles in controlling various multidrug-resistant bacteria [1,82]. According to Huq [83], biosynthesized AgNPs using *Lysinibacillus xylanilyticus* MAHUQ-40 showed strong antimicrobial activity against drug-resistant human pathogenic bacteria *Vibrio parahaemolyticus* and *Salmonella* Typhimurium. Huq and Akter [12], also discovered the potential antimicrobial activity of bacterial-mediated synthesized AgNPs against multidrug-resistant pathogenic bacteria *K. pneumoniae* and *S.* EnteritidiS. They used the disk diffusion method to investigate the zone of inhibition and microdilution assay to investigate the MICs and minimum bactericidal concentrations (MBCs) [12]. Hasnain et al. [45] reported on the panchagavya extract-mediated biosynthesis of AuNPs and investigated their antibacterial activity against *B. subtilis*, *E. coli*, and *K. pneumoniae*. They found that panchagavya extract-mediated biosynthesized AuNPs exhibited strong antibacterial activity against all these three pathogenic bacteria [84].

Silver and gold as metal exhibit toxicity even at a minimum concentration level [85]. The main properties of the bioactive polymer Ch are its nontoxicity, biodegradability, biocompatibility, low immunogenicity, and hemostatic properties [46,86–88]. By the conjugation of bioactive Ch with bioactive silver nanoparticles or gold nanoparticles, their efficacy and stability will increase significantly and the toxicity of silver and gold nanoparticles will decreasE. According to Potara et al. [89], Ch stabilizes the AgNPs and prevents agglomeration. Ch also confers a positive charge to the surface of AgNPs, which enhances their binding to the negative charges present on the cell surface of bacteria [89]. According to Saha et al. [47], Ch increases the stability and efficacy of biosynthesized AuNPS. Shinde et al. [46] investigated the antibacterial activity of biosynthesized AgNPs and Ch-AgNPs and found that the biosynthesized Ch-AgNPs show high activity against pathogenic bacteria, compared to the biosynthesized AgNPS. They also found that Ch-AgNPs do not show any toxicity in normal cellS. They used the leaf extract of *Prunus cerasus* for the biosynthesis of both AgNPs and Ch-AgNPs and evaluated their antimicrobial activity against multidrug-resistant pathogenic bacteria, such as *Enterococcus faecalis*, *E. coli*,

S. aureus, and *K. pneumonia*. The results of this study demonstrated that the Ch-AgNPs could inhibit the growth of multidrug-resistant pathogenic bacterial strains more effectively than AgNPs alone [46]. Paulkumar et al. [51] reported the antibacterial activity of plant-extract-mediated Ch silver nanocomposites against several pathogenic bacterial strains, such as *Klebsiella planticola* (MTCC 2277), *B. subtilis* (MTCC 3053), *S. faecalis* (ATCC 8043), *E. coli* (ATCC 8739), and *P. aeruginosa* (ATCC 9027). The biosynthesized Ch-AgNPs show strong antibacterial activity against all the tested pathogens, and they demonstrated that the silver-based Ch nanocomposite shows potent antibacterial activity due to the presence of small-sized silver nanoparticles on the surface of Ch [51]. Saruchi et al. [59] used the plant extract of *Saccharum officinarum* for the green synthesis of Ch-AgNPs and the synthesized nanocomposite was used to control the pathogenic *B. cereus*, *Staphylococcus*, and *E. coli*. They found that the synthesized bionanocomposites are potentially very effective against all tested pathogenic strains of bacteria and concluded that the biosynthesized Ch–silver nanocomposite could be a drug potentially used to control various pathogenic bacteria [59]. Fuster et al. [27], investigated the antibacterial activity of Ch-AuNPs against Gram-negative *E. coli* ATCC 25,922 and a clinical isolate of *E. coli* 11,046 (CI-EC) and two Gram-positive bacterial strains, methicillin-sensitive *S. aureus* ATCC 29213 and methicillin-resistant *S. aureus* ATCC 43,300. The Ch-AuNPs displayed significant antibacterial activity against all tested pathogenic strains, suggesting that Ch-AuNPs could be promising nanostructures for reducing bacterial infections [27].

Rezazadeh et al. [90] synthesized different AgNPs, including biogenic Ch-AgNPs, and investigated their antibacterial efficiency against four pathogenic bacterial strains (*E. coli*, *Proteus*, *Salmonella*, and *B. cereus*) using the disk diffusion method. The antibacterial activity of different AgNPs, such as algae-extract-mediated Ch-AgNPs (biological AgNPs), algae-extract-mediated AgNPs (algae-mediated AgNPs), only-Ch-mediated AgNPs, chemically synthesized AgNPs (chemical AgNPs), and $AgNO_3$ solution are shown in Figure 4 [90]. The results showed that the algae extract-mediated Ch-AgNPs (biological AgNPs) exhibit superior effectiveness against all four selected bacterial strains, compared to all other AgNPs, algae-Ch extract, and $AgNO_3$ precursor (Figure 4) [90]. The algae-extract-mediated Ch-AgNPs showed the largest zone of inhibition against four tested pathogenic bacterial strains, which were 21, 20, 18, and 17 mm, against *E. coli*, *Proteus*, *Salmonella*, and *B. cereus*, respectively, (Figure 4). The marine algae extract contains various biomolecules, which encompass the surface of biological AgNPS. When these bioactive AgNPs are coated by biopolymer Ch, the biological applicability and biocompatibility of Ch-AgNPs would presumably enhance, and hence increase, the antibacterial properties [90].

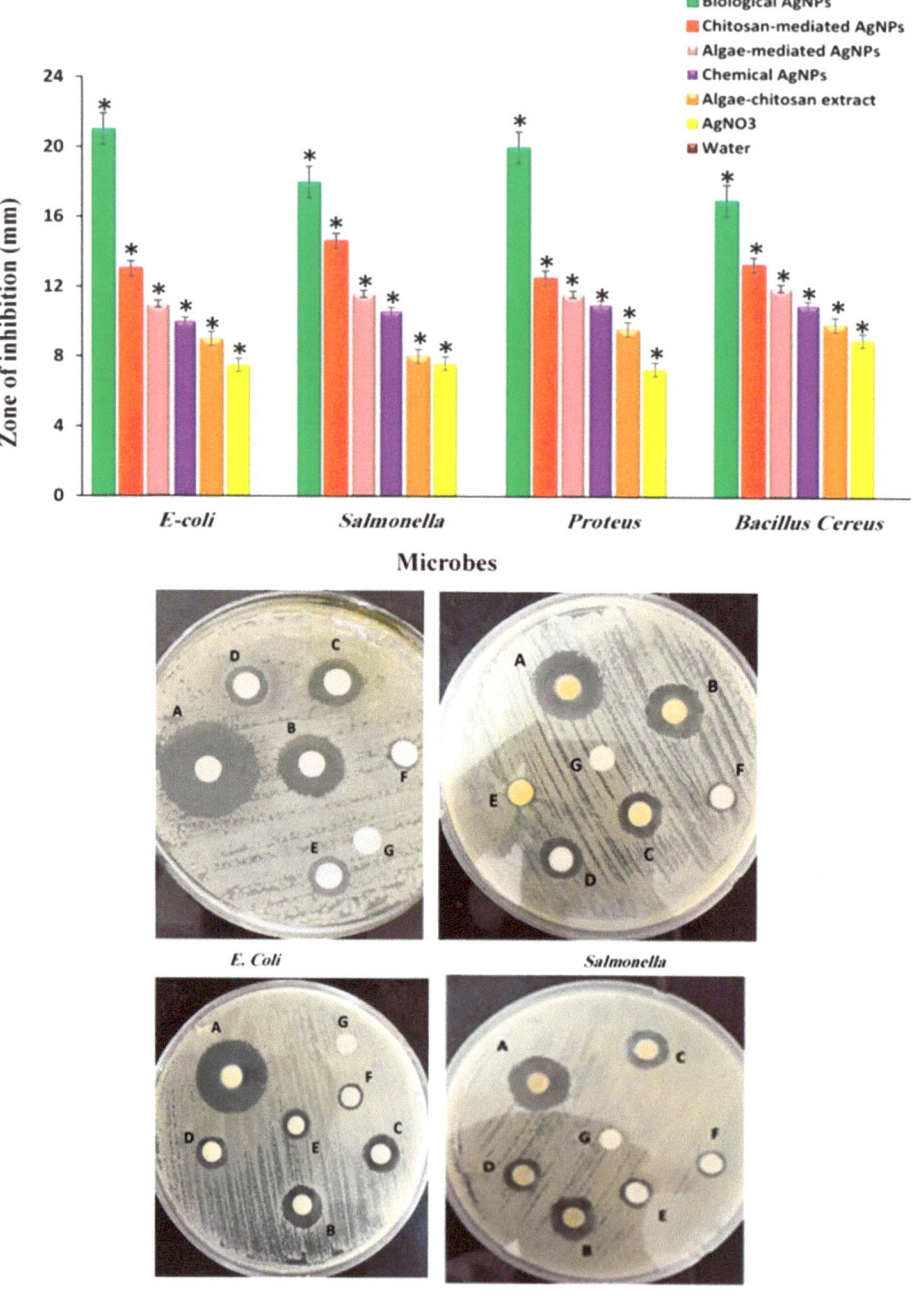

Figure 4. Comparative antibacterial effect of Ch-coated AgNPs against four selected clinical pathogenS. A, biological AgNPs; B, Ch-mediated AgNPs; C, algae-mediated AgNPs; D, chemical AgNPs; E, raw extract; F, AgNO$_3$; and G, H$_2$O. This figure has been reprinted with permission from Ref. [89], copyright 2020, Nature Portfolio.

5. Antibacterial Mechanisms of Ch-Coated Polymeric Silver and Gold Nanoparticles

The antibacterial activity of Ch-coated polymeric silver or gold nanoparticles largely depends on the type of Ch, molecular weight of Ch, type or size of silver or gold nanoparticles, molecular ratio of Ch and silver or gold nanoparticles, and the synthesis conditions, such as pH, temperature, etc. [2,27]. The positively charged Ch and silver or gold nanoparticles provide antibacterial properties because of their interaction with the negatively charged cell membranes of both Gram-negative and Gram-positive bacteria [2,27]. The complete antibacterial mechanism of Ch-coated polymeric silver or gold nanoparticles is still not fully known. There are several proposed mechanisms for the antibacterial activity of Ch against Gram-negative and Gram-positive bacteriA. The most acceptable mechanism is the interaction between positively charged Ch molecules (NH^{3+} groups) and negatively charged bacterial cell membranes, producing changes in the membrane permeability, which cause osmotic imbalances, inhibit bacterial growth and hydrolysis of the cell wall peptidoglycans of bacteria, and, finally, lead to the leakage of intracellular electrolytes, including potassium ions, as well as amino acids and low-molecular-weight proteins [27,91]. According to Sebti et al. [92], after penetrating the Ch into the nuclei of the bacteria through the cell wall, the Ch makes bonds with microbial DNA, which inhibits the synthesis of mRNA and protein and halts the normal activity of the cell. Another mechanism is the chelation of essential microbial nutrients with Ch [93]. According to Wang et al. [94], Ch has excellent metal-binding capacities, which influence the binding of different essential metallic nutrients with Ch in the bacterial cell that inhibit the growth of bacteria.

The combination of Ch and silver or gold nanoparticles seems promising because the positively charged bioactive polymer Ch potentiates interactions with bacteria, enhancing the positively charged silver or gold nanoparticles to disrupt the bacterial cell membrane more successfully. In this way, biopolymer Ch increases the biocompatibility and antibacterial activity of silver or gold nanoparticles [94,95]. The positively charged silver ions interact with the cell membrane of bacteria, disturbing the membrane permeability and respiration, as well as interacting with the negatively charged DNA and protein molecules, which could collapse the structure and function of DNA and protein [2,51]. The release of free radicals from silver might also be involved in membrane damage [51]. According to Fuster et al. [27], the antibacterial mechanism of Ch-AuNPs involves the electrostatic interactions between the Ch-AuNPs and the bacterial cell membraneS. These interactions lead to structural modification and loss of the properties of the bacterial membranE. Although the exact antibacterial mechanism of Ch-coated silver or gold nanoparticles has not been thoroughly explained, the probable antibacterial actions of Ch-coated silver or gold nanoparticles have been proposed in Figure 5. The proposed antibacterial mechanisms of Ch-coated silver and gold nanoparticles are the hydrolyses of the cell wall and cell membrane, the leakage of intracellular electrolytes and low-molecular-weight proteins, chelation of essential microbial nutrients with Ch, inhibition of the synthesis of mRNA and protein through the binding of bacterial DNA, alteration of the structure and function of the protein molecules, and the production of reactive oxygen species, which leads to the damage of ATP molecules (Figure 5). Through the above possible mechanisms, Ch-coated silver and gold nanoparticles inhibit the growth of pathogenic bacteria and finally kill them.

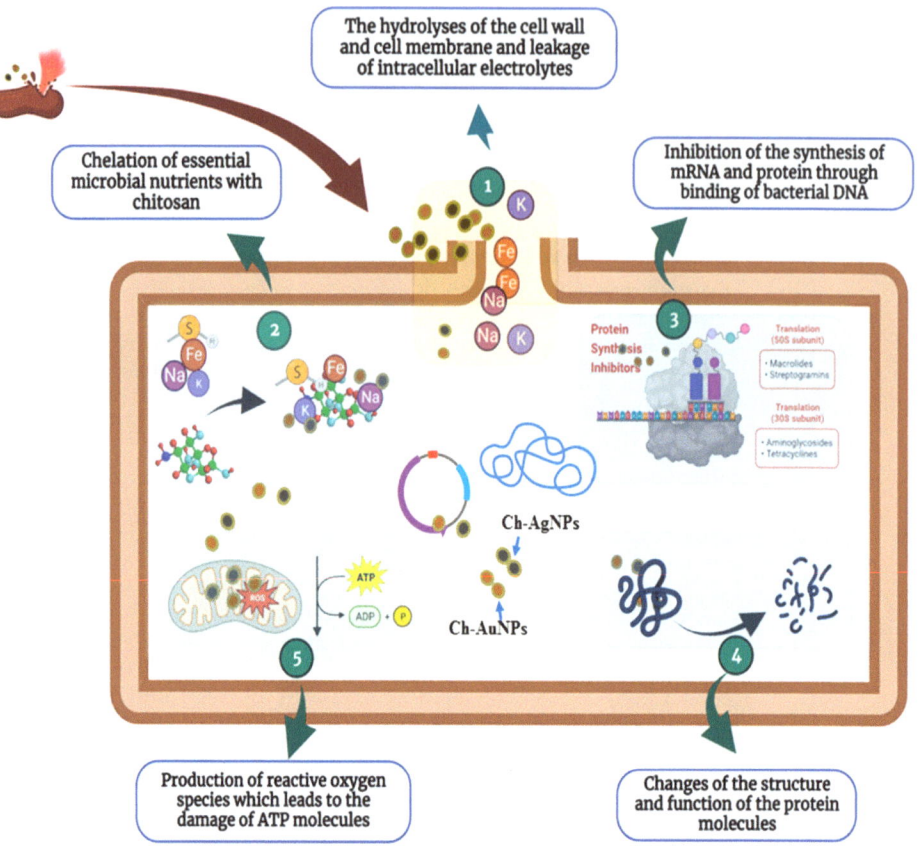

Figure 5. Probable antibacterial mechanisms of chitosan-coated silver and gold nanoparticles.

6. Conclusions and Future Perspectives

The emergence of MDR pathogenic bacteria is a serious threat to public health worldwidE. Therefore, the development of safe and effective antibacterial agents is urgently required. Ch-coated polymeric silver and gold nanoparticles represent an emerging group of bioactive hybrid materials in medical science because of their biodegradability, biocompatibility, high activity, and stability with low toxicity. Ch is a biopolymer and non-toxic polymer that shows excellent antibacterial activities against numerous pathogenic microorganismS. Similarly, biosynthesized AgNPs and AuNPs also exhibit strong antimicrobial activity against numerous pathogenic microorganisms, including multidrug-resistant bacteriA. By the conjugation of Ch with AgNPs or AuNPs, the stability of Ch-coated silver or gold nanoparticles will be increased and toxicity will be decreased, as well as the efficacy being increased significantly. Some recent studies showed the high stability and improved efficacy of Ch-coated silver and gold nanoparticles against pathogenic bacteriA. In this review, the facile, non-toxic, and eco-friendly method for the biosynthesis of Ch-coated polymeric silver or gold nanoparticles and their characterization have been comprehensively reviewed. The antibacterial applications and mechanisms of the biosynthesized Ch-coated polymeric silver or gold nanoparticles against pathogenic bacteria have also been highlighted. Although the biosynthesized Ch-coated polymeric silver or gold nanoparticles have shown great potential in controlling MDR pathogenic bacteria, several points might be considered for the future biosynthesis of Ch-coated polymeric silver or gold nanoparticles to explore their potent antibacterial activity. First, the type and molecular weight of Ch, the concentration of Ch, the concentration of silver or gold salts, and the concentration of plant or microbial extractS. These factors not only influence the synthesis process but also influence the antibacterial activity. Second, the biosynthesis of Ch-coated silver or gold nanoparticles should be conducted using potential and available plants or microbes, such as medicinal plants or other pharmacologically active plants and beneficial microbes or probioticS. Third, optimum synthesis conditions, i.e., temperature, pH, time, etc., should be maintained. Fourth, the investigation of the cytotoxic effect of biosynthesized Ch-coated polymeric silver or gold nanoparticles on human cells is essential, though some studies suggest that Ch-coated polymeric silver or gold nanoparticles are non-toxic and safe to usE. Finally, it can be said that Ch-coated polymeric silver and gold nanoparticles could be a promising tool in nanomedicine for controlling multidrug-resistant pathogenic bacteria.

Author Contributions: Conceptualization, M.A.H.; writing—original draft preparation, M.A.H. and S.A.; writing—review and editing, M.A.H., M.A., M.A.K.P., S.R.B., M.M.R. and J.H.K. All authors have read and agreed to the published version of the manuscript.

Funding: This research received no external funding.

Institutional Review Board Statement: Not applicable.

Informed Consent Statement: Not applicable.

Data Availability Statement: Not applicable.

Conflicts of Interest: The authors declare no conflict of interest.

References

1. Singh, P.; Mijakovic, I. Green synthesis and antibacterial applications of gold and silver nanoparticles from *Ligustrum vulgare* berrieS. *Sci. Rep.* **2022**, *12*, 7902. [CrossRef] [PubMed]
2. Huq, M.A.; Ashrafudoulla, M.; Rahman, M.M.; Balusamy, S.R.; Akter, S. Green synthesis and potential antibacterial applications of bioactive silver nanoparticles: A review. *Polymers* **2022**, *14*, 742. [PubMed]
3. Majdalawieh, A.; Kanan, M.C.; El-Kadri, O. Recent advances in gold and silver nanoparticles: Synthesis and applicationS. *J. Nanosci. Nanotechnol.* **2014**, *14*, 4757–4780.
4. Singh, P.; Kim, Y.J.; Singh, H. Biosynthesis, characterization, and antimicrobial applications of silver nanoparticleS. *Int. J. Nanomed.* **2015**, *10*, 2567–2577.
5. Huq, M.A.; Akter, S. Characterization and genome analysis of *Arthrobacter bangladeshi* sP. nov., applied for the green synthesis of silver nanoparticles and their antibacterial efficacy against drug-resistant human pathogenS. *Pharmaceutics* **2021**, *13*, 1691.
6. Zazo, H.; Colino, C.I.; Lanao, J.M. Current applications of nanoparticles in infectious diseaseS. *J. Control. Release* **2016**, *224*, 86–102.
7. Farhadian, N.; Mashoof, R.U.; Khanizadeh, S. *Streptococcus mutans* counts in patients wearing removable retainers with silver nanoparticles vs those wearing conventional retainers: A randomized clinical trial. *Am. J. Orthod. DentofaC. Orthop.* **2016**, *149*, 155–160.
8. Huq, M.A.; Akter, S. Bacterial mediated rapid and facile synthesis of silver nanoparticles and their antimicrobial efficacy against pathogenic microorganismS. *Materials* **2021**, *14*, 2615. [PubMed]
9. Wang, X.; Lee, S.-Y.; Akter, S.; Huq, M.A. Probiotic-mediated biosynthesis of silver nanoparticles and their antibacterial applications against pathogenic strains of *Escherichia coli* O157:H7. *Polymers* **2022**, *14*, 1834. [CrossRef] [PubMed]
10. Chaloupka, K.; Malam, Y.; Seifalian, A.M. Nanosilver as a new generation of nanoproduct in biomedical applicationS. *Trends Biotechnol.* **2010**, *28*, 580–588.
11. Ahmed, S.; Ahmad, M.; Swami, B.L.; Ikram, S. A review on plants extract mediated synthesis of silver nanoparticles for antimicrobial applications: A green expertisE. *J. Adv. Res.* **2016**, *7*, 17–28. [CrossRef] [PubMed]
12. Huq, M.A.; Akter, S. Biosynthesis, characterization and antibacterial application of novel silver nanoparticles against drug resistant pathogenic *Klebsiella pneumoniae* and *Salmonella* EnteritidiS. *Molecules* **2021**, *26*, 5996. [CrossRef]
13. Ge, L.; Li, Q.; Wang, M.; Ouyang, J.; Li, X.; Xing, M.M. Nanosilver particles in medical applications: Synthesis, performance, and toxicity. *Int. J. Nanomed.* **2014**, *16*, 2399–2407.
14. Ovais, M.; Ahmad, I.; Khalil, A.T.; Mukherjee, S.; Javed, R.; Ayaz, M.; Raza, A.; Shinwari, Z.K. Wound healing applications of biogenic colloidal silver and gold nanoparticles: Recent trends and future prospectS. *Appl. Microbiol. Biotechnol.* **2018**, *102*, 4305–4318. [CrossRef]
15. Benyettou, F.; Rezgui, R.; Ravaux, F. Synthesis of silver nanoparticles for the dual delivery of doxorubicin and alendronate to cancer cellS. *J. Mater. Chem. B* **2015**, *3*, 7237–7245. [CrossRef] [PubMed]
16. Brown, P.K.; Qureshi, A.T.; Moll, A.N.; Hayes, D.J.; Monroe, W.T. Silver nanoscale antisense drug delivery system for photoactivated gene silencing. *ACS. Nano* **2013**, *7*, 2948–2959. [CrossRef] [PubMed]

17. Zhou, W.; Ma, Y.; Yang, H.; Ding, Y.; Luo, X. A label-free biosensor based on silver nanoparticles array for clinical detection of serum p53 in head and neck squamous cell carcinomA. *Int. J. Nanomed.* **2011**, *6*, 381–386. [CrossRef]
18. Gomes, H.I.O.; Martins, C.S.M.; Prior, J.A.V. Silver Nanoparticles as Carriers of Anticancer Drugs for Efficient Target Treatment of Cancer CellS. *Nanomaterials* **2021**, *11*, 964. [CrossRef]
19. Tiwari, P.M.; Vig, K.; Dennis, V.A.; Singh, S.R. Functionalized gold nanoparticles and their biomedical applicationS. *Nanomaterials* **2011**, *1*, 31–63. [CrossRef]
20. Jeyarani, S.; Vinita, N.M.; Puja, P.; Senthamilselvi, S.; Devan, U.; Velangani, A.J.; Biruntha, M.; Pugazhendhi, A.; Kumar, P. Biomimetic gold nanoparticles for its cytotoxicity and biocompatibility evidenced by fluorescence-based assays in cancer (MDA-MB-231) and non-cancerous (HEK-293) cellS. *J. Photochem. Photobiol. B* **2020**, *202*, 111715. [CrossRef]
21. Cho, K.; Wang, X.U.; Niem, S.; Shin, D.M. Therapeutic nanoparticles for drug delivery in cancer. *Clin. Cancer Res.* **2008**, *14*, 1310–1316. [CrossRef]
22. Kumar, A.; Zhang, X.; Liang, X.J. Gold nanoparticles: Emerging paradigm for targeted drug delivery system. *Biotechnol. Adv.* **2013**, *31*, 593–606. [CrossRef] [PubMed]
23. Mieszawska, A.J.; Mulder, W.J.; Fayad, Z.A.; Cormode, D.P. Multifunctional gold nanoparticles for diagnosis and therapy of diseasE. *Mol. Pharm.* **2013**, *10*, 831–847. [CrossRef] [PubMed]
24. Zhang, L.; Gu, F.X.; Chan, J.M.; Wang, A.Z.; Langer, R.S.; Farokhzad, O.C. Nanoparticles in medicine: Therapeutic applications and developmentS. *Clin. Pharmacol. Ther.* **2007**, *83*, 761–769. [CrossRef] [PubMed]
25. Khan, A.U.; Yuana, Q.; Weia, Y. Photocatalytic and antibacterial response of biosynthesized gold nanoparticleS. *J. Photochem. Photobiol. B Biol.* **2016**, *162*, 172–177. [CrossRef]
26. Kumar-Krishnan, S.; Prokhorov, E.; Hernández-Iturriaga, M.; Mota-Morales, J.D.; Vázquez-Lepe, M.; Kovalenko, Y.; Sanchez, I.C.; Luna-Bárcenas, G. Chitosan/silver nanocomposites: Synergistic antibacterial action of silver nanoparticles and silver ionS. *Eur. Polym. J.* **2015**, *67*, 242–251. [CrossRef]
27. Fuster, M.G.; Montalbán, M.G.; Carissimi, G.; Lima, B.; Feresin, G.E.; Cano, M.; Giner-Casares, J.J.; López-Cascales, J.J.; Enriz, R.D.; Víllora, G. Antibacterial Effect of Chitosan-Gold Nanoparticles and Computational Modeling of the Interaction between Chitosan and a Lipid Bilayer Model. *Nanomaterials* **2020**, *10*, 2340. [CrossRef]
28. Sanpui, P.; Murugadoss, A.; Prasad, P.V.D.; Ghosh, S.S.; Chattopadhyay, A. The Antibacterial Properties of a Novel Chitosan-Ag-Nanoparticle CompositE. *Int. J. Food Microbiol.* **2008**, *124*, 142–146. [CrossRef]
29. Ke, C.L.; Deng, F.S.; Chuang, C.Y.; Lin, C.H. Antimicrobial Actions and Applications of Chitosan. *Polymers* **2021**, *13*, 904. [CrossRef]
30. Shih, P.Y.; Liao, Y.T.; Tseng, Y.K.; Deng, F.S.; Lin, C.H. A Potential Antifungal Effect of Chitosan Against *Candida albicans* Is Mediated via the Inhibition of SAGA Complex Component Expression and the Subsequent Alteration of Cell Surface Integrity. *Front. Microbiol.* **2019**, *10*, 602. [CrossRef]
31. El-Naggar, N.E.-A.; Saber, W.I.A.; Zweil, A.M.; Bashir, S.I. An innovative green synthesis approach of chitosan nanoparticles and their inhibitory activity against phytopathogenic *Botrytis cinerea* on strawberry leaveS. *Sci. Rep.* **2022**, *12*, 3515. [CrossRef] [PubMed]
32. Abdelhamid, H.N.; El-Ber, H.M.; Metwally, A.A.; Elshazl, M.; Hathout, R.M. Synthesis of CdS-modified chitosan quantum dots for the drug delivery of Sesamol. *Carbohdrate Polym.* **2019**, *214*, 90–99. [CrossRef] [PubMed]
33. Panda, P.K.; Sadeghi, K.; Seo, J. Recent advances in poly (vinyl alcohol)/natural polymer based films for food packaging applications: A review. *Food Packag. Shelf Life* **2022**, *33*, 100904. [CrossRef]
34. Panda, P.K.; Yang, J.-M.; Chang, Y.-H.; Su, W.-W. Modification of different molecular weights of chitosan by p-Coumaric acid: Preparation, characterization and effect of molecular weight on its water solubility and antioxidant property. *Int. J. Biol. Macromol.* **2019**, *136*, 661–667. [CrossRef] [PubMed]
35. Fernández-Pan, I.; Maté, J.I.; Gardrat, C.; Coma, V. Effect of chitosan molecular weight on the antimicrobial activity and release rate of carvacrol-enriched flmS. *Food Hydrocoll.* **2015**, *51*, 60–68. [CrossRef]
36. Hashem, A.H.; Shehabeldine, A.M.; Ali, O.M.; Salem, S.S. Synthesis of Chitosan-Based Gold Nanoparticles: Antimicrobial and Wound-Healing ActivitieS. *Polymers* **2022**, *14*, 2293. [CrossRef]
37. Corbierre, M.K.; Cameron, N.S.; Lennox, R.B. Polymer-stabilized gold nanoparticles with high grafting densitieS. *Langmuir* **2004**, *20*, 2867–2873. [CrossRef] [PubMed]
38. DeLong, R.K.; Reynolds, C.M.; Malcolm, Y.; Schaeffer, A.; Severs, T.; Wanekaya, A. Functionalized gold nanoparticles for the binding, stabilization, and delivery of therapeutic DNA, RNA, and other biological macromoleculeS. *Nanotechnol. Sci. Appl.* **2010**, *3*, 53–63. [CrossRef]
39. Chompoosor, A.; Han, G.; Rotello, V.M. Charge dependence of ligand release and monolayer stability of gold nanoparticles by biogenic thiolS. *Bioconjug. Chem.* **2008**, *19*, 1342–1345. [CrossRef]
40. Infectious Diseases Society of America (IDSA). Combating antimicrobial resistance: Policy recommendations to save liveS. *Clin. Infect. Dis.* **2011**, *52*, 397–428. [CrossRef]
41. Perez, F.; van-Duin, D. Carbapenem-resistant Enterobacteriaceae: A menace to our most vulnerable patientS. *Clevel. Clin. J. Med.* **2013**, *80*, 225–233. [CrossRef]
42. Neut, C. Carriage of Multidrug-Resistant Bacteria in Healthy People: Recognition of Several Risk GroupS. *Antibiotics* **2021**, *10*, 1163. [CrossRef]

43. Huq, M.A. Green synthesis of silver nanoparticles using *Pseudoduganella eburnea* MAHUQ-39 and their antimicrobial mechanisms investigation against drug-resistant human pathogenS. *Int. J. Mol. Sci.* **2020**, *21*, 1510. [CrossRef]
44. Hasan, K.M.F.; Wang, H.; Mahmud, S.; Genyang, C. Coloration of aramid fabric via in-situ biosynthesis of silver nanoparticles with enhanced antibacterial effect. *Inorg. Chem. Commun.* **2022**, *119*, 108115. [CrossRef]
45. Akter, S.; Huq, M.A. Biologically rapid synthesis of silver nanoparticles by *Sphingobium* sP. MAH-11^T and their antibacterial activity and mechanisms investigation against drug-resistant pathogenic microbeS. *Artif. Cells Nanomed. Biotechnol.* **2020**, *48*, 672–682. [CrossRef]
46. Shinde, S.; Folliero, V.; Chianese, A.; Zannella, C.; De Filippis, A.; Rosati, L.; Prisco, M.; Falanga, A.; Mali, A.; Galdiero, M.; et al. Synthesis of chitosan-coated silver nanoparticle bioconjugates and their antimicrobial activity against multidrug-resistant bacteriA. *Appl. Sci.* **2021**, *11*, 9340. [CrossRef]
47. Saha, S.K.; Roy, P.; Mondal, M.K.; Roy, D.; Gayen, P.; Chowdhury, P.; Babu, S.P.S. Development of chitosan based gold nanomaterial as an efficient antifilarial agent: A mechanistic approach. *Carbohydr. Polym.* **2017**, *157*, 1666–1676. [CrossRef]
48. Fouda, A.; Abdel-Maksoud, G.; Abdel-Rahman, M.A.; Salem, S.S.; Hassan, S.E.D.; El-Sadany, M.A.H. Eco-friendly approach utilizing green synthesized nanoparticles for paper conservation against microbes involved in biodeterioration of archaeological manuscript. *Int. Biodeterior. Biodegrad.* **2019**, *142*, 160–169. [CrossRef]
49. Iravani, S.; Korbekandi, H.; Mirmohammadi, S.V.; Zolfaghari, B. Synthesis of silver nanoparticles: Chemical, physical and biological methodS. *ReS. Pharm. Sci.* **2014**, *9*, 385–406.
50. Ghetas, H.A.; Abdel-Razek, N.; Shakweer, M.S.; Abotaleb, M.M.; Ahamad-Paray, B.; Ali, S.; Eldessouki, E.A.; Dawood, M.A.O.; Khalil, R.H. Antimicrobial activity of chemically and biologically synthesized silver nanoparticles against some fish pathogenS. *Saudi J. Biol. Sci.* **2022**, *29*, 1298–1305. [CrossRef]
51. Paulkumar, K.; Gnanajobitha, G.; Vanaja, M.; Pavunraj, M.; Annadurai, G. Green synthesis of silver nanoparticle and silver based chitosan bionanocomposite using stem extract of *Saccharum officinarum* and assessment of its antibacterial activity. *Adv. Nat. Sci. Nanosci. Nanotechnol.* **2017**, *8*, 035019. [CrossRef]
52. Du, J.; Sing, H.; Yi, T.H. Antibacterial, anti-biofilm and anticancer potentials of green synthesized silver nanoparticles using benzoin gum (*Styrax benzoin*) extract. *Bioprocess Biosyst. Eng.* **2016**, *39*, 1923–1931. [CrossRef]
53. Rajeshkumar, S.; Tharani, M.; Rajeswari, V.D.; Alharbi, N.S.; Kadaikunnan, S.; Khaled, J.M.; Gopinath, K.; Vijayakumar, N.; Govindarajan, M. Synthesis of greener silver nanoparticle-based chitosan nanocomposites and their potential antimicrobial activity against oral pathogenS. *Green ProcesS. Synth.* **2021**, *10*, 658–665. [CrossRef]
54. Raza, S.; Ansari, A.; Siddiqui, N.N.; Ibrahim, F.; Abro, M.I.; Aman, A. Biosynthesis of silver nanoparticles for the fabrication of non cytotoxic and antibacterial metallic polymer based nanocomposite system. *Sci. Rep.* **2021**, *11*, 10500. [CrossRef]
55. Ramesh, V.; John, S.A.; Koperuncholan, M. Impact of cement industries dust on selective green plants: A case study in Ariyalur industrial zonE. *Int. J. Pharm. Chem. Biol. Sci.* **2014**, *4*, 152–158.
56. Singh, P.; Kim, Y.J.; Zhang, D.; Yang, D.C. Biological synthesis of nanoparticles from plants and microorganismS. *Trends Biotechnol.* **2016**, *34*, 588–599. [CrossRef]
57. Sukweenadhi, J.; Setiawan, K.I.; Avanti, C.; Kartini, K.; Rupa, E.J.; Yang, D.C. Scale-up of green synthesis and characterization of silver nanoparticles using ethanol extract of *Plantago major* L. leaf and its antibacterial potential. *S. Afr. J. Chem. Eng.* **2021**, *38*, 1–8. [CrossRef]
58. Sathiyaseelan, A.; Shajahan, A.; Kalaichelvan, P.T.; Kaviyarasan, V. Fungal chitosan based nanocomposites sponges-An alternative medicine for wound dressing. *Int. J. Biol. Macromol.* **2017**, *104*, 1905–1915. [CrossRef]
59. Saruchi; Kaur, M.; Kumar, V.; Ghfar, A.A.; Pandey, S. A Green Approach for the Synthesis of silver nanoparticle-embedded chitosan bionanocomposite as a potential device for the sustained release of the itraconazole drug and its antibacterial characteristicS. *Polymers* **2022**, *14*, 1911. [CrossRef]
60. Badawy, M.E.I.; Lotfy, T.M.R.; Shawir, S.M.S. Preparation and antibacterial activity of chitosan-silver nanoparticles for application in preservation of minced meat. *Bull. Natl. ReS. Cent.* **2019**, *43*, 83. [CrossRef]
61. Mirda, E.; Idroes, R.; Khairan, K.; Tallei, T.E.; Ramli, M.; Earlia, N.; Maulana, A.; Idroes, G.M.; Muslem, M.; Jalil, Z. Synthesis of chitosan-silver nanoparticle composite spheres and their antimicrobial activitieS. *Polymers* **2021**, *13*, 3990. [CrossRef] [PubMed]
62. Kalaivani, R.; Maruthupandy, M.; Muneeswaran, T.; Hameedha-Beevi, A.; Anand, M.; Ramakritinan, C.M.; Kumaraguru, A.K. Synthesis of chitosan mediated silver nanoparticles (AgNPs) for potential antimicrobial applicationS. *Front. Lab. Med.* **2018**, *2*, 30–35. [CrossRef]
63. Kulikouskaya, V.; Hileuskaya, K.; Kraskouski, A.; Kozerozhets, I.; Stepanova, E.; Kuzminski, I.; You, L.; Agabekov, V. Chitosan-capped silver nanoparticles: A comprehensive study of polymer molecular weight effect on the reaction kinetic, physicochemical properties, and synergetic antibacterial potential. *SPE. Polymers* **2022**, *3*, 77–90. [CrossRef]
64. Dara, P.K.; Mahadevan, R.; Digita, P.A.; Visnuvinayagam, S.; Kumar, L.R.G.; Mathew, S.; Ravishankar, C.N.R.; Anandan, R. Synthesis and biochemical characterization of silver nanoparticles grafted chitosan (Chi-Ag-NPs): In vitro studies on antioxidant and antibacterial applicationS. *SN Appl. Sci.* **2020**, *2*, 665. [CrossRef]
65. Bonilla, J.J.A.; Honorato, L.; Guimarães, A.J.; Miranda, K.; Nimrichter, L. Silver Chitosan Nanocomposites are Effective to Combat SporotrichosiS. *Front. Nanotechnol.* **2022**, *4*, 857681. [CrossRef]
66. Regielfutyra, A.; Kuśliśkiewicz, M.; Sebastian, V.; Irusta, S.; Arruebo, M.; Kyzioł, A.; Stochel, G. Development of noncytotoxic silver–chitosan nanocomposites for efficient control of biofilm forming microbeS. *RSC Adv.* **2017**, *7*, 52398–52413. [CrossRef]

67. Venkatesham, M.; Ayodhya, D.; Madhusudhan, A.; Veera Babu, N.; Veerabhadram, G. A novel green one-step synthesis of silver nanoparticles using chitosan: Catalytic activity and antimicrobial studieS. *Appl. Nanosci.* **2014**, *4*, 113–119. [CrossRef]
68. Priya, K.; Vijayakumar, M.; Janani, B. Chitosan-mediated synthesis of biogenic silver nanoparticles (AgNPs), nanoparticle characterisation and in vitro assessment of anticancer activity in human hepatocellular carcinoma HepG2 cellS. *Int. J. Biol. Macromol.* **2020**, *149*, 844–852. [CrossRef]
69. Samanta, S.; Banerjee, J.; Das, B.; Mandal, J.; Chatterjee, S.; Ali, K.M.; Sinha, S.; Giri, B.; Ghosh, T.; Dash, S.K. Antibacterial Potency of Cytocompatible Chitosan-Decorated Biogenic Silver Nanoparticles and Molecular Insights towards Cell-Particle Interaction. *Int. J. Biol. Macromol.* **2022**, *219*, 919–939. [CrossRef]
70. Kanniah, P.; Chelliah, P.; Thangapandi, J.R.; Gnanadhas, G.; Mahendran, V.; Robert, M. Green synthesis of antibacterial and cytotoxic silver nanoparticles by *Piper nigrum* seed extract and development of antibacterial silver based chitosan nanocompositE. *Int. J. Biol. Macromol.* **2021**, *189*, 18–33. [CrossRef]
71. Youssef, A.M.; Mohamed, S.; Abdel-Aziz, S. Chitosan nanocomposite films based on Ag-NP and Au-NP biosynthesis by *Bacillus Subtilis* as packaging materialS. *Int. J. Biol. Macromol.* **2014**, *69*, 185–191. [CrossRef]
72. Mostafa, E.M.; Abdelgawad, M.A.; Musa, A.; Alotaibi, N.H.; Elkomy, M.H.; Ghoneim, M.M.; Badawy, M.S.E.M.; Taha, M.N.; Hassan, H.M.; Hamed, A.A. Chitosan Silver and Gold Nanoparticle Formation Using Endophytic Fungi as Powerful Antimicrobial and Anti-Biofilm PotentialitieS. *Antibiotics* **2022**, *11*, 668. [CrossRef]
73. Dananjaya, S.H.S.; Udayangani, R.M.C.; Nikapitiya, C.; Lee, J.; De-Zoysa, M. Green synthesis, physio-chemical characterization and anti-candidal function of a biocompatible chitosan gold nanocomposite as a promising antifungal therapeutic agent. *RSC Adv.* **2017**, *7*, 9182–9193. [CrossRef]
74. Tomaszewska, E.; Soliwoda, K.; Kadziola, K.; Tkacz-Szczesna, B.; Celichowski, G.; Cichomski, M.; Szmaja, W.; Grobelny, J. Detection limits of DLS and UV-Vis spectroscopy in characterization of polydisperse nanoparticles colloidS. *J. Nanomater.* **2013**, *2013*, 313081. [CrossRef]
75. Ferri, M.; Ranucci, E.; Romagnoli, P.; Giaccone, V. Antimicrobial resistance: A global emerging threat to public health systemS. *Crit. Rev. Food Sci.* **2017**, *57*, 2857–2876. [CrossRef]
76. WHO. New Report Calls for Urgent Action to Avert Antimicrobial Resistance CrisiS. Available online: https://www.who.int/news/item/29-04-2019-new-report-calls-for-urgent-action-to-avert-antimicrobial-resistance-crisis (accessed on 23 March 2022).
77. Avadi, M.R.; Sadeghi, A.M.; Tahzibi, A.; Bayati, K.H.; Pouladzadeh, M.; Zohuriaan-mehr, M.J.; Rafiee-Tehrani, M. Diethylmethyl chitosan as an antimicrobial agent: Synthesis, characterization and antibacterial effectS. *Eur. Polym. J.* **2004**, *40*, 1355–1361. [CrossRef]
78. Costa, E.M.; Silva, S.; Pina, C.; Tavaria, F.K.; Pintado, M.M. Evaluation and insights into chitosan antimicrobial activity against anaerobic oral pathogenS. *Anaerobe* **2012**, *18*, 305–309. [CrossRef]
79. Jiang, L.; Wang, F.; Han, F.; Prinyawiwatkul, W.; No, H.K.; Ge, B. Evaluation of diffusion and dilution methods to determine the antimicrobial activity of water-soluble chitosan derivativeS. *J. Appl. Microbiol.* **2013**, *114*, 956–963. [CrossRef]
80. Mansilla, A.Y.; Albertengo, L.; Rodriguez, M.S.; Debbaudt, A.; Zuniga, A.; Casalongue, C.A. Evidence on antimicrobial properties and mode of action of a chitosan obtained from crustacean exoskeletons on *Pseudomonas syringae* pv. tomato DC3000. *tomato DC3000. Appl. Microbiol. Biotechnol.* **2013**, *97*, 6957–6966. [CrossRef]
81. Ma, Z.X.; Garrido-Maestu, A.; Jeong, K.C. Application, mode of action, and in vivo activity of chitosan and its micro and nanoparticles as antimicrobial agents: A review. *Carbohydr. Polym.* **2017**, *176*, 257–265. [CrossRef]
82. Akter, S.; Lee, S.-Y.; Siddiqi, M.Z.; Balusamy, S.R.; Ashrafudoulla, M.; Rupa, E.J.; Huq, M.A. Ecofriendly synthesis of silver nanoparticles by *Terrabacter humi* sP. nov. and their antibacterial application against antibiotic-resistant pathogenS. *Int. J. Mol. Sci.* **2020**, *21*, 9746. [CrossRef]
83. Huq, M.A. Biogenic Silver Nanoparticles Synthesized by *Lysinibacillus xylanilyticus* MAHUQ-40 to Control Antibiotic-Resistant Human Pathogens *Vibrio parahaemolyticus* and *Salmonella Typhimurium*. *Front. Bioeng. Biotechnol.* **2020**, *8*, 597502. [CrossRef]
84. Sathiyaraj, S.; Suriyakala, G.; Dhanesh-Gandhi, A.; Babujanarthanam, R.; Almaary, K.S.; Chen, T.W.; Kaviyarasu, K. Biosynthesis, characterization, and antibacterial activity of gold nanoparticleS. *J. Infect. Public Health* **2021**, *14*, 1842–1847. [CrossRef] [PubMed]
85. Sabudin, S.; Derman, M.; Zainol, I.; Noorsal, K. In vitro cytotoxicity and cell seeding studies of a chitosan-silver composite for potential wound management applicationS. *J. Eng. Sci.* **2012**, *8*, 29–37.
86. Katas, H.; Moden, N.Z.; Lim, C.S.; Celesistinus, T.; Chan, J.Y.; Ganasan, P.; Suleman Ismail Abdalla, S. Biosynthesis and potential applications of silver and gold nanoparticles and their chitosan-based nanocomposites in nanomedicinE. *J. Nanotechnol.* **2018**, *2018*, 4290705. [CrossRef]
87. Panda, P.K.; Dash, P.; Chang, Y.-H.; Yang, J.-M. Improvement of chitosan water solubility by fumaric acid modification. *Mater. Lett.* **2022**, *316*, 132046. [CrossRef]
88. Katas, H.; Alpar, H.O. Development and Characterisation of Chitosan Nanoparticles for SiRNA Delivery. *J. Control. Release* **2006**, *115*, 216–225. [CrossRef]
89. Potara, M.; Jakab, E.; Damert, A.; Popescu, O.; Canpean, V.; Astilean, S. Synergistic antibacterial activity of chitosan-silver nanocomposites on *Staphylococcus aureus*. *Nanotechnology* **2011**, *22*, 135101. [CrossRef] [PubMed]
90. Rezazadeh, N.H.; Buazar, F.; Matroodi, S. Synergistic effects of combinatorial chitosan and polyphenol biomolecules on enhanced antibacterial activity of biofunctionalized silver nanoparticleS. *Sci. Rep.* **2020**, *10*, 19615. [CrossRef]

91. Devlieghere, F.; Vermeulen, A.; Debevere, J. Chitosan: Antimicrobial Activity, Interactions with Food Components and Applicability as a Coating on Fruit and VegetableS. *Food Microbiol.* **2004**, *21*, 703–714. [CrossRef]
92. Sebti, I.; Martial-Gros, A.; Carnet-Pantiez, A.; Grelier, S.; Coma, V. Chitosan Polymer as Bioactive Coating and Film against Aspergillus niger Contamination. *J. Food Sci.* **2005**, *70*, M100–M104. [CrossRef]
93. Wang, X.; Du, Y.; Fan, L.; Liu, H.; Hu, Y. Chitosan- metal complexes as antimicrobial agent: Synthesis, characterization and Structure-activity study. *Polym. Bull.* **2005**, *55*, 105–113. [CrossRef]
94. Katas, H.; Lim, C.S.; Nor Azlan, A.Y.H.; Buang, F.; Mh Busra, M.F. Antibacterial activity of biosynthesized gold nanoparticles using biomolecules from *Lignosus rhinocerotis* and chitosan. *Saudi Pharm. J.* **2019**, *27*, 283–292. [CrossRef] [PubMed]
95. Lu, B.; Lu, F.; Ran, L.; Yu, K.; Xiao, Y.; Li, Z.; Dai, F.; Wu, D.; Lan, G. Imidazole-molecule-capped chitosan–gold nanocomposites with enhanced antimicrobial activity for treating biofilm-related infectionS. *J. Colloid Interface Sci.* **2018**, *531*, 269–281. [CrossRef] [PubMed]

Article

Antimicrobial Efficacy of Quercetin against *Vibrio parahaemolyticus* Biofilm on Food Surfaces and Downregulation of Virulence Genes

Pantu Kumar Roy [1,†], Sung-Hee Park [2,†], Min Gyu Song [1] and Shin Young Park [1,*]

[1] Institute of Marine Industry, Department of Seafood Science and Technology, Gyeongsang National University, Tongyeong 53064, Korea
[2] World Institute of Khimchi, Gwangju 61755, Korea
* Correspondence: sypark@gnu.ac.kr; Tel.: +82-55-772-9143; Fax: +82-55-772-9149
† These authors contributed equally to this work.

Citation: Roy, P.K.; Park, S.-H.; Song, M.G.; Park, S.Y. Antimicrobial Efficacy of Quercetin against *Vibrio parahaemolyticus* Biofilm on Food Surfaces and Downregulation of Virulence Genes. *Polymers* 2022, 14, 3847. https://doi.org/10.3390/polym14183847

Academic Editors: Md. Amdadul Huq and Shahina Akter

Received: 15 July 2022
Accepted: 12 September 2022
Published: 14 September 2022

Publisher's Note: MDPI stays neutral with regard to jurisdictional claims in published maps and institutional affiliations.

Copyright: © 2022 by the authors. Licensee MDPI, Basel, Switzerland. This article is an open access article distributed under the terms and conditions of the Creative Commons Attribution (CC BY) license (https://creativecommons.org/licenses/by/4.0/).

Abstract: For the seafood industry, *Vibrio parahaemolyticus*, one of the most prevalent food-borne pathogenic bacteria that forms biofilms, is a constant cause of concern. There are numerous techniques used throughout the food supply chain to manage biofilms, but none are entirely effective. Through assessing its antioxidant and antibacterial properties, quercetin will be evaluated for its ability to prevent the growth of *V. parahaemolyticus* biofilm on shrimp and crab shell surfaces. With a minimum inhibitory concentration (MIC) of 220 µg/mL, the tested quercetin exhibited the lowest bactericidal action without visible growth of bacteria. In contrast, during various experiments in this work, the inhibitory efficacy of quercetin without (control) and with sub-MICs levels (1/2, 1/4, and 1/8 MIC) against *V. parahaemolyticus* was examined. With increasing quercetin concentration, swarming and swimming motility, biofilm formation, and expression levels of related genes linked to flagella motility (*flaA* and *flgL*), biofilm formation (*vp0952* and *vp0962*), and quorum-sensing (*luxS* and *aphA*) were all dramatically reduced ($p < 0.05$). Quercetin (0–110 µg/mL) was investigated on shrimp and crab shell surfaces, the inhibitory effects were 0.68–3.70 and 0.74–3.09 log CFU/cm^2, respectively ($p < 0.05$). The findings were verified using field emission scanning electron microscopy (FE-SEM), which revealed quercetin prevented the development of biofilms by severing cell-to-cell contacts and induced cell lysis, which resulted in the loss of normal cell shape. Furthermore, there was a substantial difference in motility between the treatment and control groups (swimming and swarming). According to our findings, plant-derived quercetin should be used as an antimicrobial agent in the food industry to inhibit the establishment of *V. parahaemolyticus* biofilms. These findings suggest that bacterial targets are of interest for biofilm reduction with alternative natural food agents in the seafood sector along the entire food production chain.

Keywords: *Vibrio parahaemolyticus*; quercetin; biofilm; shrimp; crab; gene expression

1. Introduction

Nutritious and tasty aquatic products are susceptible to oxidation and bacterial contamination in planned transportation. This situation damages the taste of aquatic products and seriously threatens food safety. Since seafood is freshly used, the food quality is very high, but, in time, its quality will decline and become unfit for consumption. The quality and safety of seafood is a major challenge facing its own seafood-related industry in food sciences, and mostly in fisheries and aquaculture research departments. Fish is an essential nutrient in the human diet and is also present in the global aquatic product industry for consumers. Such products are prone to oxidation and bacterial contamination in organized transportation, which not only destroys the taste of products, but also poses a very serious threat to food safety [1]. Seafood is frequently contaminated with the Gram-negative bacterium *Vibrio parahaemolyticus* [2]. During infection, it forms a biofilm, which is a layer of self-produced proteins, lipids, and polysaccharides that covers the surface of the host [3]. A crucial aspect of the pathogenesis is the production of biofilm, which might increase resistance to harmful circumstances and medications. According to studies by Han et al. [4] and Almohamad et al. [5], over 60% of outbreaks by *V. parahaemolyticus* biofilm occurred by consuming contaminated seafoods. Infections with *V. parahaemolyticus* typically have self-limiting symptoms (e.g., vomiting, diarrhea, fever, nausea, chills, headaches, and watery stools [6,7]). Although uncommon, this bacterium can cause septicemia, necrotizing fasciitis, wound infections, and even death [8,9]. Because of this, *V. parahaemolyticus* contamination poses a threat to the aquaculture industry, the food industry, and public health.

The World Health Organization reported that O3:K6 serotypes and their variants are the most common strains associated with foodborne diseases, with *V. parahaemolyticus* being the most often encountered bacterial gastroenteritis, associated with the consumption of seafood products globally [10]. One of the main issues for food safety and public health has been the prevalence of *V. parahaemolyticus* in the world. According to the CDC [11], *V. parahaemolyticus* causes 45,000 illnesses annually in the USA and is the most often reported in *vibrio* infections (https://www.cdc.gov/vibrio/faq.html, accessed on 15 June 2022) [7]. Currently, standard methods for preventing and treating *V. parahaemolyticus* contamination and infection, such as antibiotics and chemical disinfectants, are crucial [12,13]. However, studies indicate that *V. parahaemolyticus* clinical isolates and environmental isolates both show rising antibiotic resistance globally [14–16]. Because of the limitations of current control systems, other techniques of preventing bacterial contamination and infections are constantly being studied [17,18].

In comparison to their planktonic relatives, biofilms are a million times more resistant to all antimicrobial treatments [1,19]. Because of this, removing biofilm with common medicines and cleaning supplies may be challenging [4]. Aggressive chemicals, such as sodium hydroxide or sodium hypochlorite, are frequently employed in the food sector to reduce the negative impacts of biofilm [20]. However, such methods might damage the environment by corroding equipment and materials [21,22]. Therefore, it is essential to create a successful strategy that can manage and eradicate bacterial biofilm. The term "biofilm" refers to bacterial growth that defends itself by routinely embedding cells in extracellular polymeric substances (EPS) as opposed to free-living bacterial cells [7,23]. This increases the bacteria's ability to survive acquaintance to antimicrobial agents [7,24]. A number of biofilm-related genes regulate the continuous, dynamic processes that lead to the formation of biofilms, including cell attachment, EPS synthesis, resource capture, detachment, and dispersal. According to studies [14,25], *V. parahaemolyticus* can form biofilms on a variety of biotic or abiotic surfaces and interfaces, including seawater and marine organisms (shrimp, fish, crab, shellfish, etc.) [7]. This contamination of the sea and seafood leads to cross-contamination during the processing or preparation of food [7,26]. Cross-contamination may be a significant source of human diseases, according to reports [27]. The development of biofilms on or in seafood may play a significant role in the spread of *V. parahaemolyticus* and the subsequent illnesses [28]. In order to reduce contamination and infections caused by *V. parahaemolyticus*, biofilm serves as a significant target.

Bacteria are protected from physical harm, desiccation, and antibiotics by microbial biofilms [29]. Previous studies reported food-borne pathogens persist as biofilms on foods (shrimp, crab shell, and lettuce) and food contact surfaces (e.g., plastic, steel, glass, and rubber) and have an impact on the quantity, quality, and safety of food products [30–34]. Additionally, they destroy surfaces and equipment, contaminate food on a constant basis, pose a significant risk to public health, and their control is a significant barrier in the food production chain [35]. To prevent foodborne infections, natural plant extracts and antimicrobial compounds are typically regarded as secure, efficient, and environmentally friendly [36,37]. With a broad range of activity against numerous bacterial and fungal infections, plant extracts have long been used extensively for food safeguarding and disease anticipation [7,37,38].

The use of inhibitory compounds that interfere with quorum sensing (QS) is one of the preventative techniques for improving food quality and safety [39,40]. QS in a number of bacteria can be disrupted by phenolic chemicals generated from plants [39]. Plant compounds are an alternative control method against *V. parahaemolyticus* biofilms, and one of the most investigated flavonoid molecules having functional characteristics in this context is quercetin. Flavonoids have become well-known for having anti-inflammatory, antioxidant, antibacterial, and anticancer properties [41] in addition to their potential QS system inhibitory properties [42,43]. Many fruits and vegetables, including apples, tea, onions, red grapes, berries, tomatoes, and tea, contain quercetin, a flavonoid-based compound [44]. Due to its anti-inflammatory, anticancer, and neuroprotective properties, it has a wide range of applications [45,46]. The ability of flavones and flavonoids to scavenge free radicals inside cells make them recognized as having strong antioxidant activity. As a subclass of polyphenolic substances with a wide range of chemical characteristics, flavonoids are a ubiquitous component of plants. Of all the flavonoids, quercetin is one of the most effective antioxidants [47,48]. More than 4000 different flavonoids, including flavonols, flavones, flavanones, catechins, anthocyanidins, isoflavones, dihydroflavonols, and chalcones, have been classified as part of the major flavonoid group [47]. The food business uses organic antioxidants to preserve the product's color and nutritional content. Studies on the application of flavonoids in various industrial sectors have grown in number recently [47]. Similar to how they may be used in food, textiles, leather, metallurgy, medicine, and agriculture, these substances may also be used for their antioxidant capabilities [47]. Quercetin is, therefore, a common source for the food and pharmaceutical industries [47]. Therefore, there is an increasing demand for biodegradable polymers (fibers) that are effective against microorganisms. Quercetin-(Q)-loaded polylactide-based polymers (fibers) were used as antibacterial effects against *Staphylococcus aureus*, *Escherichia coli*, and *Klebsiella pneumoniae* [49]. The nanoparticle form of quercetin (nanoquercetin) is composed of a polyaspartic acid-based polymer micelle encapsulated with quercetin; colloidal in nature, they were used for inhibiting intracellular polyglutamine aggregation in cellular and animal models of Alzheimer's diseases [50]. Polymer micelle-encapsulated quercetin also has anticancer properties, and is used in the development of biodegradable nanoparticles, and enhanced delivery of quercetin by encapsulation in a surfactant polymer as an antisolvent process [51,52]. Owing to its three-ring structure with five hydroxyl groups, it possesses especially strong antioxidant capabilities [33,41,45]. Antioxidants can reduce oxidative stress and prevent biofilm formation by scavenging reactive oxygen species (ROS) accumulated in bacterial cells [33,41]. As a result, antioxidants are potent antibiofilm agents [45,53] as it is one of the primary processes by which oxidative stress induces bacteria to develop biofilm as a survival strategy. Additionally, it has already been demonstrated that quercetin has antibacterial properties against both Gram-positive and Gram-negative bacteria [44], including *S. aureus* [44,54], *E. coli* [44,55], and *Pseudomonas aeruginosa* [44,56].

The antimicrobial action of quercetin against *V. parahaemolyticus*, however, has not been specifically investigated in any investigations. In the present work, the ability of quercetin to suppress *V. parahaemolyticus* biofilm formation, including flagella motion, and its effects on virulence and QS gene expression were assessed.

2. Materials and Methods

2.1. Bacterial Strain Culture and Growth Conditions

Vibrio parahaemolyticus strain from the American Type Culture Collection (Manassas, VA, USA) (ATCC27969) was collected and used for the biofilm-forming assays. The bacteria were cultured in tryptic soy broth (TSB, BD Difco, Franklin Lakes, NJ, USA) with 2.5% NaCl at 30 °C for 24 h followed by another sub-culture at 18 h [57]. The culture was centrifuged (11,000× g for 10 min) and washed two times with phosphate buffered saline (PBS; Oxoid, Basingstoke, England). After that, peptone water (PW; Oxoid, Basingstoke, England) was added to the final bacterial solution to dilute it until it contained 10^5 log CFU/mL of bacteria. The formation of biofilms on surfaces of crab and shrimp was then accomplished using these inoculums (10^5 CFU/mL).

2.2. Preparation of Samples (Crabs and Shrimp)

With few modifications, sample preparation was done as explained in our earlier investigations [57]. The shrimp (*Penaeus monodon*) and crabs (*Corystes cassivelaunus*) were bought at a nearby grocery store at Tongyeong local market. Using a sterile scalpel, crab and shrimp shells were sliced into 2×2 cm^2 coupons. Following the removal of any leftover meat, the shells were cleaned with sterile distilled water (DW). Before being inoculated with *V. parahaemolyticus*, the coupons were sterilized by being exposed to UV-C light for 15 min on each side to remove background microflora [57]. The coupons were dipped into 10 mL of TSB, inoculated with bacteria (10^5 log CFU/mL), and then incubated for 24 h at 30 °C without shaking to test for the further experiment.

2.3. Quercetin Preparation and Determination of Minimum Inhibitory Concentration (MIC)

For our study quercetin (Q-4951) was collected from Sigma-Aldrich (St. Louis, MO, USA). After being dissolved in dimethyl sulfoxide (DMSO, Sigma-Aldrich, St. Louis, MO, USA), the product was used to create a stock solution with a concentration of 1 mg/mL. The MIC was verified and very slightly modified from the previous study [33]. A two-fold serial dilution approach using TSB was used to establish the minimum inhibitory concentration (MIC) of quercetin against *V. parahaemolyticus*. A total of 100 µL of quercetin serially diluted with TSB and 100 µL of bacterial suspension (10^5 log CFU/mL) were combined in 96-well plates (Corning Incorporated, Corning, Inc., Corning, NY, USA). Each well had a total amount of 200 µL. A microplate reader (Spectra Max 190, Sunnyvale, CA, USA) was used to measure absorbance (600 nm) while the plates were kept in a 30 °C incubator for 24 h. After an overnight incubation at 30 °C, aliquots (100 µL) taken from the wells that had no discernible growth were plated on Vibrio CHROMagar plates and the number of colonies counted. Triplicates of this experiment were run. The MIC of quercetin was 220 µg/mL. For further experiments in this study, sub-MICs of quercetin were used.

2.4. Analysis of Motility

Motility experiments were carried out in this study with minor modifications from those previously published [33]. This test was conducted to verify the effect of quercetin on the two forms of *V. parahaemolyticus* motility (swimming and swarming). Bacto agar (BD Dicfo, Franklin Lakes, NJ, USA) was mixed with TSB at a rate of 0.3% and 0.5% to provide the media for the swimming and swarming studies, respectively, and incubated at 30 °C for 13 h for swimming and 48 h for swarming. Each plate was filled with the autoclaved medium. Quercetin was added after the autoclaved medium and thoroughly mixed in before it solidified. The diameter of bacterial movement through agar was measured in millimeters (mm).

2.5. Biofilm Formation and Detachment Process

With slight adjustments, the procedure was carried out as previously described [33]. The MIC in this study was 220 µg/mL, and the inhibiting effect of biofilm was seen at sub-MIC levels, which may not have killed the bacteria, but affected their virulence factor. Control (without quercetin), 1/8, 1/4, and 1/2 MIC concentrations were used in this study. In a 50 mL conical tube 10 mL TSB with food surfaces, quercetin was added (0, 1/8, 1/4, and 1/2 MICs), and 100 µL of bacterial suspension (10^5 log CFU/mL). They were then thoroughly combined with a vortex mixer (Scientific Industries, SI-0256, Bohemia, NY, USA) and incubated for 24 h at 30 °C to form biofilms. To get rid of bacteria that had somewhat attached to the surfaces after the biofilm formation, the coupons were washed twice with distilled water (DW) [33,41]. Each washed coupon (shrimp and crab) was placed in a sterile stomacher bag with 10 mL of peptone water (PW; BD Diagnostics, Franklin Lakes, NJ, USA), and processed using a Stomacher (Bag Mixer; Interscience, Saint Nom,

France) at the highest speed of 4 for two min to release the bacteria that form biofilms on the coupons [34]. This bacterial sample was serially diluted before being placed in Vibrio CHROMagar plates as an inoculum. The number of colonies on the plates was counted after they had been kept in a 30 °C incubator for 24 h. After subtracting the populations of each concentration (0, 1/8, 1/4, and 1/2 MIC) from the populations of each group, we were able to calculate the inhibition values at log CFU/cm^2.

2.6. Confirmation of Biofilms Inhibition by FE-SEM

A *V. parahaemolyticus* strain known for its high biofilm-forming ability was used to test the biofilm-forming ability using FE-SEM. According to our earlier published work, the samples were prepared [57]. Briefly, the samples were washed in PBS and placed in 6-well dishes, fixed with 2.5% glutaraldehyde, and stored at room temperature for 4 h, and after that treated with ethanol (50, 60, 70, 80, 90% for 15 min serially) and 100% for 15 min two times. Then, the treated samples were dehydrated with soaking (33, 50, 66, and 100% hexamethyldisilazane in ethanol) for 15 min serially. The samples were platinum sputed-coated and observed by FE-SEM (Hitachi/Baltec, S-4700, Tokyo, Japan) [32].

2.7. Relative Gene Expression by Real-Time PCR (RT-PCR)

With a few minor adjustments, the experiment was carried out as previously described [33]. The test was carried out to confirm quercetin's impact on *V. parahaemolyticus* pathogenicity and quorum-sensing gene expression. After biofilm formation on coupons, biofilm cells were collected for RNA extraction. Total RNA was collected using the RNeasy Mini kit (Qiagen, Hilden, German) followed by the manufacturer's protocol. Using a Maxime RT PreMix (Random Primer) kit (iNtRON Biotechnology Co., Ltd., Seoul, Gyeonggi-do, Korea), cDNA was produced after the RNA yield and purity were assessed using a spectrophotometer at 260/280 nm and 260/230 nm (NanoDrop, Bio-Tek Instruments, Chicago, IL, USA) [58]. Table 1 lists the primers. The housekeeping gene was 16S rRNA. In a total volume of 20 µL, the cDNA sample was combined with the appropriate primers and Power SYBR Green PCR Master Mix (Applied Biosystems, Thermo Fisher Scientific, Warrington, UK). A CFX Real-Time PCR System (Bio-Rad, Hercules, CA, USA) was used to perform the RT-PCR analysis. Utilizing 2X Real-Time PCR Master Mix and 1 µL of cDNA as a template, RT-qPCR was carried out. A CFX Real-Time PCR System was used to conduct the real-time PCR. Initial denaturation for the PCR reaction took place at 95, 50, and 72 °C for 20 s each [59–61]. After PCR cycling was complete, we collected cycle threshold (Ct) values to confirm the specificity and conducted $2^{-\triangle\triangle Ct}$ method analysis [62–64].

Table 1. Primer lists used in this study for RT-qPCR. F and R stand for forward and reverse primers.

Target Gene	Sequence of Primers (5′-3′)	Product Size (bp)	NCBI Accessions No.
flaA	F: CGGACTAAACCGTATCGCTGAAA R: GGCTGCCCATAGAAAGCATTACA	128	GQ433373.1
flgL	F: CGTCAGCGTCCACCACTT R: GCGGCTCTGACTTACTGCTA	141	CP066246.1
luxS	F: GGATTTTGTTCTGGCTTTCCACTT R: GGGATGTCGCACTGGTTTTTAC	119	CP066246.1
aphA	F: ACACCCAACCGTTCGTGATG R: GTTGAAGGCGTTGCGTAGTAAG	162	CP066246.1
vp0952	F: TATGATGGTGTTTGGTGC R: TGTTTTTCTGAGCGTTTC	276	CP064041.1
vp0962	F: GACCAAGACCCAGTGAGA R: GGTAAAGCCAGCAAAGTT	358	CP064041.1
16S rRNA	F: TATCCTTGTTTGCCAGCGAG R: CTACGACGCACTTTTTGGGA	186	CP085308.1

2.8. Statistical Analysis

The experiments were performed at least three times. All data were expressed as mean ± standard error of mean (SEM). Statistical significance was set at $p < 0.05$ when Duncan's multiple-range test and an ANOVA were performed using SAS software version 9.2 (SAS Institute Inc., Cary, NC, USA) to determine the significance.

3. Results

3.1. Swimming and Swarming Motility Assays

For the formation of biofilms, bacterial flagella must be mobile. *V. parahaemolyticus* flagella can be verified by swimming and swarming assays, in particular. The impact of quercetin on inhibiting *V. parahaemolyticus* motility is depicted in Figures 1 and 2. Quercetin reduced *V. parahaemolyticus* motility by 18 and 79%, respectively, in the swimming experiment when compared to the control at 1/8 and 1/2 MIC. Figure 2 depicts the quercetin's inhibition of *V. parahaemolyticus*. Quercetin thereby reduced *V. parahaemolyticus* motility by 14 and 57% at 1/8 and 1/2 MIC, respectively. Thus, in this experiment, as quercetin concentration increased, swimming and swarming motility became more inhibited. Particularly in comparison to the control group, motility was significantly different with 1/2 MIC of quercetin.

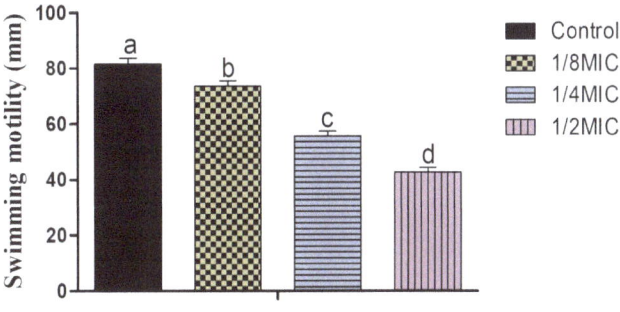

Figure 1. Swimming motility assay for *Vibrio parahaemolyticus* with sub-MICs of quercetin (μg/mL). Data represented as mean ± SEM of three independent replicates. [a–d] Values with different letters differ significantly by Duncan's multiple-range test ($p < 0.05$).

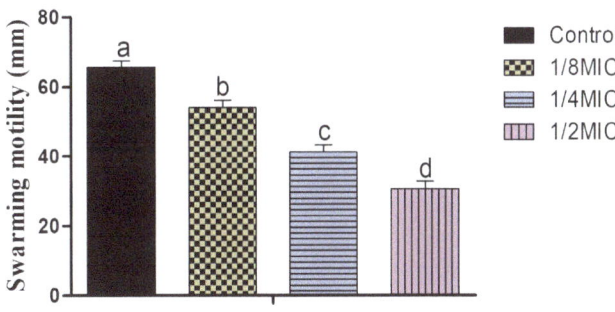

Figure 2. Swarming motility assay for *Vibrio parahaemolyticus* with sub-MICs of quercetin (μg/mL). Data represented as mean ± SEM of three independent replicates. [a–d] Values with different letters differ significantly by Duncan's multiple-range test ($p < 0.05$).

3.2. Eradication Effect of Food Additive Quercetin on Shrimp and Crab Shell Surfaces against V. parahaemolyticus

The *V. parahaemolyticus* biofilm on shrimp coupons is shown, in Figure 3, to be inhibited by quercetin. As quercetin content increased, the biofilm-inhibiting impact also grew.

The *V. parahaemolyticus* biofilm inhibition values on the shrimp surface were 0.68, 1.43, and 3.70 log CFU/cm^2, respectively, at quercetin quantities of 1/8, 1/4, and 1/2 MIC. Comparing these values to the control and other MIC groups, they were significantly ($p < 0.05$) suppressed at 1/2 MIC. On crabs, *V. parahaemolyticus* biofilm is shown, in Figure 4, to be inhibited by quercetin. The *V. parahaemolyticus* biofilm inhibitory values were 0.74, 1.40, and 3.09 log CFU/cm2 at 1/8, 1/4, and 1/2 MIC quercetin concentrations, respectively. Compared to the control and other MIC groups, 1/2 MIC significantly inhibited biofilm formation ($p < 0.05$).

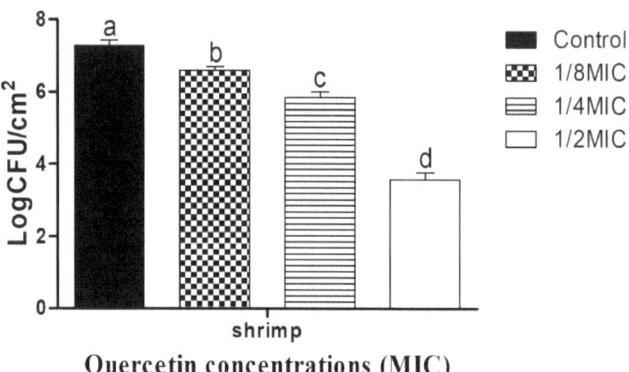

Figure 3. Inhibition of *Vibrio parahaemolyticus* biofilm formation (24 h) on shrimp surfaces by sub-MICs of quercetin (μg/mL). Data represented as mean ± SEM of three independent replicates. [a–d] Values with different letters differ significantly by Duncan's multiple-range test ($p < 0.05$).

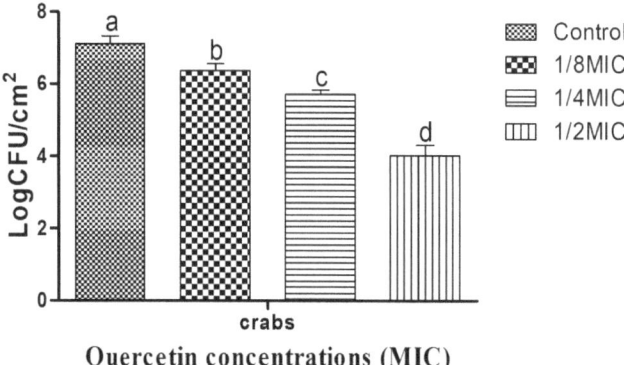

Figure 4. Inhibition of *Vibrio parahaemolyticus* biofilm formation (24 h) on crab surfaces by sub-MICs of quercetin (μg/mL). Data represented as mean ± SEM of three independent replicates. [a–d] Values with different letters differ significantly by Duncan's multiple-range test ($p < 0.05$).

3.3. Visual Confirmation of Biofilm Reduction by Quercetin under FE-SEM

The visual confirmation of biofilm inhibition by quercetin is shown in Figure 5. Biofilms were architecturally structured with intact cell-to-cell contacts in control samples. Smooth and regular cells with intact cell membranes were observed in both the control (Figure 5A,D) and the quercetin-supplemented groups (Figure 5B,C,E,F). The rough and uneven appearance of quercetin-treated bacterial cells indicated that the cells had lost their usual shape (Figure 5C,F).

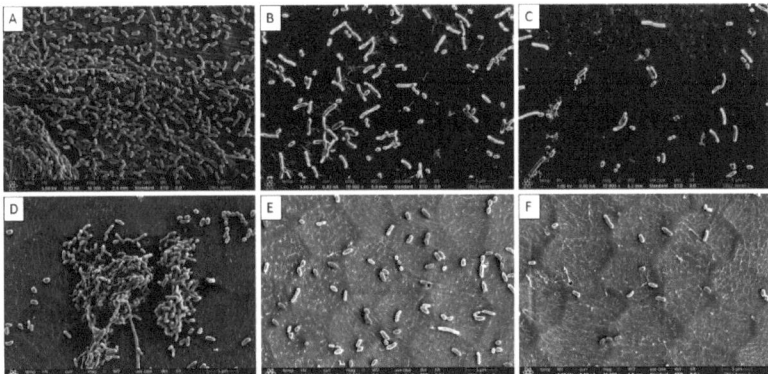

Figure 5. Representative scanning electron micrographs of *Vibrio parahaemolyticus* biofilms formation in the presence of sub-MICs of quercetin on the crab surfaces: (**A**) Control (0% quercetin); (**B**) 1/4 MIC; (**C**) 1/2 MIC and shrimp surfaces; (**D**) Control (0% quercetin); (**E**) 1/4 MIC; (**F**) 1/2 MIC.

3.4. Motility, Biofilm Forming, Virulence, and QS Sensing Relative Gene Expression Pattern

Figure 6 shows the expression of *V. parahaemolyticus* flagella motility (*flaA* and *flgL*), biofilm formation (*vp0952* and *vp0962*), and QS (*luxS* and *aphA*) determined by RT-PCR in the sub-MIC of quercetin (from 0 to 110 µg/mL). At the various sub-MIC concentrations of quercetin, gene expressio was considerably downregulated ($p < 0.05$). This section may be divided by subheadings. It should provide a concise and precise description of the experimental results, their interpretation, as well as the experimental conclusions that can be drawn.

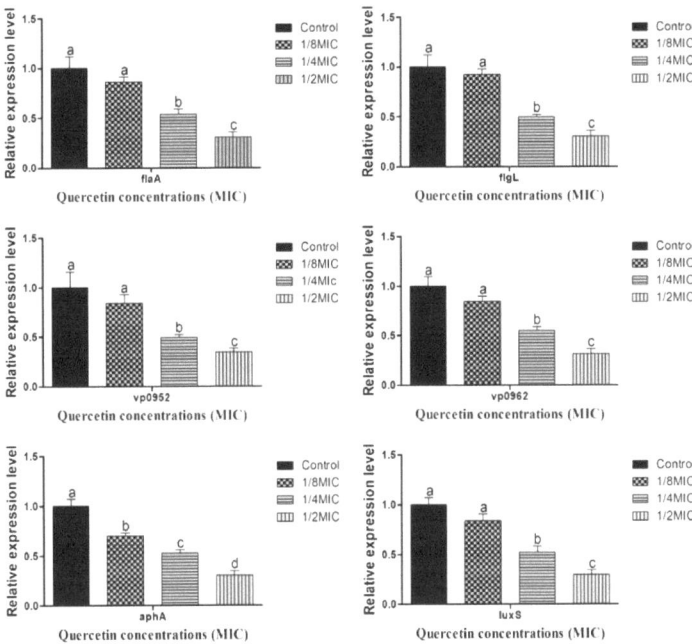

Figure 6. Relative expression levels of *flaA*, *flgL*, *vp0952*, *vp0962*, *aphA*, and *luxS* genes in *Vibrio parahaemolyticus* supplemented with sub-MICs of quercetin. [a–d] Different superscript letters indicate significant differences ($p < 0.05$) with three independent replicates.

4. Discussion

Natural chemicals originating from plants may provide a potentially feasible strategy to bypass bacterial biofilm inhibitory processes and restore quercetin efficacy. Plant extracts, which contain quercetin, could be regarded as food ingredients rather than food additives. Quercetin is a non-specific protein kinase enzyme inhibitor. The FDA approved the use of high-purity quercetin at quantities of up to 500 milligrams (mg) as an ingredient in a variety of food categories in 2010 [33]. The goal of the current investigation was to determine whether quercetin at sub-MIC levels could be used to inhibit the growth of *V. parahaemolyticus*. Against *V. parahaemolyticus*, quercetin has antibacterial efficacy, which we describe in our study. We revealed that there was a dose-dependent bactericidal effect of quercetin against *V. parahaemolyticus* as well as a considerable biofilm formation inhibition caused by quercetin using a variety of techniques, including bacterial motility and growth of biofilm. Quercetin not only inhibited bacterial growth but also suppressed pathogenicity, biofilm formation, flagella motility, and QS gene expression in response to *V. parahaemolyticus*.

The MIC of quercetin against *V. parahaemolyticus* ATCC27969 was determined to be 220 g/mL. The quantity of quercetin varies according to species. Quercetin is a pentahydroxyflavone, with the five hydroxy groups placed at the 3-, 3′-, 4′-, 5- and 7-positions. It is one of the most abundant flavonoids in edible vegetables, fruit, and wine. It has a role as an antibacterial agent, an antioxidant, a protein kinase inhibitor, and an antineoplastic agent. High concentrations of quercetin consist of more active compounds to increase the activity and bind with a specific portion to increase scavenging of ROS [48]. The MIC was established to be the lowest quantity that completely inhibited visible growth. The MIC of quercetin was determined as 80 μg/mL for *P. aeruginosa* and *K. pneumoniae*, 120 μg/mL for *Chromobacterium violaceum*, 250 μg/mL for *S.* Typhimurium, and 95 μg/mL for *Yersinia enterocolitica* [33,41,65]. By encouraging surface adhesion, swimming and swarming locomotion affect bacterial biofilm development. Our results clearly show that quercetin dramatically decreased the test pathogens' flagella-mediated motility when compared to the control (Figures 1 and 2). The outcomes are analogous to those reported by Damte et al. [66], who found that plant extracts can reduce *Pseudomonas* swarming motility by 71%. Another finding was that cinnamaldehyde prevented *E. coli* swarming by reducing biofilm development, according to Niu and Gilbert [67]. Similarly, quercetin reduced the motility at swimming (77 and 76%) and swarming (55 and 54.5%) against *S. typhimurium* [33,41]. As a result, quercetin seems to inhibit the ability of food-borne pathogens to attach to surfaces; hence, reducing the formation of biofilms. Additionally, bacterial motility, including swimming and swarming, is regarded as a key component of pathogenicity. In this case, quercetin significantly decreased the motility of the tested microorganisms.

The development of biofilms is among the most important factors in a food-borne bacteria's pathogenicity. QS is an important factor in the formation of biofilms [68]. Thus, disrupting the signal-mediated QS system may control the development of biofilms. The study's findings demonstrated that quercetin effectively decreased the biofilm development in test pathogens at all tested concentrations. Our results are in line with those previously reported [33,41], which claimed that, as compared to control, quercetin (125 μg/mL)-treated food-borne pathogens *S. typhimurium* rarely form biofilms on food and food-contact surfaces. As previously reported [69], 0.051 mg/mL of quercetin used against *Listeria monocytogenes* biofilm formation and inhibited by quercetin [33]. In order to rule out any interference from quercetin (0.051 mg/mL) on planktonic populations during the experiment, its impact on *L. monocytogenes* planktonic growth kinetics was also assessed [33]. Because planktonic cells in the bulk medium continuously deposit onto layers of attached cells throughout normal development, it is important to recognize their role in biofilm formation. The results showed that the flavonoid quercetin prevented the development of *L. monocytogenes* biofilm and suggests that quercetin affects biofilm formation mechanisms other than cell division [33,69]. However, increasing quercetin levels had an impact on the formation of biofilms, as 1.96 and 3.21 Log10 CFU/cm^2 of viable surface-associated cells were decreased at concentrations of 0.051 and 0.102 mg/mL, respectively, with a significant reduction ($p < 0.05$) in quercetin levels [33,69]. Additionally, at sub-MIC of quercetin, the biofilm was more inhibited by quercetin on food (shrimp and crab) (Figures 3 and 4). *Vibrio* can attach to surfaces and form a biofilm, making the use of plastic cutting boards and cooking raw foods extremely prone to cross-contamination [33,70,71]. Additionally, compared to glass and SS surfaces, which are hydrophilic materials, plastic is more likely to allow *Salmonella* germs to stick to them [33,41,72]. Therefore, it is crucial to avoid contaminating the plastic cutting boards used while preparing or processing food because this leads to vibriosis. Other authors looked at the efficacy of quercetin to inhibit the formation of biofilms in *S. epidermidis* [45]. Quercetin inhibited the growth of biofilms in a concentration-dependent manner. Quercetin reduced the growth of *S. epidermidis* biofilm by 90.5 and 95.3% at 250 and 500 μg/mL concentrations, respectively [45]. For this reason, quercetin is a common source for the food and pharmaceutical industries [47]. Consequently, there is a growing market for biodegradable polymers (fibers) that are effective against microbes. *S. aureus*, *E. coli*, and *K. pneumoniae* were all susceptible to the antibacterial effects of polylactide-based polymers (fibers) loaded with quercetin-(Q)- [49]. In cellular and animal models of Alzheimer's disease, nanoparticles of quercetin (nanoquercetin), which are colloidal in nature and made of polyaspartic acid-based polymer micelles encapsulated with quercetin, were utilized to prevent intracellular polyglutamine aggregation [50]. Quercetin that has been microencapsulated in a surfactant polymer capable of acting as an antisolvent process has been shown to be effective against cancer, develop biodegradable nanoparticles, and improve quercetin delivery [51,52].

We investigated biofilm generation and morphology by FE-SEM to further probe perturbations of the biofilm following quercetin application. The presence of quercetin decreased the number of connected cells and hampered the formation of biofilm, according to FE-SEM observation (Figure 5). As previously reported, quercetin has effects on *S. typhimurium* and *V. parahaemolyticus* biofilms as confirmed by FE-SEM [7,33,41]. In our study, 1/2 MIC showed inhibition was increased compared to the control and 1/4 MIC (Figure 5).

Many genes are essential for the pathogenicity, biofilm development, and physiological traits of *V. parahaemolyticus*. We observed the *V. parahaemolyticus* gene expression profiles for QS (*luxS* and *aphA*), motility (*flaA* and *flgL*), and virulence (*vp0952* and *vp0962*) in order to evaluate the potency of quercetin. There are connections between pathogenicity, QS, and virulence elements procedures. An emerging technique for preventing biofilm development, minimizing pathogenic infections, and maintaining food safety is to prevent or limit QS production. Oxidative stress develops when ROS build up occurs inside the cell [73]. By enhancing microbial population adaptation and survival protection, oxidative stress contributes significantly to the production of biofilms [53]. Not just in human cells, but also in microbes, ROS are crucial signaling molecules [41]. To keep a healthy redox cycle going and to encourage microbial adhesion, ROS can act as both intracellular and extracellular stimulants [33,46]. This will eventually result in the formation of biofilms. There may be an accumulation as a result of a disruption in the redox cycle [33]. By scavenging ROS within cells and weakening the membrane integrity of bacterial cells, the antioxidant quercetin prevents the formation of biofilms [46]. Quercetin, which acts as a potent bioactive compound, has antioxidant properties that lead to scavenging of the ROS production. Quercetin significantly reduced both forms of motility as well as the transcription of the *flaA* and *flgL* genes in the current investigation. These genes are connected to the control of flagella synthesis and structure in *V. parahaemolyticus* [7]. For instance, the *flaA* gene, which encodes polar flagellin, contributes to swimming motility, and the lateral flagellar gene system of *V. parahaemolyticus*, and allows bacteria to spread out and colonize surfaces (swarming), contains the *flgM* gene, which encodes anti-28 [7,74]. These results were in line with those of an earlier study [75], which found that thymoquinone decreased the expression of genes related to flagella production and hindered the motility of *V. parahaemolyticus*. A number of virulence factors, in addition to adhesion, are involved in the pathogenesis of *V. para-*

haemolyticus, and their expressions affect the pathogen's pathogenicity. Specifically, for the *vp0952* and *vp0962* genes, our findings showed that quercetin significantly reduced the expression of a number of virulence genes (Figure 6). On chromosome 2 of *V. parahaemolyticus*, the genes *vp0950*, *vp0952*, and *vp0962* all encode proteins that are similar to those found in biofilms [7,76]. Natural plant extracts also dramatically downregulated the transcription of the genes *ompW*, *luxS*, and *aphA*, which had previously been downregulated by citral in a prior study [7,77]. The quorum-sensing regulation, a challenging cell-to-cell process that enables bacteria to monitor their surroundings and cooperate, is primarily regulated by the two genes of *luxS* and *aphA*, which have been extensively investigated [7,78]. The regulation of thermostable direct hemolysin (TDH) and the development of *V. parahaemolyticus* biofilms have also been reported to be regulated by the *luxS* gene [7,79]. The impact of quercetin on other virulence genes, however, has to be further investigated as the current research largely focused on the alterations in biofilm-related genes.

Quercetin is not likely to enter cells and directly interact with transcriptional regulators or intracellular objectives. Our hypothesis is that quercetin might interact with specific membrane proteins, activating the bacterial signaling system and resulting in transcriptional changes that result in the downregulation of genes. In addition to microbial adhesins and cell membrane proteins, quercetin is a polyhydroxy hydrolytic chemical that has the potential to form powerful complexes with a wide range of macromolecules. By adjusting to the changes in the membrane, the bacterial cells may alter how their genes are produced using bacterial signaling processes, such two-component systems.

5. Conclusions

As a result, we were able to demonstrate that quercetin has effective antimicrobial and maybe anti-pathogenicity properties against *V. parahaemolyticus* on shrimp and crab shell surfaces. Furthermore, quercetin (1/2 MIC) significantly reduced the number of viable bacterial cells on shrimp and crab shell surfaces (3.70 and 3.09 log CFU/cm^2, respectively), disrupted cell-to-cell connections (FE-SEM images), dislodged already-formed biofilms, and significantly reduced the expression of genes associated with motility, virulence, and QS (Figure 6). Quercetin may be developed as an alternate technique to manage the biofilm of *V. parahaemolyticus* in food systems and lower the risk of foodborne illness caused by this pathogen.

Author Contributions: Conceptualization, P.K.R.; methodology, P.K.R., S.-H.P. and M.G.S.; software, P.K.R.; validation, P.K.R.; formal analysis, P.K.R.; investigation, P.K.R.; resources, P.K.R.; data curation, P.K.R. and M.G.S.; writing—original draft preparation, P.K.R.; writing—review and editing, P.K.R. and S.-H.P.; visualization, P.K.R.; supervision, S.Y.P.; project administration, S.Y.P. and S.-H.P.; funding acquisition, S.Y.P. and S.-H.P. All authors have read and agreed to the published version of the manuscript.

Funding: This study was supported by the Basic Science Research Program through the National Research Foundation of Korea (NRF) funded by the Ministry of Education, grant number 2021R1I1A3A04037468. This research was also supported by the High Value-Added Food Technology Development Program (321050-05-WT012, 321050-05-2-HD020), funded by the Korea Institute of Planning and Evaluation for Technology in Food, Agriculture, Forestry and Fisheries, Republic of Korea.

Institutional Review Board Statement: Not applicable.

Informed Consent Statement: Not applicable.

Data Availability Statement: The data presented in this study are available on request from the corresponding author.

Acknowledgments: We are grateful for the excellent support during experiments to the Food Safety Lab.

Conflicts of Interest: The authors declare no conflict of interest.

References

1. Selamoglu, M. Importance of the cold chain logistics in the marketing process of aquatic products: An update study. *Surv. Fish. Sci.* 2021, *8*, 25–29. [CrossRef]
2. Toushik, S.H.; Kim, K.; Ashrafudoulla, M.; Mizan, M.F.R.; Roy, P.K.; Nahar, S.; Kim, Y.; Ha, S.D. Korean kimchi-derived lactic acid bacteria inhibit foodborne pathogenic biofilm growth on seafood and food processing surface materials. *Food Control* 2021, *129*, 108276. [CrossRef]
3. Berlanga, M.; Guerrero, R. Living together in biofilms: The microbial cell factory and its biotechnological implications. *Microb. Cell Fact.* 2016, *15*, 165. [CrossRef] [PubMed]
4. Han, Q.; Song, X.; Zhang, Z.; Fu, J.; Wang, X.; Malakar, P.K.; Liu, H.; Pan, Y.; Zhao, Y. Removal of foodborne pathogen biofilms by acidic electrolyzed water. *Front. Microbiol.* 2017, *8*, 988. [CrossRef] [PubMed]
5. Alabdullatif, M.; Atreya, C.D.; Ramirez-Arcos, S. Antimicrobial peptides: An effective approach to prevent bacterial biofilm formation in platelet concentrates. *Transfusion* 2018, *58*, 2013–2021. [CrossRef]
6. Baker-Austin, C.; Oliver, J.D.; Alamo, M.; Ali, A.; Waldor, M.K.; Qadri, F.; Martinez-Urtaza, J. *Vibrio* spp. infections. *Nat. Rev. Dis. Prim.* 2018, *4*, 1–19. [CrossRef]
7. Liu, H.L.; Zhu, W.X.; Cao, Y.; Gao, J.Z.; Jin, T.; Qin, N.B.; Xia, X.D. Punicalagin inhibits biofilm formation and virulence gene expression of *Vibrio parahaemolyticus*. *Food Control* 2022, *139*, 109045. [CrossRef]
8. Ralph, A.; Currie, B.J. *Vibrio vulnificus* and *V. parahaemolyticus* necrotising fasciitis in fishermen visiting an estuarine tropical northern Australian location. *J. Infect.* 2007, *54*, e111–e114. [CrossRef]
9. Park, K.S.; Ono, T.; Rokuda, M.; Jang, M.H.; Okada, K.; Iida, T.; Honda, T. Functional characterization of two type III secretion systems of *Vibrio parahaemolyticus*. *Infect. Immun.* 2004, *72*, 6659–6665. [CrossRef]
10. FAO. *Advances in Science and Risk Assessment Tools for Vibrio parahaemolyticus and V. vulnificus Associated with Seafood*; Meeting Report; Food & Agriculture Organization: Rome, Italy, 2021; Volume 35.
11. Centers for Disease Control and Prevention (CDC). *National Listeria Surveillance Annual Summary, 2013*; Department of Health and Human Services, CDC: Atlanta, GA, USA, 2015.
12. Elexson, N.; Afsah-Hejri, L.; Rukayadi, Y.; Soopna, P.; Lee, H.Y.; Zainazor, T.C.T.; Ainy, M.N.; Nakaguchi, Y.; Mitsuaki, N.; Son, R. Effect of detergents as antibacterial agents on biofilm of antibiotics-resistant *Vibrio parahaemolyticus* isolates. *Food Control* 2014, *35*, 378–385. [CrossRef]
13. Reddy, P. Empiric antibiotic therapy of nosocomial bacterial infections. *Am. J. Ther.* 2016, *23*, E982–E994. [CrossRef] [PubMed]
14. Ahmed, H.A.; El Bayomi, R.M.; Hussein, M.A.; Khedr, M.H.E.; Remela, E.M.A.; El-Ashram, A.M.M. Molecular characterization, antibiotic resistance pattern and biofilm formation of *Vibrio parahaemolyticus* and *Vibrio cholerae* isolated from crustaceans and humans. *Int. J. Food Microbiol.* 2018, *274*, 31–37. [CrossRef] [PubMed]
15. Tan, C.W.; Rukayadi, Y.; Hasan, H.; Thung, T.Y.; Lee, E.; Rollon, W.D.; Hara, H.; Kayali, A.Y.; Nishibuchi, M.; Radu, S. Prevalence and antibiotic resistance patterns of *Vibrio parahaemolyticus* isolated from different types of seafood in Selangor, Malaysia. *Saudi J. Biol. Sci.* 2020, *27*, 1602–1608. [CrossRef] [PubMed]
16. Xie, T.F.; Wu, Q.P.; Zhang, J.M.; Xu, X.K.; Cheng, J.H. Comparison of *Vibrio parahaemolyticus* isolates from aquatic products and clinical by antibiotic susceptibility, virulence, and molecular characterisation. *Food Control* 2017, *71*, 315–321. [CrossRef]
17. Pisoschi, A.M.; Pop, A.; Georgescu, C.; Turcus, V.; Olah, N.K.; Mathe, E. An overview of natural antimicrobials role in food. *Eur. J. Med. Chem.* 2018, *143*, 922–935. [CrossRef]
18. Brannon, J.R.; Hadjifrangiskou, M. The arsenal of pathogens and antivirulence therapeutic strategies for disarming them. *Drug Des. Dev. Ther.* 2016, *10*, 1795–1806. [CrossRef]
19. Vazquez-Armenta, F.; Hernandez-Oñate, M.; Martinez-Tellez, M.; Lopez-Zavala, A.; Gonzalez-Aguilar, G.; Gutierrez-Pacheco, M.; Ayala-Zavala, J. Quercetin repressed the stress response factor (sigB) and virulence genes (prfA, actA, inlA, and inlC), lower the adhesion, and biofilm development of *L. monocytogenes*. *Food Microbiol.* 2019, *87*, 103377. [CrossRef]
20. Li, Y.; Dong, R.Y.; Ma, L.; Qian, Y.L.; Liu, Z.Y. Combined anti-biofilm enzymes strengthen the eradicate effect of *Vibrio parahaemolyticus* biofilm: Mechanism on cpsA-J expression and application on different carriers. *Foods* 2022, *11*, 1305. [CrossRef]
21. Meireles, A.; Borges, A.; Giaouris, E.; Simoes, M. The current knowledge on the application of anti-biofilm enzymes in the food industry. *Food Res. Int.* 2016, *86*, 140–146. [CrossRef]
22. Malone, M.; Goeres, D.M.; Gosbell, I.; Vickery, K.; Jensen, S.; Stoodley, P. Approaches to biofilm-associated infections: The need for standardized and relevant biofilm methods for clinical applications. *Expert Rev. Anti-Infect. Ther.* 2017, *15*, 147–156. [CrossRef]
23. Nahar, S.; Mizan, M.F.R.; Ha, A.J.W.; Ha, S.D. Advances and future prospects of enzyme-based biofilm prevention Approaches in the food industry. *Compr. Rev. Food Sci. Food Saf.* 2018, *17*, 1484–1502. [CrossRef] [PubMed]
24. Flemming, H.C.; Wingender, J.; Szewzyk, U.; Steinberg, P.; Rice, S.A.; Kjelleberg, S. Biofilms: An emergent form of bacterial life. *Nat. Rev. Microbiol.* 2016, *14*, 563–575. [CrossRef] [PubMed]
25. Han, N.; Mizan, M.F.R.; Jahid, I.K.; Ha, S.D. Biofilm formation by *Vibrio parahaemolyticus* on food and food contact surfaces increases with rise in temperature. *Food Control* 2016, *70*, 161–166. [CrossRef]
26. Malcolm, T.T.H.; Chang, W.S.; Loo, Y.Y.; Cheah, Y.K.; Radzi, C.W.J.W.M.; Kantilal, H.K.; Nishibuchi, M.; Son, R. Simulation of improper food hygiene practices: A quantitative assessment of *Vibrio parahaemolyticus* distribution. *Int. J. Food Microbiol.* 2018, *284*, 112–119. [CrossRef] [PubMed]

27. Liu, J.K.; Bai, L.; Li, W.W.; Han, H.H.; Fu, P.; Ma, X.C.; Bi, Z.W.; Yang, X.R.; Zhang, X.L.; Zhen, S.Q.; et al. Trends of foodborne diseases in China: Lessons from laboratory-based surveillance since 2011. *Front. Med.* **2018**, *12*, 48–57. [CrossRef]
28. Mizan, M.F.R.; Jahid, I.K.; Kim, M.; Lee, K.H.; Kim, T.J.; Ha, S.D. Variability in biofilm formation correlates with hydrophobicity and quorum sensing among *Vibrio parahaemolyticus* isolates from food contact surfaces and the distribution of the genes involved in biofilm formation. *Biofouling* **2016**, *32*, 497–509. [CrossRef]
29. Elgamoudi, B.A.; Korolik, V. *Campylobacter* biofilms: Potential of natural compounds to disrupt *Campylobacter jejuni* transmission. *Int. J. Mol. Sci.* **2021**, *22*, 12159. [CrossRef]
30. Ramić, D.; Ogrizek, J.; Bucar, F.; Jeršek, B.; Jeršek, M.; Možina, S.S. *Campylobacter jejuni* biofilm control with lavandin essential oils and by-products. *Antibiotics* **2022**, *11*, 854. [CrossRef]
31. Hossain, M.I.; Mizan, M.F.R.; Ashrafudoulla, M.; Nahar, S.; Joo, H.-J.; Jahid, I.K.; Park, S.H.; Kim, K.-S.; Ha, S.-D. Inhibitory effects of probiotic potential lactic acid bacteria isolated from kimchi against *Listeria monocytogenes* biofilm on lettuce, stainless-steel surfaces, and MBEC™ biofilm device. *LWT* **2020**, *118*, 108864. [CrossRef]
32. Hossain, M.I.; Mizan, M.F.R.; Roy, P.K.; Nahar, S.; Toushik, S.H.; Ashrafudoulla, M.; Jahid, I.K.; Lee, J.; Ha, S.D. *Listeria monocytogenes* biofilm inhibition on food contact surfaces by application of postbiotics from Lactobacillus curvatus B.67 and Lactobacillus plantarum M.2. *Food Res. Int.* **2021**, *148*, 110595. [CrossRef]
33. Roy, P.K.; Song, M.G.; Park, S.Y. Impact of Quercetin against *Salmonella typhimurium* biofilm formation on food-contact surfaces and molecular mechanism pattern. *Foods* **2022**, *11*, 977. [CrossRef] [PubMed]
34. Ashrafudoulla, M.; Mizan, M.F.R.; Ha, A.J.; Park, S.H.; Ha, S.D. Antibacterial and antibiofilm mechanism of eugenol against antibiotic resistance *Vibrio parahaemolyticus*. *Food Microbiol.* **2020**, *91*, 103500. [CrossRef] [PubMed]
35. Galie, S.; Garcia-Gutierrez, C.; Miguelez, E.M.; Villar, C.J.; Lombo, F. Biofilms in the food industry: Health aspects and control methods. *Front. Microbiol.* **2018**, *9*, 898. [CrossRef] [PubMed]
36. Baptista, R.C.; Horita, C.N.; Sant'Ana, A.S. Natural products with preservative properties for enhancing the microbiological safety and extending the shelf-life of seafood: A review. *Food Res. Int.* **2019**, *127*, 108762. [CrossRef] [PubMed]
37. Upadhyay, A.; Upadhyaya, I.; Kollanoor-Johny, A.; Venkitanarayanan, K. Combating pathogenic microorganisms using plant-derived antimicrobials: A minireview of the mechanistic basis. *BioMed Res. Int.* **2014**, *2014*, 761741. [CrossRef]
38. Durazzo, A.; Lucarini, M.; Souto, E.B.; Cicala, C.; Caiazzo, E.; Izzo, A.A.; Novellino, E.; Santini, A. Polyphenols: A concise overview on the chemistry, occurrence, and human health. *Phytother. Res.* **2019**, *33*, 2221–2243. [CrossRef]
39. Quecan, B.X.V.; Santos, J.T.C.; Rivera, M.L.C.; Hassimotto, N.M.A.; Almeida, F.A.; Pinto, U.M. Effect of Quercetin rich onion extracts on bacterial quorum sensing. *Front. Microbiol.* **2019**, *10*, 867. [CrossRef]
40. Hossain, M.I.; Kim, K.; Mizan, M.F.R.; Toushik, S.H.; Ashrafudoulla, M.; Roy, P.K.; Nahar, S.; Jahid, I.K.; Choi, C.; Park, S.H.; et al. Comprehensive molecular, probiotic, and quorum-sensing characterization of anti-listerial lactic acid bacteria, and application as bioprotective in a food (milk) model. *J. Dairy Sci.* **2021**, *104*, 6516–6534. [CrossRef]
41. Kim, Y.K.; Roy, P.K.; Ashrafudoulla, M.; Nahar, S.; Toushik, S.H.; Hossain, M.I.; Mizan, M.F.R.; Park, S.H.; Ha, S.D. Antibiofilm effects of quercetin against *Salmonella enterica* biofilm formation and virulence, stress response, and quorum-sensing gene expression. *Food Control* **2022**, *137*, 108964. [CrossRef]
42. Ortega-Vidal, J.; Cobo, A.; Ortega-Morente, E.; Galvez, A.; Alejo-Armijo, A.; Salido, S.; Altarejos, J. Antimicrobial and antioxidant activities of flavonoids isolated from wood of sweet cherry tree (*Prunus avium* L.). *J. Wood Chem. Technol.* **2021**, *41*, 104–117. [CrossRef]
43. Osonga, F.J.; Akgul, A.; Miller, R.M.; Eshun, G.B.; Yazgan, I.; Akgul, A.; Sadik, O.A. Antimicrobial activity of a new class of phosphorylated and modified flavonoids. *ACS Omega* **2019**, *4*, 12865–12871. [CrossRef] [PubMed]
44. He, Z.Y.; Zhang, X.; Song, Z.C.; Li, L.; Chang, H.S.; Li, S.L.; Zhou, W. Quercetin inhibits virulence properties of *Porphyromas gingivalis* in periodontal disease. *Sci. Rep.* **2020**, *10*, 18313. [CrossRef] [PubMed]
45. Mu, Y.Q.; Zeng, H.; Chen, W. Quercetin inhibits biofilm formation by decreasing the production of EPS and altering the composition of EPS in *Staphylococcus epidermidis*. *Front. Microbiol.* **2021**, *12*, 631058. [CrossRef] [PubMed]
46. Sreelatha, S.; Jayachitra, A. Targeting biofilm inhibition using Quercetin—Interaction with bacterial cell membrane and ROS mediated biofilm control. *Funct. Foods Health Dis.* **2018**, *8*, 292–306.
47. Ozgen, S.; Kilinc, O.K.; Selamoğlu, Z. Antioxidant activity of quercetin: A mechanistic review. *Turk. J. Agric. Food Sci. Technol.* **2016**, *4*, 1134–1138. [CrossRef]
48. Roy, P.K.; Song, M.G.; Park, S.Y. The inhibitory effect of quercetin on biofilm formation of *Listeria monocytogenes* mixed culture and repression of virulence. *Antioxidants* **2022**, *11*, 1733. [CrossRef]
49. Kost, B.; Svyntkivska, M.; Brzeziński, M.; Makowski, T.; Piorkowska, E.; Rajkowska, K.; Kunicka-Styczyńska, A.; Biela, T. PLA/β-CD-based fibres loaded with quercetin as potential antibacterial dressing materials. *Colloids Surf. B Biointerfaces* **2020**, *190*, 110949. [CrossRef]
50. Debnath, K.; Jana, N.R.; Jana, N.R. Quercetin encapsulated polymer nanoparticle for inhibiting intracellular polyglutamine aggregation. *ACS Appl. Bio Mater.* **2019**, *2*, 5298–5305. [CrossRef]
51. Fraile, M.; Buratto, R.; Gomez, B.; Martin, A.; Cocero, M.J. Enhanced delivery of quercetin by encapsulation in poloxamers by supercritical antisolvent process. *Ind. Eng. Chem. Res.* **2014**, *53*, 4318–4327. [CrossRef]
52. Baksi, R.; Singh, D.P.; Borse, S.P.; Rana, R.; Sharma, V.; Nivsarkar, M. In vitro and in vivo anticancer efficacy potential of Quercetin loaded polymeric nanoparticles. *Biomed. Pharmacother.* **2018**, *106*, 1513–1526. [CrossRef]

53. Ong, K.S.; Mawang, C.I.; Daniel-Jambun, D.; Lim, Y.Y.; Lee, S.M. Current anti-biofilm strategies and potential of antioxidants in biofilm control. *Expert Rev. Anti-Infect. Ther.* **2018**, *16*, 855–864. [CrossRef] [PubMed]
54. Amin, M.U.; Khurram, M.; Khattak, B.; Khan, J. Antibiotic additive and synergistic action of rutin, morin and quercetin against methicillin resistant *Staphylococcus aureus*. *BMC Complement. Altern. Med.* **2015**, *15*, 59. [CrossRef] [PubMed]
55. Wang, S.N.; Yao, J.Y.; Zhou, B.; Yang, J.X.; Chaudry, M.T.; Wang, M.; Xiao, F.L.; Li, Y.; Yin, W.Z. Bacteriostatic effect of Quercetin as an antibiotic alternative in vivo and its antibacterial mechanism in vitro. *J. Food Prot.* **2018**, *81*, 68–78. [CrossRef] [PubMed]
56. Ouyang, J.; Sun, F.; Feng, W.; Sun, Y.; Qiu, X.; Xiong, L.; Liu, Y.; Chen, Y. Quercetin is an effective inhibitor of quorum sensing, biofilm formation and virulence factors in *Pseudomonas aeruginosa*. *J. Appl. Microbiol.* **2016**, *120*, 966–974. [CrossRef]
57. Roy, P.K.; Mizan, M.F.R.; Hossain, M.I.; Han, N.; Nahar, S.; Ashrafudoulla, M.; Toushik, S.H.; Shim, W.B.; Kim, Y.M.; Ha, S.D. Elimination of *Vibrio parahaemolyticus* biofilms on crab and shrimp surfaces using ultraviolet C irradiation coupled with sodium hypochlorite and slightly acidic electrolyzed water. *Food Control* **2021**, *128*, 108179. [CrossRef]
58. Roy, P.K.; Ha, A.J.; Mizan, M.F.R.; Hossain, M.I.; Ashrafudoulla, M.; Toushik, S.H.; Nahar, S.; Kim, Y.K.; Ha, S.D. Effects of environmental conditions (temperature, pH, and glucose) on biofilm formation of *Salmonella* enterica serotype Kentucky and virulence gene expression. *Poult. Sci.* **2021**, *100*, 101209. [CrossRef]
59. Cho, J.; Kim, G.; Qamar, A.Y.; Fang, X.; Roy, P.K.; Tanga, B.M.; Bang, S.; Kim, J.K.; Galli, C.; Perota, A.; et al. Improved efficiencies in the generation of multigene-modified pigs by recloning and using sows as the recipient. *Zygote* **2022**, *30*, 103–110. [CrossRef]
60. Roy, P.K.; Qamar, A.Y.; Tanga, B.M.; Bang, S.; Seong, G.; Fang, X.; Kim, G.; Edirisinghe, S.L.; De Zoysa, M.; Kang, D.H.; et al. Modified *Spirulina maxima* Pectin nanoparticles improve the developmental competence of in vitro matured porcine oocytes. *Animals* **2021**, *11*, 2483. [CrossRef]
61. Roy, P.K.; Qamar, A.Y.; Tanga, B.M.; Fang, X.; Kim, G.; Bang, S.; Cho, J. Enhancing oocyte competence with Milrinone as a phosphodiesterase 3A inhibitor to improve the development of porcine cloned embryos. *Front. Cell Dev. Biol.* **2021**, *9*, 647616. [CrossRef]
62. Kim, G.; Roy, P.K.; Fang, X.; Hassan, B.M.S.; Cho, J. Improved preimplantation development of porcine somatic cell nuclear transfer embryos by caffeine treatment. *J. Veter Sci.* **2019**, *20*, e31. [CrossRef]
63. Roy, P.K.; Qamar, A.Y.; Fang, X.; Kim, G.; Bang, S.; De Zoysa, M.; Shin, S.T.; Cho, J. Chitosan nanoparticles enhance developmental competence of in vitro-matured porcine oocytes. *Reprod. Domest. Anim.* **2020**, *56*, 342–350. [CrossRef] [PubMed]
64. Roy, P.K.; Qamar, A.Y.; Fang, X.; Kim, G.; Hassan, B.M.S.; Cho, J. Effects of cobalamin on meiotic resumption and developmental competence of growing porcine oocytes. *Theriogenology* **2020**, *154*, 24–30. [CrossRef] [PubMed]
65. Gopu, V.; Meena, C.K.; Shetty, P.H. Quercetin influences quorum sensing in food borne bacteria: In-vitro and in-silico evidence. *PLoS ONE* **2015**, *10*, e0134684. [CrossRef]
66. Damte, D.; Gebru, E.; Lee, S.; Suh, J.; Park, S. Evaluation of anti-quorum sensing activity of 97 indigenous plant extracts from Korea through bioreporter bacterial strains *Chromobacterium violaceum* and *Pseudomonas aeruginosa*. *J. Microb. Biochem. Technol.* **2013**, *5*, 42–46. [CrossRef]
67. Niu, C.; Afre, S.; Gilbert, E.S. Subinhibitory concentrations of cinnamaldehyde interfere with quorum sensing. *Lett. Appl. Microbiol.* **2006**, *43*, 489–494. [CrossRef]
68. Thenmozhi, R.; Nithyanand, P.; Rathna, J.; Pandian, S.K. Antibiofilm activity of coral-associated bacteria against different clinical M serotypes of *Streptococcus pyogenes*. *FEMS Immunol. Med. Microbiol.* **2009**, *57*, 284–294. [CrossRef]
69. Vazquez-Armenta, F.J.; Bernal-Mercado, A.T.; Tapia-Rodriguez, M.R.; Gonzalez-Aguilar, G.A.; Lopez-Zavala, A.A.; Martinez-Tellez, M.A.; Hernandez-Onate, M.A.; Ayala-Zavala, J.F. Quercetin reduces adhesion and inhibits biofilm development by *Listeria monocytogenes* by reducing the amount of extracellular proteins. *Food Control* **2018**, *90*, 266–273. [CrossRef]
70. Stepanovic, S.; Cirkovic, I.; Ranin, L.; Svabic-Vlahovic, M. Biofilm formation by *Salmonella* spp. and *Listeria monocytogenes* on plastic surface. *Lett. Appl. Microbiol.* **2004**, *38*, 428–432. [CrossRef]
71. Lee, K.H.; Lee, J.Y.; Roy, P.K.; Mizan, M.F.R.; Hossain, M.I.; Park, S.H.; Ha, S.D. Viability of *Salmonella typhimurium* biofilms on major food-contact surfaces and eggshell treated during 35 days with and without water storage at room temperature. *Poult. Sci.* **2020**, *99*, 4558–4565. [CrossRef]
72. Sinde, E.; Carballo, J. Attachment of *Salmonella* spp. and *Listeria monocytogenes* to stainless steel, rubber and polytetrafluoroethylene: The influence of free energy and the effect of commercial sanitizers. *Food Microbiol.* **2000**, *17*, 439–447. [CrossRef]
73. Gambino, M.; Cappitelli, F. Mini-review: Biofilm responses to oxidative stress. *Biofouling* **2016**, *32*, 167–178. [CrossRef] [PubMed]
74. McCarter, L.L. Polar flagellar motility of the *Vibrionaceae*. *Microbiol. Mol. Biol. Rev.* **2001**, *65*, 445–462. [CrossRef] [PubMed]
75. Guo, D.; Yang, Z.Y.; Zheng, X.Y.; Kang, S.M.; Yang, Z.K.; Xu, Y.F.; Shi, C.; Tian, H.Y.; Xia, X.D. Thymoquinone inhibits biofilm formation and attachment-invasion in host cells of *Vibrio parahaemolyticus*. *Foodborne Pathog. Dis.* **2019**, *16*, 671–678. [CrossRef] [PubMed]
76. Boyd, E.F.; Cohen, A.L.V.; Naughton, L.M.; Ussery, D.W.; Binnewies, T.T.; Stine, O.C.; Parent, M.A. Molecular analysis of the emergence of pandemic *Vibrio parahaemolyticus*. *BMC Microbiol.* **2008**, *8*, 110. [CrossRef]
77. Sun, Y.; Guo, D.; Hua, Z.; Sun, H.H.; Zheng, Z.W.; Xia, X.D.; Shi, C. Attenuation of multiple *Vibrio parahaemolyticus* virulence factors by citral. *Front. Microbiol.* **2019**, *10*, 894. [CrossRef]

78. Rutherford, S.T.; van Kessel, J.C.; Shao, Y.; Bassler, B.L. AphA and LuxR/HapR reciprocally control quorum sensing in *vibrios*. *Genes Dev.* **2011**, *25*, 397–408. [CrossRef]
79. Guo, M.H.; Fang, Z.J.; Sun, L.J.; Sun, D.F.; Wang, Y.L.; Li, C.; Wang, R.D.; Liu, Y.; Hu, H.Q.; Liu, Y.; et al. Regulation of thermostable direct hemolysin and biofilm formation of *Vibrio parahaemolyticus* by quorum-sensing genes luxM and luxS. *Curr. Microbiol.* **2018**, *75*, 1190–1197. [CrossRef]

Article

Antibacterial Activity of Silver Nanoflake (SNF)-Blended Polysulfone Ultrafiltration Membrane

Gunawan Setia Prihandana [1,*], Tutik Sriani [2], Aisyah Dewi Muthi'ah [1], Siti Nurmaya Musa [3], Mohd Fadzil Jamaludin [3] and Muslim Mahardika [4]

1. Department of Industrial Engineering, Faculty of Advanced Technology and Multidiscipline, Universitas Airlangga, Jl. Dr. Ir. H. Soekarno, Surabaya 60115, Indonesia
2. Department of Research and Development, PT. Global Meditek Utama, Sardonoharjo, Ngaglik, Sleman, Yogyakarta 55581, Indonesia
3. Centre of Advanced Manufacturing & Material Processing (AMMP Centre), Department of Mechanical Engineering, Faculty of Engineering, Universiti Malaya, Kuala Lumpur 50603, Malaysia
4. Department of Mechanical and Industrial Engineering, Faculty of Engineering, Universitas Gadjah Mada, Jalan Grafika No. 2, Yogyakarta 55281, Indonesia
* Correspondence: gsprihandana@gmail.com; Tel.: +62-881-0360-00830

Citation: Prihandana, G.S.; Sriani, T.; Muthi'ah, A.D.; Musa, S.N.; Jamaludin, M.F.; Mahardika, M. Antibacterial Activity of Silver Nanoflake (SNF)-Blended Polysulfone Ultrafiltration Membrane. *Polymers* **2022**, *14*, 3600. https://doi.org/10.3390/polym14173600

Academic Editors: Md. Amdadul Huq and Shahina Akter

Received: 28 July 2022
Accepted: 27 August 2022
Published: 31 August 2022

Publisher's Note: MDPI stays neutral with regard to jurisdictional claims in published maps and institutional affiliations.

Copyright: © 2022 by the authors. Licensee MDPI, Basel, Switzerland. This article is an open access article distributed under the terms and conditions of the Creative Commons Attribution (CC BY) license (https:// creativecommons.org/licenses/by/ 4.0/).

Abstract: The aim of this research was to study the possibility of using silver nanoflakes (SNFs) as an antibacterial agent in polysulfone (PSF) membranes. SNFs at different concentrations (0.1, 0.2, 0.3 and 0.4 wt.%) were added to a PSF membrane dope solution. To investigate the effect of SNFs on membrane performance and properties, the water contact angle, protein separation, average pore size and molecular weight cutoffs were measured, and water flux and antibacterial tests were conducted. The antimicrobial activities of the SNFs were investigated using *Escherichia coli* taken from river water. The results showed that PSF membranes blended with 0.1 wt.% SNFs have contact angles of 55°, which is less than that of the pristine PSF membrane (81°), exhibiting the highest pure water flux. Molecular weight cutoff values of the blended membranes indicated that the presence of SNFs does not lead to enlargement of the membrane pore size. The rejection of protein (egg albumin) was improved with the addition of 0.1 wt.% SNFs. The SNFs showed antimicrobial activity against *Escherichia coli*, where the killing rate was dependent on the SNF concentration in the membranes. The identified bacterial colonies that appeared on the membranes decreased with increasing SNF concentration. PSF membranes blended with SNF, to a great degree, possess quality performance across several indicators, showing great potential to be employed as water filtration membranes.

Keywords: polysulfone membrane; silver nanoflake; good health; antibacterial activity; *Escherichia coli*; water treatment

1. Introduction

Clean water scarcity has become one of the greatest global challenges that need to be solved [1–3]. Different water separation technologies have been developed to tackle the problem [4]. For this purpose, economical and efficient water separation technologies are needed. Among other water separation technologies, such as flotation, skimming, ultrasonic treatment and coagulation [5–9], membrane separation has attracted more attention due to its performance, high efficiency, low-cost fabrication and operation [10–12]. To date, ceramic and polymeric membranes have been developed and used in water separation technologies [13,14]. Compared to ceramic membranes, polymeric membranes display outstanding performance, such as controllable membrane microstructures, good flexibility and easy fabrication [4,15].

Filtration utilizing membrane-based polymer technology such as microfiltration (MF), ultrafiltration (UF), nanofiltration (NF) and reverse osmosis (RO) is a potent solution to overcoming the scarcity of the clean and safe water supply [16]. The outstanding combination of water flux and rejection have made polymeric membranes widely used in various water filtration applications [17,18]. In addition, polymeric membranes possess a microporous, open spongy morphology of tiny pores ranging from 0.03 μm to 10 μm in diameter [19,20], which is suitable for all water filtration purposes.

Different types of commercial membranes have been used and developed using common polymers, such as poly (vinylidene fluoride) (PVDF) [14,21], polyether sulfone (PES) [22,23], poly (acrylonitrile) (PAN) [24,25], polyimide (PI) [26,27], polyurethane (PU) [28,29], poly (propylene) (PP) [30,31] and poly (sulfone) (PSF) [32,33].

Among those polymeric membranes, PSF is an ideal choice for membrane development due to its availability, chemical stability and high mechanical strength and can be used in a wide range of temperature and pH levels [34,35]. PSF has a tensile strength between 78 and 83 MPa; hence, it exhibits excellent mechanical properties that favor its usage as a membrane in wastewater treatment [36]. However, hydrophobicity is considered to be a disadvantage of PSF membranes, as it contributes to membrane fouling [37,38]. To further improve the properties and filtration performance of PSF membranes, many modification methods, such as polymer blending [39–41], incorporating nanoparticles [42,43], surface chemical modification [42,44], interfacial polymerization [45,46] and the incorporation of additives, have been conducted. It is known that the additives modulate the hydrophilicity of the surface of PSF membrane and control its morphology and pore formation [47–49].

Recently, various types of additives have been used to modify the PSF membrane and to attain material and filtration improvements in terms of water flux and solute rejection. Kusworo et al. prepared PSF membranes by adding TiO_2 nanoparticles to the dope solution. The study revealed that the introduction of TiO_2 nanofillers enlarged the finger-like structure of the membrane, yielded a decreased contact angle and increased the pure water flux [50].

Graphene oxide (GO) has also been evaluated as a nanofiller in PSF ultrafiltration membranes. Differently shaped GO materials, flat and crumpled GO, were used in the membrane fabrication via the phase inversion method. The experimental result shows that the shape of the GO material affects the dispersion and viscosity of the dope solution and results in changes to the membrane structures (surface hydrophilicity and porosity) and performance (permeability and rejection) [51]. Alosaimi explored the effect of the incorporation of oxidized carbon nanotubes (CNTOxi) into PSF membranes. The addition of a hybrid nanocomposite improves the thermal stability, mechanical properties and adsorption capacities of the membranes due to the good dispersion of organoclay and carbon nanomaterials [52].

Cellulose nanocrystals were used as an additive to PSF to form nanocomposite mixed-matrix membranes (NMMMs). It was reported that fabricated NMMMs have triple the pure water flux and an improvement in the methylene blue (MB) rejection efficiency [53]. MOFs have also been prepared as composite membranes. A high pure water flux membrane was obtained by incorporating metal-organic framework (MOF) MIL-100 (Fe) nanoparticles into a PSF composite membrane. In addition, fabricated PSF membranes have a 99% rejection rate of MB and excellent anti-fouling performance [54].

In water treatment, bacterial and viral penetration is the obstacle to avoid. In order to overcome this problem, silver nanoparticles (SNs) incorporated into PSF membranes have been shown to exhibit antimicrobial properties. Silver is known for its good electrical and thermal conductivity, chemical stability and antibacterial activity [55]. Its antibacterial property is due to the availability of Ag^+, which prolongs the prevention of bacterial adhesion. In addition to its antibacterial property, as an additive, SNs improved membrane hydrophilicity and reduced membrane fouling [56]. Moreover, membranes with incorporated nanosilver showed different porosities as a function of additive particle size [57]. Pal et al. [58] reported that the toxicity of SNs to bacteria is affected by SN size and shape, since in terms of the active facet, different shapes may have different effective surface areas. Nanoflakes can be defined as nanoscale particles which have three dimensions at the nanoscale [59], as shown in Figure 1. On the other hand, bacteria are susceptible to mechanical forces [60,61], and the sharp edges of nanoflakes could be beneficial in penetrating the bacterial cell membrane. In the literature, there are several works related to silver nanoflakes (SNFs); however, there is no scientific study about the characteristic properties of SNFs incorporated in PSF membranes, particularly on the effect of SNFs on the average pore size of the membrane, the separation of molecular weight protein solution (egg albumin) and the ultrafiltration performance of the fabricated membrane.

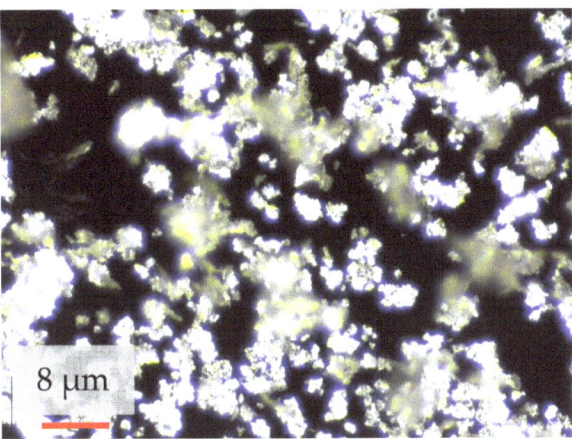

Figure 1. Image of silver nanoflake powder.

This work focuses on preparing bare and composite PSF membranes. The membranes were fabricated by introducing different quantities of SNFs (0, 0.1, 0.2, 0.3 and 0.4 wt.%) to the PSF (22 wt.%) membrane solution. Contact angle and adhesion work were the techniques used to characterize the membranes. Membrane performance was investigated by using a water flux test, a protein rejection test, an average pore size assessment and a molecular weight cutoff measurement. Finally, the antibacterial property was evaluated by quantifying the number of *Escherichia coli* attached to the membrane surface.

2. Materials and Methods

2.1. Materials

The PSF dope solution was prepared using Udel P-3500 (Solvay SA, Brussels, Belgium). Egg albumin (EA) (45 kDa) was obtained from HiMedia Laboratories Pvt Ltd., Mumbai, Maharashtra 400086, India. Chromocult® coliform agar and N-Methyl-2-pyrrolidone (NMP), used as a solvent, were obtained from Merck KGaA, Darmstadt, Germany. Silver nanoflake powder (thickness of 80–500 nm), as shown in Figure 1, was obtained from Nanostructured & Amorphous Materials, Inc. Katy, TX 77494, USA. Pure water was used for the membrane's fabrication and water flux test.

2.2. Membrane Fabrication

The PSF membrane was prepared using the phase inversion method. PSF, as a polymeric binder, was added to the solvent (NMP) until it was completely dissolved. Once a homogeneous solution was formed, SNF powder was slowly added to the dope solution at different concentrations (0, 0.1, 0.2, 0.3 and 0.4 wt.%), as presented in Table 1, and stirred with a hot plate magnetic stirrer. The dope solution of PSF/SNF was poured onto a glass plate and casted using a film applicator (Elcometer, Manchester M43 6BU, UK) at a thickness of 200 µm. The solution formed on the glass plate was gently transferred to a gelatinization bath. The bath was composed of pure water (non-solvent), and during the gelatinization process, an instantaneous liquid–liquid demixing process controlled the membrane surface and matrix phase inversion process, resulting in a membrane with a finger-like structure and a dense skin on its surface, as illustrated in Figure 2 [62].

Table 1. Composition of the casting solution of PSF/SNF membranes.

Membrane	Blend Composition, wt.%		
	PSF	SNF	NMP
SNF 0	22	0	78
SNF 0.1	22	0.1	77.9
SNF 0.2	22	0.2	77.8
SNF 0.3	22	0.3	77.7
SNF 0.4	22	0.4	77.6

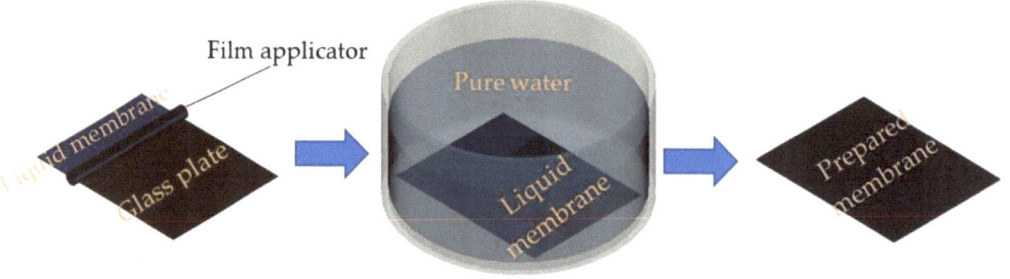

Figure 2. Membrane gelatinization process.

2.3. Membrane Characterization

Pure water at a fixed volume was deposited onto the surface of the dry membrane. The image of the deposited water was captured using an AM73915MZTL Dinolite Edge 3.0 digital microscope. The angle of the water drop was calculated using AutoCAD software. To guarantee a valid result, the contact angle for each membrane was measured three times and the values were acquired by averaging the three repetitions. From the contact angle values obtained, the work of adhesion (ω_A), i.e., the surface energy required to drag water from a membrane surface, can be determined as follows [63]:

$$\omega_A = \gamma_B (1 + \cos \theta) \tag{1}$$

where γ_B is the water surface tension and θ is the contact angle.

2.4. Equilibrium Water Content

The fabricated PSF membranes were cut to the required size, immersed in pure water for 24 h and weighed directly after wiping the free water from the membrane surface. The membranes were then dried out and weighed again. The water content of the membrane was calculated by the following [64]:

$$\%WC = \frac{W_w - W_d}{W_w} \times 100 \tag{2}$$

where W_w and W_d are the weights of the wet and dry membranes, respectively. The measured data were then analyzed using one-way ANOVA with SPSS Software version 16, SPSS Inc., Chicago, IL, USA, and the level of significance was set at 0.05.

2.5. Filtration Experiments

2.5.1. Water Flux Test

Filtration experiments were conducted in a stirred dead-end cell (Sterlitech UHP-62, Sterlitech Corp. Kent, WA, USA), as shown in Figure 3, with a diameter of 62 mm and an effective membrane area of 30 cm². Nitrogen gas at a pressure of 2 bar was delivered into the dead-end cell to provide pressure to the tested water. Data acquisition was used to record the weight of the permeate water passing through the membrane pores. The following formulas were used to calculate the volumetric flux and permeability [65]:

$$Flux\ (J_v) = \frac{Q}{A\Delta t} \quad (3)$$

$$Permeability\ (L_p) = \frac{J_v}{\Delta P} \quad (4)$$

where Q is the quantity of the permeate water (in L) during the sampling time, Δt (in h) is the time difference, A is the area of the membrane (in m²) and ΔP is the pressure difference (in bar).

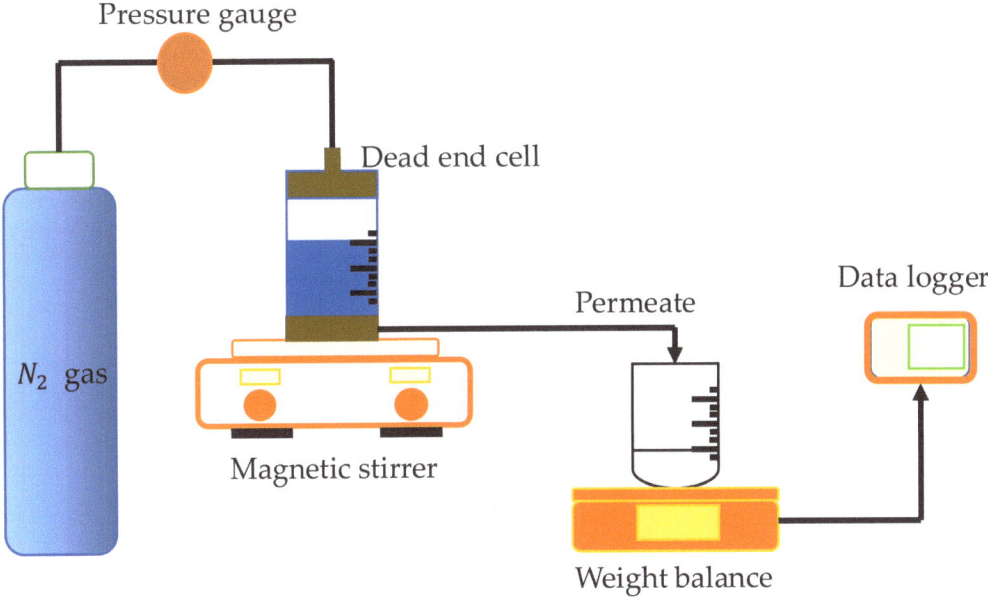

Figure 3. Experimental setup of the water flux test.

2.5.2. Protein Separation

EA solution at a concentration of 0.1 wt.% was prepared in phosphate-buffered solution (pH = 7.2). The protein rejection experiment was conducted at a fixed pressure of 2 bar using the dead-end cell test. An N4S UV-visible spectrophotometer was used to measure the concentration of the protein at a wavelength of 280 nm. Proteins contain amino acids (tryptophan) that absorb light in the UV spectrum at wavelength of 280 nm [66]. The solute rejection (SR) was determined with the following [67]:

$$\%SR = \left[1 - \frac{C_p}{C_f}\right] \times 100 \quad (5)$$

where C_p and C_f are the protein concentrations in the permeate and feed solution, respectively.

2.5.3. Measurement of Average Pore Size

The results from the ultrafiltration test were used to calculate the average pore size of the membrane surface. To do so, the molecular weight of solutes with a solute rejection (SR) of more than 80% was used in the equation below [68]:

$$\overline{R} = 100 \left[\frac{\alpha}{\%SR} \right]$$

where $\overline{R}$ is the average radius of the pore size and α is the solute radius, represented by the Stokes radius acquired from the solute molecular weight.

2.6. Molecular Weight Cutoff (MWCO)

There is a linear correlation between MWCO and the membrane's pore size. In every case, the membrane MWCO is investigated by identifying an inert solute of the smallest molecular weight that shows a protein rejection of 80–100% in an ultrafiltration test. In this study, EA was selected as a protein due to its percentage rejection of the PES/SNF-blended membrane [69].

2.7. Antibacterial Experiment

The PSF membranes were evaluated for the Gram-negative bacteria *Escherichia coli* in the untreated river water. Feng et al. [70] and Dadari et al. [71] found that heavy metals such as silver have been verified to be effective in antibacterial performance against Gram-negative *Escherichia coli* and Gram-positive *Staphylococcus aureus*. This is because the presence of silver in bacterial cells may start protein deposition in cells and may deactivate them. Either Gram-negative or Gram-positive experiments yield identical outcomes. Therefore, coliform agar was selected to grow *Escherichia coli* instead of *Staphylococcus aureus*. Untreated water, such as irrigation water or river water in populated areas—with some people having limited access to water supply still needing to fulfill their daily water intake—may have been contaminated by bacteria and viruses [72,73].

Prior to the test, coliform agar was prepared by autoclaving the liquid media (25.6 g of coliform agar in 1 L of pure water) and cooling it down in the Petri dish. During the antibacterial experiment, the fabricated membranes were used to filter out bacteria in the contaminated water (river water). The water sample was taken from the Trasi River in Sleman, Yogyakarta province, Indonesia, which is used for irrigation purposes. The tested membranes were then placed on the surface of the prepared coliform agar. The EMB agar plates were then incubated at 35 °C for 24 h. Table 2 presents the properties of the river water used in this study.

Table 2. Properties of sample water (river water).

pH	TDS (ppm)	Salinity (%)	Electrical Conductivity (µS/cm)	Temperature (°C)
8.5	143	0.01	293	28.1

The number of bacteria present on the membrane surface was analyzed with ImageJ software. For the ImageJ analysis, the images of the membrane surface were converted to 8-bit mode. Brightness and contrast were then adjusted to eliminate noise in the image background. Automatic thresholding was then utilized to convert images to binary (black and white), where the black areas represent the bacterial colonies and white areas represent the surface of the membranes [74,75].

3. Results and Discussion

3.1. Contact Angle Analysis

To examine the effect of SNF on surface hydrophilicity, the contact angles of the fabricated membranes were measured. Figure 3 presents the water contact angle of the

PSF membranes at different concentrations of SNFs. As shown in Figure 3, bare PSF membranes (0 wt.% SNFs) have the highest contact angle, whilst the most hydrophilic surface was achieved at an SNF concentration of 0.1 wt.%. The addition of SNFs to polysulfone membranes caused a reduction in the water contact angle of the modified membranes. This is in accordance with the study of Kasraei et al. [76]. The contact angle of SNF-PSF membranes slightly increased with the increase in SNF concentration. This result indicates that the introduction of SNFs could improve the hydrophilicity of the membrane in comparison to the bare PSF membrane. However, the membrane hydrophobicity increases when the SNF concentration is amplified from 0.1 to 0.4 wt.%. This is possibly due to the creation of stacked nanoflake particles at SNF concentrations that exceed 0.1 wt.%, forming clusters of SNFs on the membrane surface, thus leading to an increase in the water contact angle. Subsequently, the contact angle values were used to calculate the work of adhesion, as presented in Figure 4. The highest value of the work of adhesion was obtained for the PSF membrane with 0.1 wt.% SNFs. These results indicate that at a certain concentration, the addition of SNFs improves the surface hydrophilicity of the membrane to its maximum value.

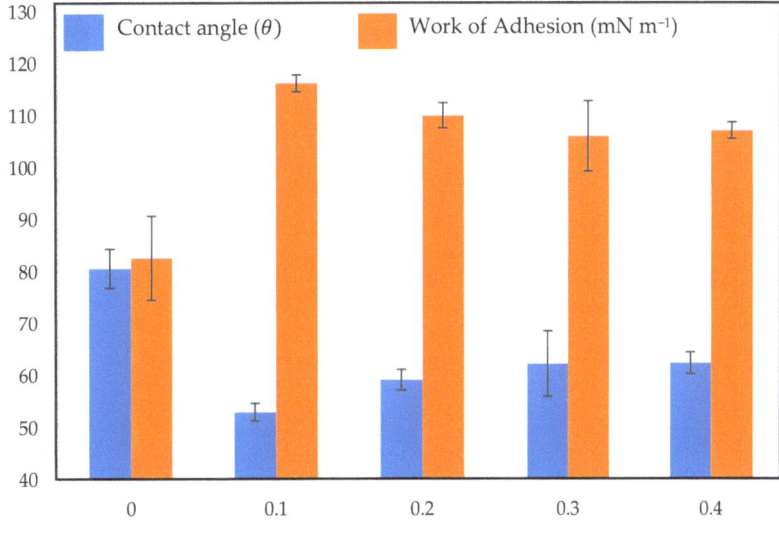

Figure 4. Contact angle and work of adhesion of the fabricated membranes at different SNF concentrations.

Contradictory to the result in this study, Bouchareb et al. [77] reported that the membrane contact angle decreases with increasing silver nanoparticles-graphene oxide (GO) concentration. They combined silver nanoparticles with graphene oxide as the additive in their study. The residual hydrophilic oxygen-based functional groups of graphene oxide and silver nanoparticles were attributed to the decrease in the superficial retention of the membrane.

3.2. Equilibrium Water Content Study

Figure 5 shows the equilibrium water content of the membranes with different concentrations of SNFs. The bare PSF membrane has a maximum water content of $65.4 \pm 0.9\%$. The water content of the membrane slightly increases with the increase in SNF concentration. At an SNF concentration of 0.1 wt.%, the membrane achieved the highest water content ($67.2 \pm 0.6\%$), as presented in Table 3, and became the most hydrophilic membrane compared to the others. When the membrane hydrophilicity increases, the membrane is capable of transporting more water through, hence increasing the water content of the membrane [67,78].

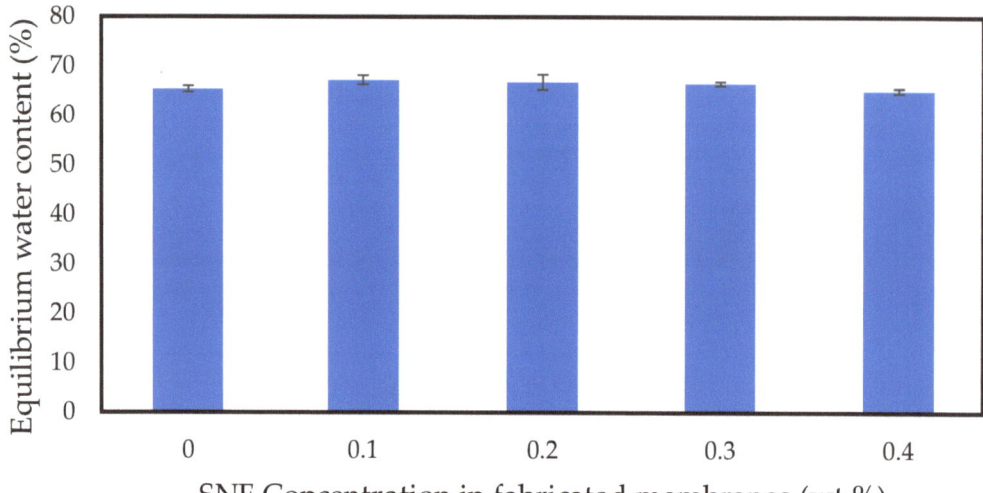

Figure 5. Equilibrium water content of the fabricated membranes at different SNF concentrations.

Table 3. Performance of PSF/SNF membranes.

Membrane Code	Water Content (%)	Average Pore Radius, R (Å)	Egg Albumin Rejection (%)
SNF 0	65.4 ± 0.6	36.9 ± 1.2	84.2 ± 2.6
SNF 0.1	67.2 ± 0.9	36.7 ± 1.6	87.4 ± 3.3
SNF 0.2	66.8 ± 1.5	38.1 ± 2.2	73.3 ± 7.7
SNF 0.3	66.5 ± 0.4	37.8 ± 3.6	78.9 ± 12.1
SNF 0.4	65.1 ± 0.5	37.3 ± 2.3	79.7 ± 10.8

A one-way ANOVA test was conducted to determine if there is a significant difference in the equilibrium water content between at least two groups. The results showed that $p = 0.023$, below the significance level of 0.05, as presented in Table 4. Therefore, we reject H_0 and conclude that there was a statistically significant difference in equilibrium water content between at least two groups.

Table 4. ANOVA statistical analysis of membrane equilibrium water content.

Equilibrium Water Content	Sum of Squares	Degree of Freedom	Mean Square	F Values	Significance
Between Groups	10.533	4	2.633	4.581	0.023
Within Groups	5.748	10	0.575		
Total	16.281	14			

3.3. Pure Water Flux Test Experiments

Figure 6 shows the pure water flux of the fabricated membranes measured at different concentrations of SNFs. As shown in Figure 5, pure water flux was increased from 88 to 116 L m^{-2} h^{-1} bar^{-1} by adding 0.1 wt.% SNFs, which is the highest amongst all the tested membranes. This is attributed to the improvement in membrane hydrophilicity, which is one of the essential factors in the enhancement of pure water flux. This result is consistent with Bilici et al. [79], who stated that the increase in pure water flux was probably due to the increase in hydrophilicity. However, at higher SNF concentrations, the pure water flux started to decline. At higher concentrations, as more SNFs cover and block the membrane pores since extra SNFs are introduced, more SNFs that are present in the membrane tend to cause extra blockage of the pores, compliant with what has been reported by Mollahosseini et al. [80] where pore blockage occurred due to higher SNF content.

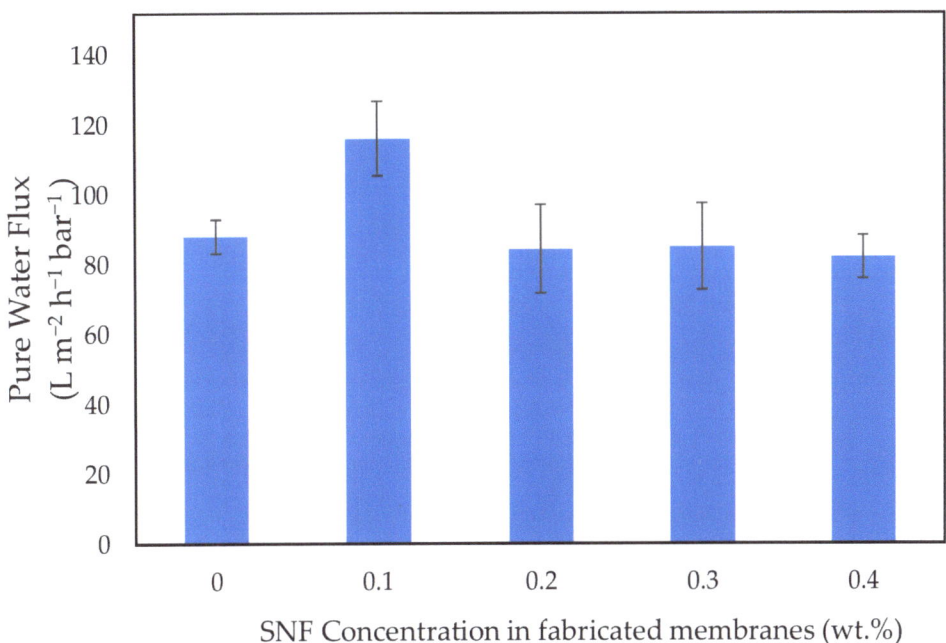

Figure 6. Pure water flux of the fabricated membranes at different SNF concentrations.

3.4. Protein Separation

Figure 7 shows the egg albumin (EA) rejection of the fabricated membranes, whose numerical rejection values are summarized in Table 3. As presented in Figure 6, the PSF membrane at 0.1 wt.% SNFs has a slightly higher EA rejection compared to the bare PSF membrane (0 wt.% SNFs). However, EA rejection of the PSF/SNF blended membrane slightly decreases at higher concentrations of SNFs. It can be observed that the EA rejection is not significantly changed at various concentrations of SNFs, since the addition of SNFs is less than 0.5 wt.% and the percentage increment is only 3.4%.

Contrary to the results of the present study, Mollahosseini et al. [80] reported that less protein macromolecules could pass through the membranes when the silver nanoparticle concentration in the membrane increased. They utilized higher concentrations of silver nanoparticles in their research, which included 0, 0.5, 2 and 4 wt.%. It is possible that the addition of more than 0.5 wt.% silver nanoparticles into the polysulfone dope solution caused pore reduction, further resulting in higher protein rejection.

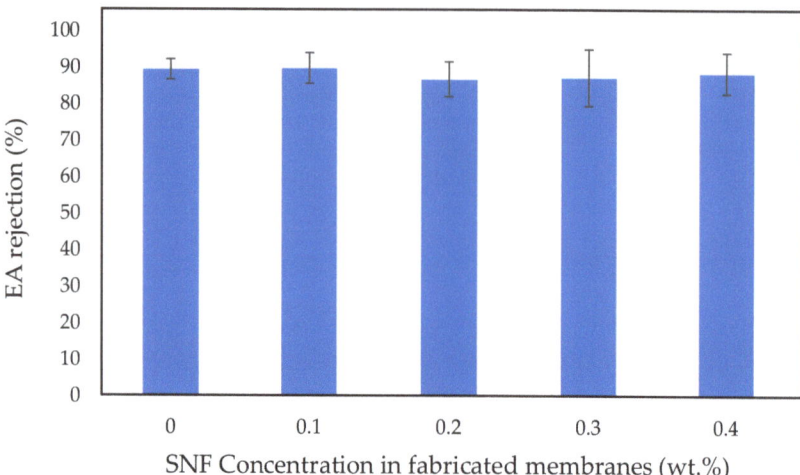

Figure 7. Effect of SNF concentration on EA rejection.

3.5. Measurement of Average Pore Size

Figure 8 presents a summary of the average pore size of the membranes. Based on the EA rejection data, all membranes have a protein rejection value between 86 and 89%, where the PSF membrane with 0.1 wt.% SNFs has the highest rejection (89.9%). These results can be further accounted for given that those membranes have an average pore size in the range of 36 Å–38 Å. There are no significant changes to the membranes' average pore size at SNF concentrations between 0.1 and 0.4 wt.%. This is due to the relatively low concentration of SNFs in the dope solution, as supported by the membrane surface morphology presented in Figure 9.

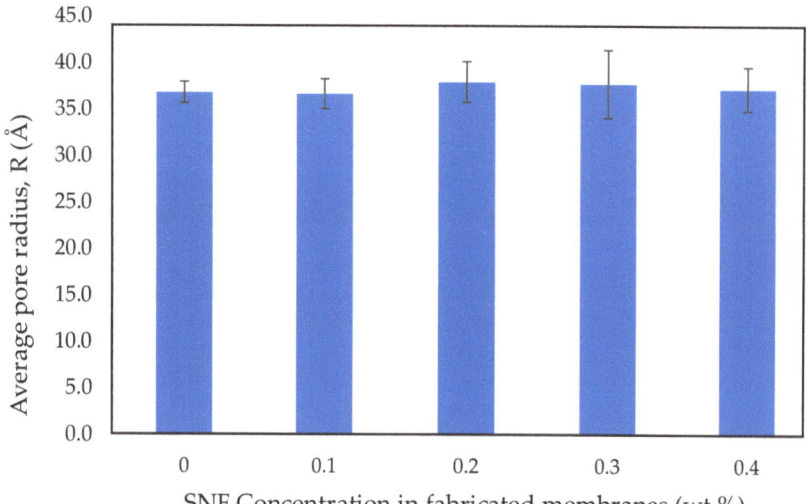

Figure 8. Effect of SNF concentration on the membrane average pore size.

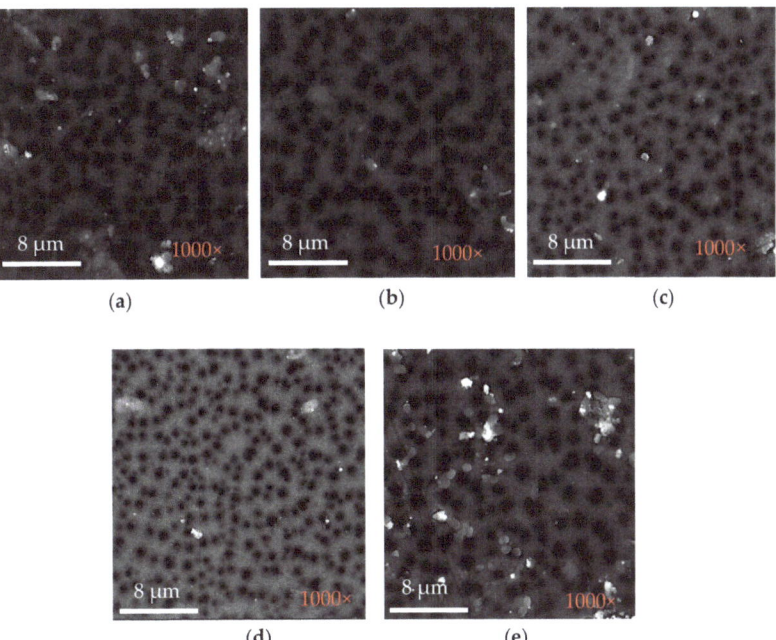

Figure 9. Surface SEM micrographs of (**a**) bare PSF (**b**) PSF-SNF 0.1 wt.%, (**c**) PSF-SNF 0.2 wt.%, (**d**) PSF-SNF 0.3 wt.% and (**e**) PSF-SNF 0.4 wt.% membranes.

In contrast to the result in this study, Li et al. [81] found that the average pore size of the membranes increased due to the increase in the total solid content (1–5 wt.%). This is possibly due to the amount of additives added to the membrane casting solution, where, to suppress defects, the small amount induced a denser skin layer, whereas the large amount of particles added led to a looser skin layer.

3.6. Molecular Weight Cutoff Measurement

In this study, EA with a molecular weight of 45 kDa was used to investigate the membrane MWCO. Based on the results obtained, the membrane MWCO is not affected by the introduction of SNFs at concentrations of 0.1–0.4 wt.% (Table 3). The size of the nanoflakes was still within the range of the permitted concentrations used in the membrane dope solution; therefore, the MWCO value remains the same.

Coincidently, similar results have previously been reported by Mahmoudi et al. [80]. Their research claimed that the addition of silver nanoparticles–graphene oxide nanoplates has no significant effect on the protein rejection of the membranes due to the low concentration of the additives (0.1–0.5 wt.%). Furthermore, the results indicated acceptable rejection values for the membranes.

3.7. Evaluation of the Antimicrobial Activity

Figure 10 presents the results of the antibacterial test for the fabricated membranes at different concentrations of SNFs. The images of the membrane area contaminated by bacterial colonies were taken and quantitatively measured using the image processing method. The images were then binarized and adjusted to detect the colonies' growth. Both the obtained images and the binarized images were compared to ensure detection validity.

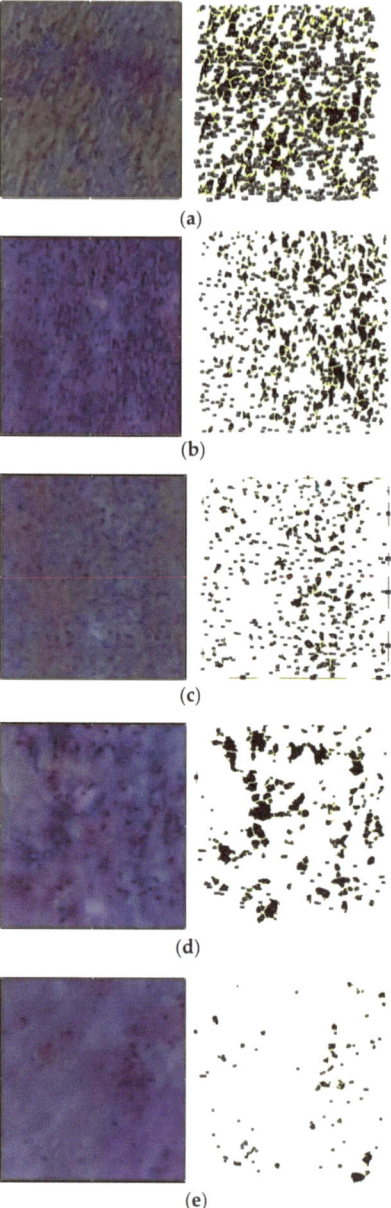

Figure 10. Results of the bacterial test on the fabricated membrane at different SNF concentrations: (**a**) 0 wt.%, (**b**) 0.1 wt.%, (**c**) 0.2 wt.%, (**d**) 0.3 wt.% and (**e**) 0.4 wt.%.

Based on these observations, the number of bacterial colonies on the PSF membrane at 0.0 wt.% SNFs was higher compared to the membranes blended with SNFs. Figure 10b–e verifies that the number of bacterial colonies decreased with the increase in SNF concentration. Quantitative observation indicated that the numbers of bacterial colonies appearing on the membrane surface were 497, 448, 377, 169 and 82 for membranes at SNF concentrations of 0, 0.1, 0.2, 0.3 and 0.4 wt.%, respectively. This result is in agreement with previous findings in which silver nanoparticles decrease the number of bacterial colonies. Silver can modify the metabolism of bacterial activity by altering their DNA so that it loses its replication ability; hence, proteins become inactivated and the bacteria enter the death phase [81,82]. In addition, a significant reduction in the colonies formed was observed at 0.4 wt.% SNFs in the membrane. This can be explained by the higher SNF concentration leading to more surface contact between the silver and the bacteria. These findings conform to the statement that exposing bacteria to a higher concentration of nanoparticles will lead to a more dramatic decrease in the extent of the colonies' growth [83,84].

4. Conclusions

In this study, we provided laboratory evidence for the efficacy of SNF-blended PSF membranes for water filtration purposes. We demonstrated that the introduction of SNFs changed the membrane surface hydrophilicity, pure water flux, protein rejection and antibacterial activity, resulting in PSF membranes with improved properties. The blended membrane with 0.1 wt.% SNFs exhibited low contact angles (55°) and the lowest antibacterial capability against *Escherichia coli*, as well as higher pure water flux and rejection against egg albumin. The best antibacterial property was related to the 0.4 wt.% SNF membrane. The blended PSF membranes developed in this study have a potential use in the water filtration process. Challenges regarding the optimal SNF concentration at their full scale and long-term usage remain and should be addressed in a pilot study.

Author Contributions: Conceptualization, G.S.P. and T.S.; methodology, G.S.P. and M.M.; validation, G.S.P., T.S., A.D.M. and M.M.; formal analysis, M.M., A.D.M. and T.S.; investigation, G.S.P. and M.M.; resources, S.N.M. and M.F.J.; writing—original draft preparation, G.S.P.; writing—review and editing, G.S.P., T.S. and M.M.; visualization, G.S.P. and T.S.; supervision, G.S.P. and M.M.; project administration, M.M., S.N.M. and M.F.J.; funding acquisition, G.S.P., S.N.M. and M.F.J. All authors have read and agreed to the published version of the manuscript.

Funding: This research and the APC were funded by the Hibah SATU Joint Research Scheme, Airlangga University (Grant No. 1307/UN3.15/PT/2021).

Institutional Review Board Statement: Not applicable.

Informed Consent Statement: Not applicable.

Data Availability Statement: The data presented in this study are available from the corresponding author upon request.

Acknowledgments: The authors thank the staff of Lembaga Penelitian Dan Pengabdian Masyarakat, Universitas Airlangga, Indonesia, for their administrative support.

Conflicts of Interest: The authors declare no conflict of interest.

References

1. Mekonnen, M.M.; Hoekstra, A.Y. Four billion people facing severe water scarcity. *Sci. Adv.* **2016**, *2*, e1500323. [CrossRef] [PubMed]
2. Peng, K.; Huang, Y.; Peng, N.; Chang, C. Antibacterial nanocellulose membranes coated with silver nanoparticles for oil/water emulsions separation. *Carbohydr. Polym.* **2022**, *278*, 118929. [CrossRef] [PubMed]
3. Prihandana, G.S.; Sururi, A.; Sriani, T.; Yusof, F.; Jamaludin, M.F.; Mahardika, M. Facile fabrication of low-cost activated carbon bonded polyethersulfone membrane for efficient bacteria and turbidity removal. *Water Pract. Technol.* **2022**, *17*, 102–111. [CrossRef]
4. Deng, Y.; Zhang, N.; Huang, T.; Lei, Y.; Wang, Y. Constructing tubular/porous structures toward highly efficient oil/water separation in electrospun stereocomplex polylactide fibers via coaxial electrospinning technology. *Appl. Surf. Sci.* **2022**, *573*, 151619. [CrossRef]
5. Gupta, R.K.; Dunderdale, G.J.; England, M.W.; Hozumi, A. Oil/water separation techniques: A review of recent progresses and future directions. *J. Mater. Chem. A* **2017**, *5*, 16025–16058. [CrossRef]
6. Zheng, W.; Huang, J.; Li, S.; Ge, M.; Teng, L.; Chen, Z.; Lai, Y. Advanced materials with special wettability toward intelligent oily wastewater remediation. *ACS Appl. Mater. Interfaces* **2021**, *13*, 67–87. [CrossRef]
7. Tanudjaja, H.J.; Hejase, C.A.; Tarabara, V.V.; Fane, A.G.; Chew, J.W. Membrane-based separation for oily wastewater: A practical perspective. *Water Res.* **2019**, *156*, 347–365. [CrossRef]
8. Ge, J.; Zhao, H.-Y.; Zhu, H.-W.; Huang, J.; Shi, L.-A.; Yu, S.-H. Advanced sorbents for oil-spill cleanup: Recent advances and future perspectives. *Adv. Mater.* **2016**, *28*, 10459–10490. [CrossRef]

9. Guha, I.; Varanasi, K. Separating nanoscale emulsions: Progress and challenges to date. *Curr. Opin. Colloid Interf. Sci.* **2018**, *36*, 110–117. [CrossRef]
10. Prihandana, G.S.; Ito, H.; Sanada, I.; Nishinaka, Y.; Kanno, Y.; Miki, N. Permeability and blood compatibility of nanoporous parylene film-coated polyethersulfone membrane under long-term blood diffusion. *J. Appl. Polym. Sci.* **2014**, *131*, 40024. [CrossRef]
11. Qiu, L.; Sun, Y.; Guo, Z. Designing novel superwetting surfaces for highefficiency oil–water separation: Design principles, opportunities, trends and challenges. *J. Mater. Chem. A* **2020**, *8*, 16831–16853. [CrossRef]
12. Prihandana, G.S.; Ito, H.; Nishinaka, Y.; Kanno, Y.; Miki, N. Polyethersulfone membrane coated with nanoporous parylene for ultrafiltration. *J Microelectromech. Syst.* **2012**, *21*, 1288–1290. [CrossRef]
13. Goswami, K.P.; Pugazhenthi, G. Credibility of polymeric and ceramic membrane filtration in the removal of bacteria and virus from water: A review. *J. Environ. Manag.* **2020**, *268*, 110583. [CrossRef] [PubMed]
14. Prihandana, G.S.; Sriani, T.; Mahardika, M. Review of surface modification of nanoporous polyethersulfone membrane as a dialysis membrane. *Int. J. Technol.* **2015**, *6*, 1025–1030. [CrossRef]
15. Wan, L.; Tian, W.; Li, N.; Chen, D.; Xu, Q.; Li, H.; He, J.; Lu, J. Hydrophilic porous PVDF membrane embedded with $BaTiO_3$ featuring controlled oxygen vacancies for piezocatalytic water cleaning. *Nano Energy* **2022**, *94*, 106930. [CrossRef]
16. Drioli, E.; Giorno, L. (Eds.) *Encyclopedia of Membranes*; Springer: Berlin/Heidelberg, Germany, 2016.
17. Al-Gamal, A.Q.; Saleh, T.A.; Alghunaimi, F.I. Nanofiltration Membrane with High Flux and Oil Rejection Using Graphene Oxide/β-Cyclodextrin for Produced Water Reuse. *Mater. Today Commun.* **2022**, *31*, 103438. [CrossRef]
18. Banerjee, P.; Das, R.; Das, P. Mukhopadhyay, Membrane Technology. In *Carbon Nanotubes for Clean Water*; Das, R., Ed.; Carbon Nanostructures; Springer: Cham, Switzerland, 2018; pp. 127–150.
19. Kang, Y.; Obaid, M.; Jang, J.; Ham, M.-H.; Kim, I.S. Novel sulfonated graphene oxide incorporated polysulfone nanocomposite membranes for enhanced-performance in ultrafiltration process. *Chemosphere* **2018**, *207*, 581–589. [CrossRef]
20. Moatmed, S.M.; Khedr, M.H.; El-dek, S.I.; Kim, H.-Y.; El-Deen, A.G. Highly efficient and reusable superhydrophobic/superoleophilic polystyrene@ Fe_3O_4 nanofiber membrane for high-performance oil/water separation. *J. Environ. Chem. Eng.* **2019**, *7*, 103508. [CrossRef]
21. Prihandana, G.S.; Sriani, T.; Mahardika, M. Effect of Polyvinylpyrrolidone on Polyvinylidene Fluoride/Hydroxyapatite Blended Nanofiltration Membranes: Characterization and Filtration Properties. *Recent Pat. Nanotechnol.* **2022**, *17*. [CrossRef]
22. Ashraf, T.; Alfryyan, N.; Ashraf, A.M.; Ahmed, S.A.; Shaban, M. Polyethersulfone Blended with Titanium Dioxide Nanoribbons/Multi-Wall Carbon Nanotubes for Strontium Removal from Water. *Polymers* **2022**, *14*, 1390. [CrossRef]
23. Zhang, Z.Y.; Gou, J.F.; Zhang, X.Y.; Wang, Z.Q.; Xue, N.; Wang, G.; Sabetvand, R.; Toghraie, D. Molecular dynamics simulation of Polyacrylonitrile membrane performance in an aqueous environment for water purification. *J. Water Process. Eng.* **2022**, *47*, 102678. [CrossRef]
24. Ren, X.; Cui, S.; Guan, J.; Yin, H.; Yuan, H.; An, S. PAN@PPy nanofibrous membrane with core-sheath structure for solar water evaporation. *Mater. Lett.* **2022**, *313*, 131807. [CrossRef]
25. Baig, U.; Faizan, M.; Dastageer, M.A. Polyimide based super-wettable membranes/materials for high performance oil/water mixture and emulsion separation: A review. *Adv. Colloid Interface Sci.* **2021**, *297*, 102525. [CrossRef] [PubMed]
26. Song, C.; Rutledge, G.C. Electrospun polyimide fiber membranes for separation of oil-in-water emulsions. *Sep. Purif. Technol.* **2021**, *270*, 118825. [CrossRef]
27. Mu, C.; Chen, H.; Sun, X.; Liu, G.; Yan, K. MoS_2@ZIF-8 doped waterborne polyurethane membranes with water vapor permeable, lubricating, and antibacterial properties. *Prog. Org. Coat.* **2021**, *161*, 106465. [CrossRef]
28. Gu, H.; Li, G.; Li, P.; Liu, H.; Chadyagondo, T.T.; Li, N.; Xiong, J. Superhydrophobic and breathable SiO_2/polyurethane porous membrane for durable water repellent application and oil-water separation. *Appl. Surf. Sci.* **2020**, *512*, 144837. [CrossRef]
29. Singh, A.K.; Kumar, S.; Bhushan, M.; Shah, V.K. High performance cross-linked dehydro-halogenated poly (vinylidene fluoride-co-hexafluoro propylene) based anion-exchange membrane for water desalination by electrodialysis. *Sep. Purif. Technol.* **2020**, *234*, 116078. [CrossRef]
30. Alsalhy, Q.F.; Ibrahim, S.S.; Khaleel, S.R. Performance of vacuum poly(propylene) membrane distillation (VMD) for saline water desalination. *Chem. Eng. Process.* **2017**, *120*, 68–80. [CrossRef]
31. Wang, T.; Gao, F.; Li, S.; Phillip, W.A.; Guo, R. Water and salt transport properties of pentiptycene-containing sulfonated polysulfones for desalination membrane applications. *J. Membr. Sci.* **2021**, *640*, 119806. [CrossRef]
32. Li, X.; Nayak, K.; Stamm, M.; Tripathi, B.T. Zwitterionic silica nanogel-modified polysulfone nanoporous membranes formed by in-situ method for water treatment. *Chemosphere* **2021**, *280*, 130615. [CrossRef]
33. Mousa, H.M.; Alfadhel, H.; Ateia, M.; Abdel-Jaber, G.T.; Gomaa, A.A. Polysulfone-iron acetate/polyamide nanocomposite membrane for oil-water separation. *Environ. Nanotechnol. Monit. MaSN* **2020**, *14*, 100314. [CrossRef]
34. Prihandana, G.S.; Sriani, T.; Muthi'ah, A.D.; Machmudah, A.; Mahardika, M.; Miki, N. Study Effect of nAg Particle Size on the Properties and Antibacterial Characteristics of Polysulfone Membranes. *Nanomaterials* **2022**, *12*, 388. [CrossRef] [PubMed]
35. Zeng, Q.; Wan, Z.; Jiang, Y.; Fortner, J. Enhanced polysulfone ultrafiltration membrane performance through fullerol addition: A study towards optimization. *Chem. Eng. J.* **2022**, *431*, 134071. [CrossRef]
36. Pouresmaeel-Selkjani, P.; Jahanshahi, M.; Peyravi, M. Mechanical, thermal, and morphological properties of nanoporous reinforced polysulfone membranes. *High Perform. Polym.* **2017**, *29*, 759–771. [CrossRef]

37. Benkhaya, S.; Lgaz, H.; Chraibi, S.; Alrashdi, A.A.; Rafik, M.; Lee, H.S.; El Harfi, A. Polysulfone/Polyetherimide Ultrafiltration composite membranes constructed on a three-component Nylon-fiberglass-Nylon support for azo dyes removal: Experimental and molecular dynamics simulations. *Colloids Surf. A Physicochem. Eng. Asp.* **2021**, *625*, 126941. [CrossRef]
38. Mondal, S.; Kumar Majumder, S. Fabrication of the polysulfone-based composite ultrafiltration membranes for the adsorptive removal of heavy metal ions from their contaminated aqueous solutions. *Chem. Eng. J.* **2020**, *401*, 126036. [CrossRef]
39. Sikorska, W.; Milner-Krawczyk, M.; Wasyłeczko, M.; Wojciechowski, C.; Chwojnowski, A. Biodegradation Process of PSF-PUR Blend Hollow Fiber Membranes Using *Escherichia coli* Bacteria—Evaluation of Changes in Properties and Porosity. *Polymers* **2021**, *13*, 1311. [CrossRef] [PubMed]
40. Barambu, N.U.; Bilad, M.R.; Bustam, M.A.; Huda, N.; Jaafar, J.; Narkkun, T.; Faungnawakij, K. Development of Polysulfone Membrane via Vapor-Induced Phase Separation for Oil/Water Emulsion Filtration. *Polymers* **2020**, *12*, 2519. [CrossRef]
41. Ibrahim, G.P.S.; Isloor, A.M.; Inamuddin; Asiri, A.M.; Ismail, A.F.; Kumar, R.; Ahamed, M.I. Performance intensification of the polysulfone ultrafiltration membrane by blending with copolymer encompassing novel derivative of poly(styrene-co-maleic anhydride) for heavy metal removal from wastewater. *Chem. Eng. J.* **2018**, *353*, 425–435. [CrossRef]
42. Dmitrieva, E.S.; Anokhina, T.S.; Novitsky, E.G.; Volkov, V.V.; Borisov, I.L.; Volkov, A.V. Polymeric Membranes for Oil-Water Separation: A Review. *Polymers* **2022**, *14*, 980. [CrossRef]
43. Deepa, K.; Arthanareeswaran, G. Influence of various shapes of alumina nanoparticle in integrated polysulfone membrane for separation of lignin from woody biomass and salt rejection. *Environ. Res.* **2022**, *209*, 112820. [CrossRef]
44. Chung, Y.T.; Ng, L.Y.; Mohammad, A.W. Sulfonated-polysulfone membrane surface modification by employing methacrylic acid through UV-grafting: Optimization through response surface methodology approach. *J. Ind. Eng. Chem.* **2014**, *20*, 1549–1557. [CrossRef]
45. Han, G.; Zhang, S.; Xue, L.; Widjojo, N.; Chung, T.-S. Thin film composite forward osmosis membranes based on polydopamine modified polysulfone substrates with enhancements in both water flux and salt rejection. *Chem. Eng. Sci.* **2012**, *80*, 219–231. [CrossRef]
46. Sutedja, A.; Josephine, C.A.; Mangindaan, D. Polysulfone thin film composite nanofiltration membranes for removal of textile dyes wastewater. *IOP Conf. Ser. Earth Environ. Sci.* **2017**, *109*, 012042. [CrossRef]
47. Nasrollahi, N.; Ghalamchi, L.; Vatanpour, V.; Khataee, A.; Yousefpoor, M. Novel polymeric additives in the preparation and modification of polymeric membranes: A comprehensive review. *J. Ind. Eng. Chem.* **2022**, *109*, 100–124. [CrossRef]
48. Rana, D.; Matsuura, T. Surface Modifications for Antifouling Membranes. *Chem. Rev.* **2010**, *110*, 2448–2471. [CrossRef] [PubMed]
49. Teli, S.B.; Benamor, A.; Nasser, M.; Hawari, A.; Zaidi, S.J.; Ba-abbad, M.; Mohammad, A.W. Effects of amphiphilic pluronic F127 on the performance of PS/SPEEK blend ultrafiltration membrane: Characterization and antifouling study. *J. Water Process Eng.* **2017**, *18*, 176–184. [CrossRef]
50. Kusworo, T.D.; Ariyanti, N.; Utomo, D.P. Effect of nano-TiO$_2$ loading in polysulfone membranes on the removal of pollutant following natural-rubber wastewater treatment. *J. Water Process. Eng.* **2020**, *35*, 101190. [CrossRef]
51. Jiang, Y.; Zeng, Q.; Biswas, P.; Fortner, J.D. Graphene oxides as nanofillers in polysulfone ultrafiltration membranes: Shape matters. *J. Membr. Sci.* **2019**, *581*, 453–461. [CrossRef]
52. Alosaimi, A.M. Polysulfone Membranes Based Hybrid Nanocomposites for the Adsorptive Removal of Hg(II) Ions. *Polymers* **2021**, *13*, 2792. [CrossRef]
53. Daria, M.; Fashandi, H.; Zarrebini, M.; Mohamadi, Z. Contribution of polysulfone membrane preparation parameters on performance of cellulose nanomaterials. *Mater. Res. Express* **2018**, *6*, 015306. [CrossRef]
54. Gnanasekaran, G.; Arthanareeswaran, G.; Mok, Y.S. A high-flux metal-organic framework membrane (PSF/MIL-100 (Fe)) for the removal of microplastics adsorbing dye contaminants from textile wastewater. *Sep. Purif. Technol.* **2021**, *277*, 119655. [CrossRef]
55. Koseoglu-Imer, D.Y.; Kose, B.; Altinbas, M.; Koyuncu, M. The production of polysulfone (PS) membrane with silver nanoparticles (AgNP): Physical properties, filtration performances, and biofouling resistances of membranes. *J. Membr. Sci.* **2013**, *428*, 620–628. [CrossRef]
56. Zodrow, K.; Brunet, L.; Mahendra, S.; Li, D.; Zhang, A.; Pedro, Q.L.; Alvarez, J.J. Polysulfone ultrafiltration membranes impregnated with silver nanoparticles show improved biofouling resistance and virus removal. *Water Res.* **2009**, *43*, 715–723. [CrossRef] [PubMed]
57. Taurozzi, J.S.; Arul, H.; Bosak, V.Z.; Burban, A.F.; Voice, T.C.; Bruening, M.L.; Tarabara, V.V. Effect of filler incorporation route on the properties of polysulfone-silver nanocomposite membranes of different porosities. *J. Membr. Sci.* **2008**, *325*, 58–68. [CrossRef]
58. Demir, D.; Ceylan, S.; Atakav, Y.; Bölgen, N. Synthesis of silver nanoflakes on chitosan hydrogel beads and their antimicrobial potential. *Int. J. Polym.* **2020**, *25*, 421–430. [CrossRef]
59. Pham, V.T.; Truong, V.K.; Quinn, M.D.; Notley, S.M.; Guo, Y.; Baulin, V.A.; Al Kobaisi, M.; Crawford, R.J.; Ivanova, E.P. Graphene Induces Formation of Pores That Kill Spherical and Rod-Shaped Bacteria. *ACS Nano* **2015**, *9*, 8458. [CrossRef] [PubMed]
60. Pal, S.; Tak, Y.K.; Song, J.M. Does the antibacterial activity of silver nanoparticles depend on the shape of the nanoparticle? A study of the Gram-negative bacterium *Escherichia coli*. *Appl. Environ. Microbiol.* **2007**, *73*, 1712–1720. [CrossRef]
61. Zhang, R.; Tang, W.; Gao, H.; Wu, C.; Gray, S.; Lu, X. In-situ construction of superhydrophobic PVDF membrane via NaCl-H$_2$O induced polymer incipient gelation for membrane distillation. *Sep. Purif. Technol.* **2021**, *274*, 117762. [CrossRef]
62. Nunes, S.P.; Peinemann, K.V. Ultrafiltration membranes of PVDF/PMMA. *J. Membr. Sci.* **1992**, *73*, 25–35. [CrossRef]

63. Tamura, M.; Uragami, T.; Sugihara, M. Studies on syntheses and permeabilities of special polymer membranes: 30. Ultrafiltration and dialysis characteristics of cellulose nitrate-poly (vinyl pyrrolidone) polymer blend membranes. *Polymer* **1981**, *22*, 829–835. [CrossRef]
64. Saraswathi, M.S.; Rana, R.; Alwarappan, S.; Gowrishankar, S.; Kanimozhia, P.; SNendran, A. Cellulose acetate ultrafiltration membranes customized with bio-inspired polydopamine coating and in situ immobilization of silver nanoparticles. *New J. Chem.* **2019**, *43*, 4216–4225. [CrossRef]
65. Pace, C.N.; Vajdos, F.; Fee, L.; Grimsley, G.; Gray, T. How to measure and predict the molar absorption coefficient of a protein. *Protein Sci.* **1995**, *4*, 2411–2423. [CrossRef] [PubMed]
66. Kanagaraj, P.; Nagendran, A.; Rana, D.; Matsuura, T.; Neelakandan, S.; Malarvizhi, K. Effects of polyvinylpyrrolidone on the permeation and fouling resistance properties of polyetherimide ultrafiltration membranes. *Ind. Eng. Chem. Res.* **2015**, *54*, 4832–4838. [CrossRef]
67. Sarbolouki, M.N. A general diagram for estimating pore size of ultrafiltration and reverse osmosis membranes. *Sep. Sci. Technol.* **1982**, *17*, 381–386. [CrossRef]
68. Mahendran, R.; Malaisamy, R.; Arthanareeswaran, G.; Mohan, D. Cellulose acetate−poly(ether sulfone) blend ultrafiltration membranes. II. Application studies. *J. Appl. Polym. Sci.* **2004**, *92*, 3659–3665. [CrossRef]
69. Feng, Q.L.; Wu, J.; Chen, G.Q.; Cui, F.Z.; Kim, T.N.; Kim, J.O. A mechanistic study of the antibacterial effect of silver ions on *Escherichia coli* and *Staphylococcus aureus*. *J. Biomed. Mater. Res. A* **2000**, *52*, 662–668. [CrossRef]
70. Dadari, S.; Rahimi, M.; Zinadini, S. Novel antibacterial and antifouling PES nanofiltration membrane incorporated with green synthesized nickel-bentonite nanoparticles for heavy metal ions removal. *Chem. Eng. J.* **2022**, *431*, 134116. [CrossRef]
71. Geldreich, E. Drinking water microbiology—New directions toward water quality enhancement. *Int. J. Food Microbiol.* **1989**, *9*, 295–312. [CrossRef]
72. Chen, L.; Peng, X. Silver nanoparticle decorated cellulose nanofibrous membrane with good antibacterial ability and high water permeability. *Appl. Mater. Today.* **2017**, *9*, 130–135. [CrossRef]
73. Collins, T.J. ImageJ for microscopy. *BioTechniques* **2007**, *43*, S25. [CrossRef] [PubMed]
74. Rasband, W.S. ImageJ. 1997–2015 National Institutes of Health, Bethesda, Maryland, USA. Available online: https://imagej.nih.gov/ij/ (accessed on 27 July 2022).
75. Kasraei, S.; Azarsina, M. Addition of silver nanoparticles reduces the wettability of methacrylate and silorane-based composites. *Braz. Oral Res.* **2012**, *26*, 505–510. [CrossRef] [PubMed]
76. Bouchareb, S.; Doufnoune, R.; Riahi, F.; Cherif-Silini, H.; Belbahri, L. High performance of polysulfone/graphene oxide-silver nanocomposites with excellent antibacterial capability for medical applications. *Mater. Today Commun.* **2021**, *27*, 102297. [CrossRef]
77. Bilici, Z.; Ozay, Y.; Yuzer, A.; Ince, M.; Ocakoglu, K.; Dizge, N. Fabrication and characterization of polyethersulfone membranes functionalized with zinc phthalocyanines embedding different substitute groups. *Colloids Surf. A Physicochem. Eng. Asp.* **2021**, *617*, 126288. [CrossRef]
78. Mollahosseini, A.; Rahimpour, A.; Jahamshahi, M.; Peyravi, M.; Khavarpour, M. The effect of silver nanoparticle size on performance and antibacteriality of polysulfone ultrafiltration membrane. *Desalination* **2012**, *306*, 41–50.
79. Li, J.; Xu, Z.; Yang, H.; Yu, L.; Liu, M. Effect of TiO_2 nanoparticles on the surface morphology and performance of microporous PES membrane. *Appl. Surf. Sci.* **2009**, *255*, 4725–4732. [CrossRef]
80. Mahmoudi, E.; Ng, L.N.; Ba-Abbad, M.M.; Mohammad, A.W. Novel nanohybrid polysulfone membrane embedded with silver nanoparticles on graphene oxide nanoplates. *Chem. Eng. J.* **2015**, *277*, 1–10. [CrossRef]
81. Mocanua, A.; Rusena, E.; Diacona, A.; Isopencua, G.; Mustăteab, G.; Şomoghic, R.; Dinescu, A. Antimicrobial properties of polysulfone membranes modified with carbon nanofibers and silver nanoparticles. *Mater. Chem. Phys.* **2019**, *223*, 39–45. [CrossRef]
82. Ng, L.Y.; Mohammad, A.W.; Leo, C.P.; Hilal, N. Polymeric membranes incorporated with metal/metal oxide nanoparticles: A comprehensive review. *Desalination* **2013**, *308*, 15–33. [CrossRef]
83. Yu, D.-G.; Teng, M.-Y.; Chou, W.-L.; Yang, M.-C. Characterization and inhibitory effect of antibacterial PAN-based hollow fiber loaded with silver nitrate. *J. Membr. Sci.* **2003**, *225*, 115–123. [CrossRef]
84. Bhardwaj, A.K.; Sundaram, S.; Yadav, K.K.; Srivastav, A.L. An overview of silver nanoparticles as promising materials for water disinfection. *Environ. Technol. Innov.* **2021**, *23*, 101721. [CrossRef]

Article

Light-Emitting-Diode-Assisted, Fungal-Pigment-Mediated Biosynthesis of Silver Nanoparticles and Their Antibacterial Activity

Nobchulee Nuanaon [1], Sharad Bhatnagar [2], Tatsuya Motoike [3] and Hideki Aoyagi [1,2,*]

[1] Life Science and Bioengineering, Graduate School of Life and Environmental Sciences, University of Tsukuba, 1-1-1, Tennodai, Tsukuba 305-8572, Ibaraki, Japan; n.nobchulee@gmail.com
[2] Faculty of Life and Environmental Sciences, University of Tsukuba, 1-1-1, Tennodai, Tsukuba 305-8572, Ibaraki, Japan; sharad.bhatnagar88@gmail.com
[3] Management Headquarters, Ushio Lighting Inc., Hatchobori, Chuo-ku, Tokyo 104-0032, Japan; t-motoike@ushiolighting.co.jp
* Correspondence: aoyagi.hideki.ge@u.tsukuba.ac.jp

Citation: Nuanaon, N.; Bhatnagar, S.; Motoike, T.; Aoyagi, H. Light-Emitting-Diode-Assisted, Fungal-Pigment-Mediated Biosynthesis of Silver Nanoparticles and Their Antibacterial Activity. *Polymers* 2022, *14*, 3140. https://doi.org/10.3390/polym14153140

Academic Editors: Md. Amdadul Huq and Shahina Akter

Received: 13 May 2022
Accepted: 23 July 2022
Published: 1 August 2022

Publisher's Note: MDPI stays neutral with regard to jurisdictional claims in published maps and institutional affiliations.

Copyright: © 2022 by the authors. Licensee MDPI, Basel, Switzerland. This article is an open access article distributed under the terms and conditions of the Creative Commons Attribution (CC BY) license (https:// creativecommons.org/licenses/by/ 4.0/).

Abstract: Nanoparticle synthesis, such as green synthesis of silver nanoparticles (AgNPs) using biogenic extracts, is affected by light, which changes the characteristics of particles. However, the effect of light-emitting diodes (LEDs) on AgNP biosynthesis using fungal pigment has not been examined. In this study, LEDs of different wavelengths were used in conjunction with *Talaromyces purpurogenus* extracellular pigment for AgNP biosynthesis. AgNPs were synthesized by mixing 10 mL of fungal pigment with $AgNO_3$, followed by 24 h exposure to LEDs of different wavelengths, such as blue, green, orange, red, and infrared. All treatments increased the yield of AgNPs. The solutions exposed to blue, green, and infrared LEDs exhibited a significant increase in AgNP synthesis. All AgNPs were then synthesized to determine the optimum precursor ($AgNO_3$) concentration and reaction rate. The results indicated 5 mM $AgNO_3$ as the optimum precursor concentration; furthermore, AgNPs-blue LED had the highest reaction rate. Dynamic light scattering analysis, zeta potential measurement, transmission electron microscopy, and Fourier transform infrared spectroscopy were used to characterize the AgNPs. All LED-synthesized AgNPs exhibited an antimicrobial potential against *Escherichia coli* and *Staphylococcus aureus*. The combination of LED-synthesized AgNPs and the antibiotic streptomycin demonstrated a synergistic antimicrobial activity against both bacterial species.

Keywords: antibacterial activity; light-emitting diodes LEDs; nanoparticle biosynthesis; silver nanoparticles; synergistic activity; *Talaromyces purpurogenus*

1. Introduction

In the past few years, increasing application of nanotechnology has ushered a progressive uptick in the field of metal nanoparticle research [1]. Metal nanoparticles have attracted interest in various fields of application, including nanosensors [2], nanocatalysts [3], textiles [4], medicine and cancer therapeutics [5,6], wound healing [7], water treatment [8,9], plant disease control [10,11], and antimicrobials [12], owing to their distinct properties. Both physical and chemical synthesis of metal nanoparticles, including top-down and bottom-up processes, are energy-intensive and adversely affect the environment via residual contaminants [1,13]. Thus, green synthesis of metal nanoparticles using biocompatible sources, such as plant [14,15], bacterial, and fungal [16] extracts, is considered an alternative and cost-effective option for metal nanoparticle production. Biological processes provide a simpler method of synthesis, use less energy, and are environmentally friendly compared with physical and chemical processes [13]. Among various nanoparticles, silver nanoparticles (AgNPs) have attracted wide interest because of their unique properties and advantages, especially their antimicrobial activity, leading to the development of biosynthetic processes wherein different plant and microbial extracts are used as reducing, capping, and stabilizing agents [17].

Several factors are known to drive AgNPs biosynthesis, with metal salt concentration, pH, temperature, and reaction time being the most important [13]. Light has also been shown to affect the biosynthesis of AgNPs when plant and microbial extracts are used as reducing and capping agents [18–20]. The study of light on the synthesis of AgNPs using extracellular polymeric substances of *Chlamydomonas reinhardtii* showed that light induced the production of AgNPs [21]. Similarly, light was used to induce fungal-mediated AgNPs biosynthesis using *Penicillium oxalicum* [22] and *Pleurotus florida* [23], whereas the use of light on size and shape control to AgNPs biosynthesis has not been widely studied. Previously, the use of blue light-emitting diode (LED) as the conversion tool to nanodecahedron AgNPs using chemical synthesis was reported [24]. Moreover, LEDs at different wavelengths (405, 590, and 720 nm) have been shown to control the shape of AgNPs to dodecahedron, triangular, and rod shape respectively during chemical synthesis using I-2959 aqueous solution. This study suggested that light at a specific wavelength can induce changes to the electromagnetic fields of AgNPs, resulting in the shape conversion of particles [25]. As the use of LEDs in AgNPs synthesis has shown a promising effect in chemical synthesis, LEDs light effect on the green synthesis of AgNPs should also be studied in the presence of light-interacting biocomponents such as fungal extracellular pigment extracts acting as a reducing agent, in order to formulate an eco-friendly size and shape control strategy. Fungal extracellular pigments have been studied for the bio-generation of metal nanoparticles owing to their high protein content and secondary metabolite components; therefore, they are considered suitable bio-factories [16]. Fungal pigment extracts from *Talaromyces purpurogenus* (*T. purpurogenus*) and *Monascus* are rich in phytochemicals, have potential for industrial pigment production, and have previously exhibited antiproliferative and antioxidant activities, especially their extracellular pigments [26,27]. This high content of secondary metabolites and proteins in fungal extracellular pigment are responsible for silver salt reduction in the formation of AgNPs. Furthermore, extracellular pigment production facilitates extraction and has been previously shown to reduce silver salts, rendering the pigments suitable for nanoparticle biosynthesis. A previous report on AgNP biosynthesis using *T. purpurogenus* showed that fungal-extracellular-pigment-mediated AgNPs are light-sensitive, with light affecting their size distribution [28]. The report also indicated the possibility of light-assisted AgNP synthesis using a fungal extracellular pigment as a reducing and capping agent. Although the effect of LED light on the size and shape of biosynthesized AgNPs using *T. purpurogenus* extracellular pigments has not yet been established, there have been reports on LED light-assisted size and shape control of AgNPs using I-2959 aqueous solution [25] and salmon DNA extract combined with $NaBH_4$ [29]. Based on data from recent studies, different LED wavelengths might affect fungal pigment extract-mediated AgNPs biosynthesis, which can perhaps be used as a novel technique for AgNP biosynthesis.

In this study, we propose a green AgNP biosynthesis method aided by different LED wavelengths, such as blue (450 nm), green (525 nm), orange (590 nm), red (660 nm), and infrared (850 nm), using *T. purpurogenus* fungal extracellular pigment as a reducing and capping agent. The AgNP production rate with different LEDs was observed using surface plasmon resonance (SPR) with a UV–Vis spectrophotometer. The effect of LEDs on the AgNP size, shape, distribution, and stability were examined. Functional groups related to pigment-mediated AgNP formation were also identified. Further, the antimicrobial potential of different AgNPs synthesized using LEDs against *Escherichia coli* and *Staphylococcus aureus* was determined as well as their synergistic effect in combination with antibiotic streptomycin against both bacteria.

2. Materials and Methods

2.1. Chemicals

Sucrose, hipolypepton, yeast extract (Nihon Seiyaku, Tokyo, Japan), magnesium sulfate heptahydrate ($MgSO_4 \cdot 7H_2O$), dipotassium hydrogen phosphate (K_2HPO_4), sodium nitrate ($NaNO_3$), potassium chloride (KCl), ferrous sulfate heptahydrate ($FeSO_4 \cdot 7H_2O$), ethanol (99%, special grade), and silver nitrate ($AgNO_3$) were purchased from Fujifilm Wako Pure Chemical Corporation (Osaka, Japan).

2.2. Fungal Extracellular Pigment Production and Extraction

T. purpurogenus was obtained from the Cell Cultivation Laboratory (Faculty of Life and Environmental Sciences, University of Tsukuba, Tsukuba, Japan). Fungal extracellular pigment production and extraction were performed as described previously [18], with minor modifications as follows. For *T. purpurogenus* extracellular pigment production, 10 mL of spore suspension was inoculated into the inoculum medium (100 mL of yeast extract 5 g/L, sucrose 30 g/L, K_2HPO_4 1 g/L, and 10 mL/L Czapek extract ($NaNO_3$ 30 g/L, KCl 5 g/L, $MgSO_4 \cdot 7H_2O$ 5 g/L, and $FeSO_4 \cdot 7H_2O$ 0.2 g/L), adjusted to pH 5.0), and incubated at 30 °C with shaking at 150 rpm in the dark for 24 h. Then, 5% (*v/v*) of the inoculum medium was transferred to the production medium (100 mL of sucrose 50 g/L, peptone 25 g/L, K_2HPO_4 2 g/L, $MgSO_4 \cdot 7H_2O$ 2 g/L, and 1% (*v/v*) of salt solution: $NaNO_3$ 1 g/L, KCl 0.05 g/L, and 0.001 g/L $FeSO_4 \cdot 7H_2O$, adjusted to pH 5.0), and incubated for 10 days at 30 °C and 150 rpm in the dark for red pigment production.

Extracellular pigment extraction was conducted by centrifuging 40 mL of the production medium at 6700× *g* and 4 °C (M-160-IV, SAKUMA, Tokyo, Japan) for 20 min, followed by separation of the supernatant from the biomass. The supernatant was collected and the extracellular pigment was extracted by mixing with 70% (*v/v*) ethanol in a 1:1 ratio at 150 rpm for 3 h. Subsequently, the mixture was evaporated using a rotary vacuum evaporator (N-1000 series, Eyela, Tokyo, Japan) to remove the ethanol and concentrate the pigment, and the extracellular pigment was filtered through a 0.45 μm filter (Advantec Toyo Kaisha, Tokyo, Japan). Finally, for AgNP biosynthesis, 1 N NaOH was used to adjust the extracellular pigment pH to pH 10.

2.3. Effect of Different Light Wavelengths on AgNP Biosynthesis

A 10 mL reaction mixture containing extracellular pigment (adjusted to pH 10) with 2 mM $AgNO_3$ was prepared, with a final pigment concentration of 0.5 g/L. The reaction mixture was then kept in a chamber connected to different LED light systems with light intensity of 100 mW/cm^2 (Advantest Optical Power Meter TQ8210, Tokyo, Japan): blue (450 nm), green (525 nm), orange (590 nm), red (660 nm), and infrared (850 nm), at 28 °C. White light and dark conditions were used as control treatments. A magnetic stirrer system was used to ensure the homogeneity of reaction mixtures (Figure 1). AgNP synthesis was evaluated by UV–Vis spectrum scanning of the samples in the range of 300–800 nm using a UV–Vis spectrophotometer (V-550, JASCO, Tokyo, Japan) at various time points over 24 h. Absorbance at a selected wavelength (412 nm) was measured at 0, 2, 4, 8, 12, and 24 h to determine the rate of AgNP synthesis using different LEDs. The colour change of the treatment mixture was visually observed to determine the AgNP synthesis.

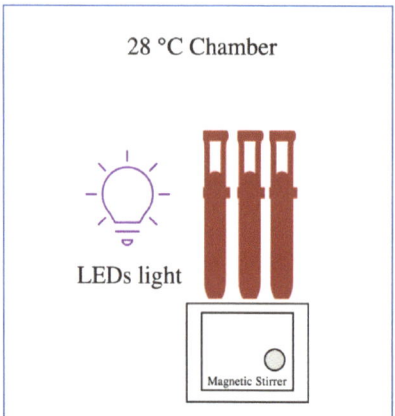

LEDs system: blue (450 nm), green (525 nm), orange (590 nm), red (660 nm) and infrared (850 nm)

Figure 1. Schematic diagram of LED-assisted *T. purpurogenus* extracellular-pigment-mediated AgNP biosynthesis.

2.4. Optimization of Metal Salt Concentration and Time Course Study for AgNP Biosynthesis

The LED light wavelengths selected from previous experiments were used to determine the optimum metal salt precursor concentration for AgNP synthesis. Different concentrations of AgNO$_3$ (2, 5, 10, 15, and 20 mM) were used to synthesize AgNP under the aforementioned reaction conditions. The reaction mixture was then exposed to the selected LEDs in a chamber at 28 °C, and homogeneity was ensured using a magnetic stirrer for 24 h. Dark condition was used as a control treatment. UV–Vis spectrophotometry and visual confirmation were used to confirm AgNP synthesis.

The optimum metal-salt precursor concentration was used to determine the time course of the selected LED light for AgNP synthesis. AgNP synthesis was evaluated at 0, 2, 4, 8, 12, and 24 h using UV–Vis spectrophotometry (300–800 nm). Absorbance at 412 nm was used to express the AgNP yield obtained using different LEDs. Three replicates were analyzed, and data were expressed as mean ± standard error.

2.5. AgNP Characterization

After 24 h of synthesis, AgNPs were centrifuged at 4800× *g* (AG 22331, Eppendorf, Hamburg, Germany) for 10 min, and the resultant pellet was washed twice with deionized water. The washed pellet was resuspended in deionized water by ultrasonication (3510-DTH, Branson, CT, USA) for 10 min before characterization. Transmission electron microscopy (TEM; H-7650, Hitachi, Tokyo, Japan) was used to analyze the size and shape of the biosynthesized AgNPs. Ten microliters of AgNPs (dilution factor = 20) were dropped onto a carbon-coated formvar copper grid, and the sample was air-dried before observation at 80 kV. Dynamic light scattering (DLS; Zetasizer Nano ZS, Malvern Panalytical, Worcestershire, UK) was used to determine the size distribution and stability (zeta potential) of AgNPs. Functional groups relevant to AgNP biosynthesis were analyzed by Fourier transform infrared spectroscopy (FTIR; JASCO FT/IR-6800, Tokyo, Japan) using lyophilized AgNPs mixed with KBr pellets.

2.6. Antimicrobial Activity of Biosynthesized AgNPs

E. coli K 12 and *S. aureus* ATCC6538P obtained from the Cell Cultivation Laboratory (Faculty of Life and Environmental Sciences, University of Tsukuba, Tsukuba, Japan) were used as representative gram-negative and gram-positive bacteria, respectively, for the antimicrobial study. The overnight grown cultures of tested microbes were diluted with autoclaved double-distilled water to reach a McFarland standard of 0.5 and used to determine the antimicrobial activity of biosynthesized AgNPs.

The disk diffusion test was used for prescreening of antimicrobial activity of LED-synthesized AgNPs against both bacterial strains. The test microbes were plated on the agar plate (hipolypepton 10 g/L, yeast extract 2 g/L, and $MgSO_4 \cdot 7H_2O$ 1 g/L). Filter paper disks loaded with AgNPs (120 µg/disk) were placed on an inoculated agar plate, and the plates were incubated at 30 °C (*E. coli* K 12) and 37 °C (*S. aureus*) for 24 h. Zone of inhibition (ZOI) was determined by measuring the area of halo around the AgNPs impregnated disk with no visible microbial growth. The minimum inhibitory concentration (MIC) and minimum bactericidal concentration (MBC) were determined using the serial broth dilution method. A 96-well microplate (Thermo Fisher Scientific, Waltham, MA, USA) containing serial dilution of LED-synthesized AgNPs (0–500 µg/mL) was prepared. Then, the test microbes were used as the inoculum to prepare the 96-well microplate. The plates were incubated at 30 °C (*E. coli* K 12) and 37 °C (*S. aureus*) for 24 h. Streptomycin was used as the standard, whereas media without any antimicrobial agent were used as the positive control. The minimum dilution that showed no growth of tested microbe was selected as the MIC. MBC was determined by plating MIC and lower dilutions of inoculum on an agar plate with the prepared media and incubating under the previously mentioned conditions for 24 h. The highest dilution that showed no microbial growth was selected as the MBC.

Antibacterial activity of the combination of LED-synthesized AgNPs and antibiotics against both bacteria were also evaluated. AgNPs (120 µg) were loaded onto the standard disk of 10 µg streptomycin (BD Sensi-Disc Streptomycin 10, NJ, USA), and the prepared disk was placed in an inoculated agar plate. The plate was incubated at 30 °C (*E. coli* K 12) and 37 °C (*S. aureus*) for 24 h, and then the ZOI was measured. The correlation graph of the logarithmic streptomycin concentration (5–80 µg/mL) and zone of inhibition against both microbes was used to determine the activity of combined treatment, as well as individual treatments of AgNPs and streptomycin [30]. ZOIs of combined treatment and AgNPs alone were expressed in terms of ZOI of equivalent streptomycin concentration using the regression equation: $y = a + b \log(x)$, where y refers to ZOI (mm), x is the streptomycin concentration (µg/mL), and a and b are constants. The synergistic effect was determined when the effect of A + B < C, where A is the AgNPs concentration; B refers to the streptomycin concentration; and C is the combined treatment concentration, expressed in terms of corresponding equivalent streptomycin concentration, calculated from the former regression equation.

The fractional inhibitory test was determined by checkerboard titration assay using a 96-well microplate. The concentration of LED-synthesized AgNPs was 2× MIC to 1/256 MIC and that of streptomycin was 2× MIC to 1/8 MIC. The fractional inhibitory concentration (FIC) index was determined as FIC Index = (A/MICA) + (B/MICB), where A is the MIC of AgNPs in combination, B is the MIC of streptomycin in combination and MICA and MICB are the MIC of each AgNPs and streptomycin individually. The following criterion was used to determine the nature of the effect: FIC $\leq$ 0.5 = synergistic, FIC > 0.5–4 = additive, and FIC > 4 = antagonistic [31].

3. Results and Discussion

3.1. Effect of Different Light Wavelengths on AgNP Biosynthesis

The effect of LEDs of different wavelengths on fungal pigment-mediated AgNP biosynthesis was confirmed by the change in colour of the reaction mixture to brown in all treatments at the 24 h mark. A change from red to brown in the reaction mixture indicated the reduction of Ag^+ to Ag^0 and the consequent formation of AgNPs after exposure to different LEDs [32]. A UV–Vis spectrophotometer was used to determine the SPR of the treatment mixture at an absorbance between 300 and 800 nm. Figure 2a–g shows the SPR of AgNPs biosynthesized by different LEDs. With increasing time, AgNP production was found to increase in all light treatments. The SPR band for all cases were observed between 409 and

430 nm. Different light exposure treatments revealed differences in the maximum wavelength (λ_{max}) and maximum absorbance (A_{max}) after 24 h, as shown in Table 1. Thus, the SPR band of all AgNPs had a wavelength shorter than 430 nm. The specific phenomenon of the SPR band with a wavelength shift near 400 nm indicated the small particle size of AgNPs. Similar results were obtained for citrate-capped AgNPs, where the decreasing SPR band signified a smaller particle size [33]. Light exposure treatments (Figure 2a–e) exhibited higher particle production than that without light (Figure 2g), indicating that light plays a crucial role in AgNP synthesis. Similarly, another report showed that LED exposure enhanced the production of biosynthesized AgNPs when fern rhizome extract was used as a reducing agent [34]. In addition, light-induced AgNP biosynthesis has previously been reported using plant [35–37], microbial [22,23,38], and algal [21,39,40] extracts as bioreducing agents. For AgNP formation, light or directly photoreduction excited metal salt ions mediate electron transfer to Ag^+ and generate Ag^0, which results in the formation of AgNPs [41]. A study on the role of light in green algal extracellular substance-mediated AgNP synthesis showed that light induced the reduction of Ag^+ to Ag^0, whereas algal substance complexes acted as reducing and stabilizing agents to form AgNPs [21]. Similarly, the biocomponents of *T. purpurogenus* extracellular pigment play a critical role in reducing the metal salt coupled with light-induced electron transfer in the formation of AgNPs.

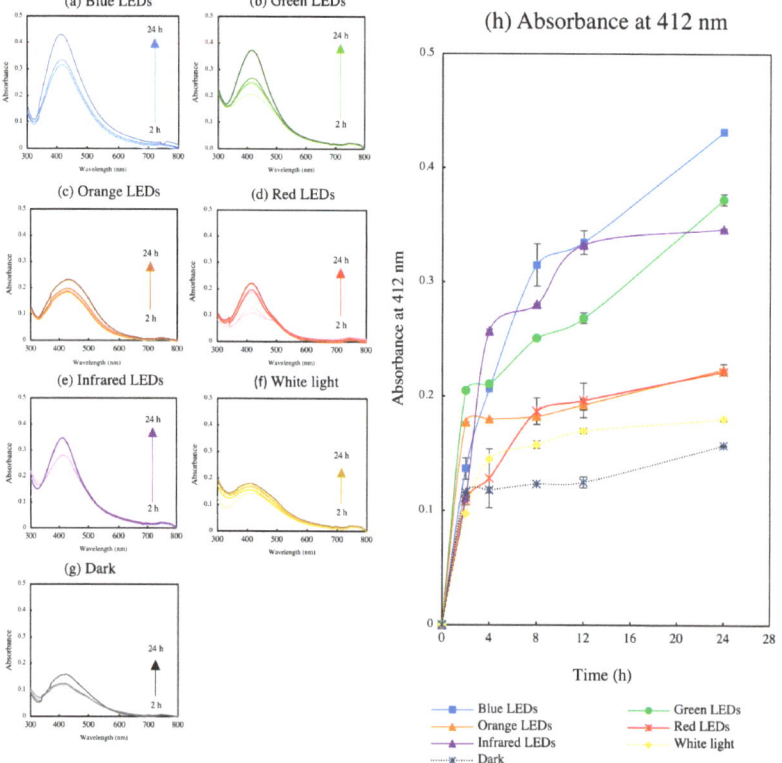

Figure 2. (**a–g**) UV–Vis spectrum of AgNPs obtained by *T. purpurogenus* extracellular pigment-mediated biosynthesis following exposure to lights at different wavelengths and (**h**) their absorbance at 412 nm for 24 h. The error bars indicate standard error (n = 3).

Table 1. The maximum wavelength (λ_{max}) and maximum absorbance (A_{max}) after 24 h during *T. purpurogenus* extracellular-pigment-mediated AgNPs biosynthesis by light exposure at different wavelengths.

Light Source	Maximum Wavelength (λ_{max})	Maximum Absorbance (A_{max})
Blue LEDs	413 nm	0.432 ± 0.010
Green LEDs	415 nm	0.373 ± 0.001
Orange LEDs	430 nm	0.231 ± 0.003
Red LEDs	415 nm	0.222 ± 0.002
Infrared LEDs	411 nm	0.347 ± 0.001
White light	409 nm	0.180 ± 0.001
Dark	415 nm	0.157 ± 0.002

Data are expressed as the means ± SE (n = 3 replicates).

Figure 2h shows the absorbance at 412 nm, representing the AgNP yield during 24 h. All light treatments exhibited an increase in AgNP production with time. After 8 h, the reaction mixture exposed to blue LED showed the highest productivity, which continued increasing with time. At the end of 24 h, the blue LED treatment demonstrated the highest AgNP yield, followed by the infrared and green LED exposure treatments. In contrast, the controls, with white light and dark conditions, showed the lowest AgNP production. The $AgNO_3$ solution exposed to light in the absence of fungal extracellular pigment did not exhibit any SPR band associated with AgNPs when examined by UV–Vis spectra (Figure S1, see Supplementary Materials). Our time course study revealed the influence of blue LED on AgNP synthesis compared with that of other light exposure treatments. Different light wavelengths induced AgNP biosynthesis in a different manner owing to variation in triggering mechanisms on diverse phytochemicals present in the bio-reducing agents [29,34]. Further, AgNPs are metal nanoparticles that respond to electromagnetic (EM) fields of light [25]. Blue LED at 450 nm represent the shortest wavelength, employing the highest energy compared with the other light wavelengths in this experiment. This high energy might be the cause for excitation and rapid reduction of Ag^+ to AgNPs in the extraction complex [34,42]. Another study has also shown that blue light irradiation enhanced AgNP biosynthesis in the presence of cherry extracts [42]. The particle size distribution determined by DLS revealed that AgNPs produced by all treatments ranged between 20 and 50 nm, except for the blue LED-irradiated AgNPs, whose size was between 2 and 15 nm (Figure S2, see Supplementary Materials). AgNPs formed by light irradiation were reported to have a size less than 60 nm [29,34], where blue light tended to increase the size distribution upon photochemical synthesis [24,25]. This work reported a smaller particle size compared with that in other previous reports of AgNPs biosynthesized in the presence of blue light [42]. TEM images revealed that the shape and appearance of AgNPs from all treatments are near-spherical. Nevertheless, a mix of non-spherical shapes was also observed in AgNPs obtained with red and orange LEDs. Previously, shape-controlled DNA-capped AgNPs were reported using LED irradiation using various combinations of DNA and $NaBH_4$ concentrations [29]. The study reported that hexagonal and truncated triangle-shaped AgNPs represented the SPR band around 495–690 nm, different from that for the near-spherical shape with the SPR presented at 416–418 nm. Although our study results indicated that different shape types, such as hexagons, can possibly be obtained with red and orange LEDs, any corresponding SPR band shift might not be observed. HR-TEM performed in another study with different LED irradiation in rhizome extract AgNP biosynthesis revealed mixed AgNPs shapes with no shifts in the corresponding SPR band. All AgNPs showed an SPR band at 410–450 nm, indicating the original SPR band for the near-spherical particles [34].

3.2. Optimization of Precursor Concentration for AgNP Biosynthesis

The optimum concentration of $AgNO_3$ was determined for blue (AgNPs-blue LED), green (AgNPs-green LED), and infrared (AgNPs-infrared LED) LEDs, which exhibited the highest yield among the tested wavelengths. Figure 3a–c shows the optimization of precursor concentration for AgNP biosynthesis. The SPR band showed that an increase in metal salt concentration from 2 to 5 mM resulted in higher AgNP production in the AgNPs-blue LED and AgNPs-green LED light exposure treatments, and a further increase in metal salt concentration (10–20 mM) resulted in decreased AgNP biosynthesis (Figure 3a,b). For the AgNPs-infrared LED treatment, 2 mM metal salt concentration showed the highest productivity, but as the concentration was increased to 5 mM, the production decreased (Figure 3c). An increase in metal salt concentration had previously been shown to be relevant for increasing AgNP production [43]. A higher metal salt concentration causes the greater reduction of Ag^+ with the light-induced formation of AgNPs [34]. After 24 h, the brown colour of AgNPs was observed in all treatments (Figure 3a–c). High metal salt concentrations (15 and 20 mM) resulted in precipitation of the precursor salt under all light exposure treatments.

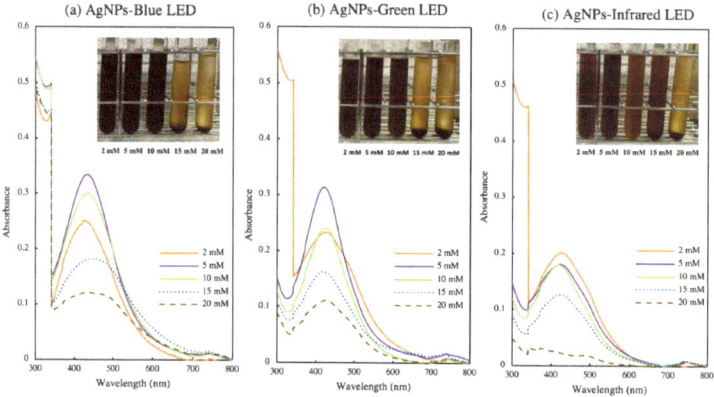

Figure 3. (**a–c**) Optimization of metal salt concentration (2–20 mM) in *T. purpurogenus* extracellular-pigment-mediated AgNP biosynthesis by blue (AgNPs-blue LED), green (AgNPs-green LED), and infrared (AgNPs-infrared LED) LEDs for 24 h.

3.3. Time Course Study of AgNP Biosynthesis

An optimum concentration of 5 mM was used to conduct fungal pigment mediated AgNPs-blue LED, AgNP-green LED, and AgNPs-infrared LED biosynthesis. A time-course study at 0, 2, 4, 8, 12, and 24 h was performed to evaluate the SPR bands of the biosynthesized AgNPs. Figure 4b shows the SPR band of the biosynthesized AgNPs at 24 h. AgNPs-blue LED exhibited the highest yield with a λ_{max} at 425 nm, whereas AgNPs-green LED and AgNPs-infrared LED exhibited λ_{max} at 426 nm and 425 nm, respectively. The sharp SPR band in each treatment indicated the presence of monodispersed spherical AgNPs [25,44]. The bathochromic (red) shift of bands compared with the lower metal salt concentration revealed larger particles [45]. To demonstrate compliance with our first experiment, the effect of different light wavelengths on AgNP biosynthesis absorbance at 412 nm (Figure 4a) was considered to represent AgNP biosynthesis in the time-course study. An increase in the reaction time readily increased AgNP production [34]. AgNPs-blue LED exhibited the highest production from the beginning, followed by AgNPs-green LED and AgNPs-infrared LED. The control treatment in the absence of light showed the lowest AgNP production with respect to time. These data indicate that blue light enhances AgNP biosynthesis. Figure 4c shows the AgNP mixture after 24 h. AgNPs-blue LED presented the darkest brown colour, indicating its high AgNPs synthesis compared with that of AgNPs-green LED and AgNPs-infrared LED [35].

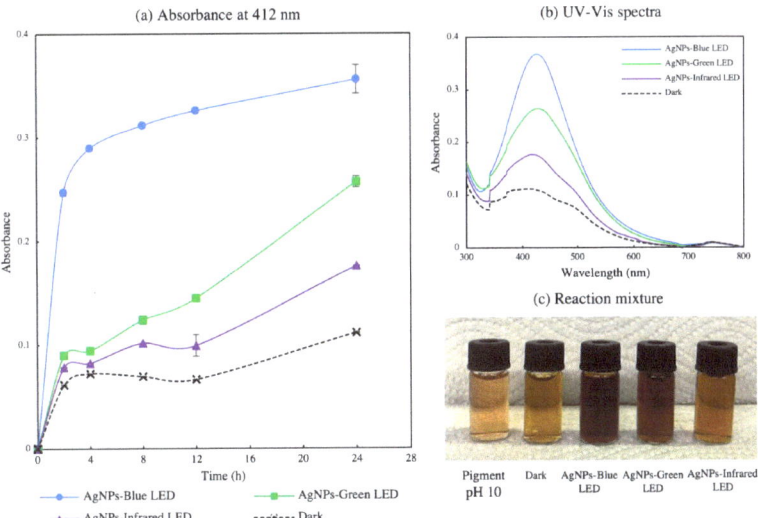

Figure 4. (**a**) Absorbance at 412 nm function of time for the metal salt at the optimized concentration for *T. purpurogenus* extracellular-pigment-mediated AgNPs' biosynthesis by blue (AgNPs-blue LED s), green (AgNPs-green LED), and infrared (AgNPs-infrared LED) LEDs. The error bars indicate standard error (n = 3). (**b**) UV–Vis spectra and (**c**) image of reaction mixture upon 24 h.

3.4. AgNP Characterization

The size distribution of AgNPs was determined using DLS analysis. AgNPs-blue LED exhibited the smallest size distribution, as shown in Figure 5a. AgNPs-blue LED showed the maximum particle size percentage at 37.84 nm, followed by AgNPs-infrared LED and AgNPs-green LED at 43.82 and 58.87 nm, respectively (Figure 5a–c). With an increase in the precursor concentration, the SPR tended to the right, exhibiting a red shift and indicating a larger particle size [45]. However, blue-light-assisted particles retained the smallest particle size. The zeta potential was then analyzed to determine particle stability, as shown in Table 2. The zeta potential of the light-biosynthesized AgNPs showed a negative charge in all treatments. AgNPs-green LED revealed the maximum negative value, followed by AgNPs-infrared LED and AgNPs-blue LED. Stable particles reportedly have zeta potential values greater than -30mV, with a higher negative charge, showing increasing particle stability [46,47]. Figure 5d–f shows the TEM images of light-biosynthesized AgNPs. All AgNPs exhibited a near-spherical shape, with a particle size of less than 50 nm. This near-spherical shape is related to the sharp SPR band, which indicates monodispersed spherical AgNPs [44]. The DLS analysis exhibited a larger particle size compared with the TEM image because of their hydrodynamic diameter related to molecules on the surface of AgNPs [18] that moved in solution owing to Brownian motion [48,49], whereas TEM presented the size of AgNPs [50].

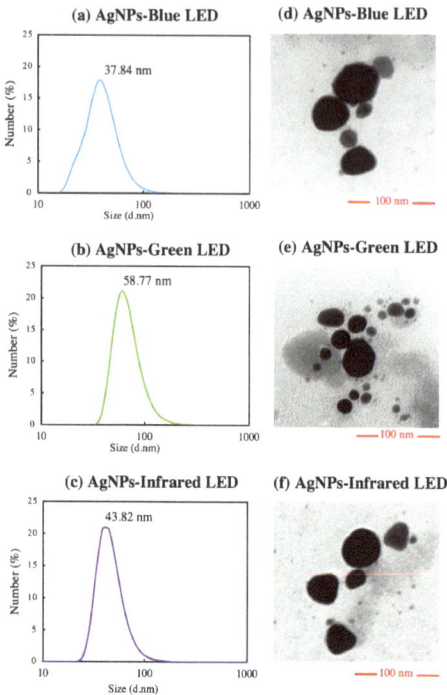

Figure 5. (**a–c**) Dynamic light scattering (DLS) analysis and (**d–f**) transmission electron microscopy (TEM) images of *T. purpurogenus* extracellular-pigment-mediated AgNP biosynthesis using blue (AgNPs-blue LED), green (AgNPs-green LED), and infrared (AgNPs-infrared LED) LEDs.

Table 2. Zeta potential of LED-assisted biosynthesized AgNPs.

Biosynthesized AgNPs	Zeta Potential (mV)
AgNPs-Blue LED	−40.60 ± 1.49
AgNPs-Green LED	−47.90 ± 1.19
AgNPs-Infrared LED	−44.40 ± 0.78

Data are expressed as the means ± SE (n = 3 replicates).

FTIR was used to determine the functional groups present in all light-biosynthesized AgNPs and fungal extracellular pigments at pH 10. Figure 6 shows the different peaks of wavelength numbers appearing in AgNPs and the pigment at pH 10, as determined by FTIR. Table 3 describes the functional groups related to the frequency ranges found in all treatments. The results indicated that light-biosynthesized AgNPs showed no particular difference in functional groups, including the OH hydroxyl group, alkenyl NH amide, and OH–bending phenol groups. The pigment at pH 10 revealed similar functional groups present in AgNPs, but also included C–H alkanes and C=O group esters. A previous study has also reported the same vibration of functional groups present in alkaline *T. purpurogenus* extracellular pigments [18]. The presence of functional groups such as OH hydroxyl, amides, and phenol groups are related to the reduction of Ag^+ to Ag^0 [18,34]. *T. purpurogenus* extracellular pigment is reported to be the source of secondary metabolites, antioxidants, and protein [51]. Thus, secondary metabolites and proteins in the extracellular pigment are correlated with silver ion reduction; metal nanoparticle binding; and nanoparticle formation, capping, and stabilization [34,52]. Similarly, AgNPs derived from exposure to different LEDs showed no significant difference in their functional groups when fern rhizome extract was used for AgNP biosynthesis [34]. This research studied the biosynthesis of AgNP using *T. purpurogenus* extracellular pigment in conjugation with LED irradiation. Biocomponents present in the fungal extracellular pigment, represented by the OH hydroxyl group, amide, and phenol group of secondary metabolites, were considered to be responsible for reduction, capping, and stabilization of Ag$^+$ to AgNP, with LED enhancing the electron transfer during AgNP formation, accelerating the synthesis.

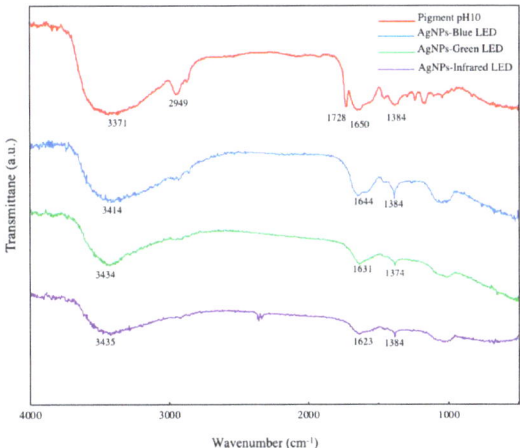

Figure 6. Fourier-transform infrared (FTIR) spectrum of *T. purpurogenus* extracellular pigment at pH 10 and fungus-mediated AgNPs biosynthesized using blue (AgNPs-blue LED), green (AgNPs-green LED), and infrared (AgNPs-infrared LED) LEDs.

Table 3. Frequency range obtained by FTIR and the corresponding functional groups.

Frequency Range (cm^{-1})	Functional Group	Treatment			
		Pigment pH 10	AgNPs-Blue LED	AgNPs-Green LED	AgNPs-Infrared LED
3570–3200	H, OH, hydroxyl group	+	+	+	+
3000–2840	C–H alkane	+	-	-	-
1730–1715	C=O group ester	+	-	-	-
1650–1600	NH amide group	+	+	+	+
1390–1310	OH bending phenol	+	+	+	+

(+): functional group is present, (-): functional group is not present in the treatment.

3.5. Antimicrobial Activity

Figure 7 shows the ZOI of LED-biosynthesized AgNPs. All AgNPs exhibited a better inhibition activity against *E. coli* than *S. aureus*, with AgNPs-green LED showing higher ZOI compared with the other AgNPs. Prescreening of antimicrobial activity of AgNPs using disk diffusion showed their potential against gram-negative and gram-positive bacteria, but MIC and MBC are also required to examine the antimicrobial activity quantitatively [53,54]. Table 4 shows that the MIC and MBC values obtained for all AgNPs against *S. aureus* and *E. coli* were less than 125 µg/mL. AgNPs-green LED and AgNPs-infrared LED showed the best MICs for both pathogens, at an AgNP concentration of 62.5 µg/mL. The MBC against *S. aureus* was 125 µg/mL and was similar for all AgNPs; 62.5 µg/mL of AgNPs-green LED inhibited the growth of gram-negative bacteria such as *E. coli* compared with AgNPs-infrared LED and AgNPs-blue LED, which were required at 125 µg/mL for the same result. Thus, the disk diffusion test and MIC and MBC showed a similar trend for antimicrobial activity of biosynthesized AgNPs, but MIC and MBC required a lower concentration of AgNPs compared with the disk diffusion test. MIC and MBC determined by the broth microdilution method have been used as a common standard for antimicrobial susceptibility test recommended by Clinical & Laboratory Standards Institute (CLSI) and The European Committee on Antimicrobial Susceptibility Testing (EUCAST) [55]. The test is recommended for studying the antimicrobial ability of metal nanoparticles because reliable and quantitative results can be obtained compared with other antimicrobial activity tests such as disk diffusion or agar well diffusion [55,56]. The disk diffusion or well diffusion test is limited to non-fastidious bacteria, gives qualitative results, and is unsuitable for NPs that slowly diffuse in the agar plate [57]. The diffusion of NPs in the agar is the major hurdle in the use of these tests, thus the broth microdilution test was also employed in this research. A study on antimicrobial activity of tea leaf mediated AgNPs on gram-negative foodborne pathogens exhibited an inhibition of bacterial growth using the disk diffusion test with a small clear zone during prescreening, but in the broth microdilution, MIC and MBC showed inhibition of bacteria at lower concentrations of AgNPs [54]. AgNPs have been known for their benefits as antimicrobial agents owing to the potential of silver ions to cause damage to the bacteria cell wall, membrane, and DNA [57]. All synthesized AgNPs showed high inhibition of pathogens. As described previously, the particle sizes of all light-synthesized AgNPs ranged between 30 and 60 nm, and this small size and large surface area influenced the bactericidal action [58,59]. A similar report has revealed that fungus-mediated AgNP particle size less than 60 nm inhibited several tested bacteria, including *E. coli* [33]. The lower effect of AgNPs on the gram-positive *S. aureus* was attributed to its higher cell wall thickness compared with that of gram-negative *E. coli* [60]. Table 2 shows that AgNPs-green LED had the highest negative charge value at -47.90 ± 1.19, whereas that for AgNPs-infrared LED and AgNPs-blue LED was -44.40 ± 0.78 and -40.60 ± 1.49, respectively. Thus, the high negative charge value showed high particle stability with a direct effect on antimicrobial activity. The most stable particles, AgNPs-green LED, exhibited better antimicrobial activity against both bacteria compared with AgNPs-infrared LED and AgNPs-blue LED. Their antimicrobial potential was related to their stability, indicating that highly stabilized particles promote greater antimicrobial activity [61]. Streptomycin exhibited a stronger antimicrobial activity against both pathogens compared with all AgNPs (MIC: 15.62 µg/mL and MBC: 31.25 µg/mL for both bacteria). The agar plates depicting the MBC of biosynthesized AgNPs against *S. aureus* and *E. coli* are shown in Supplementary Figure S3 (see Supplementary Materials). A previous study on MIC and MBC using broth microdilution against *E. coli* and *S. aureus* employing chemically synthesized AgNPs showed that AgNPs inhibited both pathogens' growth at a concentration four times higher than the antibiotic gentamicin [62]. Another study reported the MIC of bacteria-mediated AgNPs against *S. aureus* as 256 µg/mL, more than 30 times higher than the concentration of the antibiotics ampicillin (MIC 1 µg/mL), kanamycin (MIC 8 µg/mL), and tetracycline (MIC 4 µg/mL). Moreover, the antimicrobial activity against *Bacillius subtilis* was reported to be more than 100 times higher than that of antibiotics [63]. Thus, the results from the current research and literature indicate that using AgNPs could inhibit microbes, but requires a higher concentration compared with antibiotics.

Table 4. Antimicrobial activity of LED-biosynthesized AgNPs against *S. aureus* and *E. coli* K 12 evaluated by the minimum inhibitory concentration: MIC and the minimum bactericidal concentration: MBC (µg/mL).

Bacterial Strain	Antimicrobial Treatment							
	AgNPs-Blue LED		AgNPs-Green LED		AgNPs-Infrared LED		Streptomycin	
	MIC	MBC	MIC	MBC	MIC	MBC	MIC	MBC
E. coli	125	125	62.50	62.50	62.50	125	15.62	31.25
S. aureus	125	125	62.50	125	62.50	125	15.62	31.25

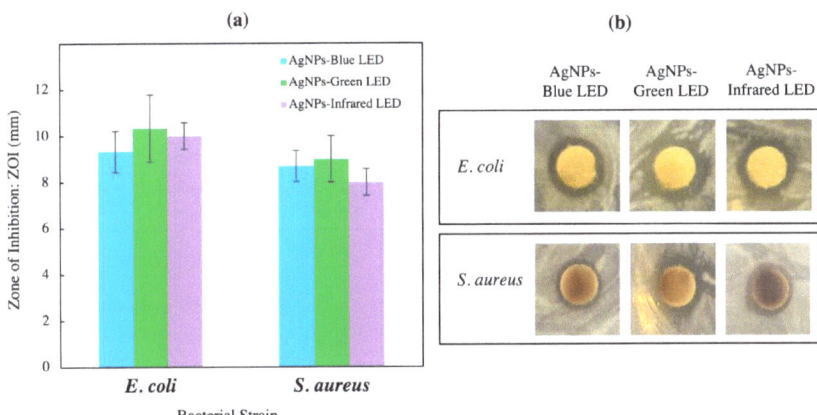

Figure 7. (a) Zone of inhibition: ZOI obtained from LED-biosynthesized AgNPs against *S. aureus* and *E. coli* K 12 evaluated by the disk diffusion method, and (b) images of developed ZOI. Data are expressed as the means ± SE (n = 3 replicates).

The combination of AgNPs and antibiotic (streptomycin) treatment against *E. coli* and *S. aureus* were evaluated by the disk diffusion test using a standard streptomycin disk loaded with LED-biosynthesized AgNPs. All combined treatments showed a synergistic effect on the antimicrobial activity against both bacterial strains compared with streptomycin alone (Figure 8). The combined treatments of AgNPs and streptomycin showed a ZOI of more than 19 mm, whereas streptomycin alone showed a ZOI at 12 ± 2.33 mm and 11.7 ± 1.67 mm against *E. coli* and *S. aureus*, respectively (Table 5). All combined treatments of AgNPs and streptomycin exhibited a slightly larger ZOI in *S. aureus* compared with *E. coli*. Previously, AgNPs and antibiotics alone showed a lower antimicrobial inhibition against gram-positive bacteria compared with gram-negative bacteria; therefore, this result demonstrated that combined treatments enhanced their antimicrobial activity against bacterial cells, especially gram-positive bacteria, *S. aureus*. Generally, AgNPs alone find it difficult to attach to the thick layer cell wall of gram-positive bacteria, but the conjugation of antibiotics with AgNPs might have promoted the bactericidal agent uptake to cells [64]. Moreover, an increase in the antimicrobial activity of the combined treatment might differ with the type of antibiotic, AgNPs, and the target organism [63,64]. The correlation graph of the logarithmic streptomycin concentration and zone of inhibition against both microbes (Figure S4, see Supplementary Materials) were used to obtain a linear relationship and regression equations, y = 9.3446x + 3.6364 (R^2 = 0.98) and y = 10.464x + 0.2859 (R^2 = 0.99), which were used to determine the effect of individual AgNP treatments and combined treatments in terms of equivalent streptomycin concentration against *E. coli* and *S. aureus*, respectively. The evaluation of combined treatment of AgNPs and streptomycin and the sum of individual treatments in terms of equivalent streptomycin concentration is shown in Table 6. All combination treatments exhibited a synergistic effect against both bacterial species. The comparison of combined treatment of AgNPs and streptomycin in terms of equivalent streptomycin concentration indicated that the combination exhibited a stronger activity than the sum of their individual parts, indicating a synergy in their action. A previous study showed that combination treatment of fungus *Trichoderma viride* mediated AgNPs and ampicillin significantly increased the ZOI of *E. coli* and *S. aureus* by up to 70-fold compared with AgNPs and antibiotic alone [64]. Similarly, the study of plant-meditates synthesis AgNPs in conjugation with two types of antibiotics, kanamycin and rifamycin, could increase the ZOI of studied bacterial strains including *B. cereus*, *E. coli*, *S. aureus*, and others [65].

Table 5. Zone of inhibition obtained from the combination of LED-biosynthesized AgNPs and streptomycin against *S. aureus* and *E. coli* K 12 was evaluated by the disk diffusion method.

Bacterial Strain	Zone of Inhibition (mm)						
	Individual Treatment				Combined Treatment with Streptomycin		
	AgNPs-Blue LED	AgNPs-Green LED	AgNPs-Infrared LED	Streptomycin	AgNPs-Blue LED	AgNPs-Green LED	AgNPs-Infrared LED
E. coli	9 ± 0.89	10 ± 1.45	10 ± 0.58	12 ± 2.33	19 ± 0.67	20 ± 1.26	19 ± 0.67
S. aureus	9 ± 0.67	9 ± 1.0	8 ± 0.58	11 ± 1.67	20 ± 0.67	21 ± 0.58	21 ± 0.33

Data are expressed as the means ± SE (n = 3 replicates).

Table 6. The evaluation of combined treatment with AgNPs and streptomycin and the sum of individual treatments expressed as equivalent streptomycin concentration calculated by the regression equation.

Bacterial Strain	Equivalent Streptomycin Concentration (µg/mL)					
	Sum of Individual Treatments (A + B)			Combined Treatment with Streptomycin (C)		
	AgNPs-Blue LED	AgNPs-Green LED	AgNPs-Infrared LED	AgNPs-Blue LED	AgNPs-Green LED	AgNPs-Infrared LED
E. coli	15 ± 2.43	16 ± 4.19	15 ± 2.59	62 ± 5.25	89 ± 17.89	62 ± 5.25
S. aureus	21 ± 5.23	22 ± 5.97	21 ± 5.22	97 ± 12.24	141 ± 21.90	97 ± 12.24

A + B < C = synergistic; A = AgNPs' concentration, B = streptomycin concentration, and C = combined treatment concentration. Data are expressed as the means ± SE (n = 3 replicates).

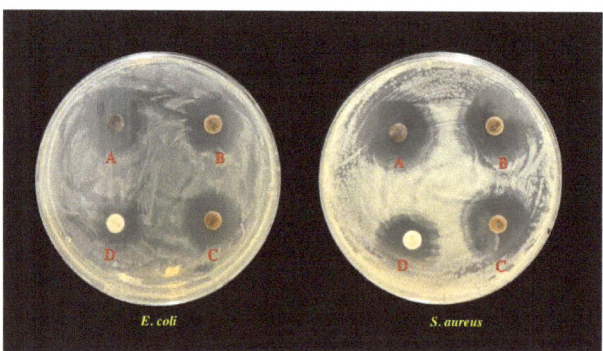

Figure 8. Effect of combination of LED-biosynthesized AgNPs and streptomycin against *S. aureus* and *E. coli* K 12 evaluated by the disk diffusion method: A is AgNPs-blue LED and streptomycin, B is AgNPs-green LED and streptomycin, C is AgNPs-infrared LED and streptomycin, and D is streptomycin alone. The diameter of the petri dish is 9 cm.

The fractional inhibitory concentration test is shown in Table 7. All combination treatments exhibited an FIC index lower than 0.5, indicating their synergistic effect against both *E. coli* and *S. aureus*, in good correlation with the previous disk diffusion assay. The small value of the FIC index showed that the effectiveness of the combined treatments of LED-synthesized AgNPs and streptomycin on the inhibition of bacterial growth of both gram-positive and gram-negative bacteria was better than an antibiotic or LED-biosynthesized AgNPs alone. A previous report on antibiotics polymyxin B and rifampicin in combination with AgNPs showed better antibacterial activity against *Acinetobacter baumannii*, a multidrug-resistant (MDR) bacterial strain, with an FIC index less than 0.5, indicating a synergistic effect of the combined treatment of AgNPs and antibiotics [31]. AgNPs plus antibiotics also showed a significant increase in antibacterial activity against several pathogens compared with AgNPs or antibiotics alone [66], showing the synergistic potential of AgNPs and antibiotics. The synergistic effect of AgNPs combined with antibiotics owes their increased antimicrobial activity to their diverse mechanisms of microbial inhibition [64,67]. An antibiotic like streptomycin causes interruption of the ribosome formation cycle, as well as inhibition and disruption of proteins synthesis in the bacterial cell [68–70], whereas AgNPs possess multiple mechanisms of action against microorganisms. The presence of Ag^+ causes cell membrane damage, and the accumulation of Ag^+ leads to the production of reactive oxygen species (ROS), causing ATP inhibition, membrane leakage, and DNA disruption [71]. The cells suffering from AgNP toxicity exhibit a depletion in oxidative stress defense, including glutathione (GSH) reduction, superoxide dismutase (SOD), and catalase (CAT) enzyme denaturation. The small size of NPs makes it easier for them to pass through the bacterial cell wall, and consequently leads to an increase in antibiotic uptake into the cell [66–68,72]. Moreover, the high surface-area-to-volume ratio of AgNPs benefits the antibiotic binding and promotes their penetrating ability against the cell membrane, leading to an easier delivery to the target site of disruption [64,68,73].

Table 7. Fractional inhibitory concentration (FIC) index of AgNPs and streptomycin combination treatments against *E. coli* and *S. aureus*.

Combined Treatment with Streptomycin	Bacterial Strains			
	E. coli		*S. aureus*	
	FIC Index	Nature of Interaction	FIC Index	Nature of Interaction
AgNPs-Blue LED	0.26 ± 0.04	Synergistic	0.25 ± 0.04	Synergistic
AgNPs-Green LED	0.24 ± 0.05	Synergistic	0.22 ± 0.05	Synergistic
AgNPs-Infrared LED	0.38	Synergistic	0.26 ± 0.01	Synergistic

FIC $\leq$ 0.5 = synergistic, FIC > 0.5–4 = additive, FIC > 4 = antagonistic. Data are expressed as the means ± SE (n = 3 replicates).

4. Conclusions

Simple biosynthesis of *T. purpurogenus* extracellular-pigment-mediated AgNPs was performed using different LED wavelengths. AgNP production varied with the type of LED exposure, wherein blue, green, and infrared LEDs enhanced the biosynthesized AgNP yields compared with the other light sources. All of the light-synthesized AgNPs showed the dominance of near-spherical particles, whereas red and orange LEDs also exhibited the possibility of non-spherical shape induction. The optimum concentration of the metal salt precursor was 5 mM. A time course study on the biosynthesis of AgNPs-blue LED, AgNPs-green LED, and AgNPs-infrared LED was performed. Characterization studies revealed the presence of a near-spherical shape particles with sizes ranging from 30 to 60 nm, with AgNPs-blue LED exhibiting the smallest size. All AgNPs exhibited a zeta potential of more than −30 mV and showed good stability, with AgNPs-green LED showing the highest values of zeta potential, followed by AgNPs-infrared LED, and AgNPs-blue LED. Furthermore, AgNPs showed good inhibitory activity against gram-positive *S. aureus* and gram-negative *E. coli* bacteria, and AgNPs-green LED exhibited better activity than AgNPs-infrared LED and AgNPs-blue LED. The combination of all LED-synthesized AgNPs and an antibiotic, streptomycin, exhibited a synergistic effect on antimicrobial activity against both gram-positive and gram-negative bacteria. Thus, blue, green, and infrared LED-assisted rapid biosynthesis of small-sized, highly stable AgNPs possessing antimicrobial action was accomplished. However, this study showed a limited effect of LED irradiation on shape induction during AgNP biosynthesis. In the future, the synergistic effect of LEDs and different factors, including the biological extract concentration, temperature, and light intensity, on the biosynthesis of AgNPs should be examined to improve particle quality, especially shape induction. Furthermore, the effect of LEDs on AgNPs biosynthesized using different bio-reducing agents, such as plant extracts, needs to be examined.

Supplementary Materials: The following supporting information can be downloaded at https://www.mdpi.com/article/10.3390/polym14153140/s1. Supplementary Figure S1: UV–Vis spectrum of AgNO$_3$ in the absence of *T. purpurogenus* extracellular pigment exposed to different LEDs at 24 h. Supplementary Figure S2: Dynamic light scattering (DLS) and transmission electron microscopy (TEM) analysis of *T. purpurogenus* extracellular-pigment-mediated AgNP biosynthesis by light exposure at different wavelengths for 24 h. Supplementary Figure S3: Minimum bactericidal concentration (MBC) of biosynthesized AgNPs and streptomycin against *S. aureus* and *E. coli*. Supplementary Figure S4: Correlation graph of the logarithm streptomycin concentration and zone of inhibition against *E. coli* and *S. aureus*. Data are expressed as the means ± SE (n = 3 replicates).

Author Contributions: N.N., S.B. and H.A. designed the research; H.A. supervised the research; N.N., S.B., T.M. and H.A. designed the experiments and N.N. performed them; N.N., S.B., T.M. and H.A. analyzed and interpreted the results; N.N., S.B. and H.A. wrote the manuscript. All authors have read and agreed to the published version of the manuscript.

Funding: This work was supported by Hirose Foundation, Japan (2021–2023) (grant to S.B.); Japan Society for the Promotion of Science (Grant-in Aid for Scientific Research B (22H02474)); Sumitomo Electric Industries Group Corporate Social Responsibility Foundation (2018–2023); and Sumitomo Foundation (Grant for Environmental Research Project 2021–2022) (grant to H.A.).

Institutional Review Board Statement: Not applicable.

Informed Consent Statement: Not applicable.

Data Availability Statement: The data presented in this study are available upon request from the corresponding authors.

Acknowledgments: N.N. was supported by a scholarship grant from the Japanese Ministry of Education, Culture, Sports, Science, and Technology and studied on the Trans-world Professional Human Resources Development Program on Food Security & Natural Resources Management (TPHRD), Graduate School of Life and Environmental Sciences, University of Tsukuba, Japan. The manuscript has been edited carefully by native-English-speaking professional editor.

Conflicts of Interest: The authors declare no conflict of interest.

References

1. Abid, N.; Khan, A.M.; Shujait, S.; Chaudhary, K.; Ikram, M.; Imran, M.; Haider, J.; Khan, M.; Khan, Q.; Maqbool, M. Synthesis of nanomaterials using various top-down and bottom-up approaches, influencing factors, advantages, and disadvantages: A review. *Adv. Colloid Interface Sci.* **2022**, *300*, 102597. [CrossRef] [PubMed]
2. Gautam, A.; Komal, P.; Gautam, P.; Sharma, A.; Kumar, N.; Jung, J.P. Recent trends in noble metal nanoparticles for colorimetric chemical sensing and micro-electronic packaging applications. *Metals* **2021**, *11*, 329. [CrossRef]
3. Jin, R. The impacts of nanotechnology on catalysis by precious metal nanoparticles. *Nanotechnol. Rev.* **2012**, *1*, 31–56. [CrossRef]
4. Giannossa, L.C.; Longano, D.; Ditaranto, N.; Nitti, M.A.; Paladini, M.P.F.; Rai, M.; Sannino, A.; Valentini, A.; Cioffi, N. Metal nanoantimicrobials for textile applications. *Nanotechnol. Rev.* **2013**, *2*, 307–331. [CrossRef]
5. Conde, J.; Doria, G.; Baptista, P. Noble metal nanoparticles applications in cancer. *J. Drug Deliv.* **2012**, *2012*, 751075. [CrossRef]
6. Yamada, M.; Foote, M.; Prow, T.W. Therapeutic gold, silver, and platinum nanoparticles. *Wiley Interdiscip Rev. Nanomed. Nanobiotechnol.* **2015**, *7*, 428–445. [CrossRef]
7. Arshad, H.; Saleem, M.; Pasha, U.; Sadaf, S. Synthesis of aloe vera-conjugated silver nanoparticles for use against multidrug-resistant microorganisms. *Electron. J. Biotechnol.* **2022**, *55*, 55–64. [CrossRef]
8. Masungaa, N.; Mmelesia, O.K.; Kefenia, K.K.; Mambaa, B.B. Recent advances in copper ferrite nanoparticles and nanocomposites T synthesis, magnetic properties and application in water treatment: Review. *J. Environ. Chem. Eng.* **2019**, *7*, 103179. [CrossRef]
9. Kim, J.; Van der Bruggen, B. The use of nanoparticles in polymeric and ceramic membrane structures: Review of manufacturing procedures and performance improvement for water treatment. *Environ. Pollut.* **2010**, *158*, 2335–2349. [CrossRef] [PubMed]
10. Jo, Y.-K.; Cromwell, W.; Jeong, H.-K.; Thorkelson, J.; Roh, J.-H.; Shin, D.-B. Use of silver nanoparticles for managing *Gibberella fujikuroi* on rice seedlings. *Crop Prot.* **2015**, *74*, 65–69. [CrossRef]
11. Nayantaraa; Kaurb, P. Biosynthesis of nanoparticles using eco-friendly factories and their role in plant pathogenicity: A review. *Biotechnol. Res. Innov.* **2018**, *2*, 63–73. [CrossRef]
12. Morones, J.R.; Elechiguerra, J.L.; Camacho, A.; Holt, K.; Kouri, J.B.; Ramirez, J.T.; Yacaman, M.J. The bactericidal effect of silver nanoparticles. *Nanotechnology* **2005**, *16*, 2346–2353. [CrossRef]
13. Singh, P.; Kim, Y.J.; Zhang, D.; Yang, D.C. Biological synthesis of nanoparticles from plants and microorganisms. *Trends Biotechnol.* **2016**, *34*, 588–599. [CrossRef]

14. Makarov, V.V.; Love, A.J.; Sinitsyna, O.V.; Makarova, S.S.; Yaminsky, I.V.; Taliansky, M.E.; Kalinina, N.O. Green nanotechnologies synthesis of metal nanoparticles using plants. *Acta Nat.* **2014**, *6*, 35–44. [CrossRef]
15. Osibe, D.A.; Chiejina, N.V.; Ogawa, K.; Aoyagi, H. Stable antibacterial silver nanoparticles produced with seed-derived callus extract of *Catharanthus roseus*. *Artif. Cells Nanomed. Biotechnol.* **2018**, *46*, 1266–1273. [CrossRef]
16. Mandal, D.; Bolander, M.E.; Mukhopadhyay, D.; Sarkar, G.; Mukherjee, P. The use of microorganisms for the formation of metal nanoparticles and their application. *Appl. Microbiol. Biotechnol.* **2006**, *69*, 485–492. [CrossRef]
17. Siddiqi, K.S.; Husen, A.; Rao, R.A.K. A review on biosynthesis of silver nanoparticles and their biocidal properties. *J. Nanobiotechnology* **2018**, *16*, 14. [CrossRef]
18. Bhatnagar, S.; Kobori, T.; Ganesh, D.; Ogawa, K.; Aoyagi, H. Biosynthesis of silver nanoparticles mediated by extracellular pigment from *Talaromyces purpurogenus* and their biomedical applications. *Nanomaterials* **2019**, *9*, 1042. [CrossRef]
19. Neethu, S.; Midhun, S.J.; Sunil, M.A.; Soumya, S.; Radhakrishnan, E.K.; Jyothis, M. Efficient visible light induced synthesis of silver nanoparticles by *Penicillium polonicum* ARA 10 isolated from *Chetomorpha antennina* and its antibacterial efficacy against *Salmonella enterica* serovar *Typhimurium*. *J. Photochem. Photobiol. B* **2018**, *180*, 175–185. [CrossRef]
20. Thatoi, P.; Kerry, R.G.; Gouda, S.; Das, G.; Pramanik, K.; Thatoi, H.; Patra, J.K. Photo-mediated green synthesis of silver and zinc oxide nanoparticles using aqueous extracts of two mangrove plant species, *Heritiera fomes* and *Sonneratia apetala* and investigation of their biomedical applications. *J. Photochem. Photobiol. B* **2016**, *163*, 311–318. [CrossRef]
21. Rahman, A.; Kumar, S.; Bafana, A.; Lin, J.; Dahoumane, S.A.; Jeffryes, C. A mechanistic view of the light-induced synthesis of silver nanoparticles using extracellular polymeric substances of *Chlamydomonas reinhardtii*. *Molecules* **2019**, *24*, 3506. [CrossRef] [PubMed]
22. Du, L.; Xu, Q.; Huang, M.; Xian, L.; Feng, J.-X. Synthesis of small silver nanoparticles under light radiation by fungus *Penicillium oxalicum* and its application for the catalytic reduction of methylene blue. *Mater. Chem. Phys.* **2015**, *160*, 40–47. [CrossRef]
23. Bhat, R.; Deshpande, R.; Ganachari, S.V.; Huh, D.S.; Venkataraman, A. Photo-irradiated biosynthesis of silver nanoparticles using edible mushroom *Pleurotus florida* and their antibacterial activity studies. *Bioinorg. Chem. Appl.* **2011**, *2011*, 650979. [CrossRef] [PubMed]
24. Anh, M.N.T.; Nguyen, D.T.D.; Thanh, N.V.K.; Phong, N.T.P.; Nguyen, D.H.; Nguyen-Le, M.-T. Photochemical synthesis of silver nanodecahedrons under blue LED irradiation and their SERS activity. *Processes* **2020**, *8*, 292. [CrossRef]
25. Stamplecoskie, K.G.; Scaiano, J.C. Light emitting diode irradiation can control the morphology and optical properties of silver nanoparticles. *J. Am. Chem. Soc.* **2010**, *132*, 1825–1827. [CrossRef]
26. Dufossé, L. Red colourants from filamentous fungi: Are they ready for the food industry? *J. Food Compos. Anal.* **2018**, *69*, 156–161. [CrossRef]
27. Pandit, S.G.; Puttananjaiah, M.H.; Harohally, N.V.; Dhale, M.A. Functional attributes of a new molecule-2-hydroxymethyl-benzoic acid 2′-hydroxy-tetradecyl ester isolated from *Talaromyces purpureogenus* CFRM02. *Food Chem.* **2018**, *255*, 89–96. [CrossRef]
28. Bhatnagar, S.; Ogbonna, C.N.; Ogbonna, J.C.; Aoyagi, H. Effect of physicochemical factors on extracellular fungal pigment-mediated biofabrication of silver nanoparticles. *Green. Chem. Lett. Rev.* **2022**, *15*, 276–286. [CrossRef]
29. Sritong, N.; Chumsook, S.; Siri, S. Light emitting diode irradiation induced shape conversion of DNA-capped silver nanoparticles and their antioxidant and antibacterial activities. *Artif. Cells Nanomed. Biotechnol.* **2018**, *46*, 955–963. [CrossRef]
30. Lalpuria, M.; Karwa, V.; Anantheswaran, R.C.; Floros, J.D. Modified agar diffusion bioassay for better quantification of Nisaplin. *J. Appl. Microbiol.* **2012**, *114*, 663–671. [CrossRef]
31. Wan, G.; Ruan, L.; Yin, Y.; Yang, T.; Ge, M.; Cheng, X. Effects of silver nanoparticles in combination with antibiotics on the resistant bacteria *Acinetobacter baumannii*. *Int. J. Nanomed.* **2016**, *11*, 3789–3800. [CrossRef]
32. Kumar, V.; Mohan, S.; Singh, D.K.; Verma, D.K.; Singh, V.K.; Hasan, S.H. Photo-mediated optimized synthesis of silver nanoparticles for the selective detection of Iron (III), antibacterial and antioxidant activity. *Mater. Sci. Eng. C Mater. Biol. Appl.* **2017**, *71*, 1004–1019. [CrossRef]
33. Sharma, A.; Sagar, A.; Rana, J.; Rani, R. Green synthesis of silver nanoparticles and its antibacterial activity using fungus *Talaromyces purpureogenus* isolated from *Taxus baccata* Linn. *Micro Nano Syst. Lett.* **2022**, *10*, 2. [CrossRef]
34. Lee, J.H.; Lim, J.M.; Velmurugan, P.; Park, Y.J.; Park, Y.J.; Bang, K.S.; Oh, B.T. Photobiologic-mediated fabrication of silver nanoparticles with antibacterial activity. *J. Photochem. Photobiol. B* **2016**, *162*, 93–99. [CrossRef]
35. Kumar, V.; Gundampati, R.K.; Singh, D.K.; Jagannadham, M.V.; Sundar, S.; Hasan, S.H. Photo-induced rapid biosynthesis of silver nanoparticle using aqueous extract of *Xanthium strumarium* and its antibacterial and antileishmanial activity. *J. Ind. Eng. Chem.* **2016**, *37*, 224–236. [CrossRef]
36. Patra, J.K.; Das, G.; Baek, K.H. Phyto-mediated biosynthesis of silver nanoparticles using the rind extract of watermelon (*Citrullus lanatus*) under photo-catalyzed condition and investigation of its antibacterial, anticandidal and antioxidant efficacy. *J. Photochem. Photobiol. B* **2016**, *161*, 200–210. [CrossRef]
37. Patra, J.K.; Das, G.; Kumar, A.; Ansari, A.; Kim, H.; Shin, H.S. Photo-mediated biosynthesis of silver nanoparticles using the non-edible accrescent fruiting calyx of *Physalis peruviana* L. fruits and investigation of its radical scavenging potential and cytotoxicity activities. *J. Photochem. Photobiol. B* **2018**, *188*, 116–125. [CrossRef]
38. Dhivahar, J.; Khusro, A.; Elancheran, L.; Agastian, P.; Al-Dhabi, N.A.; Esmail, G.A.; Arasu, M.V.; Kim, Y.O.; Kim, H.; Kim, H.-J. Photo-mediated biosynthesis and characterization of silver nanoparticles using bacterial xylanases as reductant: Role of synthesized product (Xyl-AgNPs) in fruits juice clarification. *Surf. Interfaces* **2020**, *21*, 100747. [CrossRef]

39. Bao, Z.; Lan, C.Q. Mechanism of light-dependent biosynthesis of silver nanoparticles mediated by cell extract of *Neochloris oleoabundans*. *Colloids Surf. B Biointerfaces* **2018**, *170*, 251–257. [CrossRef]
40. Bao, Z.; Cao, J.; Kang, G.; Lan, C.Q. Effects of reaction conditions on light-dependent silver nanoparticle biosynthesis mediated by cell extract of green alga *Neochloris oleoabundans*. *Environ. Sci. Pollut. Res. Int.* **2019**, *26*, 2873–2881. [CrossRef]
41. Sakamoto, M.; Fujistuka, M.; Majima, T. Light as a construction tool of metal nanoparticles: Synthesis and mechanism. *J. Photochem. Photobiol. C Photochem. Rev.* **2009**, *10*, 33–56. [CrossRef]
42. Kumar, B.; Angulo, Y.; Smita, K.; Cumbal, L.; Debut, A. Capuli cherry-mediated green synthesis of silver nanoparticles under white solar and blue LED light. *Particuology* **2016**, *24*, 123–128. [CrossRef]
43. Prathna, T.C.; Chandrasekaran, N.; Raichur, A.M.; Mukherjee, A. Biomimetic synthesis of silver nanoparticles by Citrus limon (lemon) aqueous extract and theoretical prediction of particle size. *Colloids Surf. B Biointerfaces* **2011**, *82*, 152–159. [CrossRef]
44. Steinigeweg, D.; Schlucke, S. Monodispersity and size control in the synthesis of 20–100 nm quasi-spherical silver nanoparticles by citrate and ascorbic acid reduction in glycerol–water mixtures. *Chem. Commun.* **2012**, *48*, 8682–8784. [CrossRef]
45. Mokhtari, N.; Daneshpajouh, S.; Seyedbagheri, S.; Atashdehghan, R.; Abdi, K.; Sarkar, S.; Minaian, S.; Shahverdi, H.R.; Shahverdi, A.R. Biological synthesis of very small silver nanoparticles by culture supernatant of *Klebsiella pneumonia*: The effects of visible-light irradiation and the liquid mixing process. *Mater. Res. Bull.* **2009**, *44*, 1415–1421. [CrossRef]
46. Devadiga, A.; Shetty, K.V.; Saidutta, M.B. Highly stable silver nanoparticles synthesized using *Terminalia catappa* leaves as antibacterial agent and colorimetric mercury sensor. *Mater. Lett.* **2017**, *207*, 66–71. [CrossRef]
47. Saeb, A.T.M.; Alshammari, A.S.; Al-Brahim, H.; Al-Rubeaan, K.A. Production of silver nanoparticles with strong and stable antimicrobial activity against highly pathogenic and multidrug resistant bacteria. *Sci. World J.* **2014**, *2014*, 704708. [CrossRef]
48. Chutrakulwong, F.; Thamaphat, K.; Limsuwan, P. Photo-irradiation induced green synthesis of highly stable silver nanoparticles using durian rind biomass: Effects of light intensity, exposure time and pH on silver nanoparticles formation. *J. Phys. Commun.* **2020**, *4*, 095015. [CrossRef]
49. Zhang, X.F.; Liu, Z.G.; Shen, W.; Gurunathan, S. Silver nanoparticles: Synthesis, characterization, properties, applications, and therapeutic approaches. *Int. J. Mol. Sci.* **2016**, *17*, 1534. [CrossRef]
50. Erjaee, H.; Rajaian, H.; Nazifi, S. Synthesis and characterization of novel silver nanoparticles using *Chamaemelum nobile* extract for antibacterial application. *Adv. Nat. Sci. Nanosci. Nanotechnol.* **2017**, *8*, 025004. [CrossRef]
51. Parul; Thiyam, G.; Dufossé, L.; Sharma, A.K. Characterization of *Talaromyces purpureogenus* strain F extrolites and development of production medium for extracellular pigments enriched with antioxidant properties. *Food Bioprod. Process.* **2020**, *124*, 143–158. [CrossRef]
52. Singhal, G.; Bhavesh, R.; Kasariya, K.; Sharma, A.R.; Singh, R.P. Biosynthesis of silver nanoparticles using *Ocimum sanctum* (Tulsi) leaf extract and screening its antimicrobial activity. *J. Nanoparticle Res.* **2011**, *13*, 2981–2988. [CrossRef]
53. Burt, S. Essential oils: Their antibacterial properties and potential applications in foods–a review. *Int. J. Food Microbiol.* **2004**, *94*, 223–253. [CrossRef] [PubMed]
54. Loo, Y.Y.; Rukayadi, Y.; Nor-Khaizura, M.A.; Kuan, C.H.; Chieng, B.W.; Nishibuchi, M.; Radu, S. In Vitro Antimicrobial Activity of Green Synthesized Silver Nanoparticles Against Selected Gram-negative Foodborne Pathogens. *Front. Microbiol.* **2018**, *9*, 1555. [CrossRef]
55. Balouiri, M.; Sadiki, M.; Ibnsouda, S.K. Methods for in vitro evaluating antimicrobial activity: A review. *J. Pharm. Anal.* **2016**, *6*, 71–79. [CrossRef]
56. Hoseinzadeh, E.; Makhdoumi, P.; Taha, P.; Hossini, H.; Pirsaheb, M.; Rastegar, S.O.; Stelling, J. A review of available techniques for determination of nano-antimicrobials activity. *Toxin Rev.* **2016**, *36*, 18–32. [CrossRef]
57. Lee, H.J.; Lee, S.G.; Oh, E.J.; Chung, H.Y.; Han, S.I.; Kim, E.J.; Seo, S.Y.; Ghim, H.D.; Yeum, J.H.; Choi, J.H. Antimicrobial polyethyleneimine-silver nanoparticles in a stable colloidal dispersion. *Colloids Surf. B Biointerfaces* **2011**, *88*, 505–511. [CrossRef]
58. Gomaa, E.Z. Silver nanoparticles as an antimicrobial agent: A case study on *Staphylococcus aureus* and *Escherichia coli* as models for gram-positive and gram-negative bacteria. *J. Gen. Appl. Microbiol.* **2017**, *63*, 36–43. [CrossRef]
59. Sanchez-Navarro, M.D.C.; Ruiz-Torres, C.A.; Nino-Martinez, N.; Sanchez-Sanchez, R.; Martinez-Castanon, G.A.; De Alba-Montero, I.; Ruiz, F. Cytotoxic and bactericidal effect of silver nanoparticles obtained by green synthesis method using *Annona muricata* aqueous extract and functionalized with 5-Fluorouracil. *Bioinorg. Chem. Appl.* **2018**, *2018*, 6506381. [CrossRef]
60. Rajoka, M.S.R.; Mehwish, H.M.; Zhang, H.; Ashraf, M.; Fang, H.; Zeng, X.; Wu, Y.; Khurshid, M.; Zhao, L.; He, Z. Antibacterial and antioxidant activity of exopolysaccharide mediated silver nanoparticle synthesized by *Lactobacillus brevis* isolated from Chinese koumiss. *Colloids Surf. B Biointerfaces* **2020**, *186*, 110734. [CrossRef]
61. Vanitha, G.; Rajavel, K.; Boopathy, G.; Veeravazhuthi, V.; Neelamegam, P. Physiochemical charge stabilization of silver nanoparticles and its antibacterial applications. *Chem. Phys. Lett.* **2017**, *669*, 71–79. [CrossRef]
62. Thammawithan, S.; Siritongsuk, P.; Nasompag, S.; Daduang, S.; Klaynongsruang, S.; Prapasarakul, N.; Patramanon, R. A Biological Study of Anisotropic Silver Nanoparticles and Their Antimicrobial Application for Topical Use. *Vet. Sci.* **2021**, *8*, 177. [CrossRef]
63. Wypij, M.; Czarnecka, J.; Swiecimska, M.; Dahm, H.; Rai, M.; Golinska, P. Synthesis, characterization and evaluation of antimicrobial and cytotoxic activities of biogenic silver nanoparticles synthesized from *Streptomyces xinghaiensis* OF1 strain. *World J. Microbiol. Biotechnol.* **2018**, *34*, 23. [CrossRef]

64. Fayaz, A.M.; Balaji, K.; Girilal, M.; Yadav, R.; Kalaichelvan, P.T.; Venketesan, R. Biogenic synthesis of silver nanoparticles and their synergistic effect with antibiotics: A study against gram-positive and gram-negative bacteria. *Nanomedicine* **2010**, *6*, 103–109. [CrossRef]
65. Patra, J.K.; Baek, K.H. Antibacterial activity and synergistic antibacterial potential of biosynthesized silver nanoparticles against foodborne pathogenic bacteria along with its anticandidal and antioxidant effects. *Front. Microbiol.* **2017**, *8*, 167. [CrossRef]
66. Hafez, E.H.A.; Ahmed, E.A.; Abbas, H.S.; Di, R.A.S.E. Efficacy of antibiotics combined with biosynthesized silver nanoparticles on some pathogenic bacteria. *Int. J. Sci. Res.* **2017**, *6*, 1294–1303. [CrossRef]
67. Ghaffar, N.; Javad, S.; Farrukh, M.A.; Shah, A.A.; Gatasheh, M.K.; Al-Munqedhi, B.M.A.; Chaudhry, O. Metal nanoparticles assisted revival of Streptomycin against MDRS *Staphylococcus aureus*. *PLoS ONE* **2022**, *17*, e0264588. [CrossRef]
68. Slavin, Y.N.; Asnis, J.; Hafeli, U.O.; Bach, H. Metal nanoparticles: Understanding the mechanisms behind antibacterial activity. *J. Nanobiotechnol.* **2017**, *15*, 65. [CrossRef]
69. Germovsek, E.; Barker, C.I.; Sharland, M. What do I need to know about aminoglycoside antibiotics? *Arch. Dis. Child.-Educ. Pract.* **2017**, *102*, 89–93. [CrossRef]
70. Ball, A.P.; Gray, J.A.; Murdoch, J.M. Antibacterial Drugs today: II. *Drugs* **1975**, *10*, 81–111. [CrossRef]
71. Ahmad, S.A.; Das, S.S.; Khatoon, A.; Ansari, M.T.; Afzal, M.; Hasnain, M.S.; Nayak, A.K. Bactericidal activity of silver nanoparticles: A mechanistic review. *Mater. Sci. Enegy Technol.* **2020**, *3*, 756–769. [CrossRef]
72. Yuan, Y.G.; Peng, Q.L.; Gurunathan, S. Effects of silver nanoparticles on multiple drug-resistant strains of *Staphylococcus aureus* and *Pseudomonas aeruginosa* from mastitis-infected goats: An alternative approach for antimicrobial therapy. *Int. J. Mol. Sci.* **2017**, *18*, 569. [CrossRef]
73. Cavassin, E.D.; de Figueiredo, L.F.; Otoch, J.P.; Seckler, M.M.; de Oliveira, R.A.; Franco, F.F.; Marangoni, V.S.; Zucolotto, V.; Levin, A.S.; Costa, S.F. Comparison of methods to detect the in vitro activity of silver nanoparticles (AgNP) against multidrug resistant bacteria. *J. Nanobiotechnol.* **2015**, *13*, 64. [CrossRef]

Article

The Composites of Polyamide 12 and Metal Oxides with High Antimicrobial Activity

Paulina Latko-Durałek [1,*], Michał Misiak [1], Monika Staniszewska [2], Karina Rosłoniec [2], Marta Grodzik [3], Robert P. Socha [4,5], Marcel Krzan [5], Barbara Bażanów [6], Aleksandra Pogorzelska [6] and Anna Boczkowska [1]

[1] Faculty of Materials Science and Engineering, Warsaw University of Technology, Wołoska 141 Street, 02-507 Warsaw, Poland; michal.misiak.dokt@pw.edu.pl (M.M.); anna.boczkowska@pw.edu.pl (A.B.)
[2] Centre for Advanced Materials and Technologies CEZAMAT, Poleczki 19 Street, 02-822 Warsaw, Poland; monika.staniszewska@pw.edu.pl (M.S.); karina.rosloniec@pw.edu.pl (K.R.)
[3] Institute of Biology, Warsaw University of Life Sciences, Nowoursynowska 166 Street, 02-787 Warsaw, Poland; marta_grodzik@sggw.edu.pl
[4] Research and Development Center of Technology for Industry, Ludwika Waryńskiego 3A Street, 00-645 Warsaw, Poland; robert.socha@cbrtp.pl
[5] Jerzy Haber Institute of Catalysis and Surface Chemistry, Polish Academy of Sciences, Niezapominajek 8 Street, 30-239 Kraków, Poland; marcel.krzan@ikifp.edu.pl
[6] Faculty of Veterinary Medicine, Division of Microbiology, Wrocław University of Environmental and Life Sciences, C.K. Norwida 31 Street, 50-452 Wrocław, Poland; barbara.bazanow@upwr.edu.pl (B.B.); aleksandra.pogorzelska@upwr.edu.pl (A.P.)
* Correspondence: paulina.latko@pw.edu.pl

Citation: Latko-Durałek, P.; Misiak, M.; Staniszewska, M.; Rosłoniec, K.; Grodzik, M.; Socha, R.P.; Krzan, M.; Bażanów, B.; Pogorzelska, A.; Boczkowska, A. The Composites of Polyamide 12 and Metal Oxides with High Antimicrobial Activity. *Polymers* 2022, 14, 3025. https://doi.org/10.3390/polym14153025

Academic Editors: Md. Amdadul Huq and Dimitrios Bikiaris

Received: 16 May 2022
Accepted: 20 July 2022
Published: 26 July 2022

Publisher's Note: MDPI stays neutral with regard to jurisdictional claims in published maps and institutional affiliations.

Copyright: © 2022 by the authors. Licensee MDPI, Basel, Switzerland. This article is an open access article distributed under the terms and conditions of the Creative Commons Attribution (CC BY) license (https:// creativecommons.org/licenses/by/ 4.0/).

Abstract: The lack of resistance of plastic objects to various pathogens and their increasing activity in our daily life have made researchers develop polymeric materials with biocidal properties. Hence, this paper describes the thermoplastic composites of Polyamide 12 mixed with 1–5 wt % of the nanoparticles of zinc, copper, and titanium oxides prepared by a twin-screw extrusion process and injection moulding. A satisfactory biocidal activity of polyamide 12 nanocomposites was obtained thanks to homogenously dispersed metal oxides in the polymer matrix and the wettability of the metal oxides by PA12. At 4 wt % of the metal oxides, the contact angles were the lowest and it resulted in obtaining the highest reduction rate of the *Escherichia coli* (87%), *Candida albicans* (53%), and *Herpes simplex 1* (90%). The interactions of the nanocomposites with the fibroblasts show early apoptosis (11.85–27.79%), late apoptosis (0.81–5.04%), and necrosis (0.18–0.31%), which confirms the lack of toxicity of used metal oxides. Moreover, the used oxides affect slightly the thermal and rheological properties of PA12, which was determined by oscillatory rheology, thermogravimetric analysis, and differential scanning calorimetry.

Keywords: thermoplastic composites; polyamide 12; metal oxides; pathogens; toxicity; rheology; thermal properties; microstructure

1. Introduction

Polymer and polymer-based composites have gained tremendous popularity as materials for the production of everyday appliances such as computers, pens, cases, packaging, toys, etc., due to their low price, good mechanical properties, high corrosion resistance, and easy shaping. However, one of their main drawbacks is the lack of any resistance to microorganisms. With their popularity in our everyday lives and presence in public spaces, pathogens such as fungi, viruses, and bacteria can settle on their surfaces and be spread rapidly by people. Obviously, the pathogens transmission will be dependent on the survivability of the pathogens on the surface. The longer they live, the higher risk that pathogens will be transmitted. It was examined that the virus of SARS-CoV-2 remained much longer (8 h) on a plastic surface compared to the other surfaces such as aluminum, copper, steel, rubber, cloth materials, and paper [1]. Moreover, a whole range of factors such as humidity, temperature, dust, sweat, dead skin cells or food particles contribute to rapid multiplication of pathogens. It is known that pathogens are mainly present in hospitals, and are responsible for numerous infections because they remain on plastic objects such as bedside rails, tables, floors. Besides, thermoplastic polymers are used to produce various types of small medical devices such as syringes, catheters, urine collectors, volumetric infusion pumps, packages as well as personal protective equipment including masks, gloves, and disposable aprons [2,3]. For instance, the Methicillin-Resistant *Staphylococcus aureus* (MRSA), and *Vancomycin-Resistant Enterococci* (VRE) were collected from 59% and 46% of the rooms in hospitals, respectively. Their presence was associated with insufficient surface cleaning and nurse understaffing. Moreover, the research shows that urinary catheters are responsible for more than 80% of infections of the urinary tracts [4,5]. Such a situation has prompted scientists to research the effectiveness of increasing the antimicrobial activity of the polymers and looking for new agents to combat pathogenic organisms.

The methods of creating polymeric materials which inhibit or kill microorganism are constantly being sought. These materials can be produced by a direct mixing of the biocide agent with the polymer during processing (e.g., extrusion) or by applying a coating on the surface of the protected object. The metal particles such as silver and copper are the most commonly used additives. Silver shows high antimicrobial activity, and it is widely used in the packaging industry since it improves the shelf life of food [6]. However, some research confirms that silver ions and silver nanoparticles kill bacteria, human mesenchymal stem cells, and peripheral blood mononuclear cells at the same concentration. Due to the release of ions, silver ions can interact with cell wall components, nucleic acids, or metabolic enzymes. However, some other papers show that the antimicrobial mechanism is based on not direct but indirect damage caused by reactive oxygen species and the subsequent oxidative stress [7,8]. It is important to highlight that other metal particles such as copper possess a lower price and high antimicrobial properties against many types of bacteria, and therefore it is used as a water purifier, antibacterial and antifouling agent. To avoid working with expensive metal particles, metal oxides (e.g., CuO) are commonly used instead of silver salts such as silver nitride and silver acetate. In the group of metal oxides, titanium dioxide (TiO_2) and zinc oxide (ZnO) have been tested as a filler in the polymers to obtain the biocide material. The first one has odor inhibition, a self-cleaning mechanism and low price [9], while ZnO enhances reactive oxygen species responsible for the bacterial inhibition mechanism [10]. It has been tested as an effective biocidal material for the protection of old paper documents [11]. From the other group of biocidal additives, aluminum (Al), iron oxide (FeO), magnesium oxide (MgO), calcium oxide (CaO), carbon nanotubes (CNTs), and even montmorillonite have been used as fillers in the different thermoplastic polymers. The examples of biocidal composites based on thermoplastic polymers are included in Table 1. As it can be seen from the collected data, metal-based additives are still primarily used as biocidal materials.

It should be noted that the antimicrobial activity of the metals and their oxides is not always as good as desired. The biocide composites are often resistant to only some and not all strains of bacteria, fungi, or viruses. It is so due to the too long production–consumption cycle of particles, which lowers their efficiency and relatively small surface area [12]. Hence, to improve the biocidal properties of the metals and their oxides, it has been started to synthesize them with the nanoscale using the established processes in the recently developed nanotechnology science. As a result, the nanoparticles having new physicochemical properties are formed. Compared to microparticles, their surface-area-to-volume ratio is much higher, affecting the significant increase in biological activity. Because of the higher activity of the nanoparticles themselves and resulting faster dissolving in a solution, more metal ions are formed that interact with the pathogens [2].

Many different approaches have been tested to incorporate the biocide nanoparticles in the polymers, and they are mainly dependent on the final applications of the material (nanocomposite) as well as the possibilities and limitations of the method. For instance, the biocidal polymeric fibers can be produced by immersing the pure fibers in the nanoparticles solution or through the melt-spinning process, which does not require using any solvent. The main drawback of the nanoparticles is their tendency to form agglomerates, the presence of which may reduce the final properties of the material. Therefore, they should be destroyed during manufacturing to obtain the homogenous dispersion and distribution of the nanoparticles in the whole polymer matrix. The twin-screw extrusion (melt-blending) is often applied for the mixing of thermoplastics with nanoparticles, since the applied shear rate causes agglomerates destroying, resulting in homogenous dispersion of the nanoparticles [13]. The biocidal nanocomposites in the form of pellets or powder can be further applied in the processes such as injection molding (containers, syringes), melt-spinning (fibers, yarns), single-screw extrusion (filaments for 3D printing), or melt-blown (masks, filters).

Table 1. The examples of antimicrobial thermoplastic composites described in the literature.

Polymer Matrix	Filler Type and Concentration	Pathogen	Activity	Ref.
PA12	2 wt % PHMG-DBS	*Trychophyton mentagrophytes*	qualitative activity	[14]
PA12	2–3 wt % silver nanopowder	*Escherichia. coli Staphylococcus aureus*	qualitative activity	[15]
PA12	4 wt % cuprous oxide-Cu_2O	*E. coli S. aureus*	qualitative activity	[16]
PA11	5 wt % PHMG-DBS	*E. coli Bacillus subtilis*	>99.9%	[17]
PA11	10 wt % Cu	*E. coli*	>99.9%	[18]
PLA	1–5 wt % montmorillonite	*S. aureus Enterococcus. faecalis E. coli*	lack of activity	[19]
PLA	8 wt % TiO_2	*Aspergillus fumigatus E. coli*	>94%	[17]
PANI	1 wt % and 3 wt % $AgNO_3$	*S. aureus*	over 99%	[20]
PP	TiO_2 coated with 1 wt % of nano-Ag	*S. aureus*	over 99%	[3]
PP	5 wt % TiO_2-nanotubes	*E. coli*	17%	[21]
LDPE	10 wt % Organo Modified CaO	*E. coli*	99.9%	[22]
PAN	3 wt % $AgNO_3$/3 wt %TiO_2	*S. aureus*	over 99%	[23]

PA—Polyamide, PLA—Poly(lactid acid); PANI—Poly(aniline); PP—polypropylene; PAN—poly(acrylonitrile); PHMG-DBS—poly(hexamethylene guanidine) dodecylbenzenesulfonate.

As confirmed in much research, nanoparticles are responsible for the antimicrobial properties in the polymer/metal nanocomposites. However, their effectiveness is dependent on the number of metal ions formed in the presence of water. According to the mechanism described by Palza for the bacteria [2], water molecules coming from the bacteria medium diffuse into the surface of the nanoparticles. Then, water and dissolved oxygen cause the dissolution or corrosion processes when they reach the nanoparticles' surface, leading to the formation of the ions. Moving metal ions reach the composite surface, damage the bacteria membrane and diffuse into the pathogen.

It should be noted that the biocidal properties of the polymer nanocomposites are dependent on the type of polymer used, especially its polarity and crystallinity. The highest number of metal ions is formed in more polar (hydrophilic) polymer matrices having lower crystallinity. It is associated with the ease of water absorption and movement of the macromolecules chains, which do not prevent reaching the surface of the water molecules [24].

This paper presents the characterization of the polymer/metal nanocomposites based on PA12 and the mixture of the nanoparticles of metal oxides-CuO, ZnO, and TiO_2 obtained by the melt-blending process. Such composites containing different types of metal oxides have not yet been described in the literature, which emphasizes the high novelty of the conducted research. It was examined how much the metal oxides affect the rheological and thermal properties of PA12, and the results were related to the polymer wettability and dispersion of the nanoparticles. The biological tests included the activity against representative bacteria (*Escherichia coli*), fungi (*Candida albicans*), and viruses (*Herpes simplex* virus type 1 and *Adenovirus 5*). Additionally, in vitro cytotoxicity of the nanocomposites and the programmed cell death were assessed. Because most of the polymer/metal nanocomposites described in the literature are tested against only one type of pathogens (see Table 1), the presented research is valuable for current challenges with the high activity of the microorganisms.

2. Materials and Methods

2.1. PA12 Composites Preparation

As a polymer matrix, PA12 was supplied in pellet form from the Evonik company (Essen, Germany), with the trade name Vestamid Z7321. It has the density of 1.01 g/cm^3, water absorption of 1.5%, a melting point of 179 °C, and a processing temperature between 190 and 240 °C. The pellets of PA12 were initially dried in a vacuum oven at 90 °C for at least 24 h before further usage. As an antimicrobial filler, a mixture of CuO, ZnO, and TiO_2 powders with the trade name ACRAZ-172T was used as a filler. It is in the form of a light black powder. The average grain size of these oxides was between 300 and 1000 nm, a melting point of 962 °C, and a boiling point of 2200 °C. Three oxides were intentionally applied as a filler to PA12 to strengthen antimicrobial activity of the composite. Copper oxide shows high antimicrobial activity at its surface. Zinc oxide is much less toxic, but when in contact with copper oxide, the system shows much stronger activity. Titania is added to the composite to increase wettability of the surface, which is necessary for proper and high activity of CuO and ZnO.

As presented in Figure 1, PA12 composites were produced by mixing the polymer pellets and the oxides powder using a laboratory twin-screw extruder HAAKE MiniLab (ThermoFisher Scientific, Waltham, MA, USA) equipped with a bypass cycle. The concentration of the metal oxides was 1, 2, 3, 4, and 5 wt %. The composites were extruded at 190 °C, with the rotations of the screws of 100 rpm and a mixing time (residence time) of 5 min. The fabricated pellets were afterwards processed into the round specimens with 25 mm diameter and 0.1 mm thickness using a HAAKE Mini Jet Piston Injection Molding System (ThermoFisher Scientific, Waltham, MA, USA). The parameters of the injection molding were as follows: 220 °C—the temperature of the barrel, 50 °C—the mould temperature, 700 bars and 10 s injection pressure and time, and 600 bars and 8 s—the post-processing injection pressure and time. The produced rounds were used in the rheological and antimicrobial tests as well as for the microstructure observations. Using a computer microtomography, it was also determined that the pores content in the composites was lower than 2%.

Figure 1. The approach used for the nanocomposite manufacturing.

2.2. The Analysis of the Physicochemical Properties

The wettability of the metal oxides by PA12 pellet grains was analyzed using a Drop Shape Analyser apparatus DSA100M(Kruss GmbH, Hamburg, Germany) equipped with a high-temperature cell and a digital camera (20 frames per second). The temperature cell enables the heating of samples in the range from 0 to 400 °C and controls the temperature inside the cell with the accuracy of 0.1 °C. Wetting angles were measured by tangents fit. In order to perform the measurement, the PA12 pellet grain was placed on the surface of the oxides powder filler in the center of the cell under room temperature conditions. Then, the heating of the cell was started at a rate of about 10 °C per minute. Thanks to the continuous video observation during the experiment, changes in the texture of the pellet matrix during its melting and interaction with the metal oxides were observed. The first melting effects of the PA12 pellet, the first contact angle of the molten pellet and contact angles for its full possible spread were also noticed.

The surface composition and electronic states of the elements at the surface of the filler patty before and after contact with melted PA12 were analyzed with X-ray photoelectron spectroscopy (XPS). The EA-15 (PREVAC, Rogów, Poland) hemispherical analyzer equipped with dual anode X-ray source RS 40B1 (PREVAC, Rogów, Poland) was used for the surface analysis. The spectrometer was calibrated with Ag, Au and Cu foils according to the ISO 15472:2010 standard. The area of analysis was approximately 2 mm^2 and the depth of analysis was about 10 nm.

The measuring of the quality of the composites and dispersion of the oxides in the PA12 matrix was carried out using a scanning electron microscope (SEM HITACHI TM3000, Tokyo Japan). The samples for the test were prepared by freezing the specimens after injection molding in liquid nitrogen for 10 min and then breaking them. The samples were stuck to the metal measuring tables using double-sided carbon tape so that the fractures faced upwards. The surfaces of the samples were coated with a conductive layer using an electro-deposition method to make them electrically conductive. The microstructure images were collected at magnifications of ×500, ×1000, ×1500, and ×2000. The quantitative analysis of the oxide dispersion was performed using the ImageJ software (Version: 2.0.0, National Institutes of Health and the Laboratory for Optical and Computational Instrumentation, Madison, WI, USA), which allows to count the diameters of agglomerates. For each composition, a minimum of seven images was taken for the calculation. The obtained results were shown in the form of histograms.

Thermogravimetric analysis (TGA) was performed to analyze the thermal stability of the composites. This analysis was carried out using a TGA Q500 (TA Instruments, New Castle, DE, USA) for the samples weighing 10 ± 0.5 mg placed in platinum pans. The samples were heated from 0 to 820 °C in a nitrogen atmosphere with a heating rate of 10 °C/min and a flow rate of 10 mL/min in a chamber and 90 mL/min in an oven. From the obtained curves, the degradation temperature at 5% ($T_{5\%}$) and 10% ($T_{10\%}$) weight loss, and the temperature of the maximum weight loss rate (T_d), were determined.

The effect of the addition of the metal oxides on the thermal properties of PA12 was studied using a Q1000 Differential Scanning Calorimeter (TA Instruments, New Castle, DE, USA). The samples weighing 6.5 ± 0.2 mg were placed in an aluminum hermetic pan. The used program included first heating from −80 °C to 240 °C, then cooling from 240 °C to −80 °C and second heating from −80 °C to 240 °C with a scan rate of 10 °C/min at nitrogen atmosphere. Using the Universal V4.5A TA software, the characteristic temperatures such as melting point (T_m) and crystallization temperature (T_c) were determined. The crystallinity content (X_c) of the PA12 composites were calculated from the following equation number 1:

$$X_c(\%) = \frac{\Delta H_c}{\Delta H_m^\circ (1-x)} \cdot 100\% \qquad (1)$$

where:
ΔH_c is the enthalpy of melting taken as the area under the melting peak from the second heating curve,
ΔH_m° is the melting enthalpy of 100% crystalline PA12, which is 209 J/g [25]
x is the weight fraction of metal oxides.

The processability of the PA12/oxides composites was investigated in terms of their rheological properties such as complex viscosity, storage, and loss modulus. The round specimens having a diameter of 1.5 mm and a thickness of 2 mm (Figure 1) were tested using an oscillatory ARES 4400-0107 rheometer (Rheometric Scientific Inc., TA Instruments, New Castle, DE, USA) in a parallel plate geometry mode. Firstly, the amplitude sweep test as a function of the variable strain γ (0.07–100%) at a constant frequency of 1 Hz was performed. From the linear elastic range of the storage modulus curves, the strain of 10% was selected. With that strain, a dynamic oscillatory stress-controlled rotational test was performed at 190 °C with a frequency sweep from 0.1 to 100 Hz.

2.3. Biological Tests

The prepared composites in the form of the round specimens were analyzed against representative bacteria, fungi, and viruses using the round from the injection molding machine (see Figure 1). The antibacterial activity of samples against *E. coli* was determined using two methods. The qualitative determination was made based on the modified protocol of ISO 20645 and the quantitative determination was based on the AATCC Test Method 100 protocol with modification. In the ISO 20645 method, the samples were placed between two agar layers and the plates were kept for incubation at 37 °C for 24 h. The lower layer contained 10 mL of Luria broth (LB) and the upper layer had 5 mL of LB with 5×10^5 cells/mL of the bacteria. The bacteria came from a previous LB inoculum incubated for 24 h at 37 °C. At the end of incubation, the zone of inhibition formed around the round specimen was measured in millimeters and recorded. In the AATCC Test Method 100, the tested microorganisms were grown in LB at 37 °C for 24 h, and then they were suspended in appropriate media and the cell densities were adjusted to 0.5 McFarland standards at 630 nm wavelength using a spectrophotometric method (1–1.5 × 10^8 cells/mL). The samples were then inoculated with 5 mL microbial suspension and incubated for 24 h. After that contact period, 45 mL of neutralizing solution (phosphate-buffered saline PBS, pH = 7.4) was added to the falcon tubes containing the inoculated treated swatches. After 1 min of shaking, standard serial dilution was performed and 10 μL of each solution were cultured on LB plates and incubated for 24 h at 37 °C. The reduction of microbes was calculated by the following equation: [(no. bacteria on pure PA12—no. bacteria on PA12 with oxides)/no. bacteria on pure PA12] × 100.

Antifungal activity of PA12 was assessed using the following quantitative methods: AATCC TM100-2019 (The American Association of Textile Chemists and Colorists, AATCC TM100-2019, "Test Method for Antibacterial Finishes on Textile Materials: assessment of" and PN-EN ISO 20743:2013-10E "Textiles—Determination of Antibacterial Activity of Textile Products"). Briefly, *C. albicans* ref. strain 90028 ATCC (LGC Standard, Poland) at 5×10^3 to 6×10^5 CFU/mL of YEPD was incubated with swatches at 35 °C for 24 h. Then, tenfold dilutions were plated on YEPD agar and the number of CFU was counted for tested and control (without metal oxides) swatches. Additionally, anti-*C. albicans* 90028 activity was performed using the qualitative method AATCC TM90-2011 (2016)e (The American Association of Textile Chemists and Colorists, AATCC TM90-2011(2016)e, "Test Method for Antibacterial Activity of Textile Materials: Agar Plate"). Briefly, the tested and control swatches were put on top of the solid agar inoculated with blastoconidia (2.6×10^3 CFU/mL) and incubated at 35 °C for 24 h. Zones of inhibition and growth under the tested swatches were compared to the control ones. R% was calculated using Equation: $(A - B)A \times 100\% = \%R$, where A means CFU recovered from the PA12 control inoculated with *C. albicans* and incubated over the 24 h contact period; B means CFU recovered from the PA12 with metal oxide inoculated and incubated over the 24 h contact period.

Virucidal properties were determined using Herpes simplex virus type 1 (HSV-1-ATCC® VR-1493™) and human *Adenovirus 5* (Ad-5 virus—strain Adenoid 75, ATCC VR-5™). Two types of cell lines obtained from American Type Culture Collection—ATCC (Rockville, MD, USA) (A549-human lung carcinoma (ATCC, No. CCL-185™) and HeLa -human cervix carcinoma (ATCC, No. CCL-2™)) were used in this experiment. Dulbecco's Modified Eagle's Medium—DMEM (Lonza, Basel, Switzerland) served as substrate. Media were supplemented with 10% fetal bovine serum (FBS) and 4 mM L-glutamine (Biological Industries, Kibbutz Beit-Haemek, Israel), 100 U/mL of penicillin and 100 g/mL of streptomycin (Sigma-Aldrich, Munich, Germany). The substances were tested using International Standard ISO 21702 (Measurement of antiviral activity on plastics and other non-porous surfaces, Geneva, Switzerland, 2019). This document specifies proper methods for measuring antiviral activity on plastics and other non-porous surfaces of antiviral-treated products against specified viruses. According to the standard (ISO 4.1), alternative viruses to influenza virus and feline calicivirus may be used. Therefore, the study was conducted using *Adenovirus 5* commonly found in humans. This pathogen represents non-enveloped viruses, belonging to the group of most difficult to inactivate viruses. The virus included in the test is responsible for colds. The second study was conducted using human herpesvirus type 1. This pathogen represents enveloped viruses. The virus included in the test is responsible for herpes in human. In the laboratory, these strains reach high titers, so it was possible to validate the method quickly. The virucidal efficacy was tested according to ISO 21702:2019 (ISO 7. 1-7. 4). For this purpose, a sample of the test material (ISO 4. 3. 16) treated with SCDLP neutralizer was placed in a sterile petri dish. On the sample prepared in this way, 400 µL of the test virus were spotted and covered with a 2×3 cm layer of polymer. The sample was incubated at 25 °C for 24 h at 90% humidity. After this time, the virus was collected by pipette. The study was conducted in 3 replicates. In parallel, the whole procedure was repeated using a control specimen (pure PA12) as a non-virulent material. In addition, immediately after virus inoculation onto the control round specimen, virus titers were tested (3 replicates). According to ISO 21702:2019, which allows to use the $TCID_{50}$ method instead of the plaque test, serial dilutions up to 10^{-8} of the viruses collected from each disc were prepared. In eight repeats, 50 µL of each dilution were added to the microtiter plate containing a monolayer of confluent A549 or HeLa cells. The plates were observed daily for up to 4 days for the development of viral cytopathic effect, using an inverted microscope (Olympus Corp., Hamburg, Germany; Axio Observer, Carl Zeiss MicroImaging GmbH). The calculation of the infective dose $TCID_{50}$/mL was conducted using the Spearman and Karber method with the following formula:

$$\log_{10} TCID_{50} = x_0 - 0.5 + \Sigma r/n \qquad (2)$$

where:

x_0 is $\log_{10}$ of the lowest dilution with 100% positive reaction,
r is the number of positive determinations of lowest dilution step with 100%, positive and all higher positive dilution steps,

n is the number of determinations for each dilution step.

The programmed death of the L929 ATCC cell line (LGC, Standard, Poland) treated with PA12/metal oxides composites was assessed using flow cytometer BD FACS Lyrics 2L6C with FAC Suite Software 1.4 RUO (BD Biosciences, Mississauga, ON, Canada). Liquid extracts of PA12 containing 3 wt % and 4 wt % oxides were prepared in DMEM (ATCC, LGC Standard, Poland) after 4 h and 24 h extraction according to ISO 10993-5:2009 (E). The monolayer of cells (1×10^4 cells/mL) was treated with the extracts for 18 h. Then, the treated cells were stained with annexin V and propidium iodide (FITC Annexin V Apoptosis Detection Kit I BD Biosciences, Franklin Lakes, NY, USA) [26]. Briefly, L929 were seeded in 24-well plates (10,000 cells/mL) and incubated for 24 h. Then, the medium was replaced with 100 µL of extracts (10% or 100%). The plates were incubated for 24 h at 37 ± 1 °C with 5% CO_2. Simultaneously, to assess in vitro cytotoxicity (ISO 10993-5:2009E) of PA12 containing 3 wt %, 4 wt % and 5 wt % oxides, L929 (8000 cells/100 µL) were seeded in the 96-well plates. After treatment, 11 µL PrestoBlue reagent (ThermoFisher Scientific, Waltham, MA, USA) was directly put into each well and incubated for 1.5 h. The absorbance was recorded at 570 nm (and reference at 600 nm) with Infinite M200 reader (Tecan, Durham, NC, USA). The GraphPad Prism 8.4.3 (GraphPad Software Inc., La Jolla, CA, USA) was used for the data analysis. Differences between the groups were tested using Bonferroni's test multiple comparison tests. Differences were considered statistically significant at $p < 0.05$.

3. Results

3.1. Contact Angle Measurements

The wettability of the metal oxides by melted PA12 was analyzed from the images taken during the melting process in Figure 2. In a model system, the filler was used in the form of the tablet contacted with a polymer pellet. The tablet was heated, and the shape of the pellet was analyzed. PA12 polymer starts to melt at the temperature of about 150 °C (Figure 2b) and is entirely molten at 180 °C (Figure 2c). However, the full spread of the molten polymer over the filler surface occurred only at the temperature of 224 °C (Figure 2d). Carbonization of the melted polymer was observed above 275 °C. The influence of the oxide concentration on the observed contact angles of the molten material is presented in Figure 2e. As seen, the increase of the oxide content in the filler leads to diminishing the contact angle of the molten pellet. It is worth remembering that PA12 is the example of the hydrophilic polymer which is easily wettable by water. However, in the presence of the hydrophobic metal oxides, the composites of PA12 become less hydrophilic. This effect is enhanced at higher concentration, especially at 4 wt %, for which the contact angle is 80°. The further increase of the oxide concentrations reverses the observed effect.

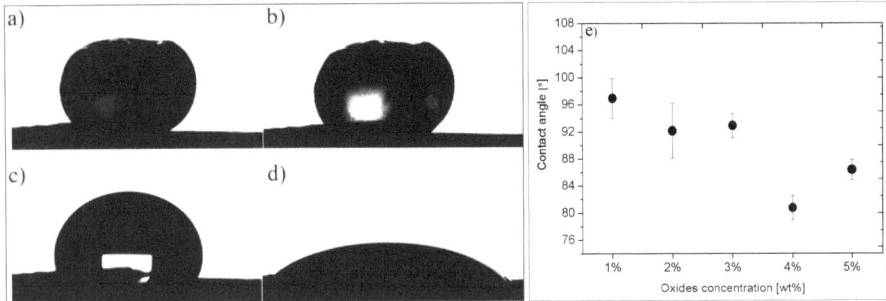

Figure 2. Images of the PA12 pellet melted at the filler surface: (**a**) original PA12 pellet; (**b**) first softening of the pellet at 150 °C; (**c**) completely melted PA12 at 180 °C; (**d**) wetting of the filler tablet by fully spread PA12 polymer at 224 °C; (**e**) dependence between contact angle of in PA12 composites and metal oxides concentration.

3.2. Surface Studies

An interaction of PA12 with the surface of oxide filler was studied. The surface of the oxide filler powder before and after contact with melted PA12 was analyzed by XPS technique. Table 2 shows concentration of the elements at both studied surfaces. The results indicate that the filler surface is strongly covered by polymer containing carbon and a small amount of nitrogen. The amount of copper (Cu), zinc (Zn) and titanium (Ti) after contact with melted PA12 at the interface filler/polymer was low. Moreover, the electronic state of the metals after reaction with melted PA12 changed, indicating reduction of the oxides at the interface (Figure 3).

Table 2. Atomic concentrations of the elements at the surface of powder oxides (filler) before and after contact with melted PA12.

State	Atomic Concentration [%]					
	C	O	Cu	Zn	Ti	N
before	8.5	45.7	32.4	10.1	3.3	0.0
after	83.1	10.7	0.4	0.2	0.1	5.6

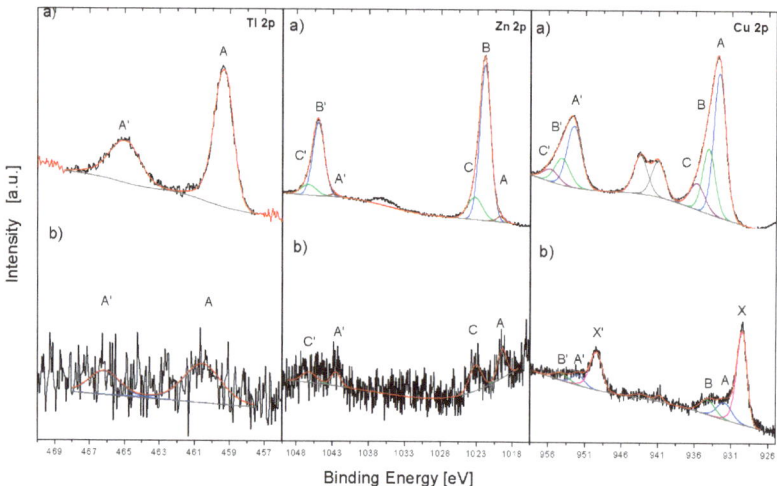

Figure 3. XPS analysis of the filler before (**a**) and after (**b**) interaction with PA12. The comparison of Cu 2p, Zn 2p and Ti 2p spectra.

In the case of Cu 2p excitation, the filler (Figure 3a) contains mainly CuO (A component), which surface is hydroxylated (B) and covered with a small amount of organic adsorbate (C) [27]. The contact with melted PA12 revealed in Cu 2p spectrum (Figure 3b) that C component disappeared and new X component appeared, which is assigned to reduction of CuO to Cu in strongly nucleophilic surrounding, and suggests strong interaction of copper oxide components with amide groups of PA12. In the case of the Zn 2p line, the filler contained mainly ZnO (B component) with hydroxylated surface (C component) and with some zinc alloy (possibly Zn-Cu bonding) surface compound (X component [27]. After the contact with melted PA12, ZnO-related components disappeared in the spectrum, showing only the alloy (X) and hydroxylated (B) ones. This can indicate strong polymer adsorption at the ZnO surface with the exposition of uncovered surface species. The described changes in the spectra of Cu, Zn and Ti in the filler indicate strong interaction of the amide groups coming from the PA12 macromolecules with metal cations. It leads to partial reduction the filler surface (metallic copper found), selective bonding with polymer (titania covered by polymer without surface change), or bonding to amide groups (zinc oxide).

3.3. Microstructure

To realize the potential of the functional fillers fully, it is necessary to obtain the well-dispersed and homogenously distribute particles in the whole polymer matrix. Therefore, the morphologies of PA12/metals oxides composites were analyzed by SEM on pellets just after extrusion and in the specimens after injection molding process. Table 3 presents the SEM images of the pellets where white areas refer to the metals oxides and they are well-dispersed in the PA12 matrix. A few agglomerates of the metals oxides can be observed in some places, which are quantitatively determined by calculation of their average grain size. It is clearly seen that increasing the content of the metals oxides causes decreasing the agglomerates diameter from 24.8 μm for 1 wt % to 4.29 μm for 5 wt % of the filler. Looking into the diameters of the initial metal oxides powder, which range from 0.3 to 1 μm, the observed agglomerates can be called as secondary ones since they are formed during extrusion from the primary agglomerates occurring in powder form. Such a scenario takes place in low viscous polymers where macromolecules move freely and when a shear rate is too low to destroy the formed agglomerates [28]. Therefore, decreasing the agglomerates sizes linearly with the oxides concentration is associated with the increase of viscosity of the PA12 matrix and higher shear force caused by rotating screws during extrusion. A similar analysis was made for the injection-molded specimens and the measured agglomerates are included also in Table 3. It can be seen that there is the same tendency as for the pellets. At higher metal concentration, the agglomerates diameters are smaller, but slightly bigger than determined for the pellets. In that process, the pellets are again melted and subject only to the pressure. Hence, the agglomerates are not able to be destroyed but they can form the new, secondary agglomerates.

Table 3. The summary of qualitative and quantitative microstructure analysis.

Metal Oxides	SEM Image of Composite Pellets	Histogram of Composite Pellets	Average Grain Size [μm]
1 wt %			24.8 ± 11.8 pellets 30.9 ± 12.3 after injection molding
2 wt %			18.2 ± 9.70 pellets 23.8 ± 6.56 after injection molding

Table 3. *Cont.*

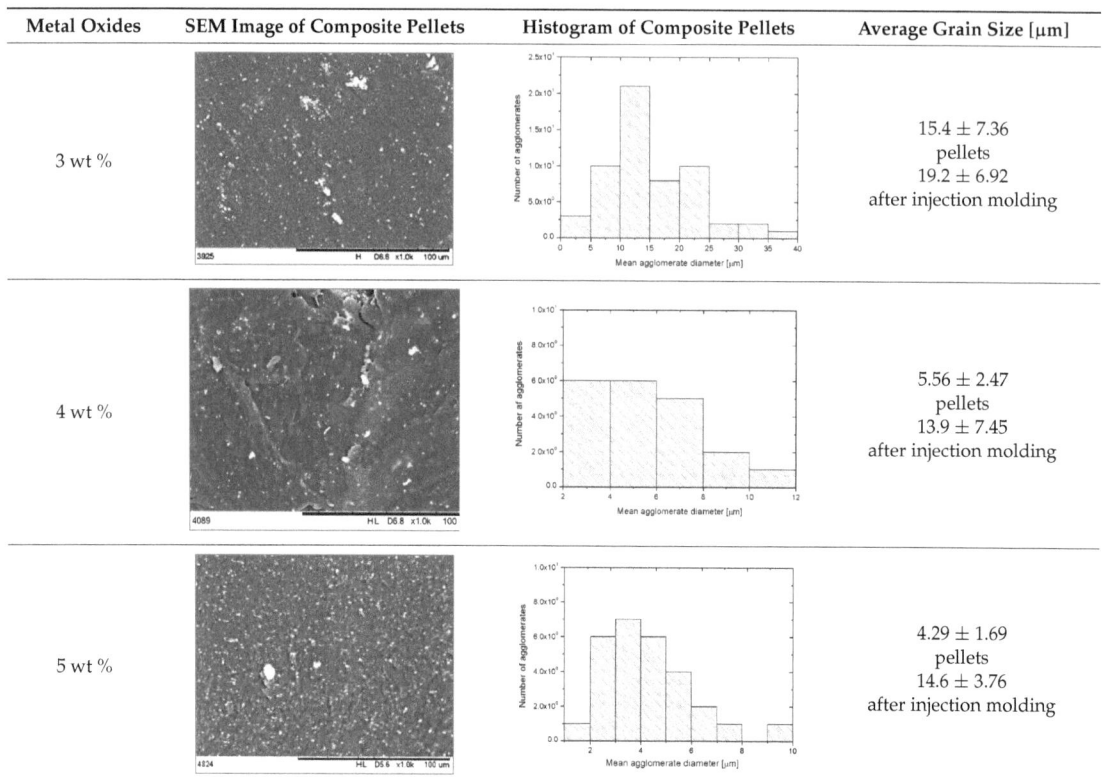

3.4. Thermal Properties

The thermal stability of the PA12 composites was determined using thermal gravimetric analysis. The results are presented in Table 4 and the raw TGA curves for each composite are collected in the Supplementary Data (Figures S1–S6). The results indicate the deterioration of the thermal properties of the composites with the increase in the amount of metal oxides. The temperature of the 5% material weight loss decreases of about a maximum of 7 °C at the highest oxides concentration in comparison to pure PA12. In the case of 10% weight loss and the fastest material degradation, the effect of the oxides is negligible. A weak impact of the oxides on the thermal stability can be connected with the presence of the oxides agglomerates. DSC test was performed to identify the effect of the metal oxides on the melting and crystallization behavior. The results were collected in Figure 4 and Table 4. For some of the composites, the glass transition decreases by about 1–5 °C without linear dependence with the oxides content. This is associated with the blocking of the polymer chains by the filler, which is more effective for the homogeneously dispersed oxides. During the 1st heating (Figure 4a), all materials show a single melting peak with a melting point at 180 °C, which refers to the presence of γ crystal forms. There is no change in the melting point indicting the weak nucleation effect of the added oxides visible also by small changes in the crystallinity content. On the thermogram obtained during the 2nd heating, a single melting peak splits to the double one (Figure 4b), suggesting the formation of the second crystalline phase-α. That crystal phase melts at the lower temperature around 165 °C and it is formed as an effect of a slow cooling subjected after the 1st heating cycle. Such a scenario favours the parallel arrangement of the polymer chains characteristic for the α phase. Because the melting peak obtained during the 2nd heating has a shoulder, it confirms the partial deformation of the α-form crystals structure to the γ-form [29,30]. That transformation is probably hindered in the presence of the metal oxides because the composites have lower crystallinity content than neat PA12. Moreover, the crystallinity changes only about few degrees with increasing amount of metal oxides, similarly to the crystallization temperature, which is not influenced by the addition of the oxides. However, it can be seen that peaks coming from the composites are narrower than for neat PA12. This fact shows the presence of more homogenous crystal phase. Similarly, the addition of copper spheres and copper flakes to PA12 do not affect characteristic temperatures of the polymer determined from the DSC thermograms [31].

Table 4. The results of thermal analysis.

Material	TGA			DSC					
	$T_{5\%}$ [°C]	$T_{10\%}$ [°C]	T_d [°C]	1st Heating			2nd Heating		Cooling
				T_g [°C]	T_m [°C]	X_c [%]	T_m [°C]	X_c [%]	T_c [°C]
Pure PA12	406	417	450	48.2	180	32.1	179	33.8	154
PA12 + 1 wt % oxides	406	420	452	43.2	180	26.0	179	27.2	156
PA12 + 2 wt % oxides	403	418	450	42.5	180	29.2	179	29.1	157
PA12 + 3 wt % oxides	402	418	451	46.1	180	26.0	178	27.9	156
PA12 + 4 wt % oxides	404	418	451	43.0	180	27.5	178	27.2	156
PA12 + 5 wt % oxides	399	417	451	47.0	180	27.9	179	28.2	156

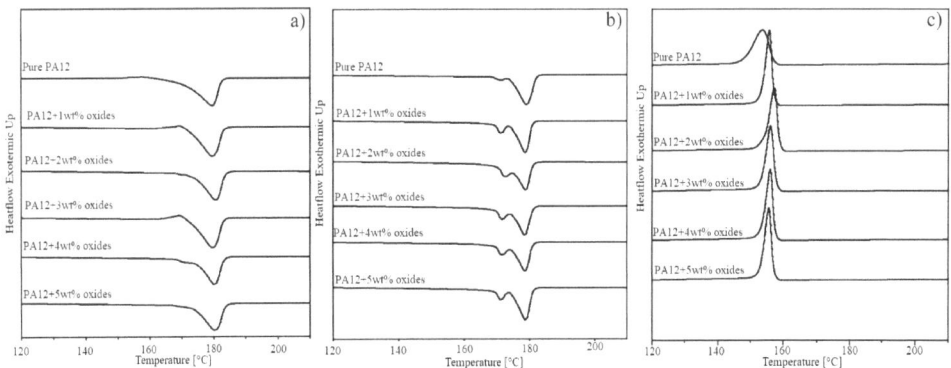

Figure 4. DSC curves for studied materials; (a) first heating; (b) second heating; (c) cooling.

3.5. Rheological Properties

The assessment of the rheological behavior of the composites is important regarding their processing performance and interactions of the polymer with the added fillers. Using rotational rheometer, the effect of the metal oxides addition on the viscosity, storage and loss modulus were analyzed. Figure 5 presents the curves obtained at 190 °C, which are presented in a logarithmic scale as a function of the angular frequency. The strain was selected as 10% from the linear viscoelastic range, and it was used in all of the tests. In the measured frequency range, for all materials including neat PA12, the viscosity curves have the linear dependence but only at low frequencies (Figure 5a). The further increase the frequency causes decreasing the viscosity, which is characteristic for the shear-thinning pseudoplastic liquids [32]. The lack of typical Newtonian curve depicted for pure PA12 can be associated with the changes in the molecular weight, the presence of the gas bubbles or additives, especially plasticizers [33]. The viscosity is higher for the composites containing 2–4 wt % of metal oxides in comparison to the viscosity of pure PA12. Because for PA12 + 1 wt % oxides the viscosity has the same characteristic for the neat PA12, it can be stated that metal oxides inhibit the movement of the polymer chains at approximately 2 wt %. Despite the fact that the composites have higher viscosity, such an increase is not significant. For instance, pure PA12 has the viscosity at 1Hz equals to 135 Pa·s, while for the composite with 5 wt % oxides, the viscosity is 190 Pa·s. In Figure 5b,c, the dependence of the storage (G′) and loss (G″) modulus refers to the elastic and viscous behavior of the material respectively. Both of the moduli increase together with the frequency and they are not dependent on the oxides concentration. So, it means that the interactions between PA12 and metal oxides are not as strong as for the reported composites of PA12 with, e.g., carbon nanotubes [34]. Obviously, such interactions are associated with a lower surface area of the used oxides, rather than with the dispersion of the oxides in the matrix. From the presented graphs, it can also be observed that for all materials, the loss modulus is larger than the storage modulus in the whole frequency range. This suggests that the studied composites exhibit more viscous than elastic properties. Looking from the practical point of view, such a behavior is desired for the processes in which PA12-containing metal oxides can be applied for the final products. The first example is a melt-blown process in which the non-woven fabric can be manufactured and used as protective masks or filters [35]. The other possibility is to use PA12 composites in the FFF (Fused Filament Fabrication) 3D printing process to print the medical components such as respiratory valves, face shields or even the implants from the medical grade of PA12 [15].

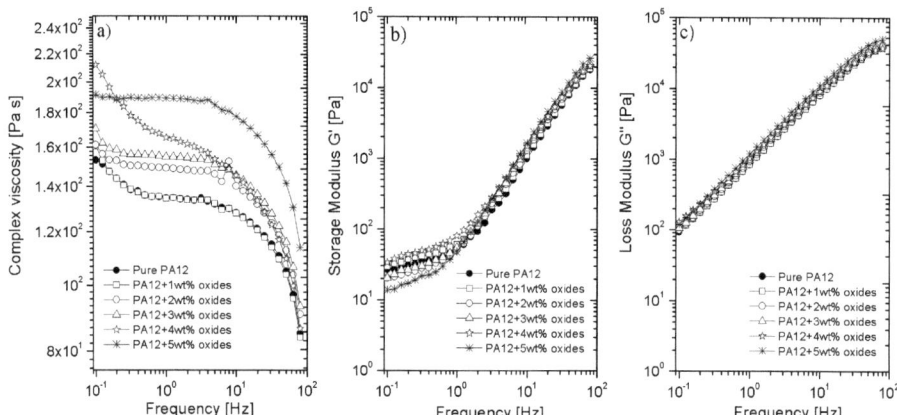

Figure 5. (**a**) Dynamic viscosity; (**b**) storage modulus G′ and (**c**) loss modulus G″ as a function of frequency for pure PA12 and PA12 with the addition of 1–5 wt % metals oxides.

3.6. Biocidal Properties

One of the objectives of this paper is to confirm the biocidal properties of PA12 as the effect of the addition of metal oxides to it. In order to meet this objective, the round specimens fabricated by injection molding (were tested against example bacteria, fungi and viruses. The qualitative activities are included in Appendix B, Figures A3 and A4, while the quantitative analysis for all pathogens is shown in Figure 6. In the case of *E. coli*, the qualitative antimicrobial activities of the pure PA12 and its composites show no inhibition zone for all of the tested samples (Appendix B, Figure A3). However, the slight growth of bacteria under all materials with oxides was observed, which indicates low antibacterial activity. The quantitative antimicrobial activities (Figure 6) of the studied materials depict the reduction of the bacteria colonies as the effect of the metal oxides addition. As it can be seen in the graph, the highest CFU reduction rate (R%) against *E. coli* was 87% in PA12 containing 4 wt % oxides, whilst 5 wt % causes decreasing of the reduction rate.

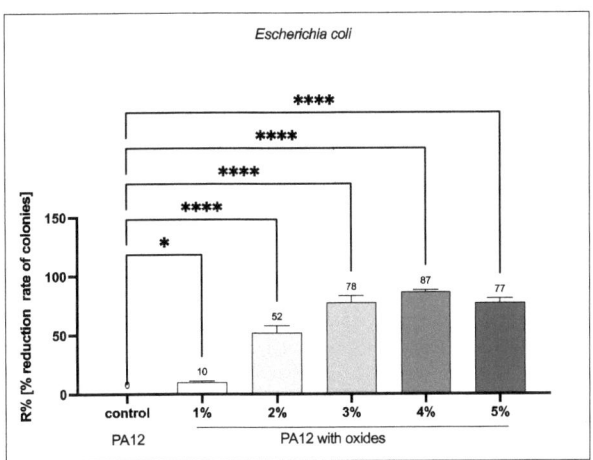

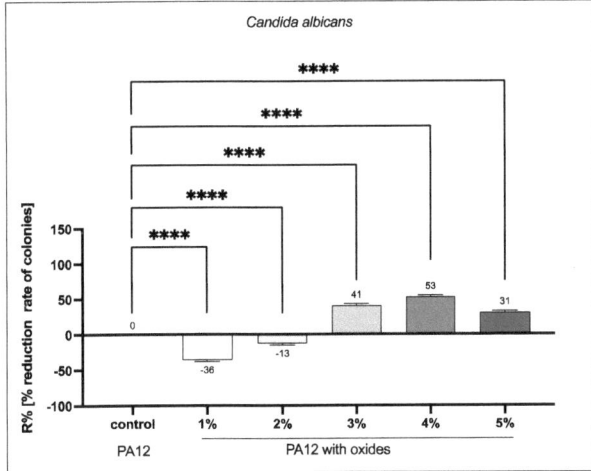

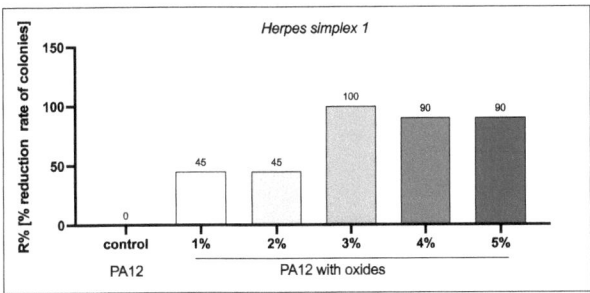

Figure 6. The antimicrobial activities in term of reduction (%) rate of colonies (CFU), against: *Escherichia coli*, *Candida albicans* and *Herpes simplex 1*. Differences with $p < 0.05$ were considered statistically significant. One asterisk (*) indicates $p < 0.05$, four asterisks (****) indicate $p < 0.0001$. Since the results for viruses are discontinuous, error bars, as well as one-way analysis of variance, cannot be added.

The standard qualitative procedure clearly demonstrates the antifungal activity of PA12 containing metal oxide vs. the control PA12 (without metal oxides), showing the lack of this activity (Figure A1 in Appendix A). As it is shown in Figure 6, the most effective reduction rate of the *C. albicans*' CFU was obtained for the composites with 3 wt % (R% = 41) and 4 wt % (R% = 53) of the metal oxides. PA12 displayed low antifungal activity in the suspension method according to PN-EN ISO 20743 2013-10E (see Figure A4 in Appendix B). Fungistatic effectiveness of the PA12 containing metal oxide was observed using AATCC TM90-2011(2016)e plate test (Figure A2 in Appendix A). The growth of *C. albicans* CFU was reduced under swatches containing metal oxides vs. the PA12 control (without metal oxides). There was no diffusion of antimicrobial agents from the swatches as indicated by the lack of a zone of inhibition (Appendix B, Figure A4). In the case of PA12 composites, the analyzed specimens displayed fungistatic activity.

The activity of PA12 composites against Ad5 and HSV-1 within 24 h was also examined. Based on the results, there was no reduction in Ad 5 titer by any of the materials tested, but in the case of HSV 1 virus, the test material showed virucidal properties (Figure 6). After the exposure time, the virus titer was reduced compared to the control material-pure PA12. The highest activity was obtained for 3 wt % of the metal oxides with the reduction R% = 100. For 4 wt % and 5 wt % the reduction was slightly lower (R% = 90) and did not vary for the oxides concentration. PA12 composites, especially at concentrations of 3 wt %, but also at 4 wt % and 5 wt %, are capable of inactivating enveloped herpes viruses. So, it can be assumed that at these concentrations PA12 composites will be equally effective in inactivating other enveloped viruses belonging to the family of *Orthomyxoviridae* (human and animal influenza viruses), *Coronaviridae* (including SARS-CoV), *Flaviviridae* and *Poxviridae*, and blood-borne viruses such as HBV, HCV and HIV.

As mentioned in the introduction, silver particles have been proven to be toxic to the human cells. Therefore, the new bioactive fillers are being researched to be used in protective materials for humans. Hence, to determine the mode of cell death generated by the PA12 composites, the Annexin V/FITC assay using flow cytometry was conducted and it was followed by the cytotoxicity assay. The treated fibroblasts displayed the features characteristic of apoptosis (Figures 7 and 8). The Annexin V binding showed the loss of membrane asymmetry and phosphatidylserine (PS) exposure in fibroblasts treated with the PA12 containing 3 wt % or 4 wt % of metal oxides (Figures 7 and 8). The fibroblasts treated with the extracts of PA12 (0–4 w% of metal oxides) underwent early apoptosis (11.85–27.79%), late apoptosis (0.81–5.04%) and necrosis (0.18–0.31%) respectively. So, the concentration of 3 wt % compared with 4 wt % generated a 2-fold higher percentage of apoptotic cells. In the case of fibroblasts, the composites containing 3 wt % of metal oxides in 4 h extract solution (Figure 7), significantly reduced fibroblast viability (52.43% of viable cells) vs. the 24 h extract (viability at 71.16% in Figure 8). For the first time PA12 (with and without metal oxides) was tested to investigate its action mode focused on the fibroblast PS externalization (apoptotic cell death). The flow cytometry results align with [34] and confirm that the viability of cells treated with PA12 remained at 71.16–72.75% (Figure 8) regardless of the concentration of metal oxides in the 24 h extract solution. So, cytometric results (Figures 7 and 8) and colorimetric data (Figure 9) confirmed that PA12 (with and without oxides) is not toxic to fibroblasts [35]. This fact leads to the conclusion that these PA12 composites induce naturally cell apoptosis, counterstain with PI is not visible in Figures 7 and 8. The antifungal concentration of 4 wt % of metal oxides in PA12 induced apoptosis and increased the early and late apoptosis population vs. untreated cells, suggesting its potential effect as a non-toxic antimicrobial agent.

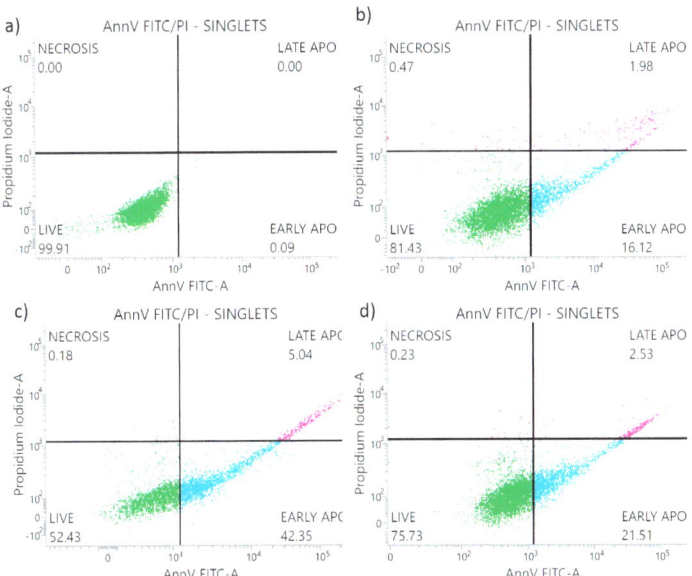

Figure 7. Percentage share of the L929 cells: alive, early apoptotic, late apoptotic and necrotic after 24 h treatment with: (**a**) DMEM—control group; (**b**) 4 h extract of pure PA12; (**c**) 4 h extract of PA12 with 3 wt % of metal oxides; (**d**) 4 h extract of PA12 with 4 wt % of metal oxides. The experiment was performed in triplicate.

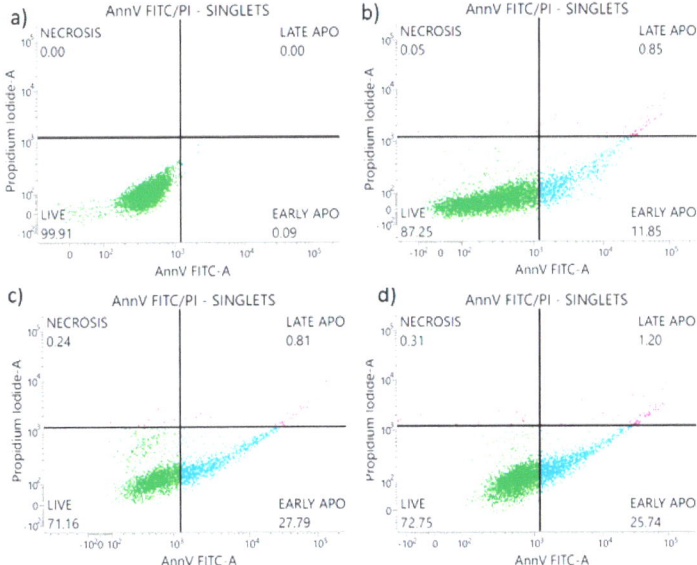

Figure 8. Percentage share of the L929 cells: alive, early apoptotic, late apoptotic and necrotic, after 24 h treatment with: (**a**) DMEM—control group; (**b**) 24 h extract of pure PA12; (**c**) 24 h extract of PA12 with 3 wt % of metal oxides; (**d**) 24 h extract of PA12 with 4 wt % of metal oxides. The experiment was performed in triplicate.

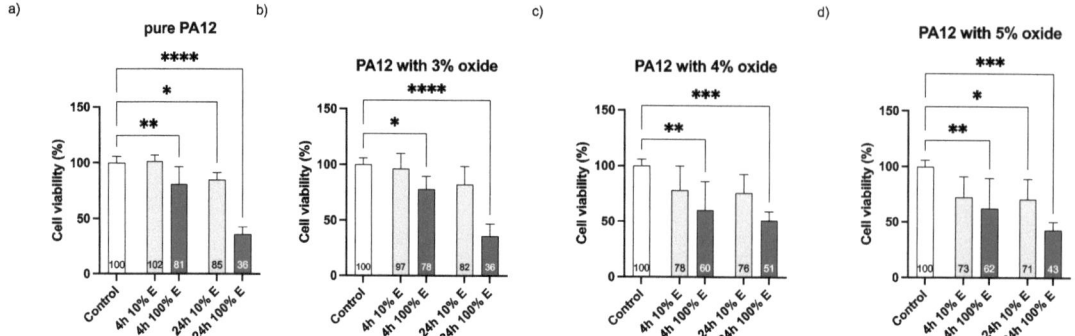

Figure 9. Cell viability: MTT assay. Histograms represent the percentage, with respect to control cells (Ctrl, 100%) of viable cells after 24 h treatment with 4 h and 24 h extracts at 10 and 100% concentration from (**a**) pure PA12; (**b**) PA12 with 3 wt % of metal oxides; (**c**) PA12 with 4 wt % of metal oxides; (**d**) PA12 with 5 wt % of metal oxides. Differences with $p < 0.05$ were considered statistically significant. One asterisk (*) indicates $p < 0.05$, two asterisks (**) indicate $p < 0.01$, three asterisks (***) indicate $p < 0.001$, four asterisks (****) indicate $p < 0.0001$.

4. Conclusions

This paper describes the thermoplastic composites of PA12 mixed with 1–5 wt % of metal oxides (CuO, ZnO, and TiO$_2$) fabricated by the melt-blending method. To obtain as high as possible biocidal activity, the used oxides occur in the nanosize, but after the extrusion process they tend to agglomerate. Therefore, SEM images show homogenously dispersed oxides in the PA12 matrix but in the form of the agglomerates. Their diameter decreases with oxides concentration, which is associated with the increased shear stress caused by higher viscosity. This growth in the viscosity of the composites was measured by rotational rheology and was the highest for 5 wt % of the oxides which were characterized by the lowest agglomerates diameter of 4.29 ± 1.69 μm. Due to the presence of the agglomerates, the interactions between polyamide macromolecules and oxides are not sufficient to affect the viscoelastic properties of PA12 since storage and loss modulus remain almost unchanged for the composites. What is more, the loss modulus is larger than the storage modulus and it leads to the conclusion that the composites of PA12 and metal oxides will behave more like viscous rather than elastic liquids. Such a property is positive from applying these polymeric composites in the standard processing methods use for the thermoplastic polymers. What is more, there are only slight changes in the temperatures and crystallinity behavior in the presence of metal oxides. It was confirmed that metal oxides are well-wetted by PA12, and the composites became less hydrophilic in the presence of hydrophobic oxides. Surface studies indicate interaction of the PA12 amide groups with metal cations resulting in the reduction of the copper surface, selective bonding with zinc, and bonding to amide groups for titania. At 4 wt % of the oxides, the contact angle was the lowest (80°). At the same time for this composite, the highest biocidal activity against the representatives of bacteria, fungi, and viruses was reported. As the effect of 4 wt % oxides, the quantitative antimicrobial activity depicts the reduction of the *E. coli* by 87%, *C. albicans* by 53%, and virus *Herpes simplex 1* by 90%. Flow cytometry also proved that cell viability remained at 71.16–72.75% in the presence of PA12 + 4 wt % oxides and that composites are not toxic for the fibroblasts. Studying the reported toxicity of the commonly used silver particles, the mixture of metal oxides seems to be a promising and safe substitute. It needs to be highlighted that PA12 composites demonstrate high biocidal activities and desired physicochemical properties. At the same time, these composites can be processed into the final products such as injected molded parts for the hospitals and nonwovens used as filters or filaments for 3D printing.

To sum up, the main achievements were highlighted below:

- PA12 composites containing CuO, TiO$_2$ and ZnO prepared by melt-blending
- Obtaining the biocidal activity against *E. coli*, *C. albicans* and *Herpes simplex 1*
- The lack of toxicity of the composites for the human cells
- Founding the highest activity and the lowest contact angle at 4 wt % oxides
- A slight effect of the oxides on the viscoelastic properties and thermal stability

Supplementary Materials: The following supporting information can be downloaded at: https://www.mdpi.com/article/10.3390/polym14153025/s1, Figure S1: TGA curve of neat PA12; Figure S2: TGA curve of PA12+1 wt % metal oxides; Figure S3: TGA curve of PA12+2 wt % metal oxides; Figure S4: TGA curve of PA12+3 wt % metal oxides; Figure S5: TGA curve of PA12+4 wt % metal oxides; Figure S6: TGA curve of PA12+5 wt % metal oxides.

Author Contributions: Conceptualization, P.L.-D., M.S. and M.G.; Formal analysis, P.L.-D.; Investigation, P.L.-D., M.M., M.S., K.R., M.G., R.P.S., M.K., B.B. and A.P.; Methodology, P.L.-D., M.S., K.R., R.P.S., M.K. and B.B.; Project administration, R.P.S. and A.B.; Software, P.L.-D., M.M., M.G. and R.P.S.; Supervision, A.B.; Validation, A.B.; Visualization, P.L.-D., M.M. and M.G.; Writing—original draft, P.L.-D., M.S., M.G. and R.P.S.; Writing—review & editing, P.L.-D. All authors have read and agreed to the published version of the manuscript.

Funding: This research was funded by the National Centre for Research and Development, within the project entitled *The development of antimicrobial and filtration layered fabrics for sanitary and medical protection and fabrication technology based on metal-polymer composites*. Call No.: POIR.01.01.01-00-1246/20.

Institutional Review Board Statement: Not applicable.

Informed Consent Statement: Not applicable.

Data Availability Statement: Not applicable.

Conflicts of Interest: The authors declare no conflict of interest.

Appendix A

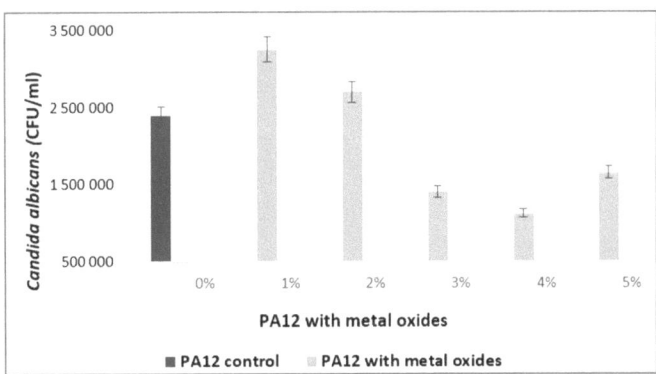

Figure A1. Antifungal activity of the PA12 with metal oxides assessed using the AATCC TM100-2019 quantitative method. Legend: The results obtained using AATCC TM100-2019 are valid due to: (1) "0" CFU of *C. albicans* recovered from the uninoculated swatches; (2) a significant increase in CFU ($\geq$1 log) recovered from the PA12 control (without metal oxides) inoculated with *C. albicans* after 24 h was noted, and it was over the CFU recovered from the PA12 control without metal oxides at "0" contact time (immediately after inoculation).

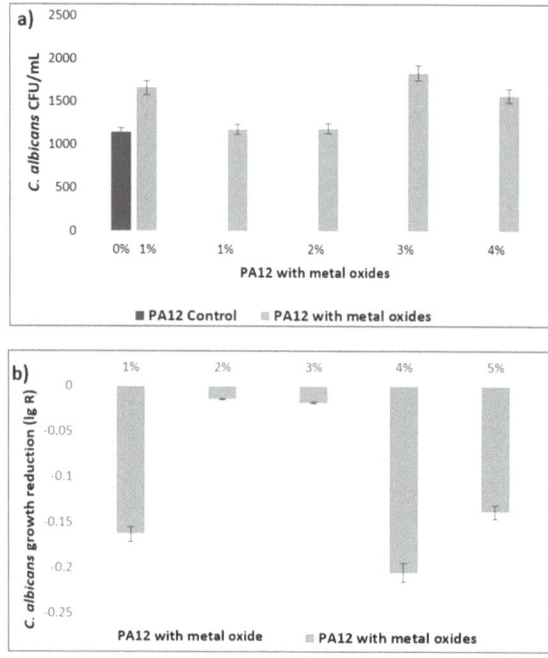

Figure A2. (a,b). Antifungal activity of the PA12 with metal oxides assessed using the PN-EN ISO 20743 2013-10E quantitative method. Legend: lg R was calculated using Equation: lg A − lg B = lg R, where A means CFU recovered from the PA12 control (without metal oxide) inoculated with *C. albicans* and incubated over the 24-h contact period; B means CFU recovered from the PA12 with metal oxide inoculated with *C. albicans* and incubated over the 24-h contact period.

Appendix B

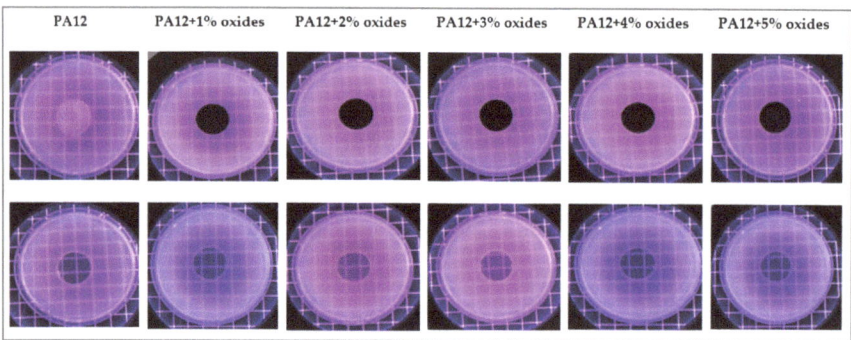

Figure A3. Comparison of *E. coli* growth in contact with PA12 and PA12 + 1–5 wt % oxides.

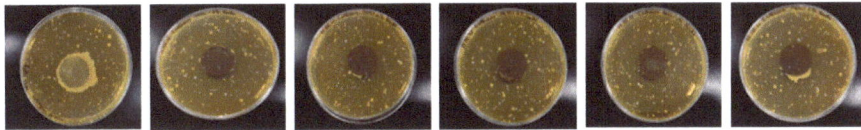

Figure A4. Fungistatic activity of the PA12 with metal oxides. *C. albicans* CFU on the YEPD agar growth medium inhibited under PA12 swatches contained 1, 2, 3, 4 and 5 wt % of metal oxides compared to the control PA12 without metal oxides. Visible confluent growth under the swatch as well as in proximity to the borders of a specimen.

References

1. Mouritz, A.P.; Galos, J.; Linklater, D.P.; Ladani, R.B.; Kandare, E.; Crawford, R.J.; Ivanova, E.P. Towards antiviral polymer composites to combat COVID-19 transmission. *Nano Sel.* **2021**, *2*, 2061–2071. [CrossRef] [PubMed]
2. Palza, H. Antimicrobial polymers with metal nanoparticles. *Int. J. Mol. Sci.* **2015**, *16*, 2099–2116. [CrossRef] [PubMed]
3. Dastjerdi, R.; Mojtahedi, M.R.M.; Shoshtari, A.M. Comparing the effect of three processing methods for modification of filament yarns with inorganic nanocomposite filler and their bioactivity against staphylococcus aureus. *Macromol. Res.* **2009**, *17*, 378–387. [CrossRef]
4. Curtis, L.T. Prevention of hospital-acquired infections: Review of non-pharmacological interventions. *J. Hosp. Infect.* **2008**, *69*, 204–219. [CrossRef] [PubMed]
5. Lourenço, J.B.; Pasa, T.S.; Bertuol, D.A.; Salau, N.P.G. An approach to assess and identify polymers in the health-care waste of a Brazilian university hospital. *J. Environ. Sci. Health—Part. A Toxic/Hazard. Subst. Environ. Eng.* **2020**, *55*, 800–819. [CrossRef] [PubMed]
6. Motelica, L.; Ficai, D.; Oprea, O.C.; Ficai, A.; Ene, V.L.; Vasile, B.S.; Andronescu, E.; Holban, A.M. Antibacterial biodegradable films based on alginate with silver nanoparticles and lemongrass essential oil–innovative packaging for cheese. *Nanomaterials* **2021**, *11*, 2377. [CrossRef]
7. Turner, R.D.; Wingham, J.R.; Paterson, T.E.; Shepherd, J.; Majewski, C. Use of silver-based additives for the development of antibacterial functionality in Laser Sintered polyamide 12 parts. *Sci. Rep.* **2020**, *10*, 892. [CrossRef]
8. Assis, M.; Simoes, L.G.P.; Tremiliosi, G.C.; Coelho, D.; Minozzi, D.T.; Santos, R.I.; Vilela, D.C.B.; Santos, J.R.; Ribeiro, L.K.; Rosa, I.L.V.; et al. Sio2-ag composite as a highly virucidal material: A roadmap that rapidly eliminates SARS-CoV-2. *Nanomaterials* **2021**, *11*, 638. [CrossRef]
9. Fonseca, C.; Ochoa, A.; Ulloa, M.T.; Alvarez, E.; Canales, D.; Zapata, P.A. Poly(lactic acid)/TiO$_2$ nanocomposites as alternative biocidal and antifungal materials. *Mater. Sci. Eng. C* **2015**, *57*, 314–320. [CrossRef]
10. Maruthapandi, M.; Saravanan, A.; Gupta, A.; Luong, J.H.T.; Gedanken, A. Antimicrobial Activities of Conducting Polymers and Their Composites. *Macromol* **2022**, *2*, 78–99. [CrossRef]
11. Motelica, L.; Popescu, A.; Răzvan, A.G.; Oprea, O.; Trușcă, R.D.; Vasile, B.S.; Dumitru, F.; Holban, A.M. Facile use of zno nanopowders to protect old manual paper documents. *Materials* **2020**, *13*, 5452. [CrossRef] [PubMed]
12. Hoseinzadeh, E.; Makhdoumi, P.; Taha, P.; Hossini, H.; Stelling, J.; Amjad Kamal, M.; Ashraf, G.M. A Review on Nano-Antimicrobials: Metal Nanoparticles, Methods and Mechanisms. *Curr. Drug Metab.* **2016**, *18*, 120–128. [CrossRef] [PubMed]
13. Li, Y.; Shimizu, H. *High-Shear Melt Processing of Polymer–Carbon Nanotube Composites*; Woodhead Publishing: Sawston, UK, 2011; pp. 133–154. [CrossRef]
14. Rogalskyy, S.; Bardeau, J.F.; Tarasyuk, O.; Fatyeyeva, K. Fabrication of new antifungal polyamide-12 material. *Polym. Int.* **2012**, *61*, 686–691. [CrossRef]
15. Vidakis, N.; Petousis, M.; Velidakis, E.; Korlos, A.; Kechagias, J.D.; Tsikritzis, D.; Mountakis, N. Medical-Grade Polyamide 12 Nanocomposite Materials for Enhanced Mechanical and Antibacterial Performance in 3D Printing Applications. *Polymers* **2022**, *14*, 440. [CrossRef] [PubMed]
16. Vidakis, N.; Petousis, M.; Michailidis, N.; Grammatikos, S.; David, C.N.; Mountakis, N.; Argyros, A.; Boura, O. Development and Optimization of Medical-Grade MultiFunctional Polyamide 12-Cuprous Oxide Nanocomposites with Superior Mechanical and Antibacterial Properties for Cost-Effective 3D Printing. *Nanomaterials* **2022**, *12*, 534. [CrossRef]
17. Rogalsky, S.; Bardeau, J.F.; Wu, H.; Lyoshina, L.; Bulko, O.; Tarasyuk, O.; Makhno, S.; Cherniavska, T.; Kyselov, Y.; Koo, J.H. Structural, thermal and antibacterial properties of polyamide 11/polymeric biocide polyhexamethylene guanidine dodecylbenzenesulfonate composites. *J. Mater. Sci.* **2016**, *51*, 7716–7730. [CrossRef]
18. Thokala, N.; Kealey, C.; Kennedy, J.; Brady, D.B.; Farrell, J.B. Characterisation of polyamide 11/copper antimicrobial composites for medical device applications. *Mater. Sci. Eng. C* **2017**, *78*, 1179–1186. [CrossRef]
19. Rapacz-Kmita, A.; Pierchała, M.K.; Tomas-Trybuś, A.; Szaraniec, B.; Karwot, J. The wettability, mechanical and antimicrobial properties of polylactide/montmorillonite nanocomposite films. *Acta Bioeng. Biomech.* **2017**, *19*, 25–33. [CrossRef]
20. Kizildag, N.; Ucar, N. Nanocomposite polyacrylonitrile filaments with electrostatic dissipative and antibacterial properties. *J. Compos. Mater.* **2016**, *50*, 4279–4289. [CrossRef]

21. Dörr, D.; Kuhn, U.; Altstädt, V. Rheological study of gelation and crosslinking in chemical modified polyamide 12 using a multiwave technique. *Polymers* **2020**, *12*, 855. [CrossRef]
22. Silva, C.; Bobillier, F.; Canales, D.; Sepúlveda, F.A.; Cament, A.; Amigo, N.; Rivas, L.M.; Ulloa, M.T.; Reyes, P.; Ortiz, J.A.; et al. Mechanical and antimicrobial polyethylene composites with CaO nanoparticles. *Polymers* **2020**, *12*, 2132. [CrossRef] [PubMed]
23. Kizildag, N.; Ucar, N.; Onen, A. Nanocomposite polyacrylonitrile filaments with titanium dioxide and silver nanoparticles for multifunctionality. *J. Ind. Text.* **2018**, *47*, 1716–1738. [CrossRef]
24. Palza, H.; Quijada, R.; Delgado, K. Antimicrobial polymer composites with copper micro- and nanoparticles: Effect of particle size and polymer matrix. *J. Bioact. Compat. Polym.* **2015**, *30*, 366–380. [CrossRef]
25. Wei, X.-F.; De Vico, L.; Larroche, P.; Kallio, K.J.; Bruder, S.; Bellander, M.; Gedde, U.W.; Hedenqvist, M.S. Ageing properties and polymer/fuel interactions of polyamide 12 exposed to (bio)diesel at high temperature. *npj Mater. Degrad.* **2019**, *3*, 28. [CrossRef]
26. Gizińska, M.; Staniszewska, A.; Kazek, M.; Koronkiewicz, M.; Kuryk, Ł.; Milner-Krawczyk, M.; Baran, J.; Borowiecki, P.; Staniszewska, M. Antifungal polybrominated proxyphylline derivative induces Candida albicans calcineurin stress response in Galleria mellonella. *Bioorganic Med. Chem. Lett.* **2020**, *30*, 127545. [CrossRef]
27. NIST. *Standard Reference Database Number 20*; National Institute of Standards and Technology: Gaithersburg, MD, USA, 2000. [CrossRef]
28. Kasaliwal, G.R.; Göldel, A.; Pötschke, P.; Heinrich, G. Influences of polymer matrix melt viscosity and molecular weight on MWCNT agglomerate dispersion. *Polymer* **2011**, *52*, 1027–1036. [CrossRef]
29. Ma, N.; Liu, W.; Ma, L.; He, S.; Liu, H.; Zhang, Z.; Sun, A.; Huang, M.; Zhu, C. Crystal transition and thermal behavior of Nylon 12. *e-Polymers* **2020**, *20*, 346–352. [CrossRef]
30. Rahim, T.N.A.T.; Abdullah, A.M.; Akil, H.M.; Mohamad, D.; Rajion, Z.A. The improvement of mechanical and thermal properties of polyamide 12 3D printed parts by fused deposition modelling. *Express Polym. Lett.* **2017**, *11*, 963–982. [CrossRef]
31. Lanzl, L.; Wudy, K.; Greiner, S.; Drummer, D. Selective laser sintering of copper filled polyamide 12: Characterization of powder properties and process behavior. *Polym. Compos.* **2019**, *40*, 1801–1809. [CrossRef]
32. Schramm, G. *A Practical Approach to Rheology and Rheometry*, 2nd ed.; Gebrueder HAAKE GmbH: Karlsruhe, Germany, 1997.
33. Majka, T.M.; Pielichowski, K.; Leszczyńska, A. Comparison of the Rheological Properties of Polyamide-6 and Its Nanocomposites with Montmorillonite Obtained by Melt Intercalation Porównanie Właściwości Reologicznych Poliamidu-6 Oraz Jego Nanokompozytów. *Czas. Tech.* **2013**, *1-Ch*, 39–46.
34. Bhattacharyya, A.R.; Pötschke, P.; Abdel-Goad, M.; Fischer, D. Effect of encapsulated SWNT on the mechanical properties of melt mixed PA12/SWNT composites. *Chem. Phys. Lett.* **2004**, *392*, 28–33. [CrossRef]
35. Kopecká, R. Modified Polypropylene Nonwoven Textile for Filter Facial Masks. *Am. J. Biomed. Sci. Res.* **2020**, *9*, 355–356. [CrossRef]

Article

Fabrication of Silver Nanoparticles Using *Cordyline fruticosa* L. Leave Extract Endowing Silk Fibroin Modified Viscose Fabric with Durable Antibacterial Property

Ngoc-Thang Nguyen [1],* and Thi-Lan-Huong Vo [2]

[1] Department of Textile Material and Chemical Processing, School of Textile-Leather and Fashion, Hanoi University of Science and Technology, 1 Dai Co Viet, Hanoi 11615, Vietnam
[2] Department of Fibre and Textile Technology, Hanoi Industrial Textile Garment University, Hanoi 12411, Vietnam; huongvtl@hict.edu.vn
* Correspondence: thang.nguyenngoc@hust.edu.vn; Tel.: +84-904309930

Citation: Nguyen, N.-T.; Vo, T.-L.-H. Fabrication of Silver Nanoparticles Using *Cordyline fruticosa* L. Leave Extract Endowing Silk Fibroin Modified Viscose Fabric with Durable Antibacterial Property. *Polymers* **2022**, *14*, 2409. https://doi.org/10.3390/polym14122409

Academic Editors: Md. Amdadul Huq and Shahina Akter

Received: 11 May 2022
Accepted: 12 June 2022
Published: 14 June 2022

Publisher's Note: MDPI stays neutral with regard to jurisdictional claims in published maps and institutional affiliations.

Copyright: © 2022 by the authors. Licensee MDPI, Basel, Switzerland. This article is an open access article distributed under the terms and conditions of the Creative Commons Attribution (CC BY) license (https://creativecommons.org/licenses/by/4.0/).

Abstract: The current work presented a green synthetic route for the fabrication of silver nanoparticles obtained from aqueous solutions of silver nitrate using *Cordyline fruticosa* L. leaf extract (Col) as a reducing and capping agent for the first time. The bio-synthesized silver nanoparticles (AgCol) were investigated using UV–visible spectroscopy (UV–vis), transmission electron microscopy (TEM), X-ray diffraction (XRD), Fourier transform infrared spectroscopy (FTIR), and thermal gravimetric analysis (TGA). The obtained data demonstrated that AgCol in spherical shape with an average size of 28.5 nm were highly crystalline and well capped by phytocompounds from the Col extract. Moreover, the bio-synthesized AgCol also exhibited the effective antibacterial activities against six pathogenic bacteria, including *Escherichia coli* (*E. coli*), *Pseudomonas aeruginosa* (*P. aeruginosa*), *Salmonella enterica* (*S. enterica*), *Staphylococcus aureus* (*S. aureus*), *Bacillus cereus* (*B. cereus*) and *Enterococcus faecalis* (*E. faecalis*). The AgCol were applied as an antibacterial finishing agent for viscose fabric using a pad-dry curing technique. The AgCol-treated viscose fabrics exhibited a good synergistic antimicrobial activity against *E. coli* and *S. aureus* bacteria. Furthermore, the silk fibroin regenerated from *Bombyx mori* cocoon waste was utilized as an ecofriendly binder for the immobilization of AgCol on the viscose fabric. Thus, the antimicrobial efficacy of the AgCol and fibroin modified viscose fabric still reached 99.99% against the tested bacteria, even after 30 washing cycles. The colorimetric property, morphology, elemental composition, and distribution of AgCol on the treated fabrics were investigated using several analysis tools, including colorimetry, scanning electron microscopy (SEM), energy-dispersive X-ray spectroscopy (EDX), atomic absorption spectroscopy (AAS), Kjeldahl, and FTIR. Because of the excellent antimicrobial efficiency and laundering durability, as well as the green synthesis method, the AgCol and fibroin modified viscose fabric could be utilized as an antibacterial material in sportswear and medical textile applications.

Keywords: *Cordyline fruticosa* L.; silver nanoparticles; antibacterial activity; silk fibroin; viscose fabric

1. Introduction

Cellulosic textiles, such as cotton and viscose, play key roles in the daily life of humans due to their excellent properties, including comfortability, water absorptivity, air permeability, and flexibility [1–3]. However, this natural material also provides a favorable medium for the adsorption, survival, and growth of hazardous microorganisms owing to its moisture, oxygen, and nutrient retaining ability [2,4–6]. Infestation by microbes can cause detrimental effects for the wearer and the textile manufacturing process, such as the generation of unpleasant odor, health concerns, loss of fabric strength properties, and discoloration [2,6]. To overcome those problems, antibacterial finishing of cellulosic textile products is often inevitable. In addition, viscose fabric absorbs more sweat than cotton when worn next to the skin, providing a convenient environment for bacterial growth.

Therefore, it is necessary to improve the antibacterial properties of viscose fabric in order to make this cellulosic material more attractive and to broaden its application fields.

Historically, numerous antimicrobial agents, such as quaternary ammonium compounds [7,8], triclosan [7,9], halogenated phenols [7], inorganic nanoparticles [9–11], chitosan and its derivatives [1,5,6,12], and their finishing techniques, have been applied for cellulosic fabrics with respect to the effective control of microbial growth [7,13,14]. However, some antimicrobial agents that are in consideration are not eco-friendly, are toxic to humans, and there is a risk that bacteria may grow resistant to the antibiotics [6,14,15]. Thus, the proper selection of antibacterial agent and application process to impart durable antibacterial efficiency to the cellulosic materials not only enables the production of highly active cellulosic products against undesirable bacterial, but also ensures minimal negative impacts on the fabric and environment. In view of these applicable and environmental concerns, the cellulosic fabrics treated with green silver nanoparticles (AgNPs) which were synthesized using bio-reductants from natural resources, i.e., plants, algae, and microorganisms, have greatly focused scientists' attention due to their wide-ranging antibacterial activity and durability to microorganisms [16–20]. Moreover, AgNPs are associated with low toxicity with regard mammalian cells, and their cytotoxicity can be reduced when they are incorporated into a polymer matrix [16–18,21].

One of the utmost important aspects for antibacterial textiles in terms of maintaining hygiene is laundering durability [1,3,17]. Thus, strategies for the immobilization of AgNPs on cellulosic textiles have been a fashion research trend in the modification of textiles. Numerous techniques for this purpose have been developed, including ex situ and in situ fabrication of AgNPs onto the fabric surfaces with the assistance of crosslinking agents or polymer binders [5,22–32]. Among polymer binders, natural polymers, such as chitosan and its derivatives, and starch, gelatin, alginate, pectin, guar gum, and rubber are reported to be very effective and ecofriendly for adhering AgNPs onto cellulosic fabric surfaces [5,27–29,31,32]. Interestingly, these polymers have great capacity to suffer bounding to the cellulose molecules of cellulosic fabrics even in very small added amounts. Therefore, they enhance the non-leaching of AgNPs during washing cycles.

Beyond the above knowledge in mind, the present work proposed a new strategy to impart antibacterial functionality to viscose fabric using bio-synthesized AgNPs and regenerated silk fibroin as an eco-friendly polymer binder. Firstly, a green synthesis of AgNPs from a water-based solution of silver nitrate utilizing *Cordyline fruticosa* L. leaf extract as a reducing and capping agent was reported for the first time. *Cordyline fruticosa* L. comprises of more than 480 species distributed mostly in tropical and subtropical regions of the world [33–35]. It can be found in gardens as an ornamental plant, as its leaves are very conspicuous and attractive. It has been used as a traditional medicinal plant for centuries in Vietnam. According to the literature, the major constituents present in the *Cordyline fruticosa* L. leaf extract are reported to contain anthocyanin, saponin, flavonoids, flavonols, chlorophyll, glycoside, etc., which possess strong reducing power [33–39]. The reduction reaction of silver ions was controlled using UV–vis spectroscopy to optimize the synthesis parameters of the AgNPs. The obtained AgNPs were examined using HR-TEM, XRD, FTIR, and TGA analyses. The antibacterial action of the bio-synthesized AgNPs was evaluated against six pathogenic bacteria, including *Escherichia coli* (*E. coli*), *Pseudomonas aeruginosa* (*P. aeruginosa*), *Salmonella enterica* (*S. enterica*), *Staphylococcus aureus* (*S. aureus*), *Bacillus cereus* (*B. cereus*), and *Enterococcus faecalis* (*E. faecalis*). Afterwards, the bio-synthesized AgNPs were applied as an antibacterial finishing agent for viscose fabrics using a pad-dry curing technique. In order to enhance the laundering durability, silk fibroin regenerated from *Bombyx mori* cocoon waste was used as an ecofriendly binder for the pre-modification of viscose fabric and subsequent loading of the AgNPs using the pad-dry curing technique, under different treatment conditions. Finally, the antibacterial activity of the functionalized viscose fabrics against both the gram-negative bacteria *E. coli* and gram-positive *S. aureus* bacteria were assessed after washing cycles. The mechanism of the modification of viscose fabric was suggested. To the best of our knowledge, the regenerated silk fibroin combined with AgNPs in the antimicrobial finishing of viscose fabric has not yet been reported.

2. Materials and Methods

2.1. Materials

Cordyline fruticosa L. leaves (Col) were collected from a locality of the Hoai Duc district, Hanoi, Vietnam, in October 2019. First of all, the Col leaves were washed with distilled water, before being dried at 60 °C for 48 h. The dried leaves were thinly chopped into small pieces (5 × 5 mm) and kept in a sealable plastic bag for further study. *Bombyx mori* cocoon waste was collected from the village of Vong Nguyet, Bac Ninh province, Vietnam. Plain woven viscose fabric (staple viscose, Ne30/1, 68 ends/cm, 68 picks/cm) was scoured and supplied by the Nam Dinh Textile Garment Co., Ltd., Nam Dinh, Vietnam. Silver nitrate (AgNO$_3$, 99.99%), sodium carbonate, ethanol, acetic acid, lithium bromide, and aluminum sulfate octadecahydrate were purchased from the Aladdin Shanghai Biochemical Technology Co. Ltd., Shanghai, China. The solvent used in all experiments was double distilled water from an EYELA Still Ace SA-2100E. Six types of bacteria, including *Escherichia coli* (ATCC 8739), *Pseudomonas aeruginosa* (ATCC 27853), *Salmonella enterica* (ATCC 2162), *Staphylococcus aureus* (ATCC 25923), *Bacillus cereus* (ATCC 11778), and *Enterococcus faecalis* (ATCC 29212) were obtained from the Center for Research and Development in Biotechnology, HUST, Hanoi, Vietnam.

2.2. Preparation of Cordyline fruticosa L. Leaf Extract

The dried Col leaf sample weighing up to 10 g was boiled in a 500 mL triangular flask containing 200 mL double distilled water for 10 min. The extract solution was filtered using a Whatman No.1 filter paper, then centrifuged at 10,000 rpm for 20 min. The obtained Col supernatant was analyzed using the pH differential method to determine the total anthocyanin content (TAC) of the Col extract [33,40]. The anthocyanin was a main component in the aqueous extraction of Col leaves [33,34], hence its concentration was adjusted to control the reducing power of the Col extract in the synthesis reaction of AgCol. The Col extract was then diluted to an anthocyanin concentration of 10 mg/L by adding double distilled water, before being using for the synthesis of AgCol, where it was kept stored at 4 °C for further use. A schematic illustration on preparation of Col leaf extract by the maceration method was represented in Figure 1.

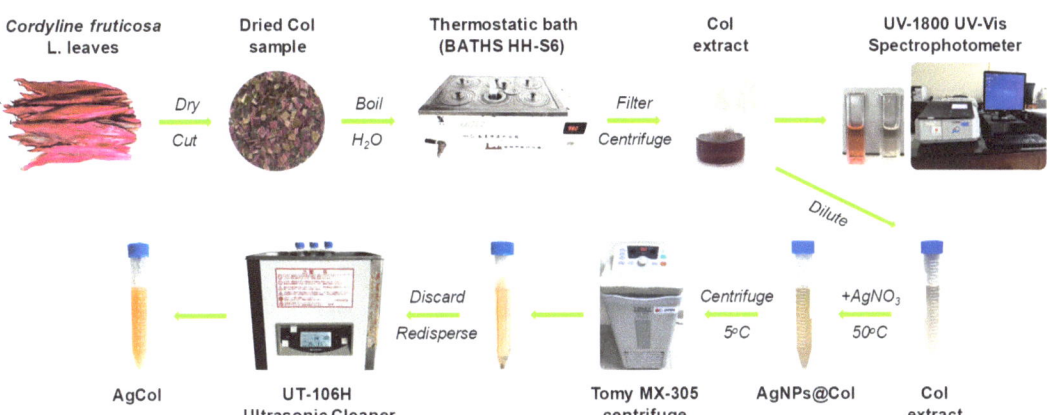

Figure 1. Schematic representation of the *Cordyline fruticosa* L. leaf extraction by the maceration method and the phytochemical-mediated synthesis of AgCol from the extract.

2.3. Synthesis of Silver Nanoparticles

The silver nanoparticles were prepared by adding 1 mL of aqueous silver nitrate solutions at varying concentrations (6, 10, 14, and 18 mM) to 10 mL of the diluted Col extract (10 mg/L anthocyanin) in Falcon tubes. The reaction mixtures were kept in the dark at 50 °C with different incubation times (2, 4, 6 and 8 h). After the passage of time, the synthesized nanoparticles were centrifuged at 15,000 rpm for 30 min at 5 °C. The solution was discarded, and the obtained pellet was re-dispersed in the double distilled water using a sonication bath (UT-106H Ultrasonic Cleaner, Japan). This step was repeated two times to remove residual reagents, and the purified nanoparticle solution (AgCol) was stored at 4 °C prior to analysis. A schematic illustration of the phytochemical-mediated synthesis of AgCol from Col extract was illustrated in Figure 1.

2.4. Dissolution of Silk Fibroin

The dissolution of silk fibroin procedure was reported in our previous work [41]. In a typical experiment, *Bombyx mori* cocoons were degummed by boiling in an aqueous solution containing 5 g/L Na_2CO_3 at a liquor ratio of 1:20 (mass in gram per volume in mL) for 30 min. The silk fibroin was rinsed several times using warm and cold distilled water, and then neutralized with acetic acid (2 mL/L). The sample was washed again in plenty of distilled water and dried at 60 °C until at a constant weight. After that, 2.8 g degummed silk fibroin was dissolved in a glass flask containing 10 mL solution of lithium bromide/ethanol/water (LiEtW) with a mass ratio of 45:44:11 at 80 °C for 60 min. The obtained fibroin solution was diluted 15 times with double distilled water to reduce the viscosity. Excess LiBr and ethanol were then removed from the fibroin solution through microfiltration and ultrafiltration systems with hollow-fiber cartridges in the QuixStand Benchtop system (GE Healthcare, Chalfont Saint Giles, UK). In the first stage of the filtering process, 0.2 μm hollow-fiber cartridge was used to remove impurities and high molecular weight fibroin segments. In the next stage, the fibroin solution was subsequently filtered through a 10,000 NMWC (nominal molecular weight cutoff) hollow fiber ultrafiltration cartridge to get the fibroin segments with molecular weight over 10 kDa retaining inside the filter tube. The solution passed through the ultrafiltration system contained low molecular weight fibroin segments, excess LiBr, ethanol, and water. The fibroin content in the obtained solution was measured using an infrared moisture analyzer (MA35, Sartorius AG, Göttingen, Germany). The scheme of degumming, dissolving, and filtering silk fibroin was illustrated in Figure 2.

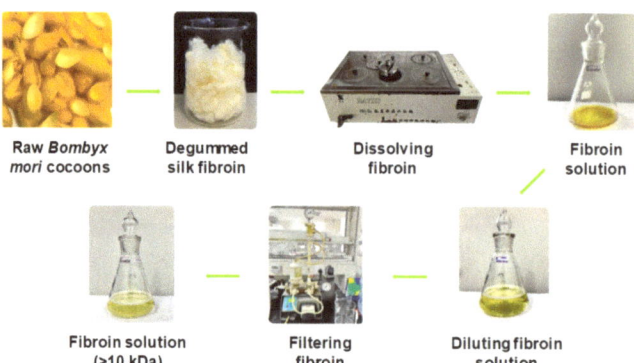

Figure 2. Schematic illustration of degumming, dissolving, and filtering silk fibroin.

2.5. Modification of Viscose Fabrics with Silk Fibroin and Silver Nanoparticles

Silk fibroin treated viscose fabric (VisFib) was prepared following our previously reported method [41]. Briefly, the viscose fabric samples with size of 35 × 35 cm were padded twice using an Atlas D394A laboratory padder to achieve a wet pickup of 80%

with the 2.5% fibroin solutions (Fib > 10 kDa). The fabrics were then dried at 110 ± 3 °C for 2 min using SDL mini-drier 398 laboratory thermo-fixation. Subsequently, the dried samples were soaked in a 10 g/L aluminum sulfate solution, padded at 80% wet pickup, and cured at 110 ± 3 °C for 5 min to regenerate and fix silk fibroin onto viscose fabrics. The samples were rinsed to remove unfixed and excess reactants and dried again.

In the next step, the untreated and treated viscose fabrics were individually immersed in serial AgCol solutions with different concentrations of 80, 40, and 20 μg/mL for 15 min at a liquor-to-fabric ratio of 5:1 (w/w) and squeezed using the Atlas D394A laboratory padder to allow various wet pickups of 70%, 80%, and 90%. After that, the padded fabrics were dried at 110 ± 3 °C for 2 min using SDL mini-drier 398 laboratory thermo-fixation. The dipping–padding–drying processes of the fabric samples were repeated two times. The name of AgCol embedded fabric samples was denoted in Table 1. The process of viscose fabrics treated with silk fibroin and AgCol was depicted in Figure 3.

Table 1. The codes of fabric samples with different treatments.

Sample	Silk Fibroin (%)	AgCo (μg/mL)	Wet Pickup (%)		
			70	80	90
Viscose fabric (Vis)	0	80	VisAg11	VisAg21	VisAg31
	0	40	VisAg12	VisAg22	VisAg32
	0	20	VisAg13	VisAg23	VisAg33
Fibroin treated viscose fabric (VisFib)	2.5	80	VisFibAg11	VisFibAg21	VisFibAg31
	2.5	40	VisFibAg12	VisFibAg22	VisFibAg32
	2.5	20	VisFibAg13	VisFibAg23	VisFibAg33

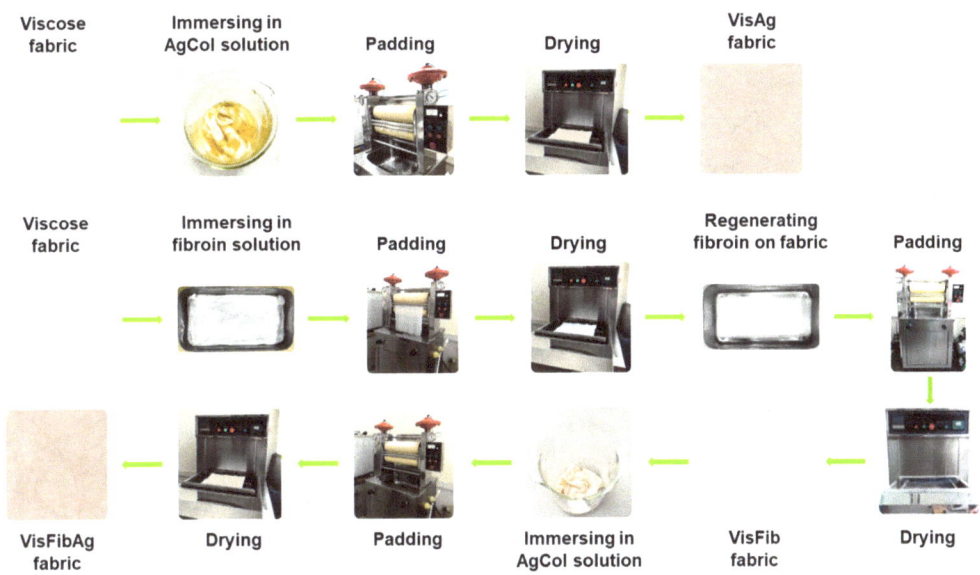

Figure 3. Scheme of modification of viscose fabrics with silk fibroin and AgCol.

2.6. Analytical Methods

2.6.1. The Total Anthocyanin Content of the Col Extract

The total anthocyanin content (TAC) of the Col extract was measured via the pH differential method using two buffer systems of pH 1.0 (KCl + HCl) and pH 4.5 ($KHC_8H_4O_4$ + HCl). The absorbance of the samples was recorded between 400 and 700 nm in a UV–vis spectrophotometer (UV-1800, Shimadzu, Japan) at a resolution of 1 nm, in 10 mm optical path length quartz cuvettes. The TAC (mg cyanidin-3-glucoside (C3G)/L) of samples was calculated via the following Equation (1) [33,40]:

$$TAC\ (mg/L) = A \times M \times DF \times 1000/(\varepsilon \times d) \quad (1)$$

where

A (absorbance) = $(A_{\lambda max} - A_{700nm})_{pH=1.0} - (A_{\lambda max} - A_{700nm})_{pH=4.5}$;
M (molecular weight) = 449.2 g/mol for C3G;
DF (dilution factor);
1000 = conversion from g to mg;
ε (molar absorptivity coefficient in L/mol/cm for C3G) = 26,900;
d (path length) = 1 cm.

2.6.2. Characterization of the AgCol

The optical absorption spectra of the AgCol were observed using a UV–vis spectrophotometer (UV-1800, Shimadzu, Japan) between 300 and 700 nm.

Transmission electron microscopy (TEM, JEOL JEM-1400, Tokyo, Japan) was employed to identify the morphology and size of the AgCol.

The XRD patterns of nanoparticles were recorded on a powder X-ray diffractometer (XRD, Oxford Instruments, Oxford, UK) using CuKα radiation (λ = 1.541 Å), operating at 40 kV and 40 mA, in the range of $10° \leq 2\theta \leq 80°$. The particle size of AgCol was calculated using the Debye–Scherrer Equation (2).

$$(d = 0.91\lambda/\beta\cos\theta) \quad (2)$$

Fourier transform infrared spectra of the Col extract and the AgCol were obtained with a FTIR spectrometer (Nicolet 6700, Thermo Scientific, Waltham, MA, USA) in the range of 4000–500 cm^{-1}.

Thermogravimetric analysis of the AgCol was carried out on a thermo gravimetric analyzer (TGA-209F, Netzsch, Selb, Germany). The sample was heated to temperature ranging from 40 °C to 800 °C under N$_2$ atmosphere, at 10 °C/min heating rate.

2.6.3. Characterization of the Modified Viscose Fabrics

A scanning electron microscope (SEM, SM-6510LV JEOL, Japan) coupled with an energy dispersive X-ray spectroscope (EDX, Oxford EDS Microanalysis System, Oxford, UK) was used to determine the morphologies and elemental compositions of the untreated and treated viscose fabrics with Fib and AgCol after platinum sputtering.

The FTIR measurements of the untreated and treated viscose fabrics with Fib and AgCol were studied via a FTIR spectrometer (Nicolet 6700, Thermo Scientific, USA).

The Kjeldahl method (Gerhardt Vapodest Kjeldahl Analysis System, Germany) was employed to detect the content of Fib in the modified viscose fabrics according to the AOAC Official Method 2001.11 with the conversion factor of 6.25 [42].

Silver contents in the modified viscose fabrics were measured using a Atomic Absorption Spectrometer (AAS, PinAAcle 900T, PerkinElmer, Waltham, MA, USA).

Color changes of the modified viscose fabrics, in terms of L*, a*, and b* values, and color differences (ΔE*) were determined using a reflectance spectrophotometer (Ci4200, X-rite, USA) with D65 illumination, and a 10° observer. In the CIELab color space, L* represents lightness, while a* and b* represent chromaticity parameters. The average color parameter values were evaluated at three positions for each sample. The overall color change was calculated based on the following Equation (3):

$$\Delta E = \sqrt{\Delta L^{*2} + \Delta a^{*2} + \Delta b^{*2}} \quad (3)$$

where ΔL^*, Δa^*, and Δb^* represent the changes in L*, a* and b* between the initial and three values, respectively.

2.7. Antibacterial Activities

2.7.1. Bio-Synthesized Silver Nanoparticles (AgCol)

The antibacterial activities of the AgCol against three gram-negative bacteria (*E. coli*, *P. aeruginosa* and *S. enterica*) and three gram-positive bacteria (*S. aureus*, *B. cereus* and *E. faecalis*) were studied by using well diffusion assay with the Clinical Laboratory Standard Institute guidelines [43]. In a typical experiment, each bacterial strain was sub-cultured in nutrient broth at 37 °C until it reached a count of approximately 10^8 colony-forming units (CFU) per mL in sterile screw-cap test tubes. Afterwards, each bacterial suspension was spread uniformly on the individual agar petri plates by using sterile cotton swabs. Seven wells with the diameter of 6 mm were aseptically punched using a sterile cork borer, and then 30 µL of Col extract (10 mg C3G/L), AgCol solutions at different concentrations (120, 60, 30 and 15 µg/mL), a standard antibiotic (Chloramphenicol, 200 µg/mL) as positive control, and double distilled water as a negative control were poured into their respective wells. These plates were incubated at 37 °C for 24 h, and the obtained zones of inhibition (ZOI) were measured using the following Equation (4):

$$W = (T - D)/2 \tag{4}$$

where

W: width of clear zone of inhibition, mm;
T: total diameter of the test specimen and clear zone, mm;
D: diameter of the test specimen, mm.

2.7.2. The Modified Viscose Fabrics with AgCol and Silk Fibroin

The antibacterial activities of the untreated and treated fabrics with AgCol and silk fibroin against both *E. coli* and *S. aureus* were evaluated qualitatively and quantitatively according to the established protocols to test the antibacterial activity of textiles. The following methods were used: AATCC 147-2004 and ASTM E2149-10 test methods.

For the parallel streak method AATCC 147-2004, the rectangular fabric samples (15 × 55 mm) were prepared and pasteurized by the autoclave at 121 °C for 20 min. About 1 mL of each strain culture grown overnight brain heart infusion broth (BHI), at approximately 10^7 CFU/mL, was mixed with 9 mL of sterile distilled water. Then, one loopful of diluted inoculum was transferred to the surface of the sterile agar plate by making five streaks, approximately 60 mm in length, spaced 10 mm apart and covering the central area of a standard petri dish. The prepared fabric samples were gently pressed by contact across the five inoculum streaks on the agar surface and incubated at 37 °C for 24 h. The width of the clear zone of inhibition was used to determine the antimicrobial activity.

For the dynamic shake flask method ASTM E2149-10, about 1 g of each fabric sample was cut into small pieces and transferred into individual sterile glass flasks containing 50 mL of bacterial suspension with a concentration of 10^5 CFU/mL. All flasks were loosely capped, placed in an incubator, and shaken at 37 °C and 120 rpm using an incubator shaker for 1 min and 1 h. Afterwards, a series of dilutions of the bacterial sample solutions using a buffer solution and 100 µL was spread over the agar plate. Bacteria recovery at after 1 min (zero contact time) and 1 h were determined. The inoculated plates were then incubated at 37 °C for 24 h and the surviving bacteria were counted. The percent reduction of bacteria was calculated using the following Equation (5):

$$R = (B - A) \times 100\%/B \tag{5}$$

where

R: the bacterial reduction percentage, %;
A and B: the number of bacteria recovered for the flasks containing test samples and without samples (blank), respectively.

Each experiment was performed in triplicates and the final values were presented as the mean ± standard deviation (SD).

2.7.3. Durability of the Antimicrobial Treatment against Washing

To evaluate the durability of the antimicrobial treatment, the washing fastness test on the modified viscose fabrics was carried out in accordance with the AATCC 61-2013 (2A) standard, which simulates five home laundering cycles. The stability of the treated viscose fabric with AgCol and Fib was examined through antibacterial tests against *E. coli* and *S. aureus* after 5, 10, 20, and 30 washes in the presence of a non-ionic detergent.

3. Results and Discussion

3.1. Synthesis and Optimization of AgCol from Col Extract

Previous reports confirmed that biological compounds existing in plant extracts, such as alkaloids, flavonoids, polysaccharides, anthocyanins, glycosides, proteins, could played a dual role of both reducing and stabilizing agents for the synthesis of AgNPs [18,20,44–46]. In addition, the size, shape, and yield of bio-synthesized AgNPs depend not only on the content of bio-compounds but also on the type of compounds present in the plant extract [45,46]. The *Cordyline fruticosa* L. leaf was reported to contain anthocyanin, saponin, flavonoids, flavonols, chlorophyll, glycoside, etc., in which anthocyanin was one of the major compounds of its aqueous extract [33–39]. The primary molecule structure of anthocyanin is cyanidin-3-glucoside, which was depicted in Figure 4a. Based on the data obtained from the UV–vis spectra of Col extract at pH = 1.0 and pH = 4.5 (Figure 4b), the TAC of the Col extract was calculated to be 109.72 ± 0.63 mg C3G/L. Anthocyanin belongs to polyphenol compounds, which have inherent reduction property. In this research, the anthocyanin content of the Col extract was adjusted to control the reducing power of the Col extract in the green synthesis reaction. For the AgNPs synthesis, the Col extract was diluted to a fixed concentration of 10 mg C3G/L, mixed with the silver nitrate solution at various concentrations, and kept at 50 °C for different periods of reaction time.

The synthesis of AgNPs was investigated in various conditions, such as $AgNO_3$ concentration and reaction time, which were identified as factors affecting the properties and yields of AgNPs. As seen from Figure 4c,d, the addition of $AgNO_3$ solution into the Col extract exhibited a color change from faint pink to yellowish-brown, indicating the excitation of surface plasmon resonance (SPR) during the reduction of silver ion to metallic silver [47]. The UV–vis absorption spectra of AgCol solutions showed the SPR peaks at around 470 nm, which confirmed the formation of the AgNPs. As depicted in Figure 4b,c, the Col extract did not show the SPR peak in the visible range, while the apparent SPR peak at 436 nm for 6 mM $AgNO_3$ concentration was exhibited. A relative red-shift of maxima of the SPR peak upon increasing the concentration of $AgNO_3$ can be noticed from 436 to 478 nm. This red-shift of the peak could be due to the enlargement of AgNPs [15]. In addition, the peak intensity reached its maximum at a wavelength of 470 nm for the 14 mM $AgNO_3$ concentration, and then it decreased with higher concentrations. The decreasing and flattening of the resonance peak at $AgNO_3$ concentration of 18 mM could be due to the lower yield and larger size of AgCol [15,19,47,48]. Based on the obtained results, 14 mM $AgNO_3$ was fixed for further optimization of other parameters. To evaluate the effect of reaction time on the AgCol synthesis, the UV–vis spectra were recorded at 2, 4, 6, and 8 h. As shown in Figure 4d, the SPR peak intensities of AgCol were noted to increase with a longer reaction time. However, a single, sharp, and highest peak absorbance was observed at 470 nm after a 6 h reaction, which provided evidence for spherical-shaped AgCol with a narrow particle size range [48–50]. In congruence with experimental observation, the optimum conditions for the bio-synthesis of AgCol were 14 mM $AgNO_3$ solution in a 1:10 ratio of the diluted Col of 10 mg C3G/L, and a reaction time of 6 h.

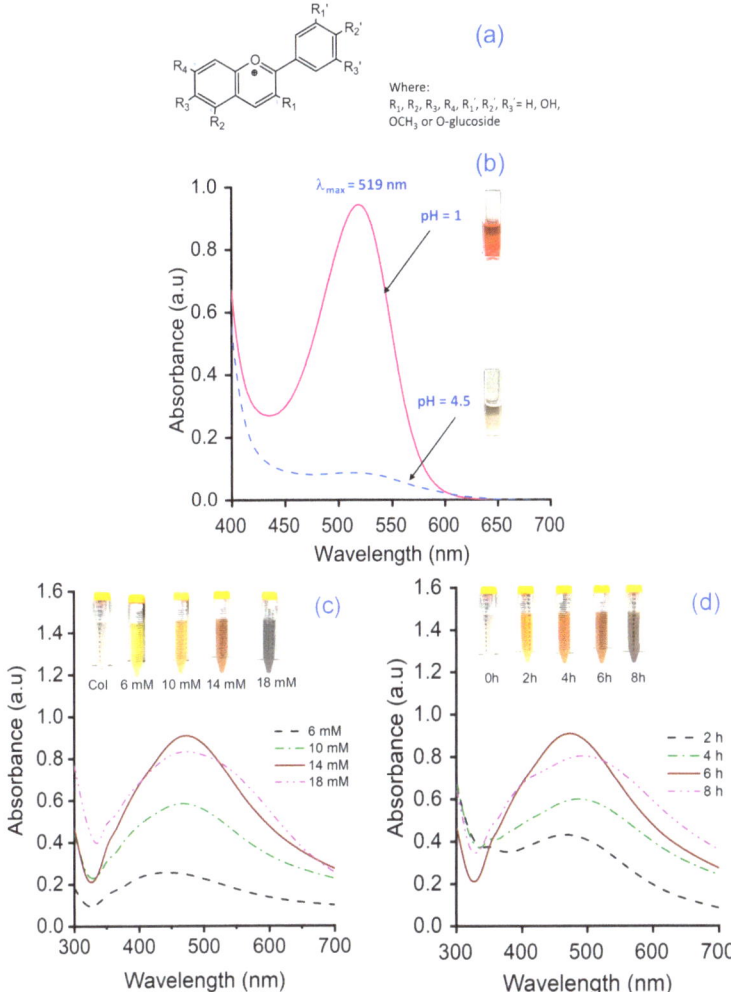

Figure 4. (**a**) The molecular structure of anthocyanin, (**b**) UV–vis spectrum of Col extract, UV–vis spectra of AgCol synthesized at (**c**) different AgNO₃ concentrations, and (**d**) reaction times. Real images of the samples are shown as insets in (**b–d**).

3.2. Morphological and Structural Characterization of AgCol

The TEM measurement was carried out to study the morphology and size of the AgCol. Figure 5a,b revealed spherical-shaped nanoparticles with a size of 10–50 nm in diameter with an average size of 28.5 nm. The spherical-shaped nanoparticles produced in the current study are in line with the expected shapes for AgCol. This result is consistent with the UV–vis observation and previous reports [44,45,48].

The crystalline structure of the AgCol was studied by recording the XRD pattern. As shown in Figure 5c, the XRD pattern of the AgCol revealed both the diffraction peaks of AgNPs and silver chloride nanoparticles. The diffraction peaks were observed at $2\theta = 38.18°$ (111), 43.72° (200), 64.45° (220), and 77.72° (311), confirming that the AgNPs were crystalline in nature and had the face center cubic (FCC) structure of metallic silver [45,51]. Other diffraction peaks at $2\theta = 27.90°$ (111), 32.30° (200), and 46.00° (220) were corresponded to silver chloride nanoparticles (JCPDS No.: 85-1355) with FCC geometry [52–54]. These silver-based nanoparticles are often found in the final product of plant-mediated green synthesis of AgNPs [52–54]. The average size of AgCol was estimated using the Debye–Scherrer equation to be about 19.3 nm [46,53]. This observation of the XRD analysis supported the UV–vis and TEM measurements that the bio-constituents in the Col extract played a crucial role in the spherical AgCol stability.

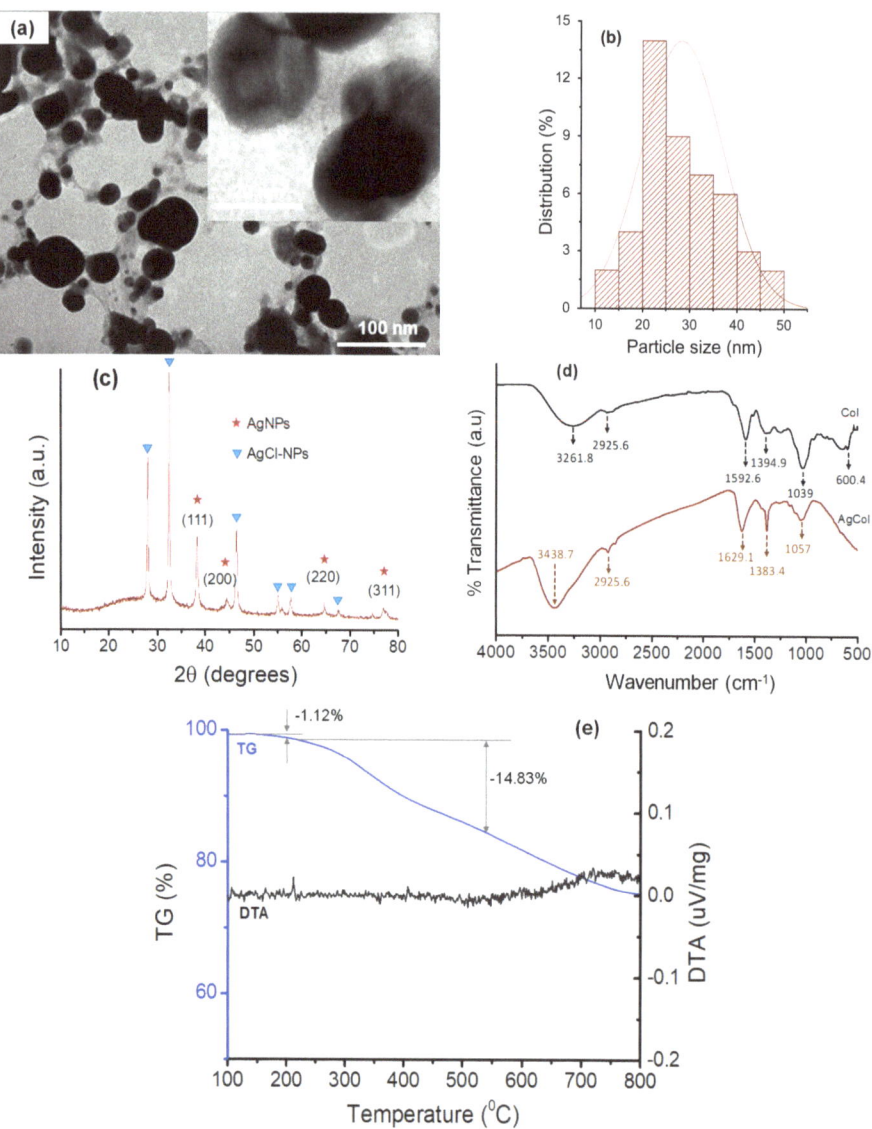

Figure 5. (a) TEM micrographs of AgCol with different magnification ×40 k and ×500 k (inset), (b) histograms of the AgCol size distribution; (c) XRD pattern of AgCol, (d) FTIR spectra of Col and AgCol, and (e) TGA of AgCol.

An FTIR analysis was used to confirm the possible interaction between AgCol and the functional groups of capping agents presented in the Col extract. The FTIR spectrum of the Col extract (Figure 5d) reflected intensive peaks of O-H (3261.8 cm^{-1}), C-H of alkanes

(2925.6 cm^{-1}), C=O of esters and ketones (1592.6 cm^{-1}), C-O of esters (1394.9 cm^{-1}), and C-O of phenolic compounds (1039 cm^{-1}) [33]. These characteristic peaks in the Col spectrum were typical of flavonoid compounds, specific anthocyanin molecules. The FTIR spectrum of the AgCol also presented similar absorption peaks at 3437.7, 2925.6, 1629.1, 1383.4, and 1057 cm^{-1}, confirming the presence of residual phytochemicals on the surface of the nanoparticles. However, shifts of peaks upon AgCol formation when compared to the Col could be attributed to the reduction of corresponding functional groups. In addition, a marked decrease in the peak intensity was observed at 1057 cm^{-1}, indicating a possible involvement of the flavonoid compounds in the reduction process. The FTIR results confirmed that different functional groups of the phytocompounds of the extract were responsible for the formation and stabilization of AgCol, as presented in the previous studies [18,20].

To further confirm the crucial role of bio-compounds in stabilizing and capping the AgCol, TGA was carried out. This authentic technique is often used to determine the thermal stability and decomposition of a sample [19,49]. The TGA thermogram of the AgCol (Figure 5e) exhibited three identical weight loss stages over a wide temperature range from room temperature to 800 °C. The initial weight loss (1.12%) at 200 °C could be involved in the evaporation process of the physically bound water in the AgCol sample. The second weight loss (16.99%) occurred between 200 °C and 600 °C, which could be correlated to decomposition of the bio-organic compounds bound on the surface of the AgCol. The last weight loss at higher temperature could be corresponded to the degradation and evaporation of the metallic silver [55]. Hence, the purity percentage of the AgCol was calculated to be about 81.89 wt% pure silver. The TGA data further confirmed that the presence of organic compounds over the AgCol surface could not only control the rate of reaction but also effectively prevented the agglomeration of nanoparticles, thus, stabilizing them.

According to the above analyses, we propose a possible mechanism for bio-synthesis of AgCol using Col leaf extract as shown in Figure 6. Anthocyanin was used as a model molecule because many of bio-molecules existing in the Col leaf extract have similar chemical structures in terms of the presence of hydroxyl groups linked with aromatic rings. The biological compounds that act as a reducing agent may be not necessarily the same which play a role as capping agents. The mechanism of AgCol formation mainly consists of the following three steps: reduction of silver ions to obtain metallic silver atoms, the nucleation of some silver atoms to build up small clusters which act as a template for AgCol formation, and the growth of these clusters and the stabilization of formed AgCol by capping agents to prevent aggregation. This proposed mechanism of AgCol bio-synthesis in the presence of Col leaf extract is in good agreement with previous reports [56–58].

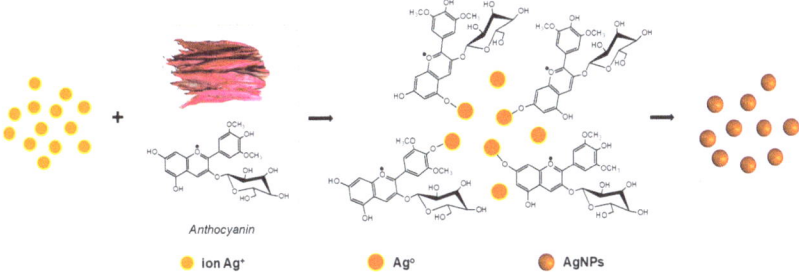

Figure 6. Proposed mechanism for bio-synthesis of AgCol using Col leaf extract.

3.3. Antimicrobial Activity of AgCol

To assess the antibacterial activity of the AgCol, six different bacterial strains including *E. coli*, *P. aeruginosa*, *S. enterica*, *S. aureus*, *B. cereus*, and *E. faecalis* were exposed to various AgCol concentrations (120, 60, 30, and 15 µg/mL) using the well diffusion method. As shown in Figure 7a–f, the double distilled water (a negative control) and Col extract did not display any antibacterial activity in terms of the ZOI, while the AgCol and the chloramphenicol (a standard antibiotic) demonstrated an obvious ZOI. It was clearly observed that the AgCol possessed promising antimicrobial activities against the tested bacterial pathogens. As presented in Figure 7g, the antibacterial activity was dependent on the tested bacteria strain and the AgCol concentration. Among the tested isolates, the ZOI was found to be the highest against *P. aeruginosa* (16.6 ± 0.27 mm) and the lowest against *S. enterica* (5.9 ± 0.37 mm) at the highest tested AgCol concentration (120 µg/mL). The antimicrobial ability of the AgCol against *E. coli* and *S. aureus* was quite good even at a low concentration (15 µg/mL), with the ZOI of 6.9 ± 0.23 and 7.6 ± 0.56 mm, respectively. Furthermore, the comparison of the ZOI against each bacterial pathogen exhibited the tendency of the enhancement in the antibacterial activity with the increase in AgCol concentration. These bactericidal properties of plant-mediated green synthesized AgNPs have been addressed in previous studies [50,59].

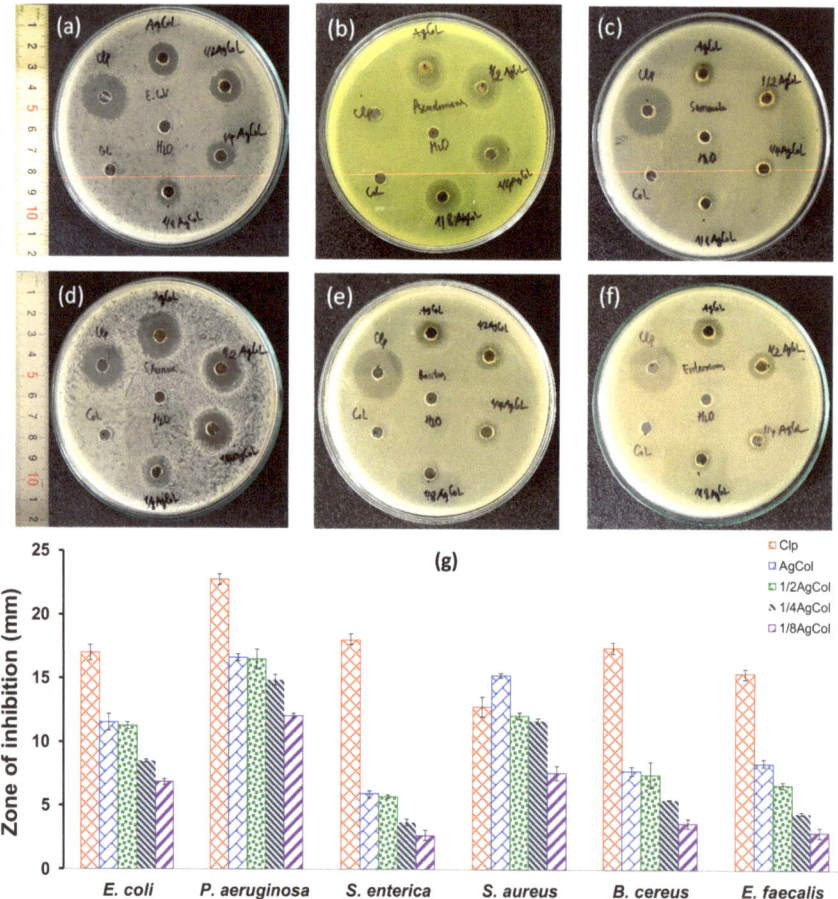

Figure 7. The photographs showing the zone of inhibition of the AgCol against (**a**) *E. coli*, (**b**) *P. aeruginosa*, (**c**) *S. enterica*, (**d**) *S. aureus*, (**e**) *B. cereus*, and (**f**) *E. faecalis*; negative control (H_2O); positive control (chloramphenicol 200 µg/mL); AgCol (120 µg/mL); 1/2 AgCol (60 µg/mL); 1/4 AgCol (30 µg/mL), 1/8 AgCol (15 µg/mL) and Col extract (10 mg C3G/L); (**g**) quantitative evaluation of antibacterial activity of the AgCol against six pathogenic bacteria (±SD, n = 3).

3.4. Coloration of the Modified Viscose Fabrics with AgCol and Silk Fibroin

In order to evaluate the adsorption capacity of AgCol on the Vis and VisFib fabrics with change in the wet pickup and AgCol concentration, the colorimetric data (L^*, a^*, b^*), color differences (ΔE^*), and color strength (K/S) of the fabric samples were determined, as shown in Table 2. The colorimetric data of both the AgCol-treated Vis and AgCol-treated VisFib fabric samples had a high L^* value, and positive a^* and b^* values, indicating the light browning effect of the AgCol on the fabrics. The ΔE^* and K/S values of all AgCol-treated fabric samples increased with the lower wet pickup and the higher AgCol concentration. Furthermore, these values of the AgCol-treated VisFib samples were higher than that of the AgCol-treated Vis samples. This could be explained by the formation of coordination bonding between AgCol and electron-rich nitrogen atoms of amine groups in the fibroin adhered on viscose fabric, which supplemented the linkage of hydroxyl groups of cellulose in viscose fabric with the nanoparticles. Therefore, the fibroin modified viscose fabric could improve the AgCol uptake to compare with the neat viscose fabric. It might enhance antibacterial efficacy and the antibacterial durability of the modified fabrics.

Table 2. Colorimetric data, color differences, K/S value, and images of the AgCol treated and untreated viscose fabric samples, with change in the wet pickup and AgCol concentration.

	Wet Pickup (%)	AgCol (µg/mL)	L^*	a^*	b^*	ΔE^*	K/S	Fabric Images
Vis	-	-	93.54	−0.96	4.32	0	0.06	
VisFib	-	-	93.47	−0.98	4.93	0.48	0.07	
VisAg11	70		85.16	3.86	8.95	8.39	0.19	
VisAg21	80	80	84.32	4.55	9.08	9.17	0.2	
VisAg31	90		85.12	4.1	8.77	8.49	0.18	
VisAg21		80	84.32	4.55	9.08	9.17	0.20	
VisAg22	80	40	85.65	3.15	6.34	6.33	0.15	
VisAg23		20	89.58	1.11	5.75	3.53	0.11	
VisFibAg21		80	79.95	4.31	6.72	8.91	0.28	
VisFibAg22	80	40	83.99	2.85	5.90	6.44	0.19	
VisFibAg23		20	87.73	1.56	5.22	4.33	0.13	

3.5. Antibacterial Efficacy of the Modified Viscose Fabrics with AgCol and Silk Fibroin

It is well known that AgNPs can endow cellulose fabrics with excellent antibacterial activity [1,3,5]. Thus, antibacterial tests were performed with the AATCC 147-2004 and ASTM E2149-10 methods to investigate the antibacterial efficacy of the AgCol-treated viscose fabrics. As shown in Figure 8, all the AgCol-treated viscose fabrics and the chloramphenicol-treated fabrics revealed an obvious ZOI against *E. coli* and *S. aureus*, while the Vis and VisFib fabrics exhibited no antibacterial activity. This confirms that AgCol-treated viscose fabrics possess antibacterial capability in this work. In addition, the ZOI of AgCol-treated fabric samples seem unchanged with the altering wet pickup, and these were substantially larger with the higher AgCol concentration. Furthermore, these values for the AgCol-treated VisFib samples were higher than those of the AgCol-treated Vis samples. These results demonstrated that the higher the AgCol uptake on fabric samples, the better the antibacterial efficiency. The antibacterial results also gave additional evidence of the enhancement of the AgCol uptake on fabric by the fibroin modifying viscose fabric. The AgCol-treated fabric conditions were chosen to get a good antibacterial efficacy with the wet pickups of 80% and AgCol concentrations of 80 µg/mL. The obtained fabrics under this treatment condition were further evaluated in terms of antibacterial durability against up to 30 home laundering cycles.

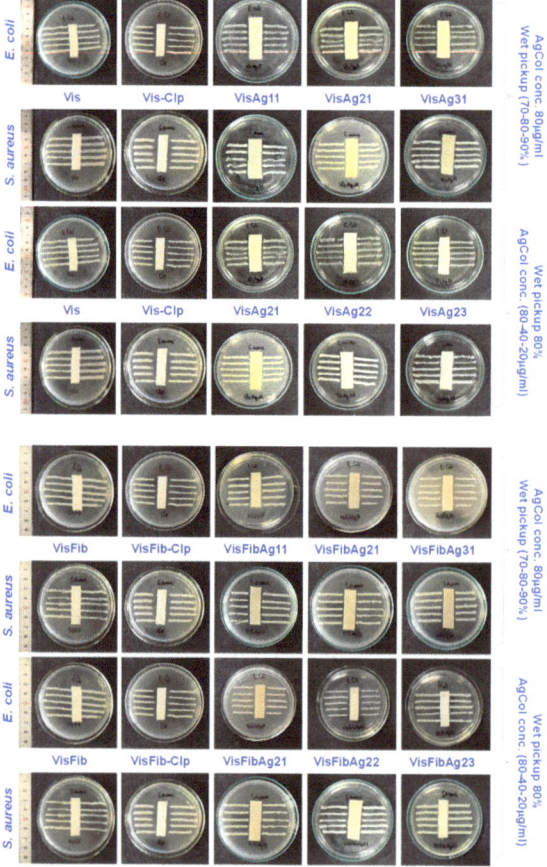

Figure 8. The photographs showing the zone of inhibition of the Vis, VisFib, chloramphenicol-treated (200 µg/mL) Vis (Vis-Clp), and VisFib (VisFib-Clp), and the AgCol-treated fabric samples against *E. coli* and *S. aureus*, with change in the wet pickup and AgCol concentration.

The durable antimicrobial activity of the AgCol-treated viscose fabrics was qualitatively visualized in Figure 9a, and then plotted in Figure 9b. As shown in Figure 9b, the antibacterial rate (BR) of the VisAg and VisFibAg fabrics for both *E. coli* and *S. aureus* after a 1 h contact period were both 99.99%. This indicated that the AgCol-treated fabrics provide excellent antibacterial properties. The BR values of the VisAg samples for *E. coli* and *S. aureus* were about 80% and 87% after 30 washing cycles, respectively, while these values for the VisFibAg samples still remained 99.99% for both *E. coli* and *S. aureus*.

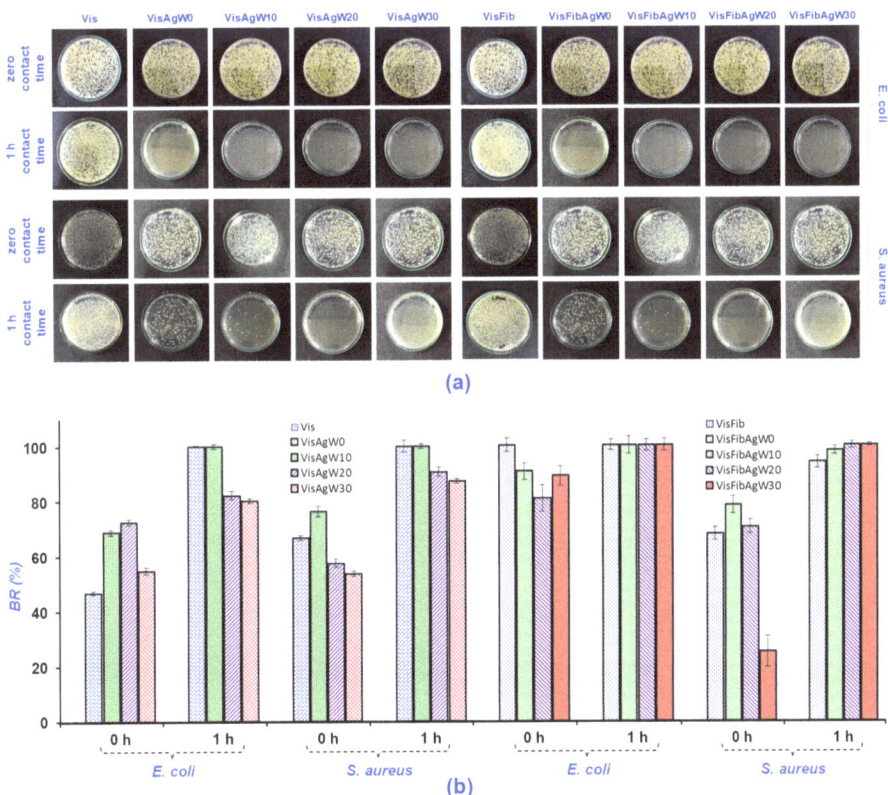

Figure 9. (a) The photographs of bacterial growth in the BHI agar plates and (b) the antibacterial rate (BR) of the Vis, VisFib, and AgCol-treated fabric samples against *E. coli* and *S. aureus* after 10, 20, and 30 repeated washing cycles.

3.6. Characteristics of the Modified Viscose Fabrics with AgCol and Silk Fibroin

To obtain more evidence of the role of fibroin as an effective binder between the AgCol and cellulose fibers in viscose fabric, the fabric surface morphology, AgCol, and fibroin content on the fabrics before and after wash were investigated. The fabric surface morphologies of the original viscose and AgCol-treated viscose fabrics before and after 30 washing tests were explored through SEM at different magnifications, as shown in Figure 10. The untreated fabric displayed the smooth surface of the viscose fibers. Rougher surfaces were observed on the fibers of AgCol-treated fabrics before and after 30 washes, indicating the occurrence of AgCol or/and Fib attached on surface of viscose fibers. The EDX spectra of those fabrics confirmed that the bright points in the SEM images were AgCol. Moreover, the bright points in the SEM micrographs of the VisFibAg fabrics were more than those of the VisAg samples, demonstrating higher AgCol uptake on the VisFibAg fabrics compared with the VisAg fabrics. After 30 washing cycles, the fibroin layer covering the surfaces of viscose fibers was still observed directly in the SEM image of the VisFibAgW30 sample. These results confirmed again that the fibroin acted as binding polymer to coat AgCol on the viscose surface.

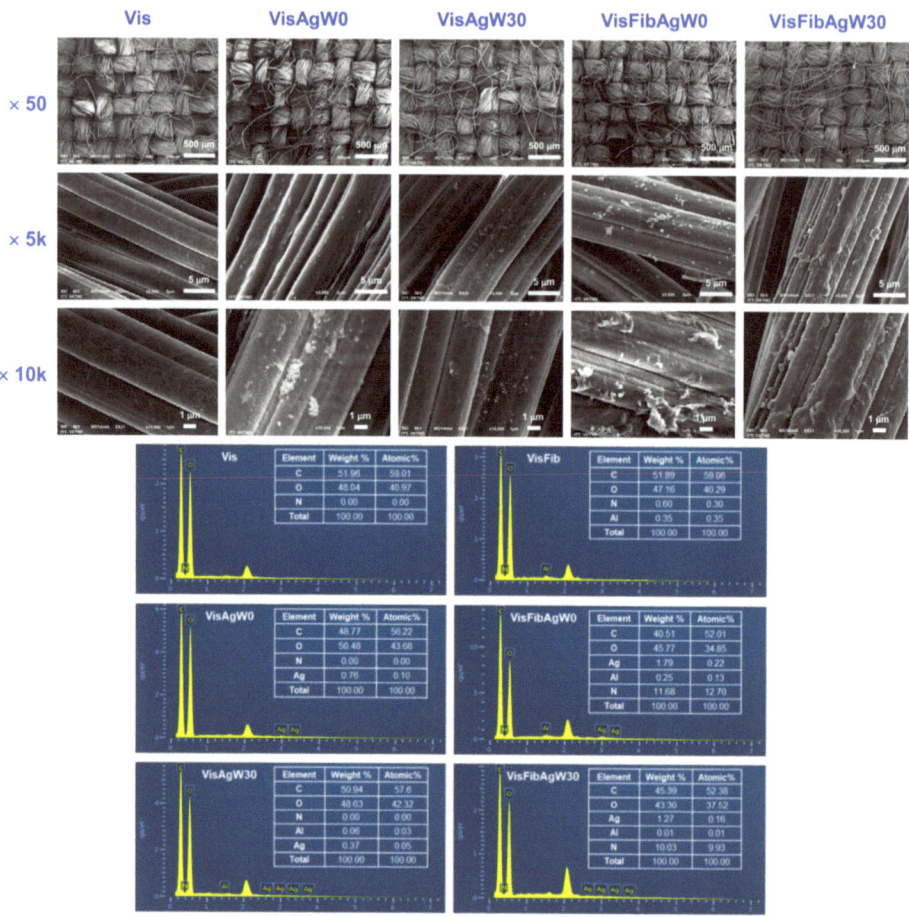

Figure 10. The SEM micrographs at different magnifications of ×50, ×5000, and ×10,000, and EDX spectra of the neat viscose fabric and the AgCol-treated fabrics before and after 30 washing cycles.

The interaction of AgCol and fibroin on the modified fabrics were analyzed by measuring silver and fibroin contents in the treated viscose fabrics before and after 30 washes (Table 3). The silver content of the VisFibAg fabric before washing was 14.3% higher than that of the VisAg fabric. After 30 washing cycles, the loss of silver content was only 25.6% for the VisFibAg fabric, while about 51.2% of silver was released from the VisAg fabric. The fibroin content of the modified fabrics decreased slightly during washing cycles. About 50% of silk fibroin still remained on both the VisFib and VisFibAg fabric after 30 washes, owing to the role of Al^{3+} ions in crosslinking between fibroin molecules and the cellulose fibers of the viscose fabric [41]. The strong linkage between the fibroin and viscose fibers lead to an enhancement in the incorporation of AgCol into the modified fabrics. Due to the presence of numerous amine groups in fibroin, it tends to bind with AgCol by coordination bonds. These results are in harmony and fit well with those illustrated in the SEM micrographs and the antibacterial ability of the AgCol-treated viscose fabrics.

Table 3. The silver content, nitrogen content, and fibroin content on the AgCol-treated viscose fabrics before and after 30 washing cycles.

No	Fabric Sample	Washing Cycles	AgCol (μg/mL)	Fibroin (%)	Silver Content (mg/kg)	Nitrogen Content (%)	Fibroin [a] Content (%)
1	Vis	-	-	-	-	0.019	-
2	VisFibW0	0	-	2.5	-	0.244	1.406
3	VisFibW30	30	-	-	-	0.121	0.638
4	VisAgW0	0	80	-	642.02	-	-
5	VisAgW30	30	80	-	313.43	-	-
6	VisFibAgW0	0	80	2.5	734.05	0.242	1.394
7	VisFibAgW30	30	80	-	545.86	0.127	0.675

[a] The protein-to-nitrogen conversion factor is 6.25.

In order to determine the possible interaction between the functional groups of fibroin on the modified fabric and AgCol, the FTIR measurements of viscose fabric (Vis), fibroin modified viscose fabric (VisFib), AgCol-treated Vis fabric (VisAg), and AgCol-treated VisFib fabric (VisFibAg) samples were carried out, and the spectra were given in Figure 11. The characteristic peaks corresponding to cellulose in the Vis at 3331.9, 2887.5, 1638.6, 1364.9, and 1017.9 cm^{-1} were assigned to the O-H, C-H, C=O, C-H, and C-O-H groups, respectively [41]. Compared to the neat viscose fabric, the VisFib spectrum revealed a fair similarity of these characteristic peaks, with the exception of the lower peak intensities at 3314.8 and 1636.5 cm^{-1} of the VisFib spectrum. It could be explained that the interactions might be occurring via the formation of hydrogen bonding between fibroin's amide groups and viscose's hydroxyl groups, and/or via the complexation of Al^{3+} ions with appropriate functional groups of fibroin and viscose [41,60]. The VisAg and VisFibAg spectra also exhibited similar characteristic peaks to those in the Vis and VisFib spectra, respectively, suggesting that the chemical structure of the cellulose was mostly unchanged. However, the peaks of the functional groups in the VisFibAg shifted toward higher wavenumbers at 3316.2, 1639.9, and 1365.3 cm^{-1}, and the intensity of these peaks were significantly decreased in comparison with the spectra of the VisFib and VisAg fabrics. The shifting of these peaks occurred due to the interaction of heavy silver atom with amino and amide groups of fibroin molecules, resulting in a decrement in the peak intensity with an ultimate red frequency shift [1,5,31,41].

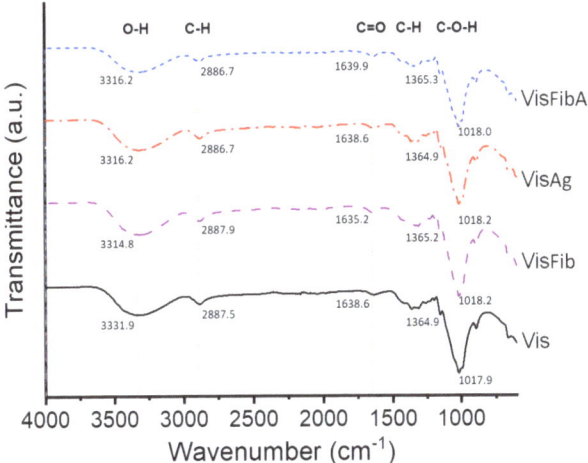

Figure 11. The FTIR spectra of the Vis, VisFib, VisAg, and VisFibAg fabrics.

On the basis of the literature review of the mechanism of silk fibroin dissolution and regeneration by metal salts [41,60], the complexation ability of fibroin and cellulose with metal ions [61], and the coordination between AgNPs and amine groups [1,5], combining with the experimental results, a proposed mechanism of fibroin regeneration and AgCol deposition onto viscose fabric was elucidated in light of Figure 12. Firstly, the silk fibroin was easily dissolved in the LiEtW solution because lithium halides solutions showed high solvency with silk fibroin [41,60]. The inter- and intra-molecular hydrogen bonds in the chains of fibroin could be broken by the electrophilic attack of the lithium ion that yielded Li-fibroin complexes [41,60]. Secondly, the silk fibroin was regenerated onto the viscose fabric by the ultrafiltration system combining with coagulation of silk fibroin using aluminum salt. Herein, Li$^+$ ions in the Li-fibroin complex were replaced by Al^{3+} ions to form a more stable Al-fibroin complex via coordination bonding between Al^{3+} ions and electron-rich nitrogen atoms of amine and amide groups in fibroin. Finally, the fibroin adhered on viscose fabric could enhance the adsorption of AgCol nanoparticles on the modified viscose fabric, and was tightly linked to the fabric via coordination bonding between the silver and the amine and amide groups in fibroin, resulting in the remarkable antibacterial efficacy and laundering durability of the VisFibAg fabric.

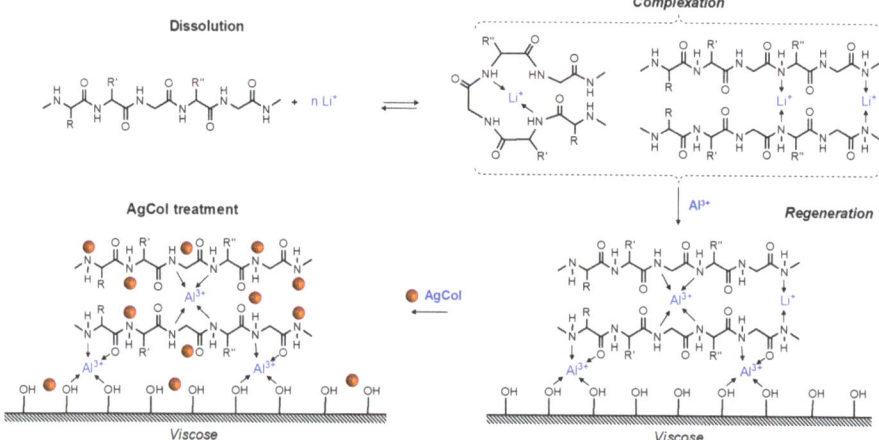

Figure 12. The proposed mechanism of fibroin regeneration and AgCol deposition onto viscose fabric.

4. Conclusions

This research provided a novel approach for the fabrication of durable antibacterial viscose fabric modified with bio-synthesized AgCol and silk fibroin as an eco-friendly binder. The AgCol were bio-synthesized using reducing and capping agents from hot water extract of *Cordyline fruticosa* L. leaves. The effect of the reaction conditions, such as the silver salt concentration and the reaction time, were investigated using several analytical techniques including UV–vis, TEM, XRD, FTIR, and TGA. The bio-synthesized AgCol were found to be effective antibacterial activities against six pathogenic bacteria. The AgCol was employed as an antibacterial agent for the functionalization of viscose fabric combined with silk fibroin. The modified viscose fabric with AgCol and fibroin exhibited a remarkable antimicrobial activity against *E. coli* and *S. aureus* bacteria even after 30 washing cycles, due to the presence of numerous amine and amide groups in fibroin, which might form coordination bonding with AgCol. The colorimetry, SEM, EDX, AAS, Kjeldahl, and FTIR analyses provided sufficient evidence for the role of the fibroin as an effective binder between the AgCol and cellulose fibers in viscose fabric. The bio-synthesized AgCol and fibroin modified viscose fabric possessed superior durable antibacterial activity, which would meet the basic criteria for medical textile applications.

Author Contributions: Conceptualization, N.-T.N.; methodology, N.-T.N. and T.-L.-H.V.; formal analysis, T.-L.-H.V.; investigation, N.-T.N. and T.-L.-H.V.; data curation, N.-T.N.; writing—original draft preparation, T.-L.-H.V.; writing—review and editing, N.-T.N.; visualization, N.-T.N. and T.-L.-H.V.; supervision, N.-T.N.; project administration, N.-T.N.; funding acquisition, N.-T.N. All authors have read and agreed to the published version of the manuscript.

Funding: This work was funded by the Vietnam Ministry of Education and Training under Grant No. B2022-BKA-23.

Institutional Review Board Statement: Not applicable.

Informed Consent Statement: Not applicable.

Data Availability Statement: Not applicable.

Acknowledgments: The authors also would like to thank colleagues from School of Textile—Leather and Fashion, Hanoi University of Science and Technology for their efforts in supporting this work.

Conflicts of Interest: The authors declare no conflict of interest.

References

1. Xu, Q.; Zheng, W.; Duan, P.; Chen, J.; Zhang, Y.; Fu, F.; Diao, H.; Liu, X. One-pot fabrication of durable antibacterial cotton fabric coated with silver nanoparticles via carboxymethyl chitosan as a binder and stabilizer. *Carbohydr. Polym.* **2019**, *204*, 42–49. [CrossRef] [PubMed]
2. Fahmy, H.M.; Eid, R.A.A.; Hashem, S.S.; Amr, A. Enhancing some functional properties of viscose fabric. *Carbohydr. Polym.* **2013**, *92*, 1539–1545. [CrossRef] [PubMed]
3. Xu, Q.; Xie, L.; Diao, H.; Li, F.; Zhang, Y.; Fu, F.; Liu, X. Antibacterial cotton fabric with enhanced durability prepared using silver nanoparticles and carboxymethyl chitosan. *Carbohydr. Polym.* **2017**, *177*, 187–193. [CrossRef] [PubMed]
4. Ibrahim, N.A.; El-Zairy, E.M.R.; Eid, B.M. Eco-friendly modification and antibacterial functionalization of viscose fabric. *J. Text. Inst.* **2017**, *108*, 1406–1411. [CrossRef]
5. Raza, Z.A.; Bilal, U.; Noreen, U.; Munim, S.A.; Riaz, S.; Abdullah, M.U.; Abid, S. Chitosan Mediated Formation and Impregnation of Silver Nanoparticles on Viscose Fabric in Single Bath for Antibacterial Performance. *Fibers Polym.* **2019**, *20*, 1360–1367. [CrossRef]
6. Mostafa, K.; El-Sanabary, A. Innovative ecological method for producing easy care characteristics and antibacterial activity onto viscose fabric using glutaraldehyde and chitosan nanoparticles. *Pigment. Resin Technol.* **2020**, *49*, 11–18. [CrossRef]
7. Emam, H.E. Antimicrobial cellulosic textiles based on organic compounds. *Biotechnology* **2019**, *9*, 29. [CrossRef]
8. Song, X.; Cvelbar, U.; Strazar, P.; Vossebein, L.; Zille, A. Antimicrobial Efficiency and Surface Interactions of Quaternary Ammonium Compound Absorbed on Dielectric Barrier Discharge (DBD) Plasma Treated Fiber-Based Wiping Materials. *ACS Appl. Mater. Interfaces* **2020**, *12*, 298–311. [CrossRef]
9. Morais, D.S.; Guedes, R.M.; Lopes, M.A. Antimicrobial Approaches for Textiles: From Research to Market. *Materials* **2016**, *9*, 498. [CrossRef] [PubMed]
10. Hassabo, A.G.; El-Naggar, M.E.; Mohamed, A.L.; Hebeish, A.A. Development of multifunctional modified cotton fabric with tri-component nanoparticles of silver, copper and zinc oxide. *Carbohydr. Polym.* **2019**, *210*, 144–156. [CrossRef]
11. Emam, H.E.; El-Hawary, N.S.; Ahmed, H.B. Green technology for durable finishing of viscose fibers via self-formation of AuNPs. *Int. J. Biol. Macromol.* **2017**, *96*, 697–705. [CrossRef] [PubMed]
12. Phan, D.-N.; Khan, M.Q.; Nguyen, N.-T.; Phan, T.-T.; Ullah, A.; Khatri, M.; Kien, N.N.; Kim, I.-S. A review on the fabrication of several carbohydrate polymers into nanofibrous structures using electrospinning for removal of metal ions and dyes. *Carbohydr. Polym.* **2021**, *252*, 117175. [CrossRef] [PubMed]
13. Edgar, K.J.; Zhang, H. Antibacterial modification of Lyocell fiber: A review. *Carbohydr. Polym.* **2020**, *250*, 116932. [CrossRef] [PubMed]
14. McQueen, R.H.; Vaezafshar, S. Odor in textiles: A review of evaluation methods, fabric characteristics, and odor control technologies. *Text. Res. J.* **2019**, *90*, 1157–1173. [CrossRef]
15. Nguyen, N.-T.; Liu, J.-H. A green method for in situ synthesis of poly(vinyl alcohol)/chitosan hydrogel thin films with entrapped silver nanoparticles. *J. Taiwan Inst. Chem. Eng.* **2014**, *45*, 2827–2833. [CrossRef]
16. Wu, Y.; Yang, Y.; Zhang, Z.; Wang, Z.; Zhao, Y.; Sun, L. Fabrication of cotton fabrics with durable antibacterial activities finishing by Ag nanoparticles. *Text. Res. J.* **2018**, *89*, 867–880. [CrossRef]
17. Zhou, Q.; Lv, J.; Ren, Y.; Chen, J.; Gao, D.; Lu, Z.; Wang, C. A green in situ synthesis of silver nanoparticles on cotton fabrics using Aloe vera leaf extraction for durable ultraviolet protection and antibacterial activity. *Text. Res. J.* **2016**, *87*, 2407–2419. [CrossRef]
18. Abdelghany, T.M.; Al-Rajhi, A.M.H.; Al Abboud, M.A.; Alawlaqi, M.M.; Ganash Magdah, A.; Helmy, E.A.M.; Mabrouk, A.S. Recent Advances in Green Synthesis of Silver Nanoparticles and Their Applications: About Future Directions. A Review. *BioNanoScience* **2018**, *8*, 5–16. [CrossRef]

19. Thi Lan Huong, V.; Nguyen, N.T. Green synthesis, characterization and antibacterial activity of silver nanoparticles using Sapindus mukorossi fruit pericarp extract. *Mater. Today Proc.* **2020**, *42*, 88–93. [CrossRef]
20. Vanlalveni, C.; Lallianrawna, S.; Biswas, A.; Selvaraj, M.; Changmai, B.; Rokhum, S.L. Green synthesis of silver nanoparticles using plant extracts and their antimicrobial activities: A review of recent literature. *RSC Adv.* **2021**, *11*, 2804–2837. [CrossRef]
21. Ovais, M.; Khalil, A.T.; Raza, A.; Khan, M.A.; Ahmad, I.; Islam, N.U.; Saravanan, M.; Ubaid, M.F.; Ali, M.; Shinwari, Z.K. Green synthesis of silver nanoparticles via plant extracts: Beginning a new era in cancer theranostics. *Nanomedicine* **2016**, *11*, 3157–3177. [CrossRef] [PubMed]
22. Heravi, M.E.M. Effects of Hydrodynamic Diameter of Nanoparticles on Antibacterial Activity and Durability of Ag-treated Cotton Fabrics. *Fibers Polym.* **2020**, *21*, 1173–1179. [CrossRef]
23. Rehan, M.; Mashaly, H.M.; Mowafi, S.; Abou El-Kheir, A.; Emam, H.E. Multi-functional textile design using in-situ Ag NPs incorporation into natural fabric matrix. *Dyes Pigments* **2015**, *118*, 9–17. [CrossRef]
24. Zhang, D.; Chen, L.; Zang, C.; Chen, Y.; Lin, H. Antibacterial cotton fabric grafted with silver nanoparticles and its excellent laundering durability. *Carbohydr. Polym.* **2013**, *92*, 2088–2094. [CrossRef]
25. Zheng, J.; Song, F.; Wang, X.-L.; Wang, Y.-Z. In-situ synthesis, characterization and antimicrobial activity of viscose fiber loaded with silver nanoparticles. *Cellulose* **2014**, *21*, 3097–3105. [CrossRef]
26. Emam, H.E.; Saleh, N.H.; Nagy, K.S.; Zahran, M.K. Functionalization of medical cotton by direct incorporation of silver nanoparticles. *Int. J. Biol. Macromol.* **2015**, *78*, 249–256. [CrossRef]
27. Su, C.-H.; Kumar, G.V.; Adhikary, S.; Velusamy, P.; Pandian, K.; Anbu, P. Preparation of cotton fabric using sodium alginate-coated nanoparticles to protect against nosocomial pathogens. *Biochem. Eng. J.* **2017**, *117*, 28–35. [CrossRef]
28. Zahran, M.; Marei, A.H. Innovative natural polymer metal nanocomposites and their antimicrobial activity. *Int. J. Biol. Macromol.* **2019**, *136*, 586–596. [CrossRef]
29. Hanh, T.T.; Thu, N.T.; Hien, N.Q.; An, P.N.; Loan, T.; Thi, K.; Hoa, P.T. Preparation of silver nanoparticles fabrics against multidrug-resistant bacteria. *Radiat. Phys. Chem.* **2016**, *121*, 87–92. [CrossRef]
30. Ibrahim, N.A.; Eid, B.M.; Abdel-Aziz, M.S. Effect of plasma superficial treatments on antibacterial functionalization and coloration of cellulosic fabrics. *Appl. Surf. Sci.* **2017**, *392*, 1126–1133. [CrossRef]
31. Arif, D.; Niazi, M.B.K.; Ul-Haq, N.; Anwar, M.N.; Hashmi, E. Preparation of antibacterial cotton fabric using chitosan-silver nanoparticles. *Fibers Polym.* **2015**, *16*, 1519–1526. [CrossRef]
32. Xu, Q.; Ke, X.; Shen, L.; Ge, N.; Zhang, Y.; Fu, F.; Liu, X. Surface modification by carboxymethy chitosan via pad-dry-cure method for binding Ag NPs onto cotton fabric. *Int. J. Biol. Macromol.* **2018**, *111*, 796–803. [CrossRef] [PubMed]
33. Nguyen, N.T.; Hoang, T.T.L. Optimization of ultrasound-assisted extraction of natural colorant from Huyet du leaves using ethanol solvent. *J. Sci. Technol. HaUI* **2019**, *51*, 109–113.
34. Lim, T.K. (Ed.) Cordyline fruticosa. In *Edible Medicinal and Non Medicinal Plants: Volume 9, Modified Stems, Roots, Bulbs*; Springer: Dordrecht, The Netherlands, 2015.
35. Nguyen Thi, D.P.; Tran, D.L.; Le Thi, P.; Park, K.D.; Hoang Thi, T.T. Supramolecular Gels Incorporating Cordyline terminalis Leaf Extract as a Polyphenol Release Scaffold for Biomedical Applications. *Int. J. Mol. Sci.* **2021**, *22*, 8759. [CrossRef]
36. Raslan, M.A.; Taher, R.F.; Al-Karmalawy, A.A.; El-Ebeedy, D.; Metwaly, A.G.; Elkateeb, N.M.; Ghanem, A.; Elghaish, R.A.; Abd El Maksoud, A.I. *Cordyline fruticosa* (L.) A. Chev. leaves: Isolation, HPLC/MS profiling and evaluation of nephroprotective and hepatoprotective activities supported by molecular docking. *New J. Chem.* **2021**, *45*, 22216–22233. [CrossRef]
37. Siddiqi, K.S.; Husen, A.; Rao, R.A.K. A review on biosynthesis of silver nanoparticles and their biocidal properties. *J. Nanobiotechnol.* **2018**, *16*, 14. [CrossRef]
38. Jadoun, S.; Arif, R.; Jangid, N.K.; Meena, R.K. Green synthesis of nanoparticles using plant extracts: A review. *Environ. Chem. Lett.* **2021**, *19*, 355–374. [CrossRef]
39. Al-Alwani, M.A.M.; Ludin, N.A.; Mohamad, A.B.; Kadhum, A.A.H.; Sopian, K. Extraction, preparation and application of pigments from *Cordyline fruticosa* and *Hylocereus polyrhizus* as sensitizers for dye-sensitized solar cells. *Spectrochim. Acta Mol. Biomol. Spectrosc.* **2017**, *179*, 23–31. [CrossRef]
40. Lee, J.; Durst, R.W.; Wrolstad, R.E. Determination of Total Monomeric Anthocyanin Pigment Content of Fruit Juices, Beverages, Natural Colorants, and Wines by the pH Differential Method: Collaborative Study. *J. AOAC Int.* **2019**, *88*, 1269–1278. [CrossRef]
41. Thang, N.N.; Huong, V.T.L. Enhancement of dye-ability of viscose fabric via modification with fibroin regenerated from waste silk cocoons. *Vlak. Text.* **2021**, *28*, 100–107.
42. Thiex, N.J.; Manson, H.; Anderson, S.; Persson, J.-Å. Determination of Crude Protein in Animal Feed, Forage, Grain, and Oilseeds by Using Block Digestion with a Copper Catalyst and Steam Distillation into Boric Acid: Collaborative Study. *J. AOAC Int.* **2019**, *85*, 309–317. [CrossRef]
43. Wikler, M.A.; Low, D.E.; Cockerill, F.R.; Sheehan, D.J.; Craig, W.A.; Tenover, F.C. *Methods for Dilution Antimicrobial Susceptibility Tests for Bacteria That Grow Aerobically: Approved Standard*, 7th ed.; Document M7-A7; CLSI: Wayne, PA, USA, 2006.
44. Ahmad, S.; Munir, S.; Zeb, N.; Ullah, A.; Khan, B.; Ali, J.; Bilal, M.; Omer, M.; Alamzeb, M.; Salman, S.M.; et al. Green nanotechnology: A review on green synthesis of silver nanoparticles-an ecofriendly approach. *Int. J. Nanomed.* **2019**, *14*, 5087–5107. [CrossRef] [PubMed]
45. Sharma, V.K.; Yngard, R.A.; Lin, Y. Silver nanoparticles: Green synthesis and their antimicrobial activities. *Adv. Colloid Interfaces Sci.* **2009**, *145*, 83–96. [CrossRef] [PubMed]

46. Hamidian, K.; Sarani, M.; Barani, M.; Khakbaz, F. Cytotoxic performance of green synthesized Ag and Mg dual doped ZnO NPs using Salvadora persica extract against MDA-MB-231 and MCF-10 cells. *Arab. J. Chem.* **2022**, *15*, 103792. [CrossRef]
47. Tran, Q.H.; Nguyen, V.Q.; Le, A.-T. Silver nanoparticles: Synthesis, properties, toxicology, applications and perspectives. *Adv. Nat. Sci. Nanosci. Nanotechnol.* **2013**, *4*, 033001. [CrossRef]
48. Kanniah, P.; Radhamani, J.; Chelliah, P.; Muthusamy, N.; Joshua Jebasingh Sathiya Balasingh Thangapandi, E.; Reeta Thangapandi, J.; Balakrishnan, S.; Shanmugam, R. Green Synthesis of Multifaceted Silver Nanoparticles Using the Flower Extract of Aerva lanata and Evaluation of Its Biological and Environmental Applications. *ChemistrySelect* **2020**, *5*, 2322–2331. [CrossRef]
49. Hemmati, S.; Rashtiani, A.; Zangeneh, M.M.; Mohammadi, P.; Zangeneh, A.; Veisi, H. Green synthesis and characterization of silver nanoparticles using Fritillaria flower extract and their antibacterial activity against some human pathogens. *Polyhedron* **2019**, *158*, 8–14. [CrossRef]
50. Vishwasrao, C.; Momin, B.; Ananthanarayan, L. Green Synthesis of Silver Nanoparticles Using Sapota Fruit Waste and Evaluation of Their Antimicrobial Activity. *Waste Biomass Valorization* **2019**, *10*, 2353–2363. [CrossRef]
51. Srikar, S.K.; Giri, D.D.; Pal, D.B.; Mishra, P.K.; Upadhyay, S.N. Green synthesis of silver nanoparticles: A review. *Green Sustain. Chem.* **2016**, *6*, 34–56. [CrossRef]
52. Ravichandran, V.; Vasanthi, S.; Shalini, S.; Shah, S.A.A.; Tripathy, M.; Paliwal, N. Green synthesis, characterization, antibacterial, antioxidant and photocatalytic activity of *Parkia speciosa* leaves extract mediated silver nanoparticles. *Results Phys.* **2019**, *15*, 102565. [CrossRef]
53. Bagherzade, G.; Tavakoli, M.M.; Namaei, M.H. Green synthesis of silver nanoparticles using aqueous extract of saffron (*Crocus sativus* L.) wastages and its antibacterial activity against six bacteria. *Asian Pac. J. Trop. Biomed.* **2017**, *7*, 227–233. [CrossRef]
54. Durán, N.; Nakazato, G.; Seabra, A.B. Antimicrobial activity of biogenic silver nanoparticles, and silver chloride nanoparticles: An overview and comments. *Appl. Microbiol. Biotechnol.* **2016**, *100*, 6555–6570. [CrossRef] [PubMed]
55. Asoro, M.; Damiano, J.; Ferreira, P.J. Size Effects on the Melting Temperature of Silver Nanoparticles: In-Situ TEM Observations. *Microsc. Microanal.* **2009**, *15*, 706–707. [CrossRef]
56. Chugh, H.; Sood, D.; Chandra, I.; Tomar, V.; Dhawan, G.; Chandra, R. Role of gold and silver nanoparticles in cancer nano-medicine. *Artif. Cells Nanomed. Biotechnol.* **2018**, *46*, 1210–1220. [CrossRef]
57. Litvin, V.A.; Galagan, R.L.; Minaev, B.F. Kinetic and mechanism formation of silver nanoparticles coated by synthetic humic substances. *Colloids Surf. Physicochem. Eng. Asp.* **2012**, *414*, 234–243. [CrossRef]
58. Alahmad, A.; Feldhoff, A.; Bigall, N.C.; Rusch, P.; Scheper, T.; Walter, J.-G. *Hypericum perforatum* L.-Mediated Green Synthesis of Silver Nanoparticles Exhibiting Antioxidant and Anticancer Activities. *Nanomaterials* **2021**, *11*, 487. [CrossRef]
59. Tripathi, D.; Modi, A.; Narayan, G.; Rai, S.P. Green and cost effective synthesis of silver nanoparticles from endangered medicinal plant Withania coagulans and their potential biomedical properties. *Mater. Sci. Eng.* **2019**, *100*, 152–164. [CrossRef]
60. Ngo, H.-T.; Bechtold, T. Surface modification of textile material through deposition of regenerated silk fibroin. *J. Appl. Polym. Sci.* **2017**, *134*, 45098. [CrossRef]
61. Wurm, F.; Rietzler, B.; Pham, T.; Bechtold, T. Multivalent Ions as Reactive Crosslinkers for Biopolymers—A Review. *Molecules* **2020**, *25*, 1840. [CrossRef]

Article

Antimicrobial Biomaterial on Sutures, Bandages and Face Masks with Potential for Infection Control

Zehra Edis [1,2,*], Samir Haj Bloukh [2,3], Hamed Abu Sara [2,3] and Nur Izyan Wan Azelee [4]

[1] Department of Pharmaceutical Sciences, College of Pharmacy and Health Sciences, Ajman University, Ajman P.O. Box 346, United Arab Emirates
[2] Center of Medical and Bio-Allied Health Sciences Research, Ajman University, Ajman P.O. Box 346, United Arab Emirates; s.bloukh@ajman.ac.ae (S.H.B.); h.abusara@ajman.ac.ae (H.A.S.)
[3] Department of Clinical Sciences, College of Pharmacy and Health Sciences, Ajman University, Ajman P.O. Box 346, United Arab Emirates
[4] Institute of Bioproduct Development (IBD), Universiti Teknologi Malaysia, Skudai 81310, Malaysia; nur.izyan@utm.my
* Correspondence: z.edis@ajman.ac.ae

Citation: Edis, Z.; Bloukh, S.H.; Sara, H.A.; Azelee, N.I.W. Antimicrobial Biomaterial on Sutures, Bandages and Face Masks with Potential for Infection Control. *Polymers* 2022, 14, 1932. https://doi.org/10.3390/polym14101932

Academic Editors: Md. Amdadul Huq and Shahina Akter

Received: 5 April 2022
Accepted: 4 May 2022
Published: 10 May 2022

Publisher's Note: MDPI stays neutral with regard to jurisdictional claims in published maps and institutional affiliations.

Copyright: © 2022 by the authors. Licensee MDPI, Basel, Switzerland. This article is an open access article distributed under the terms and conditions of the Creative Commons Attribution (CC BY) license (https://creativecommons.org/licenses/by/4.0/).

Abstract: Antimicrobial resistance (AMR) is a challenge for the survival of the human race. The steady rise of resistant microorganisms against the common antimicrobials results in increased morbidity and mortality rates. Iodine and a plethora of plant secondary metabolites inhibit microbial proliferation. Antiseptic iodophors and many phytochemicals are unaffected by AMR. Surgical site and wound infections can be prevented or treated by utilizing such compounds on sutures and bandages. Coating surgical face masks with these antimicrobials can reduce microbial infections and attenuate their burden on the environment by re-use. The facile combination of *Aloe Vera Barbadensis* Miller (AV), Trans-cinnamic acid (TCA) and Iodine (I_2) encapsulated in a polyvinylpyrrolidone (PVP) matrix seems a promising alternative to common antimicrobials. The AV-PVP-TCA-I_2 formulation was impregnated into sterile discs, medical gauze bandages, surgical sutures and face masks. Morphology, purity and composition were confirmed by several analytical methods. Antimicrobial activity of AV-PVP-TCA-I_2 was investigated by disc diffusion methods against ten microbial strains in comparison to gentamycin and nystatin. AV-PVP-TCA-I_2 showed excellent antifungal and strong to intermediate antibacterial activities against most of the selected pathogens, especially in bandages and face masks. The title compound has potential use for prevention or treatment of surgical site and wound infections. Coating disposable face masks with AV-PVP-TCA-I_2 may be a sustainable solution for their re-use and waste management.

Keywords: COVID-19; antimicrobial resistance; cinnamic acid; *Aloe Vera*; surgical site infection; iodophors

1. Introduction

The COVID-19 pandemic presented the most serious challenge for mankind in this century and exacerbated another "silent pandemic" of antimicrobial resistance (AMR) [1]. The severity of the COVID-19 crisis was elevated by insufficient health care support, supply shortages of personal protective equipment and existing antimicrobial resistance [1–5]. Resistant ESKAPE (*Enterococcus faecium, Staphylococcus aureus, Klebsiella pneumoniae, Acinetobacter baumannii, Pseudomonas aeruginosa, Enterobacter* spp., and *Escherichia coli*) pathogens augmented morbidity and mortality rates [4–9]. ESKAPE pathogens are increasingly resistant to common antibiotics, drugs and antimicrobials due to inadequate utilization of antimicrobial agents [1,3–9]. Global overuse of antimicrobials in hospital settings without proper surveillance during the COVID-19 pandemic escalated AMR dramatically [1]. Public and health sector settings are steadily encroached by such resistant, opportunistic microorganisms [1,3–9]. Morbidity and mortality rates among severely ill, immunocompromised patients skyrocketed due to nosocomial infections [6,9]. The outcomes include longer duration of treatment, delayed healing processes, exponentially growing cost of treatment, increased morbidity and surging mortality rates among patients [1,3,9]. Hospital- and community-acquired infections are caused by microbes lingering in the immediate environment [1,3–6,9–11]. Further pandemics are expected with aggravated fatality rates globally. Multiple-drug-resistant microorganisms steadily build up their defenses by adjusting to existing conventional antimicrobial agents through survival techniques [11]. Virulence factors, such as inter-kingdom biofilm formation, are examples of their "intelligent" tools to gain multidrug and AMR [11]. These methods allow them to proliferate in all settings [11]. A new generation of agents is needed to overcome AMR-related morbidity and mortality.

Iodine, a well-known microbicide, is marketed in the form of different iodophors and polymeric complexes [12–28]. Antiseptic iodophors are unaffected by microbial resistance mechanisms and can be utilized to overcome AMR [19]. Iodine readily forms a variety of polyiodide ions through donor–acceptor interactions between iodides and iodine molecules [29–36]. Therefore, there are many applications of polyiodides in many fields [30–36]. Iodine, a small, lipophilic molecule, is released and diffuses through cell membranes [18,19,36]. It acts as an oxidizing agent and denaturates enzymes, as well as proteins, through iodination [18,19,36]. However, iodine supposedly causes discoloration of skin, pain and irritation [37]. Such side effects and uncontrolled iodine release hamper its popularity among disinfectant antiseptics [37]. Triiodides are the most stable polyiodide species. Triiodides exist in the form of asymmetric or "smart" linear, symmetric units with halogen bonding character within complexes [36]. Complexed "smart" triiodides can function as slow-iodine-releasing reservoirs due to their stability [36]. These properties mitigate the adverse effects on the skin due to reduced iodine content [21,22,36]. The complexing polymer used in our investigations is polyvinylpyrrolidone (PVP). PVP is a stabilizing matrix and acts as a reservoir for I_2 molecules in the form of PVPI in medical applications [12,16–24]. We previously investigated the inhibitory actions of different formulations with silver nanoparticles, plant bio-compounds, PVP and iodine against selections of microorganisms [38–42]. Our formulations showed excellent to intermediate antimicrobial activities on discs and biodegradable polyglycolic acid (PGA) sutures. PGA sutures are multifilamented, biocompatible and non-toxic but feasible for biofilm formation due to their extended surface area [43,44]. Conventional surgical sutures and wound bandages are used for wound closure but can lead to surgical site infections (SSI) [43,44]. Incorporating medicinal plants on sutures, bandages and wound dressing materials can address AMR-related problems [44,45]. Pathogens are susceptible to biosynthesized antimicrobials and disinfectants.

Plants contain a rich spectrum of bioactive compounds, which display excellent antimicrobial activities through synergistic mechanisms in their defense against microorganisms [45–47]. Such bioactive secondary plant metabolites are the reason for the use of medicinal plants, herbs and spices throughout centuries [46,47]. Phenolic acids, polyphenols, flavonoids, hydroxycinnamic acids and other compounds are used as antimicrobial agents in an increasing number of investigations [38–56]. Sutures, bandages and wound dressings incorporating phytochemicals are promising alternatives to conventional treatment options due to targeted, on-site drug delivery [43–56].

Hydroxycinnamic acids are phytochemicals, which exert inhibitory action on a wide range of pathogens [46,47,55,56]. Trans-cinnamic acid (TCA) is a potent inhibitor of microbial proliferation [39,46,47,55–60]. TCA is abundantly available in the plant kingdom [46,47]. *Aloe Vera Barbadensis* Miller (AV) contains TCA and further biocomponents, which display synergistic mechanisms and therefore potentiate antiproliferative action on microorganisms [39,41,42,53,54,61]. AV has been known and utilized for centuries for its health-promoting, moisturizing and healing properties [39,41,42,53,54,61]. AV constituents include aloin, aloe-emodin, acemannan, hesperidin and further phytochemicals with antioxidant, antiproliferative and anti-inflammatory properties [39,41,42,53,54,61]. These compounds exert antiviral and even anti-corona-virus activities [61–71]. Such properties enable their use in wound treatment on sutures, bandages and wound dressing materials, as well as surgical face masks [72–78]. Coating or incorporating phytochemicals into surgical face masks can reduce the demand during supply shortages. The re-use of face masks will alleviate and moderate stockpiles of dumped face masks as part of a waste management strategy [73]. Recycling mitigates the economic burden for the end user, as well as the production and waste disposal processes [73]. An increasing number of investigations on antimicrobial materials showcase different approaches for sustainable, re-usable or even biodegradable face masks [73–79]. Cinnamic acid and its derivatives are increasingly used to design novel antimicrobial biomaterials within polymeric matrices [80–82]. Biodegradable, hydrophilic materials with antifouling and wound-healing abilities support tissue repair processes [83–87]. AV has stabilizing and moisturizing purposes, in addition to its anti-inflammatory qualities. TCA is a widely available bioactive phytochemical in plants with antimicrobial properties. Combining phytochemicals with iodine and PVP can potentiate the antimicrobial activities [38–42,88]. Consequently, the staining effect of iodine can be mitigated by reducing the iodine content without losing its inhibitory action. According to Kessler et al., skin irritation and cytotoxic effects are not due to molecular iodine in commercial PVP-I_2 disinfectants but are caused by other additives in the formulations [89]. The release pattern of iodine and polyiodides govern the stability of the products, skin discoloration and antimicrobial activities [89–92]. PVP serves, in general, as a stabilizing and encapsulating agent of iodine moieties with the aim to reduce iodine content in the formulation and allow a controlled release.

The main purpose of this work is to present a facile, one-pot synthesis of an antimicrobial agent out of well-known, non-toxic, sustainable and biocompatible components. We combined iodine, AV, TCA and PVP within a formulation. We investigated the formation of triiodide species within the title compound in comparison to our previous works. Disc-diffusion tests against 10 common reference strains in comparison to gentamycin and nystatin were carried out to verify the in vitro antimicrobial activity of AV-PVP-TCA-I_2. The antimicrobial action on bandages, as well as surgical sutures, suggests the potential to prevent or treat SSI and wound infections. Coating surfaces, face masks, as well as other personal protective equipment would support against AMR and envision sustainable solutions for the future. Further in vivo investigations are needed to confirm the biological activities of AV-PVP-TCA-I_2.

2. Materials and Methods

2.1. Materials

Aloe vera leaves (*Aloe barbadensis* Miller, AV) were collected from the botanical garden of Ajman University campus in December, Ajman, UAE. McFarland standard sets, disposable sterilized Petri dishes with Mueller Hinton II agar, gentamicin (9125, 30 µg/disc) and nystatin (9078, 100 IU/disc) were obtained from Liofilchem Diagnostici (Roseto degli Abruzzi (TE), Italy). *E. coli* WDCM 00013 Vitroids, *P. aeruginosa* WDCM 00026 Vitroids, *K. pneumoniae* WDCM 00097 Vitroids, *C. albicans* WDCM 00054 Vitroids and *Bacillus subtilis* WDCM 0003 Vitroids were acquired from Sigma-Aldrich Chemical Co. (St. Louis, MO, USA). The reference strains *S. pneumoniae* ATCC 49619, *S. aureus* ATCC 25923, *E. faecalis* ATCC 29212, *S. pyogenes* ATCC 19615 and *P. mirabilis* ATCC 29906 were procured from Liofilchem (Roseto degli Abruzzi (TE), Italy). Polyvinylpyrrolidone (PVP-K-30), iodine (≥99.0%), Sabouraud Dextrose broth, Mueller Hinton Broth (MHB) and ethanol (analytical grade) were purchased from Sigma Aldrich (St. Louis, MO, USA). Sterile filter paper discs (diameter of 6 mm) were obtained from Himedia (Jaitala Nagpur, Maharashtra, India). Sterile polyglycolic acid (PGA) surgical sutures (DAMACRYL, 75 cm, USP: 3-0, Metric:2, 19 mm, DC3K19) were received from General Medical Disposable (GMD), GMD Group A.S., Istanbul, Turkey. Bandages and surgical, disposable, 3-ply non-woven face masks (FOMED, Qianjiang, China) were purchased from the local pharmacy. In all experiments, absolute ethanol and ultrapure water were used. All reagents were of analytical grade and were employed as delivered.

2.2. Preparation of Aloe vera (AV) Extract

We were kindly provided with leaves of an *Aloe vera* (*Aloe barbadensis* Miller) plant from the botanical garden of Ajman University at beginning of December between 8 and 9 am. The leaves were immediately taken to the research laboratory of the College of Pharmacy and Health Sciences. The leaves with a size between 35 to 50 cm were cleaned by tissue and then washed with water to remove dust and soil. Afterward, the clean leaves were rinsed several times with distilled water, pure ethanol, ultrapure water and dried carefully. The AV leaves were cut with a sterile knife to scrap the mucilaginous gel out. This pure gel was mixed at maximum speed for 20 min and centrifuged at 4000 rpm for 40 min (3K 30; Sigma Laborzentrifugen GmbH, Osterode am Harz, Germany). The supernatant with a light-yellow color was filled immediately into a brown bottle with a screw cap and stored in darkness at 3 °C.

2.3. Preparation of AV-PVP-TCA and AV-PVP-TCA-I_2

The stock solution AV-PVP is prepared by adding 2 mL pure AV gel into 2 mL of a solution of 1 g polyvinylpyrrolidone K-30 (PVP) in 10 mL distilled water under continuous stirring at room temperature (RT). For the preparation of AV-PVP-TCA, first 0.148 g TCA is dissolved in 10 mL ethanol. Then, 2 mL of this solution is added to AV-PVP under continuous stirring at RT. After that, iodine solution is obtained by dissolving 0.05 g of iodine in 3 mL ethanol in a covered beaker at RT under stirring. An amount of 1 mL of this iodine solution is added to AV-PVP-TCA under continuous stirring at RT for the preparation of AV-PVP-TCA-I_2.

2.4. Characterization of AV Complexes

The title compounds were analyzed by SEM/EDS, Raman Spectroscopy, UV-vis, FTIR and X-ray diffraction (XRD). These investigations confirmed the composition of our biomaterials.

2.4.1. Scanning Electron Microscopy (SEM) and Energy-Dispersive X-ray Spectroscopy (EDX)

The scanning electron microscopy (SEM) and the energy-dispersive X-ray spectroscopy (EDS) analyses were performed with VEGA3 from Tescan (Brno, Czech Republic) at 15 kV by. AV-PVP-TCA-I_2 was diluted by adding one drop into distilled water and positioning it on a carbon-coated copper grid. After drying the sample, it was coated with gold through the Quorum Technology Mini Sputter Coater. SEM and EDS analyses were used to determine the morphology and elemental composition of the sample, respectively.

2.4.2. UV-Vis Spectrophotometry (UV-Vis)

The formulations AV-PVP-TCA and AV-PVP-TCA-I_2 were analyzed by UV-vis spectrophotometry. The investigation was performed by a UV-Vis spectrophotometer model 2600i from Shimadzu (Kyoto, Japan) in the wavelength range of 195 to 800 nm.

2.4.3. Raman Spectroscopy

The biohybrid AV-PVP-TCA-I_2 underwent Raman analysis on a RENISHAW (Gloucestershire, UK) equipped with an optical microscope room at RT. The sample was placed into a cuvette (1 cm × 1 cm) and put in front of the laser beam. The solid-state laser beam had an excitation of 785 nm and was directed onto the sample by the 50× objective lens of a confocal microscope with a spot diameter of 2 microns. The scattered light was collected by a CCD-based monochromator with a spectral range of between 50 and 3400 cm^{-1}. The spectral resolution was -1 cm^{-1}, the output power was 0.5%, and the integration time was -30 s.

2.4.4. Fourier Transform Infrared (FTIR) Spectroscopy

The FTIR analysis of the formulations AV-PVP-TCA and AV-PVP-TCA-I_2 was conducted on an ATR IR spectrometer equipped with a Diamond window (Shimadzu, Kyoto, Japan). Both formulations were freeze dried and analyzed in the range between 400 to 4000 cm^{-1}.

2.4.5. X-ray Diffraction (XRD)

The X-ray diffraction analysis was performed by a XRD from BRUKER (D8 Advance, Karlsruhe, Germany). The formulation AV-PVP-TCA-I_2 was analyzed by Cu radiation with a wavelength of 1.54060, coupled Two Theta/Theta, time/step of 0.5 s and a step size of 0.03.

2.5. Bacterial Strains and Culturing

The antimicrobial testing was performed with the reference microbial strains of *S. pneumoniae* ATCC 49619, *S. aureus* ATCC 25923, *E. faecalis* ATCC 29212, *S. pyogenes* ATCC 19615, *Bacillus subtilis* WDCM 0003 Vitroids, *P. mirabilis* ATCC 29906, *E. coli* WDCM 00013 Vitroids, *P. aeruginosa* WDCM 00026 Vitroids, *K. pneumoniae* WDCM 00097 Vitroids and *C. albicans* WDCM 00054 Vitroids. These reference strains were kept at -20 °C. The inoculation was performed by adding the fresh microbes to MHB. These suspensions were kept at 4 °C until needed.

2.6. Determination of Antimicrobial Properties of AV-PVP-TCA-I_2

The inhibitory action of AV-PVP-TCA-I_2 against nine reference bacterial strains (*S. pneumoniae* ATCC 49619, *S. aureus* ATCC 25923, *S. pyogenes* ATCC 19615, *E. faecalis* ATCC 29212, and *B. subtilis* WDCM 00003, *P. mirabilis* ATCC 29906, *P. aeruginosa* WDCM 00026, *E. coli* WDCM 00013 and *K. pneumoniae* WDCM 00097) was compared to the antibiotic gentamicin (positive control). The antifungal activity of the title compound was tested on *C. albicans* WDCM 00054 in comparison to the antibiotic nystatin (positive control). The negative controls of ethanol and water showed no susceptibility and were not mentioned further. The antimicrobial tests on discs, bandages, surgical face masks and sutures were repeated three times. The average of the independent experiments was presented in this investigation.

2.6.1. Procedure for Zone of Inhibition Plate Studies

The zone of inhibition plate method was used to investigate the susceptibility of the selected pathogens toward AV-PVP-TCA-I_2 [93]. We suspended the selected bacterial strains in 10 mL MHB and incubated at 37 °C for 2 to 4 h. *C. albicans* WDCM 00054 was incubated on Sabouraud Dextrose broth at 30 °C. The microbial cultures were adjusted to 0.5 McFarland standard. The disposable, sterilized Petri dishes with MHA were uniformly seeded with 100 µL microbial culture with sterile cotton swabs and dried for 10 min to be used for the antimicrobial testing.

2.6.2. Disc Diffusion Method

The antimicrobial testing was performed against the antibiotic discs of gentamycin and nystatin following the recommendations of the Clinical and Laboratory Standards Institute (CLSI) [94]. Sterile filter paper discs were coated for 24 h with 2 mL of AV-PVP-TCA-I_2 (11 µg/mL, 5.5 µg/mL, 2.75, and 1.38 µg/mL). After removing the discs from the solution, we dried the discs for 24 h under ambient conditions. *C. albicans* WDCM 00054 was incubated for 24 h at 30 °C on agar plates. The diameter of the zone of inhibition (ZOI) was measured with a ruler to the nearest millimeter. The antimicrobial properties of AV-PVP-TCA-I_2 are evident from the diameters of the clear inhibition zone around the disc. No inhibition zone confirms the resistance of the microorganisms.

2.7. Preparation and Analysis of Impregnated Sutures, Bandages and Surgical Face Masks

The uncoated, sterile, multifilamented surgical PGA sutures of 2.5 cm were impregnated with AV-PVP-TCA-I_2 for 18 h into 50 mL of AV-PVP-TCA-I_2 solution (11 µg/mL) at RT. The blue sutures became brown blue and were then dried for 24 h under ambient conditions. The bandages and surgical face masks were cut to square pieces of (5 cm × 5 cm), also being impregnated in 50 mL of our title formulation (11 µg/mL) for 18 h at RT and dried for 24 h at RT. These dip-coated, dried sutures, bandages and surgical face masks were tested in vitro by ZOI assay against our selection of 10 microbial strains (*S. pneumoniae* ATCC 49619, *S. aureus* ATCC 25923, *E. faecalis* ATCC 29212, *S. pyogenes* ATCC 19615, *Bacillus subtilis* WDCM 00003, *E. coli* WDCM 00013, *P. aeruginosa* WDCM 00026, *P. mirabilis* ATCC 29906, *K. pneumoniae* WDCM 00097 and *C. albicans* WDCM 00054).

2.8. Statistical Analysis

We utilized SPSS software (version 17.0, SPSS Inc., Chicago, IL, USA) in our statistical analysis. The data are represented as mean values. The statistical significance between groups is calculated by one-way ANOVA. Any value of $p < 0.05$ was considered statistically significant.

3. Results and Discussion

AMR is a growing concern, and it increasingly endangers the existence of mankind, in particular [1]. Worldwide, the unjustified use of antimicrobials during the COVID-19 pandemic elevated the AMR crisis [1]. ESKAPE pathogens develop resistance against common antimicrobials, exacerbating the morbidity and mortality rates [1,3–9]. Nosocomial infections in hospital settings aggravated the suffering and fatality during the recent COVID-19 pandemic [1–3,5,6]. Even community-acquired infections caused by multi-drug-resistant microorganisms impacted global health due to antibiotic overuse, increasing costs and durations of treatments [1,3,4,9]. The recent COVID-19 pandemic highlighted the need for strategic solutions to overcome AMR. Confusion, misinformation, lockdowns, disruption of transport chains and supply shortages were some of the early markers of the pandemic [3]. Resource management and alternative solutions to the shortages are eminent in times of crisis and beyond [3].

The availability of antimicrobials and personal protective equipment can be a means of survival. Disinfectants are needed for inanimate surfaces in all indoor and outdoor settings, for contact-killing surfaces and as disinfectants on skin [3,11]. Preservatives are used in different health care, pharmaceutical or cosmetic products to mitigate microbial contamination. Incorporating antimicrobials on bandages, wound dressing materials and surgical sutures leads to effective prevention and treatment of SSI, as well as wound infections [11,48–55].

Health authorities enforced the use of face masks to mitigate viral transmission and droplet movement [73–78]. However, the growing number of disposed masks presents a serious source of waste and pollution [72,73]. The COVID-19 public health measures are going to be lifted as soon as the number of cases declines globally. The recommendations of social distancing, personal hygiene and wearing of face masks must be carried on in the future during known seasonal flu outbreaks. The incorporation of inhibitory biomaterials on surgical face masks can aid the prevention and treatment of upper respiratory tract infections as well [48–55]. The antiviral, anti-coronavirus activities of plant secondary metabolites are reported in previous studies and include AV, as well as several polyphenols and hydroxycinnamic acids [67–71]. These properties can be utilized to elevate the antiviral barrier function of face masks and their re-use. Spraying or coating face masks with antimicrobial, natural biomaterials is a sustainable solution for the planet [73–78]. Antimicrobial, non-toxic agents can be utilized to disinfect and re-use masks [73–78]. At the same time, it presents a solution for low-income populations and households under economic strain. Furthermore, the re-use of face masks can effectively alleviate the supply shortage problems in different areas globally. Antimicrobial application on face mask materials can ameliorate viral transmission processes more effectively for a longer time.

Antimicrobial agents must be sustainable and impervious to microbial resistance mechanisms. AMR is a global concern and needs to be addressed by the development of new antimicrobial agents. Plant phytochemicals are suitable candidates in the AMR challenge [44,45]. Phytochemicals and their synergistic mechanisms have assisted the survival of plants against microbes since their existence [44,45]. The antimicrobial, anti-inflammatory activities of plant constituents, such as polyphenols and hydroxycinnamic acids, have been utilized since the history of mankind [46,47].

Our title compound AV-PVP-TCA-I_2 fits into the category of antimicrobial plant biomaterials and consists of TCA, PVP, AV, as well as iodine [39–42,95–99]. We used iodine, TCA and AV in our study to investigate the antimicrobial action against 10 selected pathogens. Iodine is a iodophor, which is most stable in form of "smart" triiodide species [34,36,41,42,100–104]. TCA, its derivatives and related AV bioconstituents, especially cinnamic acid, aloin, acemannan, aloe-emodin, pyrogallol, hesperidine, aloesin, 10-O-β-d-glucopyranosyl-aloenin and rhein increased biological activities due to their synergistic mechanisms [36,39,41,42,53–66,98,99]. TCA and its hydroxycinnamic derivates in AV have abilities to permeate cell membranes of Gram-positive pathogens, resulting in cell membrane disruptions and changes [46]. The -COOH group within TCA and AV polyphenols is weakly acidic [46,98]. Therefore, these compounds diffuse easily through the cell membranes and increase pH levels in the cytoplasm, leading to cell death [46,47]. The secondary and tertiary structures of proteins are unfolded due to the formation of hydrogen bonding. This can occur through the interaction between hydroxyl, carbonyl and carboxylate groups within TCA and AV components with the functional groups of proteins, outer membrane and cell wall components of the microorganisms [46]. The tertiary structure is unfolded by interactions of carboxylate groups by ion-ion–electrostatic interactions, resulting in disturbances in the salt bridges. Phenolic groups in TCA and the AV biocomponents interfere in the hydrophobic interactions between close, nonpolar phenyl and alkyl groups [46,47]. The covalent disulfide S-S linkages between the amino acids responsible for folding the tertiary structure of proteins are reduced to S-H groups by plant biomaterials in AV [12]. AV plant bio-compounds are oxidized by transforming their hydroxyl units to carbonyl groups [46,47]. We aimed to employ in our investigations sustainable, natural, plant-based solutions capable of acting as "smart" triiodide reservoirs [36,39,41,42]. Such triiodides are stable iodine moieties within the PVP matrix and are released upon contact with the microbial units. AV-PVP-TCA-I_2 formulations can be promising antimicrobial biomaterials.

The analytical results of AV-PVP-TCA-I_2 confirmed the antimicrobial properties, the formulation characteristics and sample constituents.

3.1. Elemental Composition and Morphological Examination of AV-PVP-TCA-I_2

Electron Microscope (SEM) and Energy-Dispersive X-ray Spectroscopic (EDS) Analyses.

The morphology and composition of AV-PVP-TCA-I_2 was investigated by SEM and EDS analyses (Figure 1).

The title compound AV-PVP-TCA-I_2 reveals an amorphous morphology with interesting thread- or barrel-like forms surrounded by semi-crystalline, white patches (Figure 1a). The EDS shows carbon as the main component with 81%, followed by oxygen (6.7%) and iodine (2.5%) (Figure 1b). Chlorine, potassium and copper are originating from the AV biocompounds. Aluminum and gold appear due to their use during the preparation of the samples for the SEM analysis. The samples were coated with gold.

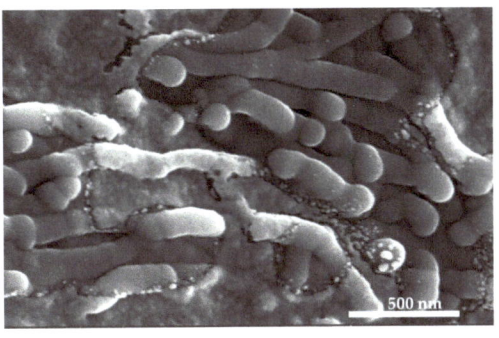 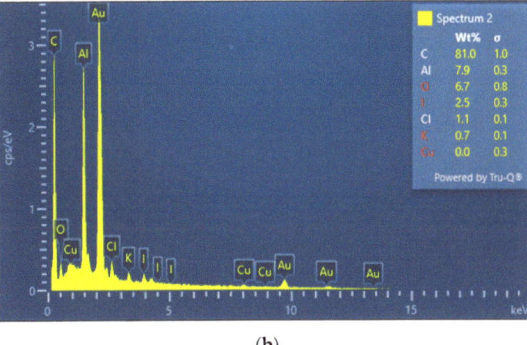

Figure 1. SEM (**a**) and EDS (**b**) of AV-PVP-TCA-I$_2$.

We investigated the antimicrobial activities of AV-PVP-TCA-I$_2$-impregnated medical sutures, bandages and face masks. Sterile, braided surgical PGA sutures were dip coated with our formulations and analyzed by SEM/EDS techniques (Figure 2).

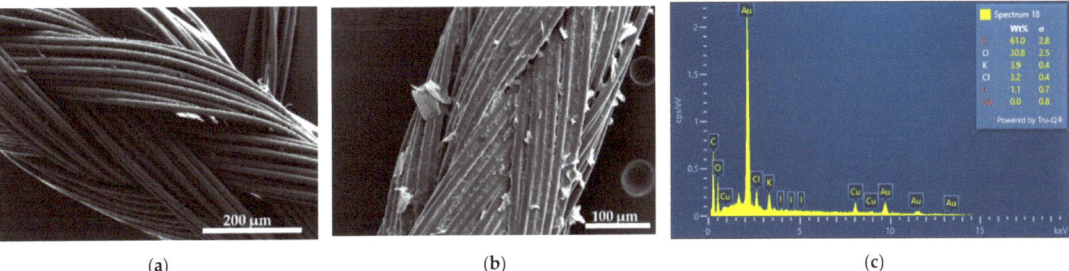

Figure 2. SEM of sutures: (**a**) plain suture [31]; (**b**) impregnated with AV-PVP-TCA-I$_2$; (**c**) EDS of coated sutures.

Figure 2a depicts the same plain PGA suture from our previous investigations [31]. Impregnating the suture with AV-PVP-Sage-I$_2$ (11 µg/mL) results in a fully coated, homogenous surface (Figure 2b). Crystalline depositions are distributed throughout the surface of the braided suture (Figure 2b). The EDS of the impregnated suture shows only the expected composition of carbon (61%), oxygen (30.8%), potassium (3.9%), chlorine (3.2%) and iodine (1.1%) (Figure 2c). This result proves the coating process and enables these sutures for their possible use in the prevention of surgical site infections.

Plain medical face masks with and without AV-PVP-TCA-I$_2$ (11 µg/mL) coating are depicted in Figure 3.

Figure 3b reveals depositions on the surface of the surgical masks. The EDS of the impregnated mask material reveals the expected compositions of carbon (82.6%), oxygen (9.8%), iodine (6,4%), calcium (0.7%) and potassium (0.5%) (Figure 3c). The plain mask depicts some rough areas, while the dip-coated mask has a smoother surface and has small aggregations on few fibers (Figure 3). These depositions formed during the drying process of the coating agent. Otherwise, there is no change in the mask material related to porosity, shape and arrangement of the fibers (Figure 3b). Such small changes will not aggravate the filtering and breathability properties of the face mask. The results may enable the use of our title material as a coating agent for face masks.

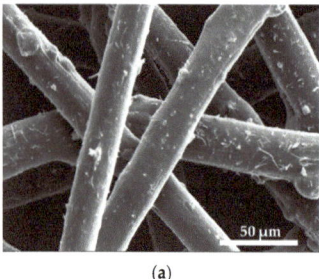 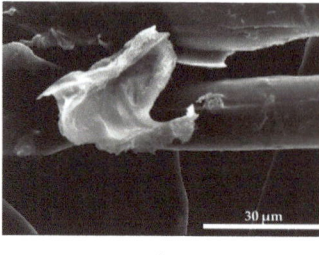 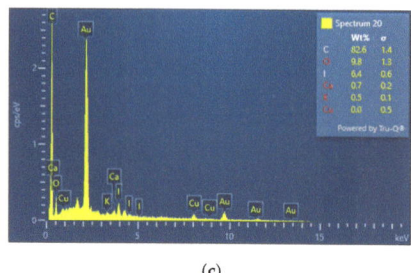

(a) (b) (c)

Figure 3. SEM of surgical face masks: (**a**) plain mask; (**b**) dip coated with AV-PVP-TCA-I_2; (**c**) EDS of coated face mask.

Medical bandage material was impregnated with AV-PVP-TCA-I_2 (11 μg/mL) and analyzed by SEM/EDS techniques (Figure 4, Supplementary Figures S1 and S2).

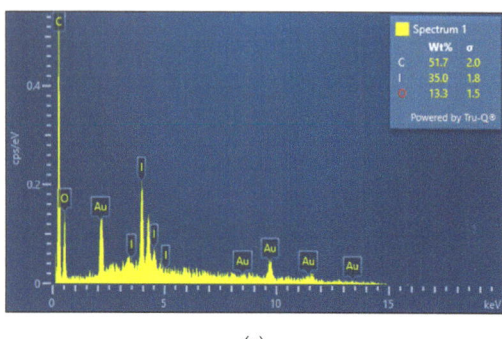 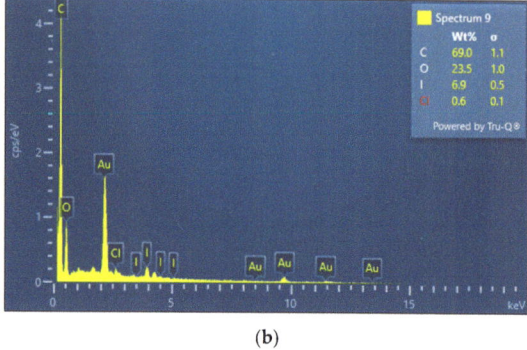

(a) (b)

Figure 4. EDS of medical bandages impregnated with AV-PVP-TCA-I_2: (**a**) area 1 coiled; (**b**) area 2 ordered.

The bandage material showed two different patterns in area1 (coiled) and area 2 (ordered) in the EDS (Figure 4, Supplementary Figures S1 and S2). The bandage was uniformly coated with the title compound and showed the expected composition of elements (Figure 4). The coiled area of the bandage adsorbed more iodine than the ordered part, with 35% and 6.9%, respectively (Figure 4). Again, carbon was the most abundant element, with 51.7 and 69% in the coiled and ordered structures of the bandage material. The SEM of the dip-coated bandage also shows small depositions due to the drying process, but no morphological changes to the fibers (Supplementary Figures S1 and S2). The results encourage the use of the impregnated bandage for the prevention and possible treatment of wound infections.

The homogenous coating of the face mask material, sutures and bandages enables its potential use as an antimicrobial coating agent on these materials (Figures 2 and 3, Supplementary Figures S1 and S2). Further investigations about the stability of the coated surfaces and the effective duration of inhibitory action are needed to judge the future applications of our title compound.

The EDS verifies the purity and composition of our title formulation AV-PVP-TCA-I_2 (Figures 1–4). Chlorine, potassium and copper were also present in our previous study with AV-PVP-I_2 and originate from the AV biocomponents [41].

The iodine moieties were available in every EDS of the analyzed samples (Figures 1–4, Supplementary Figures S1 and S2).

3.2. Spectroscopical Characterization

3.2.1. Raman Spectroscopy

Raman spectroscopic analysis of AV-PVP-TCA-I_2 is shown in Figure 5.

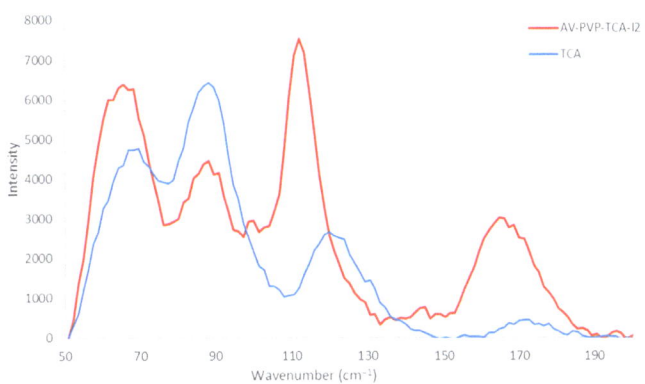

Figure 5. Raman spectroscopic analysis of AV-PVP-TCA-I_2 and TCA.

The Raman spectrum of the title compound AV-PVP-TCA-I_2 shows two broad, high-intensity absorptions around 50–130 and intermediate shifts at 150–175 cm^{-1}. A broad medium-sized absorption band is available from 200 to 455 cm^{-1} (Figure 5). Small-sized absorptions are available between 130 and 150 cm^{-1} (Figure 5). Such bands originate from a mixture of iodine moieties. The Raman spectrum is clearly dominated by the polyiodide absorption bands (Figure 5, Supplementary Figure S3). These consist of iodide ions and molecular I_2 within unsymmetrical polyiodide ions, as well as triiodide ions and, probably, pentaiodide ions (Figure 5, Table 1).

Table 1. Raman shifts in AV-PVP-TCA-I_2 (1) (cm^{-1}).

Group	AV-PVP-TCA-I_2	[41]	[90]	[91]	[23]	[15]	[92]
I_2 [$I_2 \cdots I^-$]	sh,m84* sh,m159*ν_{as} m166*ν_{as} m169*ν_{as} sh,m171*ν_s w176*ν_s w179*ν_s vw183*ν_s vw187*ν_s	sh,w80* s169*ν_{as} w189*ν	m85* m160*ν_{as}		s169 ν_s I_2		
I_3^- [I-I-I$^-$]	sh,s62ν_{2bend} s66 δ_{def} sh,m82ν_{2bend} m91$\nu_{1,s}$ w100$\nu_{1,s}$ vs112$\nu_{1,s}$ w238$^+$2$\nu_{1,s}$	sh,w61 δ_{def} sh,w70 ν_{2bend} vs110$\nu_{1,s}$ vw222$^+$2$\nu_{1,s}$	sh60 δ_{def} sh,w75ν_{2bend} s110$\nu_{1,s}$ vw221 2$\nu_{1,s}$	114$\nu_{1,s}$	vs111ν_s	vs116$\nu_{1,s}$ vw235 2$\nu_{1,s}$	m85ν_{2bend} 108ν_s 110ν_s 218$^+\nu_s$
I_3^- [I-I $\cdots$ I$^-$]	sh,w125$\nu_{3,as}$ vw145$\nu_{3,as}$ vw152$\nu_{3,as}$ sh,m162$\nu_{3,as}$ w325$^+\nu_{as}$	m144$\nu_{3,as}$ vw334$^+\nu_{as}$	m144$\nu_{3,as}$ sh,vw154$\nu_{3,as}$	144$\nu_{3,as}$	m145ν_s	m126$\nu_{3,as}$	159ν_{as} 322$^+\nu_s$ 434$^+\nu_s$ 542$^+\nu_s$

ν = vibrational stretching, $_s$ = symmetric, $_a$ = asymmetric, 1 = stretching mode 1, 3 = stretching mode 3, bend = bending, δ_{def} = deformation. * belong to the same asymmetric, nonlinear unit I_3^- = $I_2 \cdots I^-$. $^+$ overtones of triiodide ions. vw = very weak, s = strong, vs = very strong, m = intermediate, sh = shoulder.

The Raman spectrum verifies the presence of linear, "smart" triiodide ions as a major component in the AV-PVP-TCA-I$_2$ formulation, in agreement with our previous investigations (Figure 5, Table 1) [36,39,41,42].

Symmetrical, "smart" triiodide ions (I-I-I$^-$) are represented by strong Raman shifts around 91 cm^{-1}, followed by 100 and 112 cm^{-1} originating from symmetrical vibrations ν_{1s} (Figure 5, Table 1). Symmetrical triiodides appear at Raman shifts around 100–115 cm^{-1}. According to Yushina et al., triiodides appear at 100–120 cm^{-1}, bound I$_2$ at 140–180 cm^{-1} and pentaiodides at 140–160 cm^{-1} [100]. The absorption mode at 112 cm^{-1} appears in our previous works of AV-PVP-Sage-I$_2$ at 110 cm^{-1} [42]. The slight increase from 110 to 112 cm^{-1} confirms a blue shift with stronger bonds in the title formulation with TCA compared to the Sage biohybrid due to noncovalent interactions and the molecular surroundings of the compounds (Figure 5, Table 1) [42,100].

However, the Raman spectrum of AV-PVP-TCA-I$_2$ shows further medium to very weak absorption peaks related to asymmetrical triiodide ions (I-I $\cdots$ I$^-$). Savastano et al. and Xu et al. report strong symmetric stretching modes related to slightly nonlinear, symmetrical linear triiodide ions at 110 and 111 cm^{-1}, respectively [23,30]. Weak to very weak absorption signals at 145 and 152 cm^{-1} confirm asymmetric vibrations, respectively. Accordingly, a medium-sized peak at 144 cm^{-1} was also available in our previous investigation of AV-PVP-Sage-I$_2$ (Figure 5, Table 1) [42]. The peak at 152 cm^{-1} was accompanied by weak to very weak asymmetric stretching modes ν_{as} at 222 and 334 cm^{-1}, respectively [42]. The I$_3^-$ bands have overtones in the form of increasingly weak vibrational stretching modes at 238 and 325 cm^{-1} (Figure 5, Table 1). The weak symmetric stretching mode at 221 cm^{-1} supports the linear structure, according to Savastano et al. [30]. The same Raman shift is available within the broad band in our formulations with TCA and Sage [42]. Ordinartsev et al. assign Raman modes at 116 and 235 cm^{-1} to centrosymmetric, symmetrical triiodide ions (I-I-I) [15]. Our compound shows similar symmetric stretching modes confirming the linear structure of the triiodide ions (Figure 5, Table 1) [15]. The TCA compound reveals a very low absorption intensity compared to the Sage compound, which indicates that the triiodide ions are, in the majority, linear and symmetric rather than nonlinear [42]. However, both compounds also contain slightly nonlinear, symmetrical triiodides accompanied by asymmetric triiodide ions. The title compound presents a broad, weak band between 200 and 400 cm^{-1} (Figure 5, Table 1).

Unsymmetrical triiodide ions are Raman active because they are slightly nonlinear and show absorption modes at 60, 85 and 160 cm^{-1} [30]. The same absorption bands are available in the title compound with strong- to medium-sized Raman shifts at 62, 84 and 160 cm^{-1} with lower intensity in comparison to the Sage formulation (Figure 5, Table 1) [30,42]. The strong Raman shift at 62 cm^{-1} is assigned to a hot band transition related to the ν_2 symmetric stretching (Figure 5, Table 1) [30].

Further shoulders at 125 (weak) and 162 cm^{-1} (intermediate) verify unsymmetrical triiodide ions (Figure 5, Table 1). In particular, the Raman shifts at 62, 66, 84, 159, 166 and 169 cm^{-1} confirm the presence of distorted, nonlinear triiodide units (I-I $\cdots$ I$^-$) (Figure 5, Table 1). These absorptions indicate strong I-I bonds within the unsymmetrical triiodide ions due to the noncovalent interactions. Such unsymmetrical triiodides can be simplified as (I$_2 \cdots$ I$^-$). The presence of molecular iodine within such triiodide units is verified by the absorption peaks at 84, 159, 166, 169, 176, 179, 183 and 187 cm^{-1} (Figure 5, Table 1). The medium-sized shoulder at 84 cm^{-1} belongs to stretching vibrations in I$_2 \cdots$ I$^-$ moieties [30]. The I$_2$-unit within I $\cdots$ I $\cdots$ I$^-$ is confirmed by the medium shoulders at 168 and 170 cm^{-1} (Figure 5, Table 1) [30,42].

Apart from triiodide units, characteristic absorption bands for pentaiodide ions could be assigned to 148, 157 and 165 cm^{-1} (Figure 5, Table 1). Pentaiodide ions consist of [I$_2 \cdots$ I$_3^-$] units and, therefore, are detected related to molecular iodine and triiodide ions. The availability of I$_5^-$-units seems to be manifested by the high absorption intensities and broadness of these bands between 140 and 175 cm^{-1} (Figure 5, Table 1). Nevertheless, these bands are expected to be the overtones of the triiodide ions. Pentaiodide ion absorptions are usually detected around 137–147 cm^{-1} and 167 cm^{-1} [25,82,101]. The weakness of the absorption intensities is another indicator for the absence of pentaiodide ions. At the same time, free iodine is represented by a medium-sized shoulder at 171 cm^{-1} and further weak to very weak symmetric stretching modes up to 176 cm^{-1} in AV-PVP-TCA-I$_2$ (Figure 5, Table 1) [42,102–104]. The previous investigation with AV-PVP-Sage-I$_2$ was devoid of absorption modes around 172 cm^{-1} and free iodine molecules [42,102–104].

The strong presence of the iodine moieties overshadows further signals in the Raman spectrum and impedes full characterization (Figure 5, Supplementary Figure S3). AV biocompounds and TCA show different weak absorptions at higher Raman shifts in accordance to previous investigations [59,105–107]. Weak, broad bands originating from hydroxyl groups are available around the broad band between 3175 and 3400 cm^{-1} (Supplementary Figure S3). The weak bands around 3045 and 3070 cm^{-1} are due to aliphatic and aromatic, unsaturated C-H stretching modes, respectively (Supplementary Figure S3) [59]. The Raman spectrum is also overshadowed by the strong absorption intensities of TCA from 1050 to 2050 cm^{-1} in AV-PVP-TCA-I$_2$ (Supplementary Figure S3). Aromatic and aliphatic -C=C-group stretching modes are available in TCA and the title compound AV-PVP-TCA-I$_2$ around 1603 and 1645 cm^{-1} in the form of strong, sharp signals, respectively (Supplementary Figure S3) [59]. Both Raman shifts have elevated intensity in AV-PVP-TCA-I$_2$ compared to TCA alone (Supplementary Figure S3).

The deformation and stretching vibrations of unsaturated (C-H) are available in a broad band around 1200–1650 cm^{-1} with a maximum intensity at 1376 cm^{-1} (Supplementary Figure S3) [59]. The stretching vibration (C-C$_{ring}$) at 1376 cm^{-1} in AV-PVP-TCA-I$_2$ appears in TCA with higher intensity at 1371 cm^{-1} (Supplementary Figure S3) [59,105,106]. The lower intensity and blue shift in the title compound confirms the inclusion of TCA within AV-PVP-TCA-I$_2$ (Supplementary Figure S3). This Raman shift is assigned by de Souza et al. to the carboxylate ion [106]. The carbonyl groups -C=O within -COOH are more complexed in AV-PVP-TCA-I$_2$ by the interaction with PVP, iodine and AV biomolecules. This is also verified by the related Raman shifts for TCA and AV-PVP-TCA-I$_2$ at 1888 and 1848 cm^{-1}, respectively (Supplementary Figure S3) [59,105]. The -C=O stretching modes show a red shift toward 1848 cm^{-1} with increased intensity in AV-PVP-TCA-I$_2$ (Supplementary Figure S3). A sharp stretching mode at 2018 cm^{-1} is originating from C-C$_6$H$_5$ vibrations (Supplementary Figure S3) [59]. The aromatic ring breathing mode is seen around 1002 and 1029 cm^{-1} (Supplementary Figure S3) [59]. Very weak modes around 847 and 874 cm^{-1} are due to the aliphatic C-COOH vibrations (Supplementary Figure S3) [59].

AV naturally contains TCA and acemannan [41]. Acemannan is indicated in the Raman spectrum of AV-PVP-TCA-I$_2$ through the acetylation degree, with absorptions around 1740 cm^{-1} (Supplementary Figure S3) [42]. The glycosidic O-C-O stretching vibrational modes appear around 1664 cm^{-1} [41]. In this work, the Raman shifts at 1602 and 1648 cm^{-1} in TCA and AV-PVP-TCA-I$_2$ overshadow the glycosidic stretching modes of O-C-O groups within acemannan (Supplementary Figure S3). The title compound presents higher absorption intensities for the same modes. This indicates a larger size of molecules compared to pure TCA and confirms the encapsulation of TCA.

According to Hanai et al., the Raman spectrum for cis-CA and TCA shows Raman shifts at 1632 and 1637 cm^{-1} related to the C=C-stretching vibration [59]. TCA is easily changed due to UV photoisomerization to cis-CA [58,59,107]. We propose the existence of a mixture between cis-CA and TCA in our title compound AV-PVP-TCA-I$_2$ due to shifts at 1643 and 1645 cm^{-1} [58,59,107].

The Raman spectrum proves the purity of the sample AV-PVP-TCA-I$_2$ through the absence of unrelated Raman shifts (Supplementary Figure S3).

3.2.2. UV-Vis Spectroscopy

The UV-vis spectrum of the two samples AV-PVP-TCA and AV-PVP-TCA-I$_2$ is shown in Figure 6.

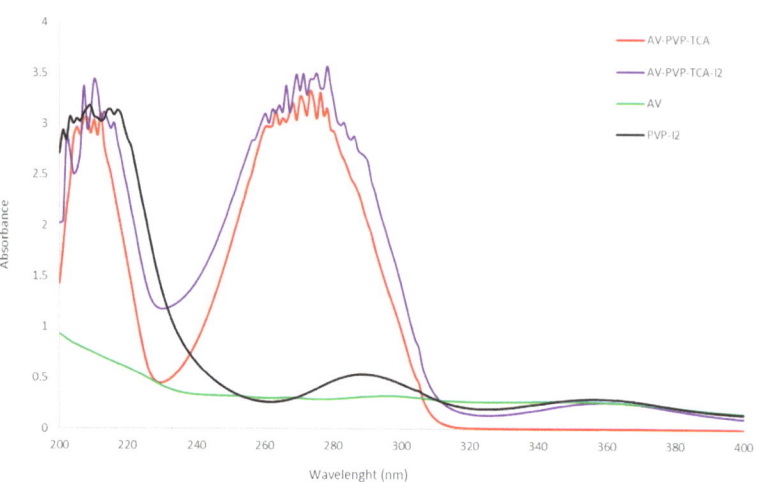

Figure 6. UV-vis analysis of AV-PVP-TCA, AV-PVP-TCA-I_2, AV and PVP-I_2 (200–400 nm). (AV-PVP-TCA: red; AV-PVP-TCA-I_2: purple; AV: green; PVP-I_2: black).

The UV-vis spectrum contains broad and high intensity absorptions in the regions around 200 to 230 nm and 240 to 310 nm (Figure 6). These broad absorption bands are the result of overlapping AV, TCA, PVP and iodine moieties [22,23,41,42,58–60]. The absorption of PVP-I_2 (black curve) determines the regions related to PVP clearly in a broad band from 200 to 250 nm with a λ_{max} at 209 nm and a shoulder at 305 nm (Figure 6) [22,23,41,42]. The complexed iodine moieties in PVP-I_2 absorb in the same broad region around 200 to 250 nm. They reveal further signals around λ_{max} = 288 nm and 356 nm (Figure 6). These absorptions are compliant with similar peaks in the UV-vis spectrum of AV-PVP-I_2 and can be used to interpret the spectrum of AV-PVP-TCA-I_2 (purple curve) (Figure 6, Table 2).

The UV-vis spectrum of AV-PVP-TCA-I_2 confirms the composition of the sample. The UV-visible spectrum of AV-PVP-TCA-I_2 reveals absorption peaks of molecular I_2 (207 and 210 nm), iodide ions (202 nm) and triiodide ions (290 and 359 nm) (Figure 6, Table 2). Iodide ions, iodine molecules and triiodide ions are detected in accordance with previous investigations (Figure 6, Table 2) [15,22,23,25,27,30,33,34,36,41,42,89–91,100–104]. The bands at 290 and 359 nm depict a higher availability of symmetrical, linear triiodide ions (I-I-I$^-$) compared to asymmetric triiodide ions and iodine. The predominance of "smart" triiodide ions was confirmed by the Raman spectrum (Figure 5, Table 1). The very weak absorption intensity at 359 nm is expected to be an overtone to triiodide ions instead of being an indicator of pentaiodide units.

The biohybrids AV-PVP-Sage-I_2 and AV-PVP-I_2 reveal comparable UV-vis absorptions to the title compound (Table 2) [41,42]. The iodine moieties appear in the UV-vis spectrum at similar wavelengths. The triiodides absorb in the Sage formulation at 291 and 359 nm (Table 2) [42]. The basic biohybrid AV-PVP-I_2 has absorption peaks at 291 and 358 nm [41]. These triiodides are "smart" triiodides with [I-I-I] units consisting of pure halogen bonding [33,34,36,41,42]. "Smart" triiodides are more stable and have enhanced antimicrobial activities [33,34,36,41,42]. They release iodine molecules in a controlled manner when the complexes are deformed due to electrostatic interaction with the microbial cell membranes [33,34,36,41,42]. Hence, the PVP in the biohybrid AV-PVP-TCA-I_2 complexes iodine moieties similar to the previous investigations [41,42]. The formula PVP-I_2 can be used as PVP-I_3^- according to previous reports [16–21,23,41,42]. After one-month storage of AV-PVP-TCA-I_2, the resulting UV-vis spectrum did not show any difference compared to the UV-vis spectrum of the fresh sample.

Table 2. UV-vis absorption signals in the samples AV-PVP-TCA (1), AV-PVP-TCA-I$_2$ (2), AV-PVP (3), AV-PVP-I$_2$ (4), PVP-I$_2$ (5) and [58] (nm).

Group	1	2	3	4	5	[58]
I$_2$		207 vs 210 vs		205	206 vs	
I$_3^-$		290 s,sh 359 w,br 440 vw,sh		290 358	288 m,br 356 w,br 440 vw,br	
I$^-$		202 vs		202	201 vs	
AV/Aloin	208 **	207 **	207 vs			
PVP	203–215 ** 210 vs 216 sh	203–219 ** 210 vs 218 s,sh	201 vs 202 vs 209 vs 211 br 213 sh 217 sh	222	201 vs 203 vs 210 vs,br 215 vs,br 217 vs,br 221 s,sh	
PVP-I$_2$	305 w,sh	305 m,sh	305	305	305 w,sh	
TCA	203–215 ** 278 vs 283 s,sh 286 s,sh	203–219 ** 278 vs 283 vs 286 vs				209 218 276

** The broad bands overlap, and several peaks related to AV compounds, iodine moieties and TCA cannot be observed. vw = very weak, br = broad, s = strong, vs = very strong, m = intermediate, sh = shoulder.

Enhanced interactions between the iodine moieties, AV biocomponents, TCA molecules and the PVP are expressed by the UV-vis spectrum of the title compound. The broadening also refers to the increased hydrogen bonding within the sample AV-PVP-TCA-I$_2$. In comparison to the PVP-I$_2$ (black curve), the title compound shows increased absorption intensities for triiodides at 290 nm, while the band at 359 nm is decreased (Figure 6, Table 2). TCA molecules have strong absorptions in the region between 203 to 219, as well as 270 to 290 nm [58–60]. These overlap with the absorption bands of the iodine moieties and impede further clarifications (Figure 6, Table 2). The absorption bands of TCA are represented by strong bands at 210, 212, 278, 283 and 286 nm [58–60]. The addition of iodine leads to a hyperchromic effect, which coincides with increased availability of chromophores and conjugated systems. The mentioned bands increase in intensity in the title compound AV-PVP-TCA-I$_2$, indicating the liberation of TCA and further AV components from the encapsulation by PVP (Figure 6, Table 2). The freed molecules perform hydrogen bonding, resulting in higher intensity and broadened bands in the spectrum of AV-PVP-TCA-I$_2$ (Figure 6, Table 2).

PVP shows strong absorption at 203–219 and 305 nm (Figure 6, Table 2) [22,23,41,42]. Iodine moieties and TCA overshadow the region between 203 to 219 nm, but the shoulder at 305 nm can be observed easily in the UV-visible spectrum (Figure 6, Table 2). Starting with PVP-I$_2$ (black curve), the intensity of the shoulder at 305 nm increases from AV-PVP-TCA to AV-PVP-TCA-I$_2$ (Figure 6, Table 2).

The vibrational mode for –C=O can be observed in PVP-I$_2$ at 221 nm and undergoes a blue shift toward 215 nm in AV-PVP-TCA (Figure 6, Table 2) [22,23,41,42]. The addition of iodine into the latter compound results in a red shift, with increased absorption intensity at 217 nm (Figure 6, Table 2). In comparison with AV-PVP-TCA, this bathochromic effect means for AV-PVP-TCA-I$_2$ an increase in chromophores and conjugated systems related to –C=O, as well as less encapsulation and decreased hydrogen bonding. Adding iodine into the system liberates TCA and the AV components from the encapsulation by PVP. Their –C=O bonds absorb at higher intensities, while PVP encapsulates the iodine units.

The biohybrid, as well as AV-PVP-TCA-I$_2$, shows higher absorbance between the wavelength range of 240 to 320 nm compared to our previously reported complex AV-PVP-

Sage-I_2 [42]. This hyperchromic effect reveals the existence of more π-electrons in the title compound. This means an increase in conjugation, less complexation, less encapsulation and less hydrogen bonding. TCA aromatic rings absorb around 277 nm and are, together with the AV aromatic biocomponents, the reason for the broad signal between 260 to 320 nm in the UV-vis spectrum. TCA has an absorption maximum at λ-max = 274 nm, with additional absorption peaks at 215 and 204 nm in accordance with previous investigations (Figure 6, Table 2) [58–60]. However, TCA overshadows the UV-vis spectrum and makes further clarifications about AV biocomponents difficult (Figure 6, Table 2). According to Saleh et al., cis-CA absorbs at 255 nm [58]. The additional shoulder at 256 nm in AV-PVP-TCA-I_2 may be related to cis-CA as a UV-induced photoisomerization of TCA (Figure 6) [58,59,107]. Therefore, we may confirm a mixture between TCA and cis-CA in the sample after iodination. The same result was obtained during the discussion of the Raman spectra (Supplementary Figure S3). The broadness of the absorption band in the UV-vis around 260 to 320 nm for both of the compounds AV-PVP-TCA and AV-PVP-TCA-I_2 may even suggest the availability of cis-CA before iodination.

3.2.3. Fourier Transform Infrared (FTIR) Spectroscopy

The FTIR analysis of AV-PVP-TCA and AV-PVP-TCA-I_2 augments the small differences between the two samples clearly (Figure 7, Supplementary Figure S2).

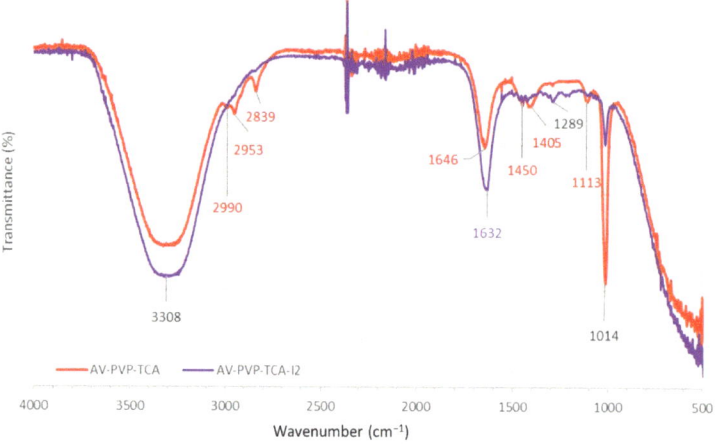

Figure 7. FTIR spectroscopic analysis from AV-PVP-TCA (red) and AV-PVP-TCA-I_2 (purple).

The title compounds AV-PVP-TCA-I_2 and AV-PVP-TCA reveal a similar pattern in the transmission spectrum (Figure 7). The iodinated biomaterial shows intense broadening and higher absorption intensity in two regions. These are evident around 3308, 1632, 1639, 1649 and 1655 cm^{-1}, being related to stretching vibrations of –COOH, –C=O (-COOH) and C=C groups, respectively (Figure 7, Table 3) [39,41,42,59,108–110].

The addition of iodine into AV-PVP-TCA produces shifts in the vibrational bands from 1646, 1651 and 1656 cm^{-1} toward 1632 cm^{-1}, 1639 cm^{-1} and 1655 cm^{-1}, respectively (Figure 7, Table 3) [59]. The blue shift from 1646 to 1632 cm^{-1} is related to the C=C bonds within TCA after iodine addition (Table 3). The C=O of TCA underwent a small shift from 1656 to 1655 cm^{-1} after iodination (Table 3). The carbonyl group in PVP, which is responsible for complexing iodine moieties, showed a blue shift from 1651 to 1639 cm^{-1} after the addition of iodine (Table 3) [19,22,59]. The blue shifts are due to the interaction of iodine with the C=O groups of PVP, verifying the encapsulation of the polyiodide ions by the PVP polymer matrix. This is also confirmed by a blue shift of the vibrational band related to PVP amide groups from 1292 to 1288 cm^{-1} after iodination (Table 3) [19]. The broadening, coupled with the increase in absorption intensities, indicates a higher availability of -C=C-, carboxyl and carbonyl groups originating from the released TCA and AV biocomponents. These molecules interact with the light and engage in hydrogen bonding. Adding molecular iodine also triggers oxidation reduction reactions. Iodine is reduced to iodide-ions, while the hydroxyl groups of AV biocomponents are oxidized to carbonyl groups. The encapsulation of PVP increases with the addition of iodine (Figure 7, Table 3).

Table 3. FTIR analysis [cm^{-1}] of AV-PVP-TCA (A) and AV-PVP-TCA-I$_2$ (B).

	$\nu_{1,2}$ (O–H)$_{s,a}$ ν (COO<u>H</u>)$_a$	ν (C–H)$_a$	ν (CH$_2$)$_{s,a}$	ν (C-H)$_s$ ν (O-H) *	ν (C=O)$_a$ ν (C=C)	δ (C-H)$_a$ δ (CH$_2$) δ (O-H)	ν (C-C)	ν (C-O) $\underline{\nu\ (N-H)}$	ν (C-O)
A	3308	2990	2953	2839	1646 (C=C) TCA 1651 (C=O) PVP 1656 (C=O) TCA	1405 1417 δ (O-H) TCA 1450 1465	1314 1343	1290 TCA <u>1292</u>	1014 1065 1074 1113 1205 1223 1275
B	3308	2989	2953	2839	1632 (C=C) TCA 1639 (C=O) PVP 1649 1655 (C=O) TCA	1405 1417 δ (O-H) TCA 1425 1450 1465 1495	1314 1343	1289 TCA <u>1288</u>	1016 1064 1073 1113 1205 1223 1275

* ν = vibrational stretching, δ = deformation, $_s$ = symmetric, $_a$ = asymmetric.

The FTIR analysis reveals striking similarities to our previous investigations (Figure 7, Table 3) [41,42]. AV-PVP-TCA and its iodinated formulation AV-PVP-TCA-I$_2$ absorb in the same wavelength regions as in our previous reports (Figure 7, Table 3) [41,42]. AV biocomponents are available in the FTIR spectrum (Figure 7). Acemannan is one of the major components in AV and can be detected by the O-C-O acetyl stretching at 1275 cm^{-1} in both FTIR spectra (Figure 7, Supplementary Figure S4) [41]. The small intensity peak at 1113 cm^{-1} is originating from aloin and is reduced in intensity after adding I$_2$ because it is also more encapsulated by the PVP complex [41,42]. The complexation of iodine moieties also releases other molecules previously complexed by the PVP backbone (Figure 7, Table 3) [41,42]. This finding is manifested in AV-PVP-TCA-I$_2$ by the increased band intensities and broadening between 3100 and 3600 cm^{-1} (Figure 7, Table 3) [41,42]. This proves a higher interaction of the light with more released –COOH-groups, which connect after their release from the PVP carbonyl oxygen atoms with other biomolecules through hydrogen bonding. The same increase in intensity occurs for the bands of the carbonyl-C=O stretching vibration (Figure 6). The band appears in AV-PVP-TCA at 1646 cm^{-1} with lower intensity and undergoes a red shift, coupled with an increase in intensity toward 1632 cm^{-1} in the iodinated title compound. The increased intensity is proof of the release from encapsulation from PVP after adding iodine (Figure 7, Table 3) [41,42]. At the same time, this indicates the increase in –C=O groups as a result of oxidation (Figure 7, Table 3) [41,42]. Hydroxyl groups within the AV biocompounds are oxidized, while iodine is reduced into iodide ions. The red shift toward 1632 cm^{-1} underlines the process of more hydrogen bonding and increased interactions between the newly formed –C=O groups and other moieties within the sample (Figure 7, Table 3) [41,42].

As previously reported, adding iodine leads to a hypsochromic effect because of complexation of triiodide ions into the PVP matrix [16–21,23,41,42]. This complexation reduces the number of π-electrons and conjugation systems, which leads to more hydrogen bonding [41,42]. Once molecular iodine is added into the formulation, it is reduced by AV biocomponents to iodide, while the phytochemical functional groups are oxidized and partly released from the PVP matrix [39,41,42]. Triiodide ions form hydrogen bonding to carbonyl groups within PVP [16–21,23,41,42]. The released AV phytochemicals result in higher absorption intensities for –COOH, –OH, –C=O, -C=C- and –C–O in AV-PVP-TCA-I_2 (Figure 7, Table 3) [41,42].

The FTIR spectrum reveals bands for PVP carbonyl groups at 1651 and 1639 cm^{-1} for AV-PVP-TCA and AV-PVP-TCA-I_2, respectively (Figure 7, Table 3) [41,42]. This red shift indicates a stronger encapsulation of the carbonyl groups of PVP after addition of iodine, resulting in less hydrogen bonding with the AV biocomponents and TCA. Again, this proves a partial release of those components by replacement with iodine moieties. The same peak is observed for the Sage formulations at 1650 cm^{-1} for PVP [41,42]. The stretching vibration of -C-O appears for TCA at 1290 cm^{-1} before and at 1289 cm^{-1} with higher intensity after adding iodine. This also confirms the partial release of TCA molecules after adding iodine (Figure 7, Table 3) [41,42]. The methylene groups of PVP are usually located around 2700 to 2900 cm^{-1} for PVP [41,42,108,109]. In our title compounds, the bands are available at 2990/2989 cm^{-1} (C-H)$_a$, 2953 (CH$_2$) and 2839 (C-H)$_s$ (Figure 7, Table 3) [41,42,108,109]. However, their intensities are reduced after iodine addition. This proves higher involvement of PVP entities in complexation in AV-PVP-TCA-I_2, in accordance with previous investigations (Figure 7, Table 3) [41,42]. The same results can be observed for C-C and CH$_2$ rocking vibrational bands at 1014 cm^{-1} (Figure 7, Table 3). The absorption intensity of this band is highly reduced in the title compound AV-PVP-TCA-I_2, confirming the higher coiled structure of the PVP (Figure 7, Table 3). Our previously investigated compounds AV-PVP-Sage-I_2, AV-PVP-I_2, AV-PVP-I_2-Na and pure PVP show the same band at 1017 cm^{-1} [41,42,108]. The red shift from 1017 to 1016/1014 cm^{-1} proves higher complexation in our title compounds AV-PVP-TCA/AV-PVP-TCA-I_2 in comparison to pure PVP and our previous investigations [41,42,108]. The encapsulation of iodine moieties leads to a coiled structure of PVP, which reduces the absorption intensities of the backbone carbon chain of the polymer.

According to Hanai et al., cis-CA shows an absorption peak for the C=C-stretching vibrations at 1631 cm^{-1}, which is comparable to our result of 1632 cm^{-1} in Figure 7 (Table 3) [59]. This can confirm the availability of cis-CA in our sample after iodination due to UV-induced photoisomerization (Figure 7, Table 3).

In general, incorporating iodine into the system partially releases TCA and AV biocompounds from the PVP matrix. TCA and AV components appear in the FTIR spectrum of the iodinated title compound with higher absorbance. These molecules show higher intensities in the spectrum through their available functional groups, chromophores and increased abilities to form hydrogen bonding. The increase in absorption intensities is manifested by the AV-PVP-TCA-I_2 FTIR spectrum around 3308 (-O-H and -COOH), 1632 (asymmetric C=O from -COOH) and 1641 cm^{-1} (C=C) (Figure 7, Table 3). The bands originate from TCA and the main AV component aloin. Further indicators are the vibrational stretching bands around 1343 and 1314 cm^{-1} (both C-C) (Figure 7, Table 3). The C-O vibrational stretching bands at 1289, 1275, 1223, 1205, 1073 and 1064 cm^{-1} show higher absorbances compared to their counterparts in the AV-PVP-TCA compound.

As a result, iodine addition intensifies the -OH, -COOH, -C-O and -C=O interaction of de-complexed TCA and AV biocomponents. At the same time, higher encapsulation of iodine by the PVP backbone leads to lower intensity bands of -CH and -CH$_2$ groups. The FTIR spectrum proves the purity of the synthesized biohybrids AV-PVP-TCA and AV-PVP-TCA-I_2 (Figure 7, Table 3).

3.2.4. X-ray Diffraction (XRD)

The XRD analysis of AV-PVP-TCA-I_2 reveals sharp diffraction peaks originating from AV, TCA, PVP and iodine (Figure 8).

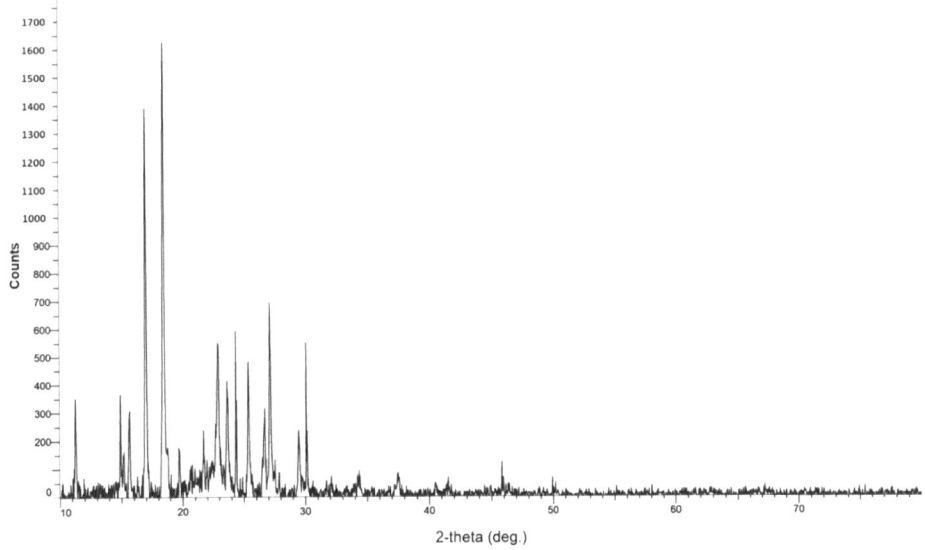

Figure 8. XRD analysis of AV-PVP-TCA-I$_2$.

The two main peaks at 2θ = 17, 19° and smaller peaks at 11 and 15° belong to TCA, while others at 11 and 20° are due to PVP [41,95–100]. Further peaks reveal crystalline behavior at 2θ = 10, 15, 16, 19, 20, 21 and 22° [41,95–100]. The small broad bands from 2θ = 20° to 30° indicate the formation of amorphous phases after the deposition of polyiodide species into the PVP matrix (Figure 8). Iodide moieties are encapsulated into the PVP backbone and cause small but broad amorphous regions below the sharp peaks at 22.8, 23.8, 25.5, 27 and 30 (Figure 8, Table 4).

Table 4. XRD analysis of the samples AV-PVP-TCA-I$_2$, AV-PVP-Sage-I$_2$ (1), AV-PVP-I$_2$ (2) [41], AV-PVP-I$_2$-NaI (3) [41] and in previous reports (2Theta°).

Group	AV-PVP-TCA-I$_2$	1	2	3	[95]	[96]	[13]
I$_2$	22.8 m 23.8 m 25.5 m 27 m 30 m 37 vw 46 w	-	-	28 m 40 w	25 29 36	-	24.5 s 25 s 28 s 37 w 38 w 43 w 46 m
PVP	11 m 20 m	13 s	10 s 19 s,br	11 s,br 20 s,br	-	-	-
TCA	11 m 15 w 17 s 19 vs 24 m 26 m	-	-	-	-	10 vs 19 w 20 w 23 s 25 m 27 w 29 w 32 w 34 w	-
AV	16 m 22 m	14 s	14 s 21 s,br 22 s,br	-	-	-	-

w = weak, br = broad, s = strong, m = intermediate.

The decrease in peak intensities related to PVP and their shift toward 2Theta values of 11° and 20° confirm the higher encapsulation of polyiodides in the title compound compared to AV-PVP-I$_2$ alone. The addition of TCA into the system clearly induced a higher crystallization degree, lower intensities and sharp peaks (Figure 8, Table 4).

The purity of our title compound AV-PVP-TCA-I$_2$ is confirmed by the XRD analysis (Figure 8, Table 4).

Iodine and TCA are represented according to our previous studies and other investigations around 24–46 2Theta degrees [41,95–97]. The intermediate peaks around 23, 24, 26, 27 and 30 degrees (2Theta) are followed by very weak to weak signals at 37 and 46 degrees, respectively (Figure 8, Table 4). These signals appear in previous investigations and confirm the iodine within our sample [41,95,97]. Figure 8 and Table 4 indicate the related strong and intermediate peaks of the crystalline TCA at 2Theta degrees of 10, 23 and 25, 29, respectively (Figure 8, Table 4) [96]. These are accompanied by several other weak peaks at 2Theta 19, 20, 27, 32, 34 (Figure 8, Table 4) [96]. AV biocomponents are indicated by intermediate peaks at 2Theta degrees of 16 and 22 in AV-PVP-TCA-I$_2$ (Figure 8, Table 4), which are all available in previous investigations as well [41,98,99]. The title compound AV-PVP-TCA-I$_2$ has a higher crystalline character in comparison to our previous investigation of AV-PVP-I$_2$, AV-PVP-NaI and AV-PVP-Sage-I$_2$ [41,42]. The higher crystallinity in AV-PVP-TCA-I$_2$ is due to the higher presence of TCA in the formulation compared to the other bio-hybrids [41,42]. The sample shows crystalline character in the SEM of the surgical suture (Figure 2). The same result was obtained in the XRD analysis with sharp, distinct peaks for AV, TCA, iodine and PVP (Figure 8, Table 4) [41,98–100].

3.3. Antimicrobial Activities of AV-PVP-TCA and AV-PVP-TCA-I$_2$

Disc diffusion assay against 10 reference microorganisms was used to investigate the antimicrobial activities of AV-PVP-TCA and AV-PVP-TCA-I$_2$. The utilized Gram-positive bacteria were *S. pneumonia* ATCC 49619, *S. aureus* ATCC 25923, *S. pyogenes* ATCC 19615, *E. faecalis* ATCC 29212 and *B. subtilis* WDCM0003. Gram-negative bacteria included the strains *E. coli* WDCM 00013 Vitroids, *P. mirabilis* ATCC 29906, *P. aeruginosa* WDCM 00026 Vitroids and *K. pneumonia* WDCM00097 Vitroids. *C. albicans* WDCM 00054 Vitroids, which were used to test the antifungal activities of our biomaterial. The antimicrobial properties of our bio-compounds were compared to the positive controls, gentamicin and nystatin. Methanol, ethanol and water were the negative controls. They showed no zone of inhibition (ZOI) and were excluded from Table 5.

Table 5. Antimicrobial testing of antibiotics (A), AV-PVP-TCA-I$_2$ by disc dilution studies (1,2,3), suture (S), bandage (B) and mask (M). ZOI (mm) against microbial strains by diffusion assay.

Strain	Antibiotic	A	1 [+]	2 [+]	3 [+]	S	B	M
S. pneumoniae ATCC 49619	G	18	14	12	0	3	24	20
S. aureus ATCC 25923	G	28	17	13	11	7	40	35
S. pyogenes ATCC 19615	G	25	16	12	9	4	24	19
E. faecalis ATCC 29212	G	25	15	13	0	3	23	20
B. subtilis WDCM 00003	G	21	14	0	0	4	23	16
P. mirabilis ATCC 29906	G	30	0	0	0	0	0	0
P. aeruginosa WDCM 00026	G	23	13	12	0	2	20	20
E. coli WDCM 00013	G	23	13	0	0	0	26	16
K. pneumoniae WDCM 00097	G	30	13	10	0	2	28	21
C. albicans WDCM 00054	NY	16	46	25	13 *	17	55	54

[+] Disc diffusion studies (6 mm disc impregnated with 2 mL of 11 µg/mL (1), 2 mL of 5.5 µg/mL (2) and 2 mL of 2.75 µg/mL (3) of AV-PVP-TCA-I$_2$. A = G Gentamicin (30 µg/disc). NY (Nystatin) (100 IU). Suture (S), bandage (B) and mask (M) impregnated with 2 mL of 11 µg/mL AV-PVP-TCA-I$_2$. Gray shaded area represents Gram-negative bacteria. 0 = Resistant. * Further dilution to 1.38 µg/mL yielded ZOI = 9 mm. No statistically significant differences ($p > 0.05$) between row-based values through Pearson correlation.

Our title biohybrid AV-PVP-TCA-I$_2$ shows remarkable antimicrobial properties of impregnated discs, sutures, bandages and face masks (Table 4). The tests were performed on discs at concentrations of 11, 5.5 and 2.75 μg/mL. Sutures, bandages and masks were treated with a concentration of 11 μg/mL. The highest inhibition zones were achieved, in general, against the fungal reference strain *C. albicans* WDCM 00054, followed by Gram-positive and, finally, Gram-negative bacterial strains.

C. albicans WDCM 00054, *S. aureus* ATCC 25923, *S. pneumoniae* ATCC 49619 and *E. coli* WDCM 00013 were more susceptible toward AV-PVP-TCA-I$_2$ than their respective antibiotic positive controls (Table 3). AV-PVP-TCA-I$_2$ inhibits *C. albicans* WDCM 00054 at concentrations of 11 and 5.5 μg/mL with 46 and 25 mm stronger than nystatin (Table 5, Figure 9a).

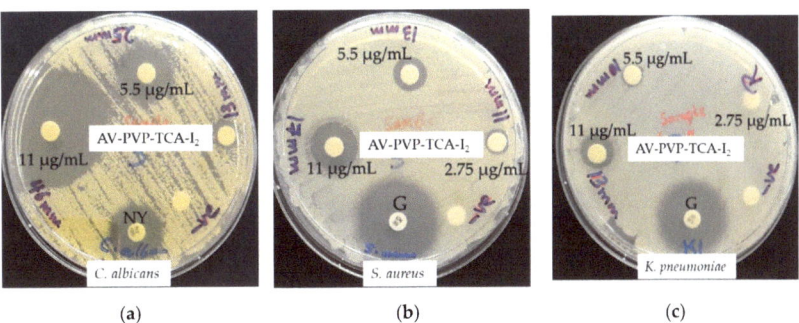

Figure 9. Disc diffusion assay of AV-PVP-TCA-I$_2$ (with concentrations of 11, 5.5 and 2.75 μg/mL) with positive control antibiotic nystatin (NY = 100 IU) and gentamicin (G = 30 μg/disc). From left to right: AV-PVP-TCA-I$_2$ against (**a**) *C. albicans* WDCM 00054; (**b**) *S. aureus* ATCC 25932; (**c**) *K. pneumoniae* WDCM 00097.

S. aureus ATCC 25923 (Figure 9b) and *K. pneumoniae* WDCM 00097 reveal a ZOI of 17 and 13 mm, respectively. The susceptibility toward AV-PVP-TCA-I$_2$ decreases in the order of the Gram-positive pathogens *S. pyogenes* ATCC 19615 (16 mm), *E. faecalis* ATCC 29212 (15 mm), *S. pneumoniae* ATCC 49619 and *B. subtilis* WDCM 00003 (both 14 mm) (Table 5, Figure 9c). The Gram-negative reference strains follow with *P. aeruginosa* WDCM 00026 (13/12 mm), *K. pneumoniae* WDCM 00097 (13/10 mm) and *E. coli* WDCM 00013 (13/0 mm) at concentrations of 11 and 5.5 μg/mL, respectively (Table 5, Figure 9).

The disc diffusion tests revealed strong antifungal activity of AV-PVP-TCA-I$_2$ against *C. albicans* WDCM 00054. The next most susceptible microorganism is the Gram-positive bacteria *S. aureus* ATCC 25932. Intermediate antibacterial activity against further Gram-positive and Gram-negative species *K. pneumoniae* WDCM 00097, *P. aeruginosa* WDCM 00026 and *E. coli* WDCM 00013 are manifested as well (Table 5, Figure 9).

The promising results in the disc diffusion studies tempted us to investigate AV-PVP-TCA-I$_2$ on surgical sutures as potential preventive agents against surgical site infections. We coated braided PGA surgical sutures with the title compound. The inhibitory activity on surgical sutures was tested against the same 10 reference strains (Table 5, Figure 10).

The inhibitory zones around the sutures follow the same trends displayed by the disc diffusion studies on sterile discs. The dip-coated sutures showed extraordinary antifungal activity against *C. albicans* WDCM 00054, with an inhibition zone of 17 mm (Table 5, Figure 10). The Gram-positive pathogen, *S. aureus* ATCC 25,932 (7 mm), is followed by *S. pyogenes* ATCC 19615 (4 mm), *B. subtilis* WDCM 00003 (4 mm), *S. pneumoniae* ATCC 49619 and *E. faecalis* ATCC 29212 (both 3 mm) (Figure 10, Table 5). The two Gram-negative pathogens, *P. aeruginosa* WDCM 00026 and *K. pneumoniae* WDCM 00097, revealed inhibition zones of 2 mm against AV-PVP-TCA-I$_2$ (Figure 10, Table 5).

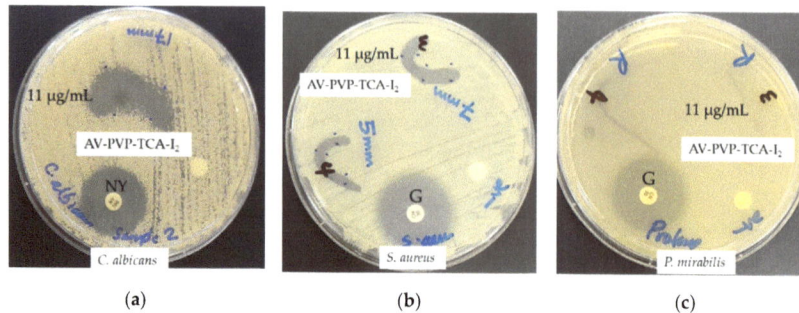

Figure 10. Impregnated, sterile PGA sutures with AV-PVP-TCA-I_2. with positive control antibiotic nystatin (NY = 100 IU) and gentamicin (G = 30 μg/disc). From left to right: AV-PVP-TCA-I_2 (11 μg/mL) against (**a**) *C. albicans* WDCM 00054; (**b**) *S. aureus* ATCC 25932; (**c**) *P. mirabilis* ATCC 29906.

We impregnated sterile face masks with the biomaterial AV-PVP-TCA-I_2 to investigate the potential antimicrobial action on masks. The process can potentiate the re-use of masks during any pandemic. This will ensure safety, sustainability and waste reduction. Communities without adequate resources could solve this issue by re-using face masks after treatment with our biocompatible, non-toxic title compound.

The antimicrobial tests on face masks against the same 10 reference strains delivered the same trends. AV-PVP-TCA-I_2 is a strong antifungal agent against *C. albicans* WDCM 00054 (54 mm) (Table 5, Figure 11).

The highest susceptibility against our title compound is shown by the Gram-positive *S. aureus* ATCC 25932 (35 mm), followed by *S. pneumoniae* ATCC 49619 (20 mm), *E. faecalis* ATCC 29212 (20 mm), *S. pyogenes* ATCC 19615 (19 mm) and *B. subtilis* WDCM 00003 (16 mm) (Figure 11, Table 5). The three Gram-negative pathogens, *K. pneumoniae* WDCM 00097, *P. aeruginosa* WDCM 00026 and *E. coli* WDCM 00013, displayed against AV-PVP-TCA-I_2 inhibition zones of 21, 20 and 16 mm, respectively (Figure 10, Table 5).

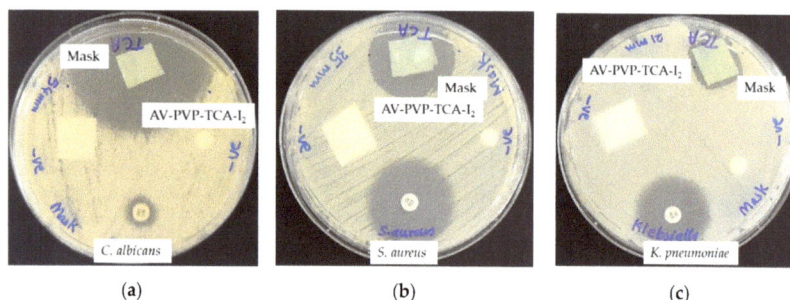

Figure 11. Impregnated sterile masks with AV-PVP-TCA-I_2. with positive control antibiotic nystatin (NY = 100 IU) and gentamicin (G = 30 μg/disc). From left to right: AV-PVP-TCA-I_2 (11 μg/mL) against (**a**) *C. albicans* WDCM 00054; (**b**) *S. aureus* ATCC 25932; (**c**) *K. pneumoniae* WDCM 00097.

We tested the impregnated sterile cotton bandages with AV-PVP-TCA-I_2 and tested them against the same 10 microorganisms. Our compound can be potentially used as a strong microbial agent in the treatment of wounds and wound infections. The results show strong antifungal activities against *C. albicans* WDCM 00054, with inhibition zone of 55 mm (Table 5, Figure 12).

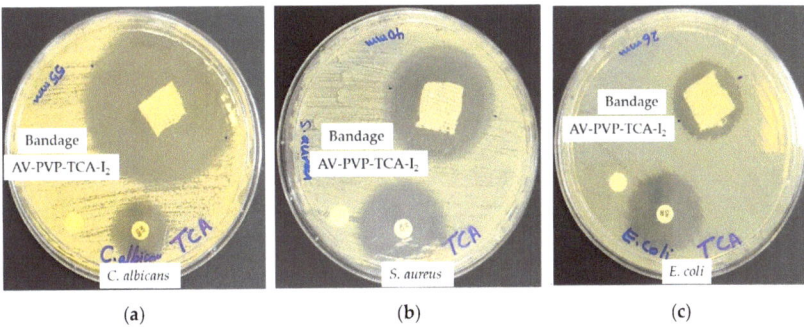

Figure 12. Impregnated sterile bandages with AV-PVP-TCA-I_2. with positive control antibiotic nystatin (NY = 100 IU) and gentamicin (G = 30 µg/disc). From left to right: AV-PVP-TCA-I_2 (11 µg/mL) against (**a**) *C. albicans* WDCM 00054; (**b**) *S. aureus* ATCC 25932; (**c**) *E. coli* WDCM 00013.

The Gram-positive *S. aureus* ATCC 25932 (40 mm) is followed in decreasing inhibitory action by *S. pneumoniae* ATCC 49619 and *S. pyogenes* ATCC 19615 (both 24 mm), *E. faecalis* ATCC 29212 and *B. subtilis* WDCM 00003 (both 23 mm) (Figure 11, Table 5).

The three Gram-negative bacteria, *K. pneumoniae* WDCM 00097, *E. coli* WDCM 00013 and *P. aeruginosa* WDCM 00026, revealed inhibition zones of 28, 26 and 20 mm, respectively (Figure 12, Table 5).

The highest inhibitory action was achieved by AV-PVP-TCA-I_2 against the fungal strain *C. albicans* WDCM 00054 with a ZOI = 46 mm compared to nystatin with 16 mm (Table 5). We used the same method and concentration in our previous work with AV-PVP-Sage-I_2 and reported an alleviated susceptibility of *C. albicans* WDCM 00054 with 52 mm [42]. The inhibitory action is similar or slightly weaker when the formulation contains TCA instead of Sage. The same result is seen in slightly weaker inhibition zones against *S. aureus* ATCC 25923 with 17, 13 and 11 mm (Table 5). In comparison, the Sage formulation showed inhibition zones of 20, 15 and 14 mm against the same pathogen [42]. The spore forming *B. subtilis* WDCM 0003 is slightly more inhibited by the title formulation with TCA in comparison to the Sage formulation (Table 5) [42]. When TCA replaces Sage in the formulation at a concentration of 11 µg/mL, the susceptibility of the Gram-positive *S. pyogenes* ATCC 19615 is slightly higher with 16 mm compared to 14 mm [42]. The Gram-negative, multi-drug-resistant *P. aeruginosa* WDCM 00026 is inhibited by AV-PVP-TCA-I_2 with 13 and 12 mm, while it is resistant against AV-PVP-Sage-I_2 (Table 5) [42,46]. The same pathogen shows augmented susceptibility against AV-PVP-TCA-I_2 on dip-coated sutures as well (Table 5) [42]. All Gram-positive pathogens and *C. albicans* WDCM 00054 were more inhibited on AV-PVP-TCA-I_2-coated sutures compared to AV-PVP-Sage-I_2-coated ones (Table 5) [42]. *C. albicans* WDCM 00,054 was inhibited by AV-PVP-TCA-I_2-coated suture with 17 mm, while the Sage-formulation-coated suture resulted in 15 mm (Table 5) [42]. We also soaked the bandages and surgical face masks with our formulation AV-PVP-TCA-I_2. The impregnated cotton bandages and face masks showed high antimicrobial activity against *C. albicans* WDCM 00054, followed by the Gram-positive pathogen *S. aureus* ATCC 25932 and the Gram-negative *K. pneumoniae* WDCM 00097 (Table 5, Figures 11 and 12).

In conclusion, the synergistic action of the Sage biocompounds in the biohybrid AV-PVP-Sage-I_2 achieves slightly higher inhibition zones for *C. albicans* WDCM 00054 and *S. aureus* ATCC 25932 in comparison to AV-PVP-TCA-I_2 at the same concentration in disc diffusion studies [42]. However, the title compound AV-PVP-TCA-I_2 is effective against *P. aeruginosa* WDCM 00026, while the Sage formulation fails to inhibit this multi-drug-resistant pathogen on disc diffusion studies [42]. The AV-PVP-TCA-I_2 impregnated sutures were more effective than the ones coated with AV-PVP-Sage-I_2 against Gram-positive bacteria and *C. albicans* WDCM 00054 [42]. These results encourage the use of AV-PVP-TCA-I_2 on sutures, bandages and face masks against *C. albicans* WDCM 00054,

Gram-positive *S. aureus* ATCC 25932, Gram-negative *P. aeruginosa* WDCM 00026 and *K. pneumoniae* WDCM 00097.

The susceptibility of the tested microorganisms against our biomaterial AV-PVP-TCA-I_2 reveals similar patterns to previously reported biomaterials AV-PVP-I_2 and AV-PVP-Sage-I_2 [41,42]. The highest susceptibility was showcased by *C. albicans* WDCM 00054, followed by high to intermediate inhibition of the Gram-positive pathogens (Table 5). Gram-negative microorganisms displayed the lowest inhibition zones due to their higher negatively charged, complicated outer cell membrane with porin channels. This confirms the antimicrobial action of molecular iodine, which does not go through porin channels like its hydrophilic iodide and triiodide counterparts. Iodine is released from triiodide ions, which are encapsulated by the PVP complex [41,42,46]. The release occurs through deformation of the PVP complex through dipol–dipol interactions of TCA and AV biocomponents [41,42,46]. Remarkably, the results show higher susceptibility of the multi-drug-resistant *P. aeruginosa* WDCM 00026 against AV-PVP-TCA-I_2 in comparison to the Sage formulation (Table 5) [42,46]. The title formulation seems to be able to avoid the efflux pumps within Gram-negative, motile bacilli. *P. mirabilis* ATCC 29906 is a swarming bacillus with several flaggellae. *P. mirabilis* ATCC 29906 shows resistance against both of our formulations, AV-PVP-TCA-I_2 and AV-PVP-Sage-I_2 (Table 5) [42].

In general, the fungi *C. albicans* WDCM 00054 is highly susceptible, followed by Gram-positive and lastly Gram-negative bacteria. The bacterial morphology rules susceptibility of the studied microorganisms against AV-PVP-TCA-I_2 and our previous compounds in the same patterns [41,42]. Cocci, non-motile, Gram-positive, higher agglomerated clusters (*S. aureus*), chains (*E. faecalis*, *S. pyogenes*) and pairs (*S. pneumoniae*) are more susceptible than motile, Gram-negative pathogens (Table 5) [41,42]. Fungus and Gram-positive microorganisms of our selection have less complicated outer cell membrane structures and are therefore more vulnerable toward AV-PVP-TCA-I_2 [8,11,41,42,44,47,110–112]. The thick peptidoglycan layers of Gram-positive pathogens are crosslinked with peptide bridges, which possibly form hydrogen bonding, dipol–dipol and electrostatic interactions with components in our title formulation through TCA, AV and PVP [8,11,36,39,41,42,44,47,110–112]. These interactions deform the complex AV-PVP-TCA-I_2 and result in the release of molecular iodine from the "smart" triiodides reservoir [36,39,41,42]. Iodine compromises the cell membranes, leads to cell leakage, diffuses easily through the cell membranes and causes cell death by inhibiting important cell functions [17,18,36,39,41,42].

In conclusion, the biomaterial is a strong antifungal agent on bandages, discs, sutures and masks against *C. albicans* WDCM 00054. All the tested materials exhibited highest inhibitory action against the Gram-positive pathogen *S. aureus* ATCC 25932 and other tested bacterial strains. Intermediate to high inhibitory action is recorded with the Gram-negative pathogens, *K. pneumoniae* WDCM 00097, *E. coli* WDCM 00013 and *P. aeruginosa* WDCM 00026. *P. mirabilis* ATCC 29906 showed resistance in all the investigations.

Our title compound AV-PVP-TCA-I_2 is a combination of affordable, commonly used and well-known non-toxic antimicrobial agents. Coating common gauze bandages and surgical sutures with AV-PVP-TCA-I_2 can mitigate or even prevent surgical site and wound infections. The work is based on our previous investigations [36,39,41,42]. Impregnating face masks with AV-PVP-TCA-I_2 can potentially reduce the viral load in the area between the mask and the face during its use. AV-PVP-TCA-I_2 can deactivate common microorganisms, moisturize skin and tissues and alleviate the filtering properties of the mask fibers. The formulation AV-PVP-TCA-I_2 may potentially support the community during pandemics, crises, shortages of antimicrobial disinfectants and reduce the need for disposing face masks after single use. The components are readily available for any individual and can be produced in households. Further investigations are needed to confirm the suggested uses and other biological activities of the formulation.

4. Conclusions

Iodine-based disinfectants and antiseptics are impervious to AMR and can play a strategic role against resistant, opportunistic pathogens. Biomaterials consisting of iodine and plant biocompounds have enhanced inhibitory action against a variety of microorganisms. These microbicidal formulations can be employed as disinfectants on inanimate surfaces to mitigate community- and hospital-acquired infections, as well as antiseptics on skin tissues. Coating sutures, wound dressing materials and gauze bandages can alleviate the development of SSI and wound infections. Surgical face masks are needed in the prevention of microbial infections through air droplets. Single-use, disposable face masks already place a huge burden on the environment, resources and low-income populations. Impregnating face masks with antimicrobial formulations can alleviate viral transmission and ameliorate sustainability. The title compound AV-PVP-TCA-I_2 showed excellent antifungal and good to intermediate antibacterial activities against our selection of Gram-positive- and Gram-negative reference strains, respectively. AV-PVP-TCA-I_2 inhibits *C. albicans* WDCM 00054 remarkably on sutures, bandages and face masks with a ZOI of 17, 55 and 54 mm, respectively. Gram-positive *S. aureus* ATCC 25932, Gram-negative *P. aeruginosa* WDCM 00026 and *K. pneumoniae* WDCM 00097 are also strongly susceptible on bandages and face masks. Therefore, the title compound has the potential to be used as an infection control agent against the mentioned pathogens. The composition and morphology of the title compound was verified by analytical methods. Our results confirm AV-PVP-TCA-I_2 as a strong antifungal agent against *C. albicans*. The title compound is an alternative in the prevention or treatment of infectious diseases on inanimate surfaces, face masks, sutures and bandages. Further investigations are needed to confirm the intended uses and other biological activities.

Supplementary Materials: The following supporting information can be downloaded at: https://www.mdpi.com/article/10.3390/polym14101932/s1, Figure S1: EDS of Bandage (coiled part), Figure S2: EDS of Bandage (ordered fibers part), Figure S3: Raman analysis full spectrum of AV-PVP-TCA-I2, Figure S4: FTIR spectrum of AV, Figure S5: FTIR spectrum of PVPI.

Author Contributions: Conceptualization, Z.E., S.H.B. and N.I.W.A.; methodology, S.H.B. and H.A.S.; software, Z.E.; validation, Z.E., H.A.S. and S.H.B.; formal analysis, S.H.B. and H.A.S.; investigation, Z.E., S.H.B., N.I.W.A. and H.A.S.; resources, Z.E. and S.H.B.; data curation, Z.E. and H.A.S.; writing—original draft preparation, Z.E.; writing—review and editing, Z.E., N.I.W.A. and S.H.B.; visualization, Z.E.; supervision, S.H.B. and Z.E.; project administration, S.H.B. and Z.E.; funding acquisition, Z.E. All authors have read and agreed to the published version of the manuscript.

Funding: This work was kindly supported by Ajman University, Deanship of Research and Graduate Studies, Ajman, United Arab Emirates (Project ID No: Ref. # 2021-IRG-PH-2).

Institutional Review Board Statement: Not applicable.

Informed Consent Statement: Not applicable.

Data Availability Statement: Not applicable.

Acknowledgments: The authors deeply appreciate and thank Sohaib Naseem Khan from Ajman University for providing the FTIR analysis, his continuous support and valuable input during the investigations. We are grateful to Hussain Alawadhi, Mohammad Shameer, Muhammed Irshad and Fatima Mohammed Abla from Sharjah University for providing the SEM, XRD, Raman and EDS analyses. We thank Irina Yushina for her valuable input regarding the UV-vis and Raman analysis. We thank Iman Haj Bloukh from Ajman University and Ibrahim Haj Bloukh from Ajman University for their support and input during the investigations. We are grateful to the artist "@art_by_amie" and Iman Haj Bloukh for preparing digital art images as graphical abstracts.

Conflicts of Interest: The authors declare no conflict of interest.

References

1. Mahoney, A.R.; Safaee, M.M.; Wuest, W.M.; Furst, A.L. The silent pandemic: Emergent antibiotic resistances following the global response to SARS-CoV-2. *iScience* **2021**, *24*, 102304. [CrossRef] [PubMed]
2. Ranney, M.L.; Griffeth, V.; Jha, A.K. Critical supply shortages—The need for ventilators and personal protective equipment during the COVID-19 pandemic. *N. Engl. J. Med.* **2020**, *382*, e41. [CrossRef] [PubMed]
3. Bloukh, S.H.; Edis, Z.; Shaikh, A.A.; Pathan, H.M. A Look behind the Scenes at COVID-19: National Strategies of Infection Control and Their Impact on Mortality. *Int. J. Environ. Res. Public Health* **2020**, *17*, 5616. [CrossRef] [PubMed]
4. Mulani, M.S.; Kamble, E.E.; Kumkar, S.N.; Tawre, M.S.; Pardesi, K.R. Emerging Strategies to Combat ESKAPE Pathogens in the Era of Antimicrobial Resistance: A Review. *Front. Microbiol.* **2019**, *10*, 539–563. [CrossRef] [PubMed]
5. Chen, X.; Liao, B.; Cheng, L.; Peng, X.; Xu, X.; Li, Y.; Hu, T.; Li, J.; Zhou, X.; Ren, B. The microbial coinfection in COVID-19. *Appl. Microbiol. Biotechnol.* **2020**, *104*, 7777–7785. [CrossRef] [PubMed]
6. Rawson, T.M.; Moore, L.S.P.; Zhu, N.; Ranganathan, N.; Skolimowska, K.; Gilchrist, M.; Satta, G.; Cooke, G.; Holmes, A. Bacterial and fungal co-infection in individuals with coronavirus: A rapid review to support COVID-19 antimicrobial prescribing. *Clin. Infect. Dis.* **2020**, *71*, 2459–2468. [CrossRef]
7. Avire, N.J.; Whiley, H.; Ross, K. A Review of *Streptococcus pyogenes*: Public health risk factors, prevention and control. *Pathogens* **2021**, *10*, 248. [CrossRef]
8. Mahalingam, S.S.; Jayaraman, S.; Pandiyan, P. Fungal Colonization and Infections—Interactions with Other Human Diseases. *Pathogens* **2022**, *11*, 212. [CrossRef]
9. Cassini, A.; Hogberg, L.D.; Plachouras, D.; Quattrocchi, A.; Hoxha, A.; Simonsen, G.S.; Colomb-Cotinat, M.; Kretzschmar, M.E.; Devleesschauwer, B.; Cecchini, M.; et al. Attributable deaths and disability-adjusted life-years caused by infections with antibiotic-resistant bacteria in the EU and the European Economic Area in 2015: A population-level modelling analysis. *Lancet Infect. Dis.* **2019**, *19*, 56–66. [CrossRef]
10. Kampf, G.; Todt, D.; Pfaender, S.; Steinmann, E. Persistence of coronaviruses on inanimate surfaces and their inactivation with biocidal agents. *J. Hosp. Infect.* **2020**, *104*, 246–251. [CrossRef]
11. Rao, H.; Choo, S.; Rajeswari Mahalingam, S.R.; Adisuri, D.S.; Madhavan, P.; Md. Akim, A.M.; Chong, P.P. Approaches for Mitigating Microbial Biofilm-Related Drug Resistance: A Focus on Micro- and Nanotechnologies. *Molecules* **2021**, *26*, 1870. [CrossRef] [PubMed]
12. Kaiho, T. *Iodine Chemistry and Applications*, 1st ed.; Kaiho, T., Ed.; John Wiley & Sons, Inc.: Hoboken, NJ, USA, 2015; pp. 15–410, ISBN 78-1-118-46629-2.
13. Zhang, Q.; Wu, Z.; Liu, F.; Liu, S.; Liu, J.; Wang, Y.; Yan, T. Encapsulating a high content of iodine into an active graphene substrate as a cathode material for high-rate lithium–iodine batteries. *J. Mater. Chem. A* **2017**, *5*, 15235–15242. [CrossRef]
14. Yao, L.; Xu, P.; Gao, W.; Li, J.; Gao, L.; Niu, G.; Li, D.; Chen, S.; Tang, J. A chain-type diamine strategy towards strongly anisotropic triiodide of DMEDA·I_6. *Sci. China Mat.* **2020**, *63*, 566–574. [CrossRef]
15. Ordintsev, A.A.; Petrov, A.A.; Lyssenko, K.A.; Petrov, A.V.; Goodilin, E.A.; Tarasov, A.B. Crystal structure of new formamidinium triiodide jointly refined by single-crystal XRD, Raman scattering spectroscopy and DFT assessment of hydrogen-bond network features. *Acta Cryst.* **2021**, *E77*, 692–695. [CrossRef] [PubMed]
16. Moulay, S. Molecular iodine/polymer complexes. *J. Polym. Eng.* **2013**, *33*, 389–443. [CrossRef]
17. Moulay, S. Macromolecule/Polymer-Iodine Complexes: An Update. *Recent Innov. Chem. Eng.* **2019**, *12*, 174. [CrossRef]
18. Vasudevan, P.; Tandon, M. Antimicrobial properties of iodine based products. *J. Sci. Ind. Res.* **2010**, *69*, 376–383.
19. Gillam, T.A.; Goh, C.K.; Ninan, N.; Bilimoria, K.; Shirazi, H.S.; Saboohi, S.; Al-Bataineh, S.; Whittle, J.; Blencowe, A. Iodine complexed poly(vinylpyrrolidone) plasma polymers as broad spectrum antiseptic coating. *Appl. Surf. Sci.* **2021**, *537*, 147866. [CrossRef]
20. Bigliardi, P.L.; Alsagoff, S.A.L.; El-Kafrawi, H.Y.; Pyon, J.-K.; Wa, C.T.C.; Villa, M.A. Povidone iodine in wound healing: A review of current concepts and practices. *Int. J. Surg.* **2017**, *44*, 260–268. [CrossRef]
21. Gao, T.; Fan, H.; Wang, X.; Gao, Y.; Liu, W.; Chen, W.; Dong, A.; Wang, Y.J. Povidone-Iodine-Based Polymeric Nanoparticles for Antibacterial applications. *ACS Appl. Mater. Interfaces* **2017**, *9*, 25738–25746. [CrossRef]
22. Ma, J.; Zhang, Y.; Zhang, D.; Niu, X.; Lin, Z. Insights into the molecular interaction between poly(vinylpyrrolidone)-iodine disinfection system and polypropylene microplastics in aquatic environment. *Chem. Eng. J.* **2022**, *430*, 132276. [CrossRef]
23. Xu, X.; Guan, Y. Investigating the Complexation and Release Behaviors of Iodine in Poly(vinylpyrrolidone)-Iodine Systems through Experimental and Computational Approaches. *Ind. Eng. Chem. Res.* **2020**, *59*, 22667–22676. [CrossRef]
24. Viswanathan, K.; Babu, D.B.; Jayakumar, G.; Raj, G.D. Anti-microbial and skin wound dressing application of molecular iodine nanoparticles. *Mater. Res. Express* **2017**, *4*, 104003. [CrossRef]
25. Edis, Z.; Raheja, R.; Bloukh, S.H.; Bhandare, R.R.; Sara, H.A.; Reiss, G.J. Antimicrobial Hexaaquacopper(II) Complexes with Novel Polyiodide Chains. *Polymers* **2021**, *13*, 1005. [CrossRef] [PubMed]
26. Bloukh, S.H.; Edis, Z. Structure and Antimicrobial properties of bis(1,4,7,10-tetraoxacyclododecane-$\kappa^4 O,O',O'',O'''$)cesium pentaiodide, $C_{16}H_{32}CsI_5O_8$. *Z. Krist. NCS* **2020**, *235*, 759–761. [CrossRef]
27. Wang, M.; Takahama, T.; Tashiro, K. Crystalline Iodine Complexes of Amorphous Poly(vinyl acetate) as Studied by X-ray Diffraction, Vibrational Spectroscopy, and Computer Simulation. *Macromolecules* **2020**, *53*, 4395–4406. [CrossRef]

28. Tashiro, K.; Takahama, T.; Wang, M.F. X-ray study of Poly(vinyl Alcohol)-Iodine complex prepared from the dilute iodine solution as a hint to know the inner structure of polarizer. *Polymer* **2021**, *233*, 124180. [CrossRef]
29. Blake, A.J.; Li, W.S.; Lippolis, V.; Schröder, M.; Devillanova, F.A.; Gould, R.O.; Parsons, S.; Radek, C. Template self-assembly of polyiodide networks. *Chem. Soc. Rev.* **1998**, *27*, 195–206. [CrossRef]
30. Savastano, M.; Bazzicalupi, C.; Gellini, C.; Bianchi, A. Genesis of Complex Polyiodide Networks: Insights on the Blue Box/I$^-$/I$_2$ Ternary System. *Crystals* **2020**, *10*, 387. [CrossRef]
31. Edis, Z.; Bloukh, S.H. Preparation and structural and spectroscopic characterization of a pentaiodide [Rb(12-crown-4)$_2$]I$_5$. *Z. Nat.* **2013**, *68*, 1340–1346.
32. Okuda, M.; Hiramatsu, T.; Yasuda, M.; Ishigaki, M.; Ozaki, Y.; Hayashi, M.; Tominaga, K.; Chatani, E. Theoretical Modeling of Electronic Structures of Polyiodide Species Included in α-Cyclodextrin. *J. Phys. Chem. B* **2020**, *124*, 4089–4096. [CrossRef] [PubMed]
33. Haj Bloukh, S.; Edis, Z. Halogen bonding in Crystal structure of bis(1,4,7,10-tetraoxacyclododecane-κ^{4}O,O',O'',O''')cesium triiodide, C16H32CsI3O8. *Z. Krist. New Cryst. Struct.* **2020**, *235*, 717–719. [CrossRef]
34. Edis, Z.; Bloukh, S.H. Preparation and structural and spectroscopic characterization of triiodides [M(12-crown-4)2]I3 with M = Na and Rb. *Z. Nat.* **2014**, *69*, 995–1002.
35. Van Mengen, M.; Reiss, G.J. I$_6^{2-}$ Anion composed of two asymmetric triiodide moieties: A competition between halogen and hydrogen bond. *Inorganics* **2013**, *1*, 3–13. [CrossRef]
36. Edis, Z.; Haj Bloukh, S.; Abu Sara, H.; Bhakhoa, H.; Rhyman, L.; Ramasami, P. "Smart" triiodide compounds: Does halogen bonding influence antimicrobial activities? *Pathogens* **2019**, *8*, 182. [CrossRef]
37. Schmitz, G.; Rosenblatt, L.; Salerno, N.; Odette, J.; Ren, R.; Emanuel, T.; Michalek, J.; Liu, Q.; Du, L.; Jahangir, K.; et al. Treatment data using a topical povidone-iodine antiseptic in patients with superficial skin abscesses. *Data Brief* **2019**, *23*, 103715. [CrossRef]
38. Haj Bloukh, S.; Edis, Z.; Abu Sara, H.; Alhamaidah, M.A. Antimicrobial Properties of Lepidium sativum L. Facilitated Silver Nanoparticles. *Pharmaceutics* **2021**, *13*, 1352. [CrossRef]
39. Bloukh, S.H.; Edis, Z.; Ibrahim, M.R.; Abu Sara, H. "Smart" antimicrobial nanocomplexes with potential to decrease surgical site infections (SSI). *Pharmaceutics* **2020**, *12*, 361. [CrossRef]
40. Reda, M.; Ashames, A.; Edis, Z.; Bloukh, S.; Bhandare, R.; Abu Sara, H. Green Synthesis of Potent Antimicrobial Silver Nanoparticles Using Different Plant Extracts and Their Mixtures. *Processes* **2019**, *7*, 510. [CrossRef]
41. Edis, Z.; Bloukh, S.H. Facile Synthesis of Antimicrobial Aloe Vera-"Smart" Triiodide-PVP Biomaterials. *Biomimetics* **2020**, *5*, 45. [CrossRef]
42. Edis, Z.; Bloukh, S.H. Facile Synthesis of Bio-Antimicrobials with "Smart" Triiodides. *Molecules* **2021**, *26*, 3553. [CrossRef] [PubMed]
43. Joseph, B.; George, A.; Gopi, S.; Kalarikkal, N.; Thomas, S. Polymer sutures for simultaneous wound healing and drug delivery—A review. *Int. J. Pharm.* **2017**, *524*, 454–466. [CrossRef] [PubMed]
44. Bhatia, P.; Sharma, A.; George, A.J.; Anvitha, D.; Kumar, P.; Dwivedi, V.P.; Chandra, N.S. Antibacterial activity of medicinal plants against ESKAPE: An update. *Heliyon* **2021**, *7*, e06310. [CrossRef] [PubMed]
45. Zhao, Q.; Luan, X.; Zheng, M.; Tian, X.-H.; Zhao, J.; Zhang, W.-D.; Ma, B.-L. Synergistic Mechanisms of Constituents in Herbal Extracts during Intestinal Absorption: Focus on Natural Occurring Nanoparticles. *Pharmaceutics* **2020**, *12*, 128. [CrossRef]
46. Godlewska-Żyłkiewicz, B.; Świsłocka, R.; Kalinowska, M.; Golonko, A.; Świderski, G.; Arciszewska, Ż.; Nalewajko-Sieliwoniuk, E.; Naumowicz, M.; Lewandowski, W. Biologically Active Compounds of Plants: Structure-Related Antioxidant, Microbiological and Cytotoxic Activity of Selected Carboxylic Acids. *Materials* **2020**, *13*, 4454. [CrossRef]
47. Malheiro, J.F.; Maillard, J.-Y.; Borges, F.; Simões, M. Biocide Potentiation Using Cinnamic Phytochemicals and Derivatives. *Molecules* **2019**, *24*, 3918. [CrossRef]
48. Souto, E.B.; Ribeiro, A.F.; Ferreira, M.I.; Teixeira, M.C.; Shimojo, A.A.M.; Soriano, J.L.; Naveros, B.C.; Durazzo, A.; Lucarini, M.; Souto, S.B.; et al. New Nanotechnologies for the Treatment and Repair of Skin Burns Infections. *Int. J. Mol. Sci.* **2020**, *21*, 393. [CrossRef]
49. Bianchera, A.; Catanzano, O.; Boateng, J.; Elviri, L. The place of biomaterials in wound healing. In *Therapeutic Dressings and Wound Healing Applications*; John Wiley & Sons: Hoboken, NJ, USA, 2020; pp. 337–366.
50. Rezvani Ghomi, E.; Khalili, S.; Nouri Khorasani, S.; Esmaeely Neisiany, R.; Ramakrishna, S. Wound dressings: Current advances and future directions. *J. Appl. Polym. Sci.* **2019**, *136*, 47738. [CrossRef]
51. Reczyńska-Kolman, K.; Hartman, K.; Kwiecień, K.; Brzychczy-Włoch, M.; Pamuła, E. Composites Based on Gellan Gum, Alginate and Nisin-Enriched Lipid Nanoparticles for the Treatment of Infected Wounds. *Int. J. Mol. Sci.* **2022**, *23*, 321. [CrossRef]
52. Schulte-Werning, L.V.; Murugaiah, A.; Singh, B.; Johannessen, M.; Engstad, R.E.; Škalko-Basnet, N.; Holsæter, A.M. Multifunctional Nanofibrous Dressing with Antimicrobial and Anti-Inflammatory Properties Prepared by Needle-Free Electrospinning. *Pharmaceutics* **2021**, *13*, 1527. [CrossRef]
53. Tummalapalli, M.; Berthet, M.; Verrier, B.; Deopura, B.L.; Alam, M.S.; Gupta, B. Composite wound dressings of pectin and gelatin with *Aloe vera* and curcumin as bioactive agents. *Int. J. Biol. Macromol.* **2016**, *82*, 104–113. [CrossRef] [PubMed]
54. Goudarzi, M.; Fazeli, M.; Azad, M.; Seyedjavadi, S.S.; Mousavi, R. Aloe vera gel: Effective Therapeutic Agent against Multidrug-Resistant *Pseudomonas aeruginosa* Isolates Recovered from Burn Wound Infections. *Chemother. Res. Pract.* **2015**, *2015*, 639806. [CrossRef] [PubMed]

55. Ahmed, J.; Altun, E.; Aydogdu, M.O.; Gunduz, O.; Kerai, L.; Ren, G.; Edirisinghe, M. Anti-fungal bandages containing cinnamon extract. *Int. Wound J.* **2019**, *16*, 730–736. [CrossRef] [PubMed]
56. Ahmed, J.; Gultekinoglu, M.; Bayram, C.; Kart, D.; Ulubayram, K.; Edirisinghe, M. Alleviating the toxicity concerns of antibacterial cinnamon-polycaprolactone biomaterials for healthcare-related biomedical applications. *MedComm* **2021**, *2*, 236–246. [CrossRef]
57. Wang, S.; Kang, O.-H.; Kwon, D.-Y. Trans-Cinnamaldehyde Exhibits Synergy with Conventional Antibiotic against Methicillin-Resistant *Staphylococcus aureus*. *Int. J. Mol. Sci.* **2021**, *22*, 2752. [CrossRef]
58. Saleh, N.; Bufaroosha, M.S.; Moussa, Z.; Bojesomo, R.; Al-Amodi, H.; Al-Ahdal, A. Encapsulation of Cinnamic Acid by Cucurbit[7]uril for Enhancing Photoisomerization. *Molecules* **2020**, *25*, 3702. [CrossRef]
59. Hanai, K.; Kuwae, A.; Takai, T.; Senda, H.; Kunimoto, K.-K. A comparative vibrational and NMR study of cis-cinnamic acid polymorphs and trans-cinnamic acid. *Spectrochim. Acta A* **2001**, *57*, 513–519. [CrossRef]
60. Huang, X.; Wang, P.; Li, T.; Tian, X.; Guo, W.; Xu, B.; Huang, G.; Cai, D.; Zhou, F.; Zhang, H.; et al. Self-Assemblies Based on Traditional Medicine Berberine and Cinnamic Acid for Adhesion-Induced Inhibition Multidrug-Resistant *Staphylococcus aureus*. *ACS Appl. Mater. Interfaces* **2020**, *12*, 227–237. [CrossRef]
61. Sánchez, M.; González-Burgos, E.; Iglesias, I.; Gómez-Serranillos, M.P. Pharmacological Update Properties of *Aloe Vera* and its Major Active Constituents. *Molecules* **2020**, *25*, 1324. [CrossRef]
62. Liu, C.; Cui, Y.; Pi, F.; Cheng, Y.; Guo, Y.; Qian, H. Extraction, Purification, Structural Characteristics, Biological Activities and Pharmacological Applications of Acemannan, a Polysaccharide from *Aloe vera*: A Review. *Molecules* **2019**, *24*, 1554. [CrossRef]
63. Donkor, A.; Donkor, M.N.; Kuubabongnaa, N. Evaluation of anti-infective potencies of formulated aloin A ointment and aloin A isolated from *Aloe barbadensis* Miller. *BMC Chem.* **2020**, *14*, 8. [CrossRef] [PubMed]
64. Kuntić, V.; Pejic, N.; Mićić, S. Direct Spectrophotometric Determination of Hesperidin in Pharmaceutical Preparations. *Acta Chim. Slov.* **2012**, *59*, 436–441. [PubMed]
65. Xiang, H.; Cao, F.; Ming, D.; Zheng, Y.; Dong, X.; Zhong, X.; Mu, D.; Li, B.; Zhong, L.; Cao, J.; et al. Aloe-emodin inhibits *Staphylococcus aureus* biofilms and extracellular protein production at the initial adhesion stage of biofilm development. *Appl. Microbiol. Biotechnol.* **2017**, *101*, 6671–6681. [CrossRef] [PubMed]
66. Meza-Valle, K.Z.; Saucedo-Acuña, R.A.; Tovar-Carrillo, K.L.; Cuevas-González, J.C.; Zaragoza-Contreras, E.A.; Melgoza-Lozano, J. Characterization and Topical Study of Aloe Vera Hydrogel on Wound-Healing Process. *Polymers* **2021**, *13*, 3958. [CrossRef]
67. Sun, Z.; Yu, C.; Wang, W.; Yu, G.; Zhang, T.; Zhang, L.; Zhang, J.; Wei, K. Aloe Polysaccharides Inhibit Influenza A Virus Infection—A Promising Natural Anti-flu Drug. *Front. Microbiol.* **2018**, *9*, 2338. [CrossRef] [PubMed]
68. Borges-Argáez, R.; Chan-Balan, R.; Cetina-Montejo, L.; Ayora-Talavera, G.; Sansores-Peraza, P.; Gómez-Carballo, J.; Mirbella Cáceres-Farfán, M. In vitro evaluation of anthraquinones from *Aloe vera* (*Aloe barbadensis* Miller) roots and several derivatives against strains of influenza virus. *Ind. Crop. Prod.* **2019**, *132*, 468–475. [CrossRef]
69. Paraiso, I.L.; Revel, J.S.; Stevens, J.F. Potential use of polyphenols in the battle against COVID-19. *Curr. Opin. Food Sci.* **2020**, *32*, 149–155. [CrossRef]
70. Piccolella, S.; Crescente, G.; Faramarzi, S.; Formato, M.; Pecoraro, M.T.; Pacifico, S. Polyphenols vs. Coronaviruses: How Far Has Research Moved Forward? *Molecules* **2020**, *25*, 4103. [CrossRef]
71. Mani, J.S.; Johnson, J.B.; Steel, J.C.; Broszczak, D.A.; Neilsen, P.M.; Walsh, K.B.; Naiker, M. Natural product-derived phytochemicals as potential agents against coronaviruses: A review. *Virus Res.* **2020**, *284*, 197989. [CrossRef]
72. Tuñón-Molina, A.; Takayama, K.; Redwan, E.M.; Uversky, V.N.; Andrés, J.; Serrano-Aroca, Á. Protective Face Masks: Current Status and Future Trends. *ACS Appl. Mat. Interfaces* **2021**, *13*, 56725–56751. [CrossRef]
73. Ray, S.S.; Lee, H.K.; Huyen, D.T.T.; Chen, S.-S.; Young-Nam Kwon, Y.-N. Microplastics waste in environment: A perspective on recycling issues from PPE kits and face masks during the COVID-19 pandemic. *Environ. Technol. Innov.* **2022**, *26*, 102290. [CrossRef] [PubMed]
74. Jung, S.; Yang, J.-Y.; Byeon, E.-Y.; Kim, D.-G.; Lee, D.-G.; Ryoo, S.; Lee, S.; Shin, C.-W.; Jang, H.W.; Kim, H.J.; et al. Copper-Coated Polypropylene Filter Face Mask with SARS-CoV-2 Antiviral Ability. *Polymers* **2021**, *13*, 1367. [CrossRef] [PubMed]
75. Chowdhury, M.A.; Shuvho, M.B.A.; Shahid, M.A.; Haque, A.K.M.M.; Kashem, M.A.; Lam, S.S.; Ong, H.C.; Uddin, M.A.; Mofijur, M. Prospect of Biobased Antiviral Face Mask to Limit the Coronavirus Outbreak. *Environ. Res.* **2021**, *192*, 110294. [CrossRef] [PubMed]
76. Lu, W.-C.; Chen, C.-Y.; Cho, C.-J.; Venkatesan, M.; Chiang, W.-H.; Yu, Y.-Y.; Lee, C.-H.; Lee, R.-H.; Rwei, S.-P.; Kuo, C.-C. Antibacterial Activity and Protection Efficiency of Polyvinyl Butyral Nanofibrous Membrane Containing Thymol Prepared through Vertical Electrospinning. *Polymers* **2021**, *13*, 1122. [CrossRef] [PubMed]
77. Takayama, K.; Tuñón-Molina, A.; Cano-Vicent, A.; Muramoto, Y.; Noda, T.; Aparicio-Collado, J.L.; Sabater i Serra, R.; Martí, M.; Serrano-Aroca, Á. Non-Woven Infection Prevention Fabrics Coated with Biobased Cranberry Extracts Inactivate Enveloped Viruses Such as SARS-CoV-2 and Multidrug-Resistant Bacteria. *Int. J. Mol. Sci.* **2021**, *22*, 12719. [CrossRef] [PubMed]
78. Borojeni, I.A.; Gajewski, G.; Riahi, R.A. Application of Electrospun Nonwoven Fibers in Air Filters. *Fibers* **2022**, *10*, 15. [CrossRef]
79. Rahman, M.Z.; Hoque, M.E.; Alam, M.R.; Rouf, M.A.; Khan, S.I.; Xu, H.; Ramakrishna, S. Face Masks to Combat Coronavirus (COVID-19)—Processing, Roles, Requirements, Efficacy, Risk and Sustainability. *Polymers* **2022**, *14*, 1296. [CrossRef]
80. Fonseca, A.C.; Lima, M.S.; Sousa, A.F.; Silvestre, A.J.; Coelho, J.F.J.; Serra, A.C. Cinnamic acid derivatives as promising building blocks for advanced polymers: Synthesis, properties and applications. *Polym. Chem.* **2019**, *10*, 1696–1723. [CrossRef]

81. Panda, P.K.; Yang, J.-M.; Chang, Y.-H. Preparation and characterization of ferulic acid-modified water soluble chitosan and poly (γ-glutamic acid) polyelectrolyte films through layer-by-layer assembly towards protein adsorption. *Int. J. Biol. Macromol.* **2021**, *171*, 457–464. [CrossRef]
82. Alfuraydi, R.T.; Alminderej, F.M.; Mohamed, N.A. Evaluation of Antimicrobial and Anti-Biofilm Formation Activities of Novel Poly (vinyl alcohol) Hydrogels Reinforced with Crosslinked Chitosan and Silver Nano-Particles. *Polymers* **2022**, *14*, 1619. [CrossRef]
83. Chandel, A.K.S.; Shimizu, A.; Hasegawa, K.; Ito, T. Advancement of Biomaterial-Based Postoperative Adhesion Barriers. *Macromol. Biosci.* **2021**, *21*, 2000395. [CrossRef] [PubMed]
84. Chandel, A.K.S.; Kumar, C.U.; Jewrajka, S.K. Release Performance of Biodegradable/Cytocompatible Agarose–Polyethylene Glycol–Polycaprolactone Amphiphilic Co-Network Gels. *ACS Appl. Mater. Interfaces* **2016**, *8*, 3182–3192. [CrossRef] [PubMed]
85. Bera, A.; Chandel, A.K.S.; Kumar, C.U.; Jewrajka, S.K. Degradable/cytocompatible and pH responsive amphiphilic conetwork gels based on agarose-graft copolymers and polycaprolactone. *J. Mater. Chem. B* **2015**, *3*, 8548–8557. [CrossRef] [PubMed]
86. Ohta, S.; Mitsuhashi, K.; Chandel, A.K.S.; Qi, P.; Nakamura, N.; Nakamichi, A.; Yoshida, H.; Yamaguchi, G.; Hara, Y.; Sasaki, R.; et al. Silver-loaded carboxymethyl cellulose nonwoven sheet with controlled counterions for infected wound healing. *Carbohydr. Polym.* **2022**, *286*, 119289. [CrossRef] [PubMed]
87. Chandel, A.K.S.; Nutan, B.; Raval, I.H.; Suresh, K.; Jewrajka, S.K. Self-Assembly of Partially Alkylated Dextran-graft-poly[(2-dimethylamino)ethyl methacrylate] Copolymer Facilitating Hydrophobic/Hydrophilic Drug Delivery and Improving Conetwork Hydrogel Properties. *Biomacromolecules* **2018**, *19*, 1142–1153. [CrossRef] [PubMed]
88. Prundeanu, M.; Brezoiu, A.-M.; Deaconu, M.; Gradisteanu Pircalabiuru, G.; Lincu, D.; Matei, C.; Berger, D. Mesoporous Silica and Titania-Based Materials for Stability Enhancement of Polyphenols. *Materials* **2021**, *14*, 6457. [CrossRef] [PubMed]
89. Freeman, C.; Duan, E.; Kessler, J. Molecular iodine is not responsible for cytotoxicity in iodophors. *J. Hosp. Inf.* **2022**, in press. [CrossRef]
90. Savastano, M.; Bazzicalupi, C.; García, C.; Gellini, C.; López de la Torre, M.D.; Mariani, P.; Pichierri, F.; Bianchi, A.; Melguizo, M. Iodide and triiodide anion complexes involving anion–π interactions with a tetrazine-based receptor. *Dalton Trans.* **2017**, *46*, 4518. [CrossRef]
91. Lundin, J.G.; McGann, C.L.; Weise, N.K.; Estrella, L.A.; Balow, R.B.; Streifel, B.C.; Wynne, J.H. Iodine binding and release from antimicrobial hemostatic polymer foams. *React. Funct. Polym.* **2019**, *135*, 44–51. [CrossRef]
92. Shestimerova, T.A.; Mironov, A.V.; Bykov, M.A.; Grigorieva, A.V.; Wei, Z.; Dikarev, E.V.; Shevelkov, A.V. Assembling Polyiodides and Iodobismuthates Using a Template Effect of a Cyclic Diammonium Cation and Formation of a Low-Gap Hybrid Iodobismuthate with High Thermal Stability. *Molecules* **2020**, *25*, 2765. [CrossRef]
93. Bauer, A.W.; Perry, D.M.; Kirby, W.M.M. Single-disk antibiotic-sensitivity testing of staphylococci: An analysis of technique and results. *AMA Int. Intern. Med.* **1959**, *104*, 208–216. [CrossRef]
94. Clinical and Laboratory Standards Institute (CLSI). *Performance Standards for Antimicrobial Disk Susceptibility Testing*, 28th ed.; M100S; CLSI: Wayne, PA, USA, 2018; Volume 38.
95. Sai, M.; Zhong, S.; Tang, Y.; Ma, W.; Sun, Y.; Ding, D. Research on the preparation and antibacterial properties of 2-N-thiosemicarbazide-6-O-hydroxypropyl chitosan membranes with iodine. *J. Appl. Polym. Sci.* **2014**, *131*, 1–8. [CrossRef]
96. Laus, M.; Li, W.; Zhao, X.; Sun, X.; Zu, Y.; Liu, Y.; Ge, Y. Evaluation of Antioxidant Ability In Vitro and Bioavailability of *trans*-Cinnamic Acid Nanoparticle by Liquid Antisolvent Precipitate. *J. Nanomater.* **2016**, *2016*, 9518362. [CrossRef]
97. Mamatha, G.; Rajulu, A.V.; Madhukar, K. In Situ Generation of Bimetallic Nanoparticles in Cotton Fabric Using Aloe Vera Leaf Extract, as a Reducing Agent. *J. Nat. Fibers* **2020**, *17*, 1121–1129. [CrossRef]
98. Aziz, S.B.; Abdullah, O.G.; Hussein, S.A.; Ahmed, H.M. Effect of PVA Blending on Structural and Ion Transport Properties of CS: AgNt-Based Polymer Electrolyte Membrane. *Polymers* **2017**, *9*, 622. [CrossRef] [PubMed]
99. Aghamohamadi, N.; Sanjani, N.S.; Majidi, R.F.; Nasrollahi, S.A. Preparation and characterization of *Aloe vera* acetate and electrospinning fibers as promising antibacterial properties materials. *Mater. Sci. Eng. C* **2019**, *94*, 445–452. [CrossRef] [PubMed]
100. Yushina, I.; Tarasova, N.; Kim, D.; Sharutin, V.; Bartashevich, E. Noncovalent Bonds, Spectral and Thermal Properties of Substituted Thiazolo [2,3-b][1,3]thiazinium Triiodides. *Crystals* **2019**, *9*, 506. [CrossRef]
101. Zhang, S.; Kai, C.; Liu, B.; Zhang, S.; Wei, W.; Xu, X.; Zhou, Z. Facile fabrication of cellulose membrane containing polyiodides and its antibacterial properties. *Appl. Surf. Sci.* **2020**, *500*, 144046. [CrossRef]
102. Yang, Y.; Zheng, Z.; Lin, J.; Zhou, L.; Chen, G. Effect of KI Concentration in Correcting Tank on Optical Properties of PVA Polarizing Film. *Polymers* **2022**, *14*, 1413. [CrossRef]
103. Yushina, I.D.; Kolesov, B.A.; Bartashevich, E.V. Raman spectroscopy study of new thia- and oxazinoquinolinium triodides. *New J. Chem.* **2015**, *39*, 6163–6170. [CrossRef]
104. Bartashevich, E.V.; Grigoreva, E.A.; Yushina, I.D.; Bulatova, L.M.; Tsirelson, V.G. Modern level for the prediction of properties of iodine-containing organic compounds: Iodine forming halogen bonds. *Russ. Chem. Bull. Int. Ed.* **2017**, *66*, 1345–1356. [CrossRef]
105. Kowczyk-Sadowy, M.; Świsłocka, R.; Lewandowska, H.; Piekut, J.; Lewandowski, W. Spectroscopic (FT-IR, FT-Raman, ^{1}H- and ^{13}C-NMR), Theoretical and Microbiological Study of *trans o*-Coumaric Acid and Alkali Metal *o*-Coumarates. *Molecules* **2015**, *20*, 3146–3169. [CrossRef] [PubMed]

106. de Souza, M.L.; Otero, J.C.; López-Tocón, I. Comparative Performance of Citrate, Borohydride, Hydroxylamine and β-Cyclodextrin Silver Sols for Detecting Ibuprofen and Caffeine Pollutants by Means of Surface-Enhanced Raman Spectroscopy. *Nanomaterials* **2020**, *10*, 2339. [CrossRef] [PubMed]
107. Vanholme, B.; El Houari, I.; Boerjan, W. Bioactivity: Phenylpropanoids' best kept secret. *Curr. Opin. Biotechnol.* **2019**, *56*, 156–162. [CrossRef]
108. Rahma, A.; Munir, M.M.; Khairurrijal, K.; Prasetyo, A.; Suendo, V.; Rachmawati, H. Intermolecular Interactions and the Release Pattern of Electrospun Curcumin-Polyvinyl (pyrrolidone) Fiber. *Biol. Pharm. Bull.* **2016**, *39*, 163–173. [CrossRef] [PubMed]
109. D'Amelia, R.P.; Gentile, S.; Nirode, W.F.; Huang, L. Quantitative Analysis of Copolymers and Blends of Polyvinyl Acetate (PVAc) Using Fourier Transform Infrared Spectroscopy (FTIR) and Elemental Analysis (EA). *World J. Chem. Edu.* **2016**, *4*, 25–31. [CrossRef]
110. Geetanjali, R.; Sreejit, V.; Sandip, P.; Preetha, R. Preparation of *Aloe vera* mucilage-ethyl vanillin Nano-emulsion and its characterization. *Mater. Today Proc.* **2021**, *43*, 3766–3773. [CrossRef]
111. Carroll, K.C.; Morse, A.M.; Mietzner, T.; Miller, S. *Jawetz, Melnick and Adelberg's Medical Microbiology*, 27th ed.; International Edition; Mc Graw-Hill Education: New York, NY, USA, 2016; ISBN 978-1-25-925534-2.
112. Wiley, J.M.; Sherwood, L.M.; Woolverton, C.J. *Prescott's Microbiology*, 10th ed.; Mc Graw-Hill Education: New York, NY, USA, 2016; ISBN 978-1259281594.

Article

Synthetic Polypeptides with Cationic Arginine Moieties Showing High Antimicrobial Activity in Similar Mineral Environments to Blood Plasma

Kuen Hee Eom [1], Shuwei Li [1], Eun Gyeong Lee [1], Jae Ho Kim [2], Jung Rae Kim [1] and Il Kim [1,*]

1. School of Chemical Engineering, Pusan National University, Busandaehag-ro 63-2, Geumjeong-gu, Busan 46241, Korea; reimreim@pusan.ac.kr (K.H.E.); lishuwei0325@pusan.ac.kr (S.L.); rud6063@pusan.ac.kr (E.G.L.); j.kim@pusan.ac.kr (J.R.K.)
2. Department of Physiology, School of Medicine, Pusan National University, Busandaehak-ro, Mulgeum-eup, Yangsan-si 50612, Korea; jhkimst@pusan.ac.kr
* Correspondence: ilkim@pusan.ac.kr

Citation: Eom, K.H.; Li, S.; Lee, E.G.; Kim, J.H.; Kim, J.R.; Kim, I. Synthetic Polypeptides with Cationic Arginine Moieties Showing High Antimicrobial Activity in Similar Mineral Environments to Blood Plasma. *Polymers* **2022**, *14*, 1868. https://doi.org/10.3390/polym14091868

Academic Editors: Md. Amdadul Huq and Shahina Akter

Received: 22 March 2022
Accepted: 28 April 2022
Published: 2 May 2022

Publisher's Note: MDPI stays neutral with regard to jurisdictional claims in published maps and institutional affiliations.

Copyright: © 2022 by the authors. Licensee MDPI, Basel, Switzerland. This article is an open access article distributed under the terms and conditions of the Creative Commons Attribution (CC BY) license (https:// creativecommons.org/licenses/by/ 4.0/).

Abstract: Translocation of cell-penetrating peptides is promoted by incorporated arginine or other guanidinium groups. However, relatively little research has considered the role of these functional groups on antimicrobial peptide activity. A series of cationic linear-, star- and multi-branched-poly(L-arginine-*co*-L-phenylalanine) have been synthesized via the ring-opening copolymerizations of corresponding *N*-carboxyanhydride monomers followed by further modifications using the *N*-heterocyclic carbene organocatalyst. All the polymers are characterized by the random coiled microstructure. Antibacterial efficacy, tested by the gram-positive *B. subtilis* bacteria and the gram-negative *E. coli* bacteria, was sensitive to the structure and relative composition of the copolymer and increased in the order of linear- < star- < multi-branched structure. The multi-branched-p[(L-arginine)$_{23}$-*co*-(L-phenylalanine)$_7$]$_8$ polymer showed the best antibacterial property with the lowest minimum inhibitory concentration values of 48 μg mL^{-1} for *E. coli* and 32 μg mL^{-1} for *B. subtilis*. The efficacy was prominent for *B. subtilis* due to the anionic nature of its membrane. All of the resultant arginine moiety-containing polypeptides showed excellent blood compatibility. The antibiotic effect of the copolymers with arginine moieties was retained even in the environment bearing Ca^{2+}, Mg^{2+}, and Na$^+$ ions similar to blood plasma. The cationic arginine-bearing copolypeptides were also effective for the sterilization of naturally occurring sources of water such as lakes, seas, rain, and sewage, showing a promising range of applicability.

Keywords: amino acids; antimicrobial peptides; arginine; *N*-carboxyanhydrides; polypeptides; ring-opening polymerization

1. Introduction

Antibiotics are used to prevent and treat bacterial infections. Antimicrobial resistance (AMR) is caused by mutations in bacteria due to the use of antibiotics; this is a significant concern because the resistant subject is bacteria, not humans or animals. Antibiotic-resistant bacteria are more difficult to treat than non-resistant bacteria, leading to increased medical costs, extended hospitalization, and increased mortality [1]. Infectious diseases that can be treated with general treatment courses of antibiotics have become more difficult to treat due to the spread of AMR. The World Health Organization states that the emergence of antibiotic-resistant bacteria is causing the world to once again enter a post-antibiotic era in which humans can die from common infections [2]. There is an urgent need for the research and development of new antibiotics.

Antimicrobial peptides (AMPs) are emerging as alternatives to traditional antibiotics, and AMPs such as nisin, gramicidin, polymyxins, daptomycin, vancomycin, and melittin have been used [3,4]. Existing antibiotic mechanisms interfere with cell wall synthesis, protein synthesis, or DNA replication [5] and can stimulate the immune system or interact with bacterial membranes to kill bacteria [6]. Among them, the mechanism of interaction with the bacterial membrane selectively binds to the negatively charged bacterial cell membrane with cationic amino acids like lysine, histidine and arginine, and hydrophobic amino acids like valine, alanine, phenylalanine, and leucine [7].

AMPs are attracting attention as next-generation antibiotics as biodegradable polymers that do not remain in the environment and are difficult for bacteria to develop resistance to because they primarily have mechanisms that interact with bacterial cell membranes. AMP is obtained only from gene-modified microorganisms; therefore, the production rate is low and the cost is high [8].

To circumvent this problem, AMP-mimetic polypeptides (AMPPs), which reproduce AMP-like structures with synthetic polymers, are emerging as alternatives [9]. Among them, the ring-opening polymerization (ROP) of α-amino acid N-carboxyanhydrides (NCAs) can synthesize polypeptides with cationic amino acids and hydrophobic amino acids on a large scale [10]. Therefore, AMPPs synthesized by the ring-opening polymerization of NCAs have the advantage of low cost. Because of lysine NCA's storability, well-understood mechanism, and high yield, many AMPPs utilizing lysine have been reported because lysine is the easiest to synthesize of the cationic amino acids [11–13]. AMPPs with arginine moieties have been reported less because they are relatively difficult to synthesize [14–21]. In addition, the synthesis of AMPs with arginine is important in antibiotic studies. AMPs with arginine units have been reported to have stronger membrane interactions and perturbation properties than AMPs with lysine counterparts [22–28].

Lysine has been reported to have a significant decrease in antibiotic efficacy due to the presence of divalent cations, but arginine has been reported to have a relatively low antibiotic effect [29,30]. There are divalent cations such as Ca^{2+} and Mg^{2+} in biological matrices, which means that the antibiotic effect may decrease when AMP enters the body. Thus, we have studied AMPPs that were less disturbed by divalent cations and maintained antibacterial properties by synthesizing polypeptides with arginine moieties as positive charges.

In our previous study, we reported an imidazolium hydrogen carbonate (IHC) catalyst that synthesizes topological nanoengineered antimicrobial polypeptides of various structures through the ROP of NCAs [10]. IHC-mediated NCA polymerizations have many advantages, including metal-free procedure, absence of any toxic impurities, and a living pathway [10]. It was also possible to synthesize polypeptides of various structures by only modifying the initiator. Here, several proportions of linear- (*l*-), star- (*s*-), and multibranched- (*mb*-) poly(L-lysine-*co*-L-phenylalanine) [p(Lys-*co*-Phe)] have been synthesized and the L-lysine units were further modified to L-arginine units to yield poly(L-arginine-*co*-poly(L-phenylalanine) [p(Arg-*co*-Phe)] counterparts. According to the antibacterial tests made for the gram-positive *B. subtilis* bacteria and the gram-negative *E. coli* bacteria, the efficacy was sensitive to the to the structure and compositions of the copolymer and increased in the order *l*- < *s*- < *mb*-structure. All of the p(Arg-*co*-Phe) copolymers showed excellent blood compatibility. Notably, the antibiotic effect was retained in the presence of Ca^{2+}, Mg^{2+}, and Na^+ ions and the copolymers show their efficiency for the sterilization of naturally occurring sources of water such as lakes, seas, rain, and sewage.

2. Materials and Methods

Anhydrous *N,N*-dimethylformamide (DMF), trifluoroacetic acid (TFA), triphosgene, trimethylamine (TEA), $KHCO_3$, calcium chloride, magnesium chloride, and solvents like tetrahydrofuran (THF), diethyl ether, hexane, and methanol were purchased from Fisher Scientific Korea (Incheon, Korea) and used without purification. *L*-phenylalanine (Phe), 1,3-Diisopropylimidazolium chloride, and ε-benzyloxycarbonyl-*L*-lysine (Cbz-Lys) were purchased from Fisher Scientific Korea and stored under N_2. Hexylamine (HA); tris-(2-aminoethyl)amine (TREN); polyethylenimine, branched (PEI; number average molecular weight = 600); 1*H*-pyrazole-1-carboxamide monohydrochloride; and 33 wt% hydrogen bromide solution in acetic acid were purchased from Merck Korea (Seoul, Korea) and stored under nitrogen. All reactions and polymerization were carried out under nitrogen atmosphere using the Schlenk techniques. *Escherichia coli* BL21 (*E. coli*) and *Bacillus subtilis* ATCC 6633 (*B. subtilis*) were obtained from the Korean Collection for Type Cultures (Jeollabuk-do, Korea). Sewage was provided by the East Busan Sewage Treatment Plant (Busan, Korea).

2.1. Synthesis

The 1,3-diisopropylimidazolium hydrogen carbonate ($[^i\text{PrNHC(H)}]^+[\text{HCO}_3]^-$) catalyst precursor was synthesized by reacting 1,3-diisopropylimidazolium chloride with $KHCO_3$ following a reported procedure [10].

Poly(ε-carbobenzoxy-L-lysine), p(Cbz-Lys), and p(Cbz-Lys-*co*-Phe) with different arms were synthesized using HA, TREN, and PEI as initiators according to methods previously described [10]. Figure 1 illustrates general synthetic procedures. For the deprotection, 5 mmol of p(Cbz-Lys) and p(Cbz-Lys-*co*-Phe) were dissolved in TFA in a round-bottom flask equipped with glass stoppers and stirrer bars. Two equivalents of HBr (33% v/v) per carboxyl group were added and the mixture was stirred for 4 h at 25 °C. The solution was then poured into an excess amount of cold diethyl ether and purified in a dialysis bag (1 kDa) for 2 d. After freeze-drying, white solid p(Lys) and p(Lys-*co*-Phe) were obtained.

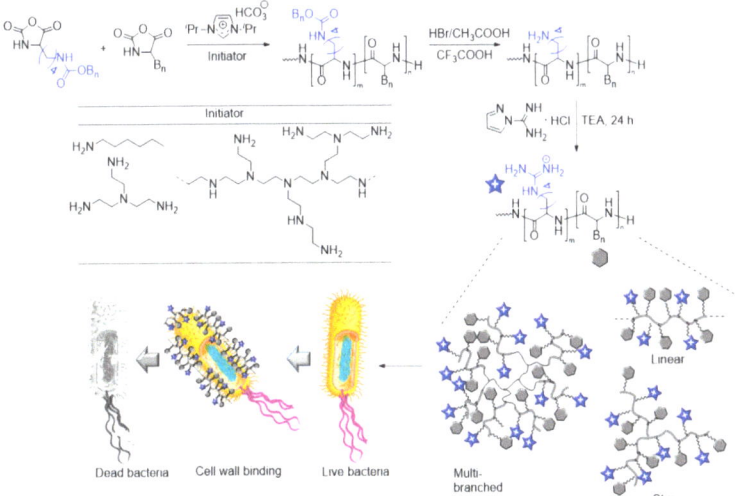

Figure 1. General synthetic procedure p(Arg-*co*-Phe) with different topologies for the design of polypetides with bactericidal properties.

The guanidination of δ-NH_2 was achieved with 1*H*-pyrazole-1-carboxamide monohydrochloride as a guanidylating agent at pH 9.5 in aqueous medium [14–17]. In a typical experiment, p(Lys-*co*-Phe) was dissolved in DMF in a Schlenk flask under nitrogen. PCH (1.2 times the number of lysine units) and TEA (the same number of moles as PCH) were dissolved in DMF, added to a syringe, administered dropwise to the polypeptide solution, and reacted for 3 d. This solution was then purified in a dialysis bag (1000 Da) for 4 d and lyophilized. Finally, p(Arg-*co*-Phe) was obtained in a white solid state and used for the antibacterial activity assay.

2.2. Antibacterial Test

A growth inhibition assay was performed to determine the antibacterial effects of p(Lys-*co*-Phe) and p(Arg-*co*-Phe) with different structures. The model bacteria *E. coli* (a gram-negative bacillus) and *B. subtilis* (a gram-positive, catalase-positive bacterium) cells were cultured overnight in Luria-Bertani (LB) medium at 37 °C. Sterilized LB broth (50 mL) was added to a sterilized culture dish. Sterilized LB broth was added to a bacterial broth and 3.6 mL of the broth containing the bacterial suspension diluted at a concentration of 5×10^7 CFU mL^{-1} was added to 5 mL of sterile quartz cuvette. The standard solution of p(Lys-co-Phe) or p(Arg-co-Phe) was prepared in distilled water at a concentration of 5.12 mg mL^{-1}, and then the standard solutions were diluted in series. A mixture of bacteria and copolymer was obtained by adding a standard stock solution (400 µL) to each quartz cuvette. 400 µL of distilled water was added to 3.6 mL of a diluted stock containing bacteria suspended in solution to use as a control group. The bacteria in the cuvette were cultured overnight in a rocking bed at 80 rpm and 37 °C. In the suspended fluid, the optical density (OD) value of the bacteria was monitored at a wavelength of 600 nm every 3 h using UV-vis spectroscopy. The minimum inhibitory concentration (MIC) value was defined as the lowest concentration of an antibiotic at which bacterial growth is completely inhibited. All analyses were performed three times at different dates.

Plate counting was performed to determine whether antibacterial performance was observed in natural water gathered from different sources such as lake, rain, sewage, and sea. The copolymer standard solution was prepared in distilled water at a concentration of 5.12 mg mL^{-1}, and the standard solution was diluted in series. Natural water and copolymer solution were added in a ratio of 9:1 to create an experimental group sample; natural water and distilled water were added as controls in a ratio of 9:1. The samples were diluted to different concentration gradients and 10 µL of the diluted samples were spread evenly on solid LB agar medium at 30 °C overnight for colony growth and observation.

2.3. Hemolysis Test

The hemolytic activity of p(Lys-co-Phe) and p(Arg-co-Phe) was investigated in the range between 100 µg mL^{-1} and 3000 µg mL^{-1}. Blood from mice was washed three times with phosphate buffered saline (PBS) and then diluted to 10% (v/v) with PBS. Blood (90 µL) was first added to a 1.5 mL microfuge tube. Then, 10 µL of each standard stock solution of copolymer was added. Blood suspensions were also added to blank PBS as the negative control. PBS containing 0.2% Triton-X was used as a positive control. After incubation at 37 °C for 1 h, each mixture was centrifuged at 5000 rpm for 5 min. An aliquot of 70 µL of the supernatant was transferred to each well of a 96-well plate and the OD value was read at 350 nm. The hemolysis ratio relative to the TX-100 control group was calculated as $[(A_{sample} - A_{negative})/(A_{sample} - A_{positive})] \times 100\%$, where A_{sample}, $A_{negative}$, and $A_{positive}$ are the OD values of the supernatant from the incubated sample, negative control, and positive control, respectively (Figure S14 in Supplementary Materials).

2.4. Characterization

The ^{1}H and ^{13}C nuclear magnetic resonance (NMR) spectra were recorded on a Varian Unity 400 MHz spectrometer; shifts were reported from tetramethylsilane in a downfield of 1 millionth and referenced residual solvent peaks. Fourier transform infrared (FT-IR) spectral data were collected for film samples cast on KBr disks, which were measured using a Shimadzu IRPrestige-21 spectrophotometer (Shimadzu, Kyoto, Japan) with 32 scans per experiment at a resolution of 1 cm^{-1}. UV-vis spectrometer analysis was performed at a scan rate of 300 nm·min^{-1} using a Shimadzu UV-1650 PC (Shimadzu, Kyoto, Japan). The absorbance and transmittance spectra of the copolymer/bacteria solution were recorded at 600 nm. Circular dichroism (CD) analysis was performed using a Jasco J-1500 spectrometer (Jasco, Easton, MD, USA) with 1 cm quartz cells at 25 °C. The copolymers were dissolved in PBS to 0.5 mg mL^{-1}. Wavelengths between 190 nm and 260 nm were analyzed; the integration time was 1 s and the wavelength step was 0.2 nm. Distilled water was used as a reference solvent and five scans were recorded for all the copolymers. The secondary structure of copolymers was analyzed using the CD Multivariate Calibration Creation Program in Spectra Manager™ Version 2 (Jasco, Easton, MD, USA).

The zeta potential was measured using a Malvern Zetasizer Nano ZS device (Malvern Pananalytical, Malvern, UK), equipped with a monochromatic coherent He-Ne laser (633 nm) as the light source and a detector that detects light scattered at an angle of 173° and a constant temperature of 25 °C. Zeta potential was measured at a concentration of 1 mg mL^{-1} in PBS. All samples were filtered through a 0.45 μm nylon filter before measurement and performed in triplicate. Scanning electron microscopy (SEM) was used to detect the morphology of the pathogen on a JCM-5700 Scanning Electron Microscope (JEOL USA, Peabody, MA, USA). Bacteria were treated with the copolymer at a 3 × MIC value and cultured for 4 h in a shaking bed at 37 °C, and the control group was prepared under the same conditions without adding copolymer. Both the treated bacterial group and the control group were washed three times with PBS, and the samples were fixed with 2.5% (v/v) glutaraldehyde in phosphate buffer (0.1 M) overnight at 4 °C. The fixed bacterial suspension was washed three times with PBS to remove excess fixers and dehydrated in an ethanol series (30%, 50%, 70%, 80%, 95%, and 100% in PBS (0.01 M)—100% was repeated three times). The solution was dried by moving it to a cover slide in a 100% ethanol suspension state. Finally, the samples were ready for SEM analysis.

3. Results and Discussion

3.1. Synthesis and Characterization of Copolypeptides

For the synthesis of p(Arg-*co*-Phe) with different topologies, a series of ROPs of Cbz-Lys NCA and Phe NCA were performed using three different amine initiators—HA, TREN, and PEI that result in *l*-, *s*-, and *mb*-copolymers, respectively—in the presence of an organocatalyst, [^{i}PrNHC(H)]$^{+}$[HCO$_3$]$^{-}$ (Figure 1). The [^{i}PrNHC(H)]$^{+}$[HCO$_3$]$^{-}$-mediated ROP of amino acid NCA permitted the achievement of rapid and efficient synthesis of well-defined polypeptides and provided control over the polypeptide architecture simply by tuning the type of amine initiators due to its living nature [10]. Table 1 shows the results of polymerizations performed by controlling the initial [monomer]/[initiator]/[catalyst] ratio to 120:1:0.2, 90:1:0.2, and 60:1:0.2, respectively, by varying the relative Cbz-Lys NCA/Phe NCA ratio. According to the molecular weight of the obtained p(Cbz-Lys) and p(Cbz-Lys-*co*-Phe) polymers, the targeted homo- and copolymers were successfully achieved within 20 min of polymerization at 25 °C. As the incorporation of Phe units increases, the solubility of resultant copolymers sharply decreases. However, the deprotected counterparts, p(Lys) and p(Lys-*co*-Phe), were soluble in common NMR solvents. ^{1}H NMR spectra of *l*-, *s*-, and *mb*-polypeptides are in Figures S1–S9 (in Supplementary Materials).

The transformation of ornithine into arginine to prepare arginine-containing peptides by guanylation has long been recognized [21,31]. Our preliminary studies using commercially available guanylating reagents cyanamide, O-methylisourea hydrogen sulfate [32], 2-ethyl-2-thiopseudourea hydrobromide [33], and 3,5-dimethylpyrazole-l-carboxamidine nitrate [34] showed insufficient reactivity for practical use. We found that 1H-pyrazole-1-carboxamide monohydrochloride was the most efficient and chemically specific guanylation of sterically unhindered δ-NH$_2$ (Scheme 1).

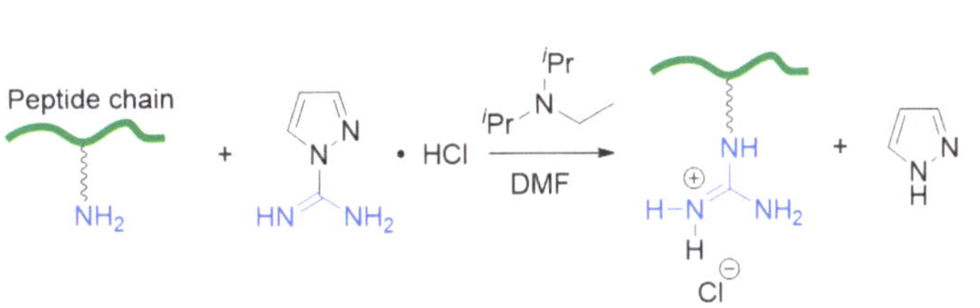

Scheme 1. Guanylation of 1H-pyrazole-1-carboxamide monohydrochloride with a primary amine in Lys units.

Table 1. Synthesis and characterization of various homo- and copolypeptides with different microstructures.

Entry	No. Repeat Unit in Feed		Initiator [1]	No. Repeat Unit in Copolymer [2]		M_n [3] (kg mol^{-1})	ζ [4] (mV)	MIC [5] (µg mL^{-1})	
	Cbz-Lys NCA	Phe NCA		Lys	Phe			E. coli	B. substilis
1	120	0	TREN	128	0	16.5	42.5	64	64
2	96	24	TREN	94	25	15.9	37.7	96	64
3	72	48	TREN	74	45	16.2	36.7	>128	>128
4	90	0	TREN	92	0	11.9	42.2	96	64
5	72	18	TREN	71	20	12.2	37.2	48	48
6	54	36	TREN	57	34	12.4	36.9	64	128
7	60	0	TREN	63	0	8.2	43.0	>128	>128
8	48	12	TREN	49	12	8.2	39.2	>128	>128
9	36	24	TREN	38	27	9.0	38.3	>128	>128
10	24	6	HA	23	7	4.1	41.1	>64	>64
11	200	50	PEI [6]	195	57	34.0	38.0	32	48

[1] Polymerization conditions: DMF = 5 mL; [Cbz-Lys NCA] + [Phe NCA] = 0.2 M; [monomer]/[amino group in initiator] = 20, 30, and 40; [catalyst] = 0.02 M; T = 25 °C; and time = 20 min. [2] Number of repeat unit in p(Cbz-Lys) or p(Cbz-Lys-co-Phe) estimated by ^{1}H NMR spectroscopy. [3] Number average molecular weight of p(Cbz-Lys) or p(Cbz-Lys-co-Phe) estimated by ^{1}H NMR spectroscopy. [4] Zeta potential (ζ) of the p(Arg) and p(Arg-co-Phe) determined by DLS. [5] The minimum inhibitory concentration from antibacterial tests performed after guanylation of p(Lys) and p(Lys-co-Phe). [6] Number of amine group in a PEI molecule is assumed to be 8.3.

Figure 2a–c shows ^{1}H NMR spectra of s-[p(Cbz-Lys)$_{21}$]$_3$ (Entry 7 in Table 1), s-[p(Lys)$_{21}$]$_3$, and its guanylated counterpart (s-[p(Arg)$_{21}$]$_3$), respectively, and Figure 2d–f shows ^{1}H NMR spectra of l-[p(Cbz-Lys)$_{23}$-co-p(Phe)$_7$] (Entry 10), l-p[(Lys)$_{23}$-co-(Phe)$_7$], and l-p[(Arg)$_{23}$-co-(Phe)$_7$], respectively. The methyne group of Cbz-Lys (−NH−**CH**−; c) and methylene proton (−**CH$_2$**−NH$_2$; g) of the of Cbz-Lys side chain in s-p(Cbz-Lys)$_{21}$ are detected at 3.73 ppm and 3.08 ppm, respectively; they appeared at 4.23 ppm and 2.87 ppm, respectively, in s-p(Lys)$_{21}$; and 4.25 ppm and 3,07 ppm, respectively, in s-p(Lys)$_{21}$. The methylene protons (−**CH$_2$**−; d, e, and f) of the other side chain appeared as a multiplet in between 1.18 and 2.05 ppm for s-p(Cbz-Lys)$_{21}$, between 1.23 and 1.76 ppm for s-p(Lys)$_{21}$, and between 1.20 and 1.75 ppm for s-p(Arg)$_{21}$. It is not surprising to observe similar results from the ^{1}H NMR spectra of l-p[(Cbz-Lys)$_{23}$-co-(Phe)$_7$], l-p[(Lys)$_{23}$-co-(Phe)$_7$], and l-p[(Arg)$_{23}$-co-(Phe)$_7$] except the appearance of the peaks corresponding to Phe fragments. A similar peak and shift pattern was reported during the transformation of poly(Lys-co-Val) to poly(Arg-co-Val) [31]. Successful transformations of Cbz-Lys to Lys and to Arg units were also confirmed by the FTIR spectra of s-p(Lys)$_{21}$, s-p(Arg)$_{21}$, l-p[(Lys)$_{23}$-co-(Phe)$_7$], and l-p[(Arg)$_{23}$-co-(Phe)$_7$] by observing a characteristic stretching band at 1542 cm^{-1} (Figure S10) for the guanidine functionality.

The secondary structures of s-p(Arg)$_{21}$ and l-p[(Arg)$_{23}$-co-(Phe)$_7$] were analyzed by CD spectroscopy in PBS at 25 °C. The spectra of both samples had a strong negative band near 200 nm and a few positive shoulder bands above 220 nm (Figure 3), indicating that both samples consist of similar secondary structures. Each absorption band gives rise to different characteristic bands that can be deconvoluted to estimate the secondary structure components of the polymers using the CD Multivariate Calibration Creation Program in Spectra Manager^{TM} Version 2. Through curve-fitting procedures based on a set of reference spectra with known secondary structure components, the four secondary structure components (α-helical, β-sheet, turn, and random coil) could be estimated (see inset in Figure 3). Both s-p(Arg)$_{21}$ and l-p[(Arg)$_{23}$-co-(Phe)$_7$] consist of predominantly random coil structures with small portions of β-sheet and turn structures. Both samples show no absorption bands indicating α-helical structures.

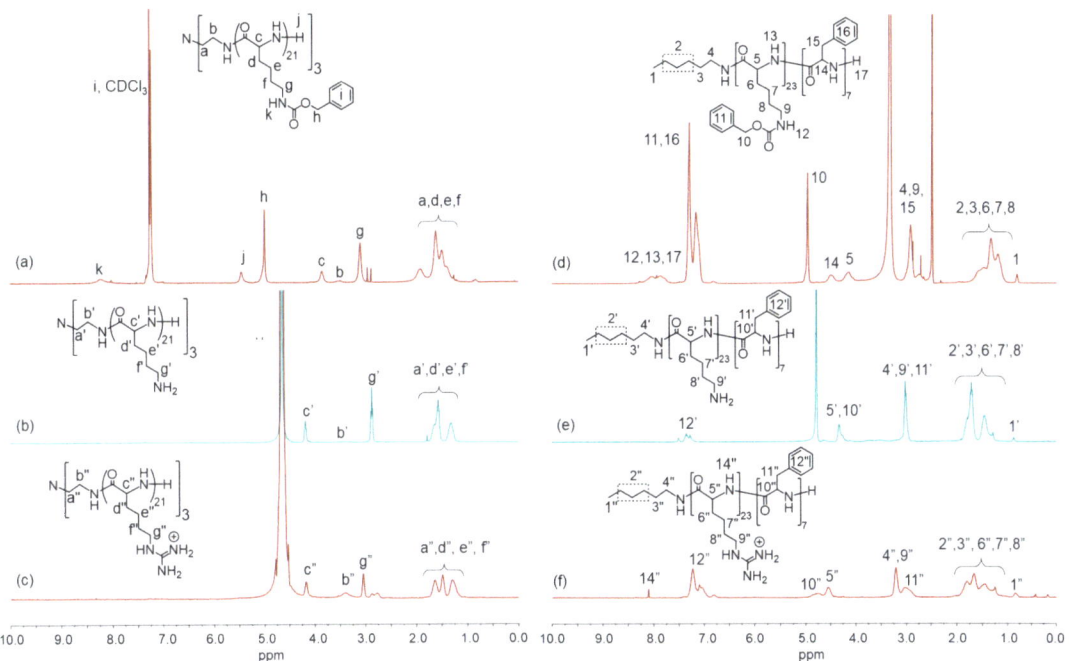

Figure 2. ^{1}H NMR spectra of *s*-p(Cbz-Lys)$_{21}$ (Entry 7 in Table 1) in CDCl$_3$ (**a**), *s*-p(Lys)$_{21}$ in D$_2$O (**b**), *s*-p(Arg)$_{21}$ in D$_2$O (**c**), *l*-p[(Cbz-Lys)$_{23}$-*co*-(Phe)$_7$] (Entry 10 in Table 1) in DMSO-*d$_6$* (**d**), *l*-p[(Lys)$_{23}$-*co*-(Phe)$_7$] in D$_2$O (**e**), and *l*-p[(Arg)$_{23}$-*co*-(Phe)$_7$] in TFA (**f**). Note that the NMR solvents of samples are different from each other due to the solubility variation.

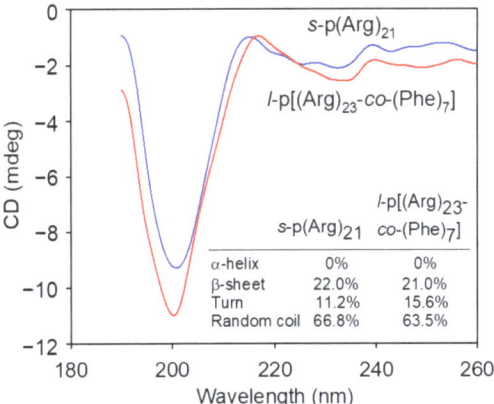

Figure 3. CD spectra of *s*-p(Arg)$_{21}$ and *l*-p[(Arg)$_{23}$-*co*-(Phe)$_7$] recorded at the same concentration (0.5 mg mL^{-1}) in PBS at 25 °C. Inset: summary of the secondary structure analyzed using CD Multivariate Calibration Creation Program in Spectra Manager^{TM} Version 2.

3.2. Antimicrobial Activities

Prior to performing antimicrobial activities, the potential stability of the p(Arg) and p(Arg-*co*-Phe) colloidal systems was measured at a concentration of 1 mg mL^{-1} in PBS using zeta potential measurement. As summarized in Table 1, the zeta potentials of all polymer particles are more positive than +30 mV, demonstrating their intrinsic stability in buffer solution.

The antibacterial activity of all guanylated polymers including s-p(Arg), s-p[(Arg)-co-(Phe)], l-p[(Arg)-co-(Phe)], and mb-p[(Arg)-co-(Phe)] was evaluated by measuring MICs using B. subtilis as the gram-positive bacteria and E. coli as the gram-negative bacteria (Figures S11 and S12). In the antibacterial ability evaluation conducted on LB broth, bacterial proliferation was clearly observed in the control group containing only LB broth and bacteria, but bacteria did not proliferate in the experimental group containing LB broth, bacteria, and polymer. As summarized in Table 1, comparing three s-(Arg) polymers (Entry 1, 4, and 7 in Table 1) with different numbers of Arg units, the s-(Arg) polymers with the largest number of Arg units shows the lowest MIC value. Comparing s-p[(Arg)-co-(Phe)] copolymers with different numbers and compositions of each repeat unit, the MIC value tend to decrease as the repeat unit increases and the relative composition of Phe units decreases, demonstrating that the Arg unit is a vital component in identifying antibacterial activity. However, s-p[(Arg)$_{24}$-co-(Phe)$_7$]$_3$ polymer shows the lowest MIC value of 48 µg mL^{-1} in E. coli and 48 µg mL^{-1} in B. subtilis, indicating that the hydrophobic portion affected antibacterial activity to some extent. This polymer shows better antibacterial activity than its linear counterpart, l-p[(Arg)$_{23}$-co-(Phe)$_7$]. The PEI-initiated multi-branching copolymer with similar repeat units in each arm (mb-p[(Arg)$_{23}$-co-(Phe)$_7$]$_8$; Entry 11) shows the lowest MIC values: 48 µg mL^{-1} in E. coli and 32 µg mL^{-1} in B. subtilis. Therefore, the topology of polymers also affects antibacterial efficacy because the topology is important when developing antibacterial agents with membrane destruction mechanisms. This multi-branching copolypeptide also showed the low hemolysis of 32.4% at 3000 µg·mL^{-1}, which is 93.75 times higher than its MIC in E. coli, demonstrating acceptable blood suitability (Figure S13).

Blood plasma consists of about 90% water and transports nutrients, wastes, antibodies, ions, hormones, etc. Even though ions make up only about 1% by weight of blood plasma, they are the major contributors to plasma molarity, since their molecular weights are much less than those of proteins. NaCl constitutes more than 65% of the plasma ions. Bicarbonate, potassium, calcium, phosphate, sulfate, and magnesium are other plasma ions. Antibacterial effects were tested on s-[p(Lys)$_{23}$-co-p(Phe)$_7$]$_3$ (Entry 5, Table 1) and s-[p(Arg)$_{23}$-co-p(Phe)$_7$]$_3$ polymers in the absence or presence of additional Ca^{2+} and Mg^{2+} (Figure 4). LB broth contains nutrients and ions necessary for microbial growth, which reduces the likelihood of antibiotic capacity being hindered by nutrients and ions in performance. Additionally, it has lower concentrations of Ca^{2+} and Mg^{2+} ions than plasma. Therefore, the concentrations of Ca^{2+} and Mg^{2+} ions were set to 2 mM, considering the ion concentration in the blood plasma is 2.5 mM for Ca^{2+} and 1.5 mM for Mg^{2+}. Both polymers showed high antibiotic efficacy at 64 ug/mL in pure water with a higher efficiency for s-[p(Arg)$_{23}$-co-p(Phe)$_7$]$_3$. Specifically, polymer shows a high efficacy for B. subtilis. The efficacy decreased in the media bearing Ca^{2+} and Mg^{2+} ions. The decrease in efficacy was apparent for s-[p(Lys)$_{23}$-co-p(Phe)$_7$]$_3$, most probably because the amine groups in Lys units may serve as anchors for metal ion chelation with amine groups.

Considering the percentage of salt in blood is about 0.9 percent by weight, antibacterial activity was tested in 1.0% NaCl solution. As illustrated in Figure 4, both s-[p(Lys)$_{23}$-co-p(Phe)$_7$]$_3$ and s-[p(Arg)$_{23}$-co-p(Phe)$_7$]$_3$ polymers showed better antibacterial activity in salt solution than in pure water. s-[p(Arg)$_{23}$-co-p(Phe)$_7$]$_3$ polymer showed better efficacy than s-[p(Lys)$_{23}$-co-p(Phe)$_7$]$_3$ polymer and no bacterial proliferation was apparently observed for B. subtilis. The antibacterial efficacy of s-[p(Arg)$_{23}$-co-p(Phe)$_7$]$_3$ polymer was retained in the presence of additional cations, Ca^{2+} or Mg^{2+}. It is worth noting that s-[p(Arg)$_{23}$-co-p(Phe)$_7$]$_3$ polymer showed much better activity than s-[p(Lys)$_{23}$-co-p(Phe)$_7$]$_3$ polymer in all media. These results indicate that the copolypetides bearing cationic Arg side chains show better antibiotic efficacy than those bearing neutral Lys side chains. The cationic Arg moieties would more easily bind to the negatively charged bacterial cell membrane with relatively lower interference of the various metal cations existing in the media, which is favorable in vivo applications. The low antimicrobial activity in *E. coli* may be due to the fact that a strong interaction between the Lys or Arg moieties and the phospholipid head group prevents translocation of the peptide into the inner leaflet of the membrane [35]. In addition, the hydrophobic Phe residues govern the extent to which the water-soluble p(Arg)-*co*-p(Phe) and p(Lys)-*co*-p(Phe) polymers will be able to partition into the membrane lipid bilayer, and excessive levels of Phe unit can lead to cell toxicity and loss of antimicrobial selectivity [36].

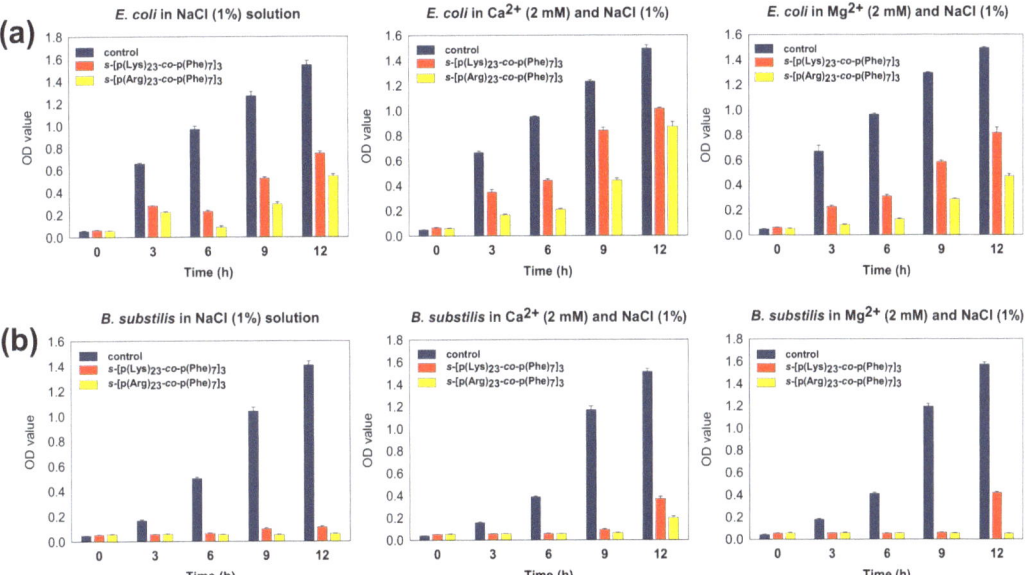

Figure 4. Growth inhibition assay of (**a**) *E. coli* and (**b**) *B. substilis* with 64 ug/mL of s-[p(Lys)$_{23}$-*co*-p(Phe)$_7$]$_3$ and s-[p(Arg)$_{23}$-*co*-p(Phe)$_7$]$_3$ polymers in salt solution (1% NaCl) in the presence or absence of divalent cations (2 mM Ca^{2+} or Mg^{2+}).

In all cases, p(Arg)-*co*-p(Phe) showed lower efficacy for the growth inhibition of *E. coli* than for the growth inhibition of *B. subtilis*. These results can be inferred from the membrane compositions of both bacterial cells. *E. coli* accumulates three major membrane phospholipids: zwitterionic phosphatidylethanolamine (PE; ~75% of membrane lipids), the anionic lipid phosphatidylglycerol (PG; ~20%), and cardiolipin [37]. The *B. subtilis* lipidome comprises 70% PG, 12% PE, 5% phosphoglycolipid, 4% cardiolipin, 4% diglycosyldiacylglycerol, 2% monoglycosyl diacylglycerol, 2% aminoacyl phosphatidylglycerol, and other components [38]. The cationic Arg moieties in p(Arg)-*co*-p(Phe) strongly interact with anionic PG while displaying repulsive action with zwitterionic PE, as illustrated in Figure 5. Accordingly, p(Arg)-*co*-p(Phe) can be adsorbed onto the *B. subtilis* membrane more strongly than onto the *E. coli* membrane.

Mechanistically, copolypeptides interact with the cytoplasmic membrane first and then accumulate intracellularly, blocking critical cellular processes. Peptide-based antimicrobial action has been studied extensively since it was discovered [6]. The mechanism of action can be divided into two major classes: direct killing and immune modulation. As illustrated in Figure 5, p(Arg)-*co*-p(Phe) accumulates at the surface and self-assembles on the bacterial cell membrane after the initial electrostatic and hydrophobic interactions that are dependent on the composition of the membranes. The direct killing action can be observed by using SEM images (Figure 6). Even though it is difficult to propose the mechanism of action with only the data collected here, the polymer is expected to bind to the surface of the membrane. The membrane-bound peptides then orientate themselves so that the cationic

Arg residues face the polar lipid headgroups and the hydrophobic Phe residues face toward the lipid tails. Once a threshold concentration of bound peptide is reached, the hydrophobic Phe residues will penetrate the lipophilic phospholipid layer, eventually destroying the membrane and leading to a formation wrinkle surface tangled with other destroyed bacterial cell membranes.

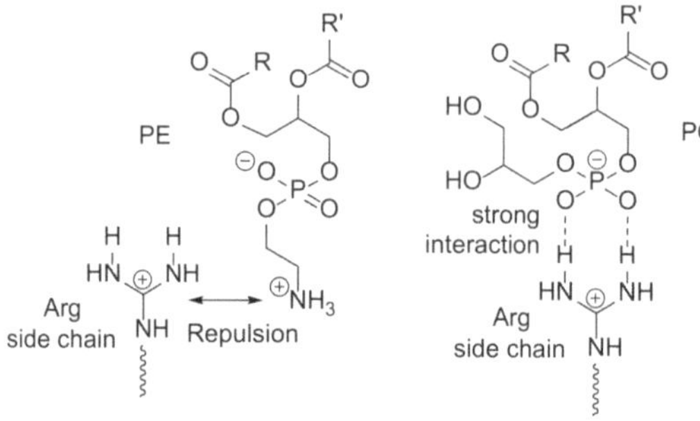

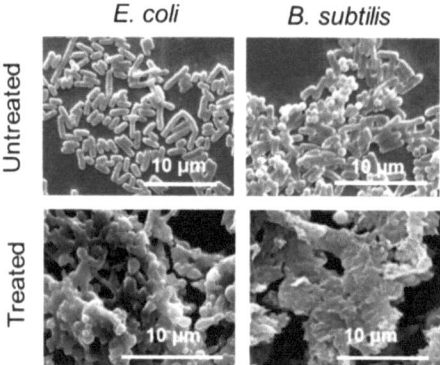

Figure 5. Proposed interactions of Arg moiety with zwitterionic phosphatidylethanolamine (PE) and with the anionic lipid phosphatidylglycerol (PG).

Figure 6. SEM images of untreated and s-[p(Arg)$_{23}$-co-p(Phe)$_7$]$_3$-polymer-treated (3 × MIC) groups of *E. coli* and *B. substilis*.

To evaluate the practical utilization, antibacterial capacity was determined at the s-[p(Arg)$_{23}$-co-p(Phe)$_7$]$_3$ concentration of MIC×3 in natural waters (Figure 7). The results of the colony formation assays clearly show that s-[p(Arg)$_{23}$-co-p(Phe)$_7$]$_3$ is highly effective for the inhibition of bacterial growth in various natural water sources, except rainwater. This material shows particularly high antibacterial ability for seawater bearing ~3.5% NaCl. This result is in line with the previously obtained results that the antibacterial activity of s-[p(Arg)$_{23}$-co-p(Phe)$_7$]$_3$ was retained even in the presence of various metal cations, which may broaden the scope of application of the materials of this study.

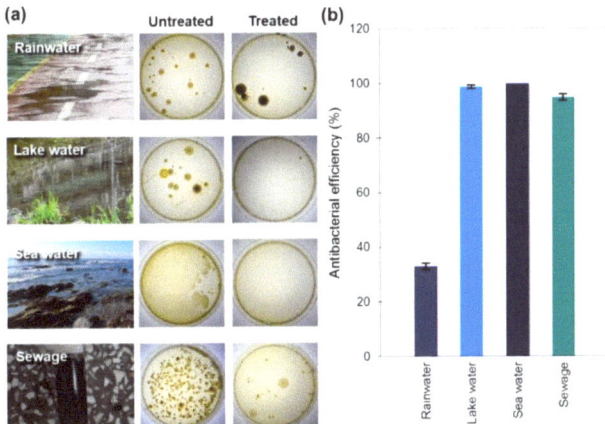

Figure 7. Antibacterial effect of natural water samples collected, including: rainwater, lake water (from Hoedong Suwonji, Busan, Korea), seawater (from Haeundae beach, Busan, Korea), and sewage (from East Busan Sewage Treatment Plant, Busan, Korea): (**a**) colony formation images of the untreated and s-[p(Arg)$_{23}$-co-p(Phe)$_7$]$_3$-polymer-treated (3 × MIC) groups and (**b**) antibacterial efficiency calculated by colony forming units.

4. Conclusions

The l-, s-, and mb-p[(Cbz-Lys)-co-(Phe)] copolypeptides were synthesized via the random ring-opening copolymerization of corresponding NCAs using organic [^{i}PrNHC(H)]$^+$[HCO$_3$]$^-$ as a catalyst. The Cbz-Lys units were deprotected to Lys units and further modified to cationic Arg units to yield p[(Arg)-co-(Phe)] counterparts. The living character of the copolymerization made it possible to synthesize various copolypeptides with targeted MWs and topologies having predominantly random coil microstructure. The p[(Arg)-co-(Phe)] copolymers were intrinsically stable in buffer solution and showed acceptable blood compatibility.

The antibacterial activity of all the guanylated polymers including s-p(Arg), s-p[(Arg)-co-(Phe)], l-p[(Arg)-co-(Phe)], and mb-p[(Arg)-co-(Phe)] evaluated by measuring MICs using *B. subtilis* and *E. coli* showed that the Arg unit was a vital component to show antibacterial activity. The s-p[(Arg)$_{24}$-co-(Phe)$_7$]$_3$ polymer shows the lowest MIC values of 48 μg mL^{-1} in *E. coli* and 48 μg mL^{-1} in *B. subtilis*. The multi-branching mb-p[(Arg)$_{23}$-co-(Phe)$_7$]$_8$ copolymer showed the lowest MIC values: 48 μg mL^{-1} in *E. coli* and 32 μg mL^{-1} in *B. subtilis*.

s-[p(Arg)$_{23}$-co-p(Phe)$_7$]$_3$ copolymer was much more effective for inhibiting *B. subtilis* rather than *E. coli* due to the differences of interaction capability between polymer and each bacterial membrane; because it did not lose its antibacterial activity even in the presence of various metal cations, Ca^{2+} or Mg^{2+}, and/or Na$^+$; and because it showed better activity than its Lys counterpart, s-[p(Lys)$_{23}$-co-p(Phe)$_7$]$_3$. s-[p(Arg)$_{23}$-co-p(Phe)$_7$]$_3$ was also effective to treat natural water sources such as lake, sea, rain, and sewage. Antimicrobial copolypeptides are unique materials that have shown great promise in treating bacteria while causing them to develop no or only low resistance.

Supplementary Materials: The following supporting information can be downloaded at: https://www.mdpi.com/article/10.3390/polym14091868/s1, Figures S1–Figure S9: ^{1}H NMR spectra of linear-, star-, and multi-branched-polypeptides, Figure S10: FTIR spectra of s-p(Lys)21, s-p(Arg)21, l-p[(Lys)23-co-(Phe)7], and l-p[(Arg)23-co-(Phe)7], Figure S11: Determination of minimal inhibitory concentration on *E. coli* cells with l-, s-, and mb-polypeptides, Figure S12: Determination of minimal inhibitory concentration on *B. substilis* cells with l-, s-, and mb-polypeptides, Figure S13: Hemolytic activity of mb-p[(Arg)23-co-(Phe)7]8 in fresh red blood cells of mice.

Author Contributions: Conceptualization, I.K., J.H.K. and J.R.K.; Methodology, K.H.E. and S.L.; Validation, K.H.E. and S.L.; formal analysis, K.H.E. and S.L.; investigation, K.H.E., S.L. and E.G.L.; resources, K.H.E. and S.L.; data curation, I.K., J.H.K., J.R.K., K.H.E. and S.L.; writing—original draft preparation, K.H.E. and S.L.; writing—review and editing, I.K.; visualization, I.K.; supervision, I.K., J.H.K. and J.R.K.; project administration, K.H.E.; funding acquisition, I.K. All authors have read and agreed to the published version of the manuscript.

Funding: This work was supported by the National Research Foundation of Korea grant funded by the Korean government (MSIT) (2021R1A2C2003685) and the Korea Institute of Energy Technology Evaluation and Planning (KETEP) grant funded by the Korea government (MOTIE) (20208401010080).

Institutional Review Board Statement: Not applicable.

Informed Consent Statement: Not applicable.

Data Availability Statement: The data presented in this study are available on request from the corresponding author.

Conflicts of Interest: The authors declare no conflict of interest.

References

1. Friedman, N.D.; Temkin, E.; Carmeli, Y. The negative impact of antibiotic resistance. *Clin. Microbiol. Infect.* **2016**, *22*, 416–422. [CrossRef] [PubMed]
2. Ventola, C.L. The antibiotic resistance crisis: Part 1: Causes and threats. *Pharm. Ther.* **2015**, *40*, 277–283.
3. Rima, M.; Rima, M.; Fajloun, Z.; Sabatier, J.-M.; Bechinger, B.; Naas, T. Antimicrobial peptides: A potent alternative to antibiotics. *Antibiotics* **2021**, *10*, 1095. [CrossRef]
4. Dijksteel, G.S.; Ulrich, M.M.W.; Middelkoop, E.; Boekema, B.K.H. Lessons learned from clinical trials using antimicrobial peptides (AMPs). *Front. Microbiol.* **2021**, *12*, 616979. [CrossRef]
5. Aslam, B.; Wang, W.; Arshad, M.I.; Khurshid, M.; Muzammil, S.; Rasool, M.H.; Nisar, M.A.; Alvi, R.F.; Aslam, M.A.; Qamar, M.U.; et al. Antibiotic resistance: A rundown of a global crisis. *Infec. Drug Resist.* **2018**, *11*, 1645–1658. [CrossRef]
6. Kumar, P.; Kizhakkedathu, J.N.; Straus, S.K. Antimicrobial peptides: Diversity, mechanism of action, and strategies to improve the activity and biocompatibility in vivo. *Biomolecules* **2018**, *8*, 4. [CrossRef]
7. Wang, G. Structures of human host defense cathelicidin LL-37 and its smallest antimicrobial peptide KR-12 in lipid micelles. *J. Biol. Chem.* **2008**, *283*, 32637–32643. [CrossRef] [PubMed]
8. Magana, M.; Pushpanathan, M.; Santos, A.L.; Leanse, L.; Fernandez, M.; Ioannidis, A.; Giulianotti, M.A.; Apidianakis, Y.; Bradfute, S.; Ferguson, A.L.; et al. The value of antimicrobial peptides at the age of resistance. *Lancet Infect. Dis.* **2020**, *20*, e216–e230. [CrossRef]
9. Ambrosio, P.J.S.; Tronnet, A.; Verhaeghe, P.; Bonduelle, C. Synthetic polypeptide polymers as simplified analogues of antimicrobial peptides. *Biomacromolecules* **2021**, *22*, 57–75. [CrossRef] [PubMed]
10. Zhang, Y.; Song, W.; Li, S.; Kim, D.; Kim, J.H.; Kim, J.R.; Kim, I. Facile and scalable synthesis of topologically nanoengineered polypeptides with excellent antimicrobial activities. *Chem. Commun.* **2020**, *56*, 356–359. [CrossRef] [PubMed]
11. Su, X.; Zhou, X.; Tan, Z.; Zhou, C. Highly efficient antibacterial diblock copolypeptides based on lysine and phenylalanine. *Biopolymers* **2017**, *107*, e23041. [CrossRef] [PubMed]
12. Zhou, X.; He, J.; Zhou, C. Strategies from nature: Polycaprolactone-based mimetic antimicrobial peptide block copolymers with low cytotoxicity and excellent antibacterial efficiency. *Polym. Chem.* **2019**, *10*, 945–953. [CrossRef]
13. Xi, Y.; Song, T.; Tang, S.; Wang, N.; Du, J. Preparation and antibacterial mechanism of polypeptide-based micelles with excellent antibacterial activities. *Biomacromolecules* **2016**, *17*, 3922–3930. [CrossRef] [PubMed]
14. Andreev, K.; Bianchi, C.; Laursen, J.S.; Citterio, L.; Hein-Kristensen, L.; Gram, L.; Kuzmenko, I.; Olsen, C.A.; Gidalevitz, D. Guanidino groups greatly enhance the action of antimicrobial peptidomimetics against bacterial cytoplasmic membranes. *Biochim. Biophys. Acta Biomembr.* **2014**, *1838*, 2492–2502. [CrossRef]
15. Albertshofer, K.; Siwkowski, A.M.; Wancewicz, E.V.; Esau, C.C.; Watanabe, T.; Nishihara, K.C.; Kinberger, G.A.; Malik, L.; Eldrup, A.B.; Manoharan, M.; et al. Structure—Activity relationship study on a simple cationic peptide motif for cellular delivery of antisense peptide nucleic acid. *J. Med. Chem.* **2005**, *48*, 6741–6749. [CrossRef]
16. Exley, S.E.; Paslay, L.C.; Sahukhal, G.S.; Abel, B.A.; Brown, T.D.; McCormick, C.L.; Heinhorst, S.; Koul, V.; Choudhary, V.; Elasri, M.O.; et al. Antimicrobial peptide mimicking primary amine and guanidine containing methacrylamide copolymers prepared by raft polymerization. *Biomacromolecules* **2015**, *16*, 3845–3852. [CrossRef]
17. Mbizana, S.; Hlalele, L.; Pfukwa, R.; Du Toit, A.; Lumkwana, D.; Loos, B.; Klumperman, B. Synthesis and cell interaction of statistical l-arginine–glycine–l-aspartic acid terpolypeptides. *Biomacromolecules* **2018**, *19*, 3058–3066. [CrossRef]
18. Tsogas, I.; Theodossiou, T.; Sideratou, Z.; Paleos, C.M.; Collet, H.; Rossi, J.C.; Romestand, B.; Commeyras, A. Interaction and transport of poly(l-lysine) dendrigrafts through liposomal and cellular membranes: The role of generation and surface functionalization. *Biomacromolecules* **2007**, *8*, 3263–3270. [CrossRef]

19. Carlson, P.M.; Schellinger, J.G.; Pahang, J.A.; Johnson, R.N.; Pun, S.H. Comparative study of guanidine-based and lysine-based brush copolymers for plasmid delivery. *Biomater. Sci.* **2013**, *1*, 736–744. [CrossRef]
20. Sideratou, Z.; Sterioti, N.; Tsiourvas, D.; Tziveleka, L.; Thanassoulas, A.; Nounesis, G.; Paleos, C.M. Arginine end-functionalized poly(l-lysine) dendrigrafts for the stabilization and controlled release of insulin. *J. Colloid Interface Sci.* **2010**, *351*, 433–441. [CrossRef]
21. Bernatowicz, M.S.; Wu, Y.; Matsueda, G.R. 1H-Pyrazole-1-carboxamidine hydrochloride an attractive reagent for guanylation of amines and its application to peptide synthesis. *J. Org. Chem.* **1992**, *57*, 2497–2502. [CrossRef]
22. Tencza, S.B.; A Miller, M.; Islam, K.; Mietzner, T.A.; Montelaro, R.C. Effect of amino acid substitutions on calmodulin binding and cytolytic properties of the LLP-1 peptide segment of human immunodeficiency virus type 1 transmembrane protein. *J. Virol.* **1995**, *69*, 5199–5202. [CrossRef] [PubMed]
23. Mitchell, D.J.; Steinman, L.; Kim, D.T.; Fathman, C.G.; Rothbard, J.B. Polyarginine enters cells more efficiently than other polycationic homopolymers. *J. Pept. Res.* **2000**, *56*, 318–325. [CrossRef]
24. Phadke, S.M.; Lazarevic, V.; Bahr, C.C.; Islam, K.; Stolz, D.B.; Watkins, S.; Tencza, S.B.; Vogel, H.J.; Montelaro, R.C.; Mietzner, T.A. Lentivirus lytic peptide 1 perturbs both outer and inner membranes of Serratia marcescens. *Antimicrob. Agents Chemother.* **2002**, *46*, 2041–2045. [CrossRef]
25. Kalia, V.; Sarkar, S.; Gupta, P.; Montelaro, R.C. Rational site-directed mutations of the LLP-1 and LLP-2 lentiviral lytic peptide domains in the intracytoplasmic tail of human immunodeficiency virus type 1 gp41 indicate common functions in cell-cell fusion but distinct roles in virion envelope incorporation. *J. Virol.* **2003**, *77*, 3634–3646. [PubMed]
26. Phadke, S.M.; Islam, K.; Deslouches, B.; Kapoor, S.A.; Stolz, D.B.; Watkins, S.C.; Montelaro, R.C.; Pilewski, J.M.; Mietzner, T.A. Selective toxicity of engineered lentivirus lytic peptides in a CF airway cell model. *Peptides* **2003**, *24*, 1099–1107. [CrossRef]
27. Phadke, S.M.; Deslouches, B.; Hileman, S.E.; Montelaro, R.C.; Wiesenfeld, H.C.; Mietzner, T.A. Antimicrobial peptides in mucosal secretions: The importance of local secretions in mitigating infection. *J. Nutr.* **2005**, *135*, 1289–1293. [CrossRef] [PubMed]
28. Su, Y.; Doherty, T.; Waring, A.J.; Ruchala, P.; Hong, M. Roles of arginine and lysine residues in the translocation of a cell-penetrating peptide from ^{13}C, ^{31}P, and ^{19}F solid-state NMR. *Biochem.* **2009**, *48*, 4587–4595. [CrossRef]
29. Deslouches, B.; Hasek, M.L.; Craigo, J.K.; Steckbeck, J.D.; Montelaro, R.C. Comparative functional properties of engineered cationic antimicrobial peptides consisting exclusively of tryptophan and either lysine or arginine. *J. Med. Microbiol.* **2016**, *65*, 554–565. [CrossRef]
30. Lee, J.-K.; Park, Y. All d-lysine analogues of the antimicrobial peptide HPA3NT3-A2 increased serum stability and without drug resistance. *Int. J. Mol. Sci.* **2020**, *21*, 5632. [CrossRef]
31. Sell, D.R.; Monnier, V.M. Conversion of arginine into ornithine by advanced glycation in senescent human collagen and lens crystallins. *J. Biol. Chem.* **2004**, *279*, 54173–54184. [CrossRef] [PubMed]
32. Granier, C.; Muller, E.P.; Van Rietschoten, J. Use of synthetic analogs for a study on the structure-activity relationship of apamin. *Eur. J. Biochem.* **1978**, *82*, 293–299. [CrossRef] [PubMed]
33. Jen, T.; Van Hoeven, H.; Groves, W.; McLean, R.A.; Loev, B. Amidines and related compounds. 6. Structure-activity relations of antihypertensive and antisecretory agents related to clonidine. *J. Med. Chem.* **1975**, *18*, 90–99. [CrossRef] [PubMed]
34. Bannard, R.A.B.; Casselman, A.A.; Cockburn, W.F.; Brown, G.M. Guanidine compounds. II. preparation of Mono-and N, N-di-alkylguanidines. *Can. J. Chem.* **1958**, *36*, 1541–1549. [CrossRef]
35. Yeaman, M.R.; Yount, N.Y. Mechanisms of antimicrobial peptide action and resistance. *Pharmacol. Rev.* **2003**, *55*, 27–55. [CrossRef] [PubMed]
36. Yin, L.M.; Edwards, M.A.; Li, J.; Yip, C.M.; Deber, C.M. Roles of hydrophobicity and charge distribution of cationic antimicrobial peptides in peptide-membrane interactions. *J. Biol. Chem.* **2012**, *287*, 7738–7745. [CrossRef] [PubMed]
37. Raetz, C.R.; Dowhan, W. Biosynthesis and function of phospholipids in *Escherichia coli*. *J. Biol. Chem.* **1990**, *265*, 1235–1238. [CrossRef]
38. Clejan, S.; Krulwich, T.A.; Mondrus, K.R.; Seto-Young, D. Membrane lipid composition of obligately and facultatively alkalophilic strains of *Bacillus* spp. *J. Bacteriol.* **1986**, *168*, 334–340. [CrossRef]

Article

Probiotic-Mediated Biosynthesis of Silver Nanoparticles and Their Antibacterial Applications against Pathogenic Strains of *Escherichia coli* O157:H7

Xiaoqing Wang [1,†], Sun-Young Lee [1], Shahina Akter [2,†] and Md. Amdadul Huq [1,*]

1. Department of Food and Nutrition, College of Biotechnology and Natural Resource, Chung-Ang University, Anseong 17546, Gyeonggi-do, Korea; wxq2016@naver.com (X.W.); nina6026@cau.ac.kr (S.-Y.L.)
2. Department of Food Science and Biotechnology, Gachon University, Seongnam 461701, Gyeonggi-do, Korea; shahinabristy16@gmail.com
* Correspondence: amdadbge@gmail.com or amdadbge100@cau.ac.kr
† These authors contributed equally to this work.

Abstract: The present study aimed to suggest a simple and environmentally friendly biosynthesis method of silver nanoparticles (AgNPs) using the strain *Bacillus sonorensis* MAHUQ-74 isolated from kimchi. Antibacterial activity and mechanisms of AgNPs against antibiotic-resistant pathogenic strains of *Escherichia coli* O157:H7 were investigated. The strain MAHUQ-74 had 99.93% relatedness to the *B. sonorensis* NBRC 101234[T] strain. The biosynthesized AgNPs had a strong surface plasmon resonance (SPR) peak at 430 nm. The transmission electron microscope (TEM) image shows the spherical shape and size of the synthesized AgNPs is 13 to 50 nm. XRD analysis and SAED pattern revealed the crystal structure of biosynthesized AgNPs. Fourier transform infrared spectroscopy (FTIR) data showed various functional groups associated with the reduction of silver ions to AgNPs. The resultant AgNPs showed strong antibacterial activity against nine *E. coli* O157:H7 pathogens. Minimum inhibitory concentration (MIC) values of the AgNPs synthesized by strain MAHUQ-74 were 3.12 µg/mL for eight *E. coli* O157:H7 strains and 12.5 µg/mL for strain *E. coli* ATCC 25922. Minimum bactericidal concentrations (MBCs) were 25 µg/mL for *E. coli* O157:H7 ATCC 35150, *E. coli* O157:H7 ATCC 43895, *E. coli* O157:H7 ATCC 43890, *E. coli* O157:H7 ATCC 43889, and *E. coli* ATCC 25922; and 50 µg/mL for *E. coli* O157:H7 2257, *E. coli* O157: NM 3204-92, *E. coli* O157:H7 8624 and *E. coli* O157:H7 ATCC 43894. FE-SEM analysis demonstrated that the probiotic-mediated synthesized AgNPs produced structural and morphological changes and destroyed the membrane integrity of pathogenic *E. coli* O157:H7. Therefore, AgNPs synthesized by strain MAHUQ-74 may be potential antibacterial agents for the control of antibiotic-resistant pathogenic strains of *E. coli* O157:H7.

Keywords: probiotic-mediated biosynthesis; AgNPs; antibacterial applications; *E. coli* O157:H7

Citation: Wang, X.; Lee, S.-Y.; Akter, S.; Huq, M.A. Probiotic-Mediated Biosynthesis of Silver Nanoparticles and Their Antibacterial Applications against Pathogenic Strains of *Escherichia coli* O157:H7. *Polymers* **2022**, *14*, 1834. https://doi.org/10.3390/polym14091834

Academic Editor: Tzu-wei Wang

Received: 15 April 2022
Accepted: 28 April 2022
Published: 29 April 2022

Publisher's Note: MDPI stays neutral with regard to jurisdictional claims in published maps and institutional affiliations.

Copyright: © 2022 by the authors. Licensee MDPI, Basel, Switzerland. This article is an open access article distributed under the terms and conditions of the Creative Commons Attribution (CC BY) license (https://creativecommons.org/licenses/by/4.0/).

1. Introduction

Silver nanoparticles (AgNPs) refer to nanoparticles with sizes ranging from 1 nm to 100 nm. AgNPs have been widely used in various applications in biomedical science such as in antibiotics, biosensors, drug delivery systems, antimicrobial, anticancer, and anti-inflammatory agents, etc. [1–7]. Many devices, including textiles, keyboards, and medical devices, now contain AgNPs that continuously release small amounts of silver ions to provide antibacterial protection [8]. AgNPs also have been implemented in the development of various bioactive materials, including polymer composites, because of their high antimicrobial activity. The polymer-based nanoparticles are usually a combination of organic polymers or a specific biomolecule in which an inorganic nanoparticle is embedded. Polymer-based metallic nanoparticles are widely explored due to their versatility, biodegradability, being environmentally friendly, and being biocompatible in nature [9,10]. The successful capping of polymers on AgNPs may prevent the agglomeration of synthesized nanocomposite as well as facilitate the stabilization process [11].

It was reported that AgNPs-containing polymer composite exhibits significant antimicrobial activity against pathogenic bacteria [12]. AgNPs have been used extensively in recent times to create antibacterial coatings [13–15]. Depending on the application, various shapes of nanoparticles can be constructed. Spherically shaped AgNPs are commonly used, but diamond, octagonal, and flake shapes are also used [16,17]. For synthesizing nanoparticles, various methods have been utilized, including physical, chemical, and biological methods [18]. The most commonly used physical approaches for nanoparticle synthesis are laser ablation and evaporation–condensation [19]. The advantages of these approaches are controllable chemical purity and the generation of clusters with free surfaces; in addition, the free choice of size has proven to be feasible [20]. However, a tubular-reactor-mediated physical synthesis at atmospheric pressure has some drawbacks. For example, a tube furnace requires a long time, large space, and enormous energy [19]. The chemical method provides a simple method to synthesize AgNPs in solution using organic and inorganic reducing agents. Typically, the chemical synthesis of AgNPs solution generally consists of the following three main components: (1) a metal precursor, (2) a reducing agent, and (3) a stabilizing/blocking agent. Silver nitrate is often used as the metal precursor, and sodium citrate and ascorbate, sodium borohydride ($NaBH_4$), polyol process, elemental hydrogen, hydrazine, N, N-dimethylformamide (DMF), ascorbic acid, poly (ethylene glycol)-block copolymers, and ammonium formate are used as different reducing agents [21–23]. However, clinical and food applications avoid the use of chemical processes to produce nanoparticles, because these toxic chemicals are used in the process. In contrast, biosynthesis has many advantages, such as high stability, low toxicity to healthy cells, and no synthesis of toxic by-products [14]. Therefore, the use of biological pathways to synthesize nanoparticles is the preferred method, which eliminates the use of toxic chemicals, and is cost-effective and environmentally friendly [24].

Regarding the biosynthesis of AgNPs, the most commonly used materials are bacteria, fungi, algae, plants, and their components, etc. [14]. Two methods can be used for the biosynthesis of silver nanoparticles using microorganisms. According to the location of the synthesis of nanoparticles, they can be classified as intracellular or extracellular. Contrary to an intracellular synthesis that requires complicated purification steps, the extracellular synthesis of nanoparticles is more convenient, easy, and cost-effective. Therefore, extra-cellular synthesis is preferred because the removal of bacterial cells in advance simplifies the recovery of nanoparticles and can be easy and rapid purification of synthesized nanoparticles. Various bacteria including *Bacillus* sp., *Escherichia coli*, *Lactobacillus kimchicus*, *Pseudomonas deceptionensis*, *Solibacillus isronensis*, *Cedecea* sp., *Novosphingobium* sp., and *Sporosarcina koreensis* strains can reduce Ag+ ions to Ag and form AgNPs [25–32]. Biosynthesized AgNPs show superior properties such as hydrophilicity, stability, and large surface area. Compared with chemical synthesis routes, the bacterial-based synthesis method is economical, simple, reproducible, and requires less energy.

Fermented food is a type of food manufactured by humans ingeniously using beneficial microorganisms. Therefore, there are abundant microorganisms in fermented food, which provides a good source of microbial diversity. Various biomolecules are produced by bacteria including biopolymers such as polysaccharides, proteins, enzymes, and DNA, where the monomer units are sugars, amino acids, and nucleotides, respectively. These extracellular polymeric substances may play important roles in the synthesis and stabilization of nanoparticles and may enhance the efficacy of NPs [33]. Bacteria secrete various enzymes in the culture supernatant, including nitrate reductase responsible for nanoparticle biosynthesis. Different *Bacillus* sp. can secrete NADH- and NADH-dependent enzymes that involve in the reduction of silver ions in the form of AgNPs [25]. Members of the genus *Bacillus* have significant microbiological uses [34]. Numerous enzymes, antibiotics, and other metabolites have medical, agricultural, pharmaceutical, and other industrial applications [35]. For the biosynthesis of AgNPs, plant products such as fruit extract and soil bacteria are commonly used. There are a few studies to synthesize AgNPs using microorganisms isolated from foods. At present, drug-resistant pathogenic bacteria are a serious threat to humans. *Escherichia coli* O157:H7 is a notorious pathogen, and this pathogen is associated with a broad spectrum of illnesses in humans ranging from mild diarrhea and hemorrhagic colitis to the potentially fatal hemolytic uremic syndrome [36]. In this study, we report a newly isolated *B. sonorensis* MAHUQ-74 from kimchi, which we used for the synthesis of AgNPs using an extracellular method. The culture supernatant of *B. sonorensis* MAHUQ-74 was used to quickly and easily synthesize AgNPs without adding any reducing agent. Then, resultant AgNPs were characterized, and their antimicrobial activity at the mechanism level was investigated against various pathogenic strains of *E. coli* O157:H7.

2. Materials and Methods

2.1. Materials and Bacterial Strains

The culture medium was purchased from Difco, MB Cell (Seoul, South Korea). Analytical grade AgNO$_3$ (silver nitrate) was purchased from Sigma-Aldrich Chemicals (St. Louis, MO, USA). Nine pathogenic *E. coli* O157:H7 strains were used in this study—namely, *E. coli* O157:H7 ATCC 43895, ATCC 35150, ATCC 43889, ATCC 43890, ATCC 43894, ATCC 25922; *E. coli* O157:H7 8624, 2257; and *E. coli* O157:NM 3204-92—which were collected from the bacterial culture collection of Chung-Ang University (Anseong-si, Korea). All strains were maintained at $-80\ °C$ in tryptic soy broth (TSB) containing 20% glycerol (v/v) and were activated by cultivation on LB agar for 24 h at 37 °C.

2.2. Isolation, Identification, and Characterization of AgNP-Producing Probiotics

Kimchi, a traditional, fermented food product, was collected from the retail market in Anseong, Korea. Briefly, 1 g of kimchi was dissolved in 9 mL of 0.85% (w/v) NaCl solution, serially diluted up to 10^{-5}, and 100 µL of each dilution was spread on both nutrient agar (NA) and Reasoner's 2A agar (R$_2$A agar) plates [37]. Then, the plates were incubated at 37 °C for 3 days. Single colonies were purified by successive transferring to new NA or R$_2$A agar plates. AgNP synthesis ability was screened by culturing all isolates in 5 mL R2A broth media for 48 h at 37 °C. Then, 1 mM AgNO$_3$ solution was added to the culture supernatant and incubated again for 48 h. Among all of these isolated strains, strain MAHUQ-74 showed strong AgNP synthesis ability. Then, the strain was identified by 16S rRNA gene sequence analysis [38]. The 16S rRNA gene sequence of strain MAHUQ-74 was submitted to GeneBank of NCBI. The 16S rRNA gene sequences of related taxa were obtained from the EzBioCloud server [39]. The phylogenetic tree was created using the MEGA6 program [40] and the neighbor-joining method, to discover the phylogenetic location of isolated strain MAHUQ-74 [41]. The optimum growth conditions of strain MAHUQ-74 including media, temperature and pH, gram-staining reaction, catalase, and oxidase activities were examined according to the previous description [42]. According to the manufacturer's instructions, commercial API ZYM and API 20 NE kits (bioMérieux) were used to further determine enzyme activity and carbon source utilization. The probiotic strain MAHUQ-74 was deposited to the Korean Agricultural Culture Collection (KACC).

2.3. Biosynthesis of AgNPs Using Probiotic Strain MAHUQ-74

For the biosynthesis of AgNPs, the *B. sonorensis* MAHUQ-74 strain was cultured in 100 mL of R$_2$A broth and incubated at 37 °C for 3 days in a shaking incubator at 180 rpm. After three days of incubation, the culture supernatant was collected by centrifugation at 10,000 rpm, 4 °C for 10 min in a centrifuge. Then, 0.1 mL (1 M concentration) filter-sterilized AgNO$_3$ solution was added to 100 mL supernatant and incubated again in an orbital shaker at 180 rpm and 30 °C for 24 h. The synthesis of AgNPs was confirmed by visual observation of the color change. Finally, the biosynthesized AgNPs were collected by centrifugation at 13,000 for 30 min. The precipitate of synthesized AgNPs was washed with distilled water. Then, the air-dried samples were used for characterization and antibacterial studies.

2.4. Characterization of Biosynthesized AgNPs

The biosynthesis of AgNPs was monitored by a UV–Vis spectrophotometer in the range of 300–800 nm. By using field-emission transmission electron microscopy (FE-TEM), energy-dispersive X-ray (EDX) spectroscopy, element map, and selective area diffraction (SAED) mode, the morphology, element composition, and purity of the synthesized AgNPs were studied. A drop of AgNP solution was kept on a copper mesh, dried at room temperature, and finally transferred to a microscope for analysis. X-ray diffraction (XRD) analysis was conducted with an X-ray diffractometer (D8 Advance, Bruker, Germany) in the range of 30–90° over 2θ value, using CuKα radiation, at 40 kV, 40 mA with 6°/min scanning rate. For XRD analysis, air-dried AgNP samples were used. Fourier transform-infrared (FTIR) spectroscopy showed biomolecules related to the biosynthesis and stability of AgNPs. FTIR analysis was performed by using a PerkinElmer Fourier Transform Infrared Spectrometer (PerkinElmer Inc., Waltham, MA, USA), with a resolution of 4 cm^{-1} and a range of 400–4000 cm^{-1}. A Malvern Zetasizer Nano ZS90 (Otsuka Electronics, Osaka, Japan) was used to determine the particle size of the green synthetic AgNPs by dynamic light scattering (DLS) at 25 °C. Pure water was used as a dispersion medium with a dielectric constant of 78.3, a viscosity of 0.8878 cP, and a refractive index of 1.3328.

2.5. Antimicrobial Activity of Probiotic-Mediated Synthesized AgNPs

The tested pathogenic microbes (nine different *E. coli* O157:H7 strains) were grown overnight in LB broth. Briefly, 100 μL of the bacterial culture sample of the tested pathogen was spread on the LB agar plate. Sterile paper discs containing AgNPs 50 μL and 100 μL (1000 μg/mL) were placed on the LB agar plates. Then, the plates were incubated in an incubator at 37 °C for 24 h. Similarly, the antibacterial activity of some commercial antibiotics such as erythromycin (15 μg/disc), vancomycin (30 μg/disc), and penicillin G (10 μg/disc) was tested against nine *E. coli* O157:H7 strains. The inhibition zones were calculated after 24 h of incubation. The test was performed twice.

2.6. Determination of MIC and MBC

The minimum inhibitory concentration (MIC) of probiotic-mediated synthesized AgNPs was measured using the broth microdilution technique. Nine *E. coli* O157:H7 strains were grown in LB broth overnight at 37 °C, and the turbidity was fixed at around 1×10^6 CFUs/mL. Then, 100 mL of test bacterial (1×10^6 CFUs/mL) suspension were added to a 96-well plate, after which an equal volume of AgNP solution with various concentrations (3.12–200 μg/mL) was added, and finally, the plates were incubated in a 37 °C incubator for 24 h. Every 3 h of the interval, the absorbance (at 600 nm) was measured in a microplate reader. Minimum bactericidal concentration (MBC) was determined by streaking 8 μL of each set on an agar plate and again incubated for 24 h at 37 °C. The culture plates were observed by direct visualization to determine the MBC that blocked bacterial growth [24].

2.7. Morphological Evaluation of Treated Cells Using FE-SEM

The morphological alterations of *E. coli* O157:H7 (ATCC 35150) were examined using FE-SEM. Logarithmic growth phase cells (1×10^7 CFU/mL) were treated with probiotic-mediated synthesized AgNPs at a concentration of $1 \times$ MBC for 6 h. In the control, the bacterial culture was treated without AgNP solution. The treated cells were washed with phosphate-buffered saline (PBS). The cells were fixed for 4 h using 2.5% glutaraldehyde solution and then, washed several times with PBS. Again, with 1% tetroxide cells were fixed and washed with PBS buffer solution. The fixed cells were dehydrated using different concentrations of ethanol (30 to 100%, every 10% interval) at room temperature for 10 min. Then, the samples were dried with a dryer. Finally, the samples were placed on the SEM metal grid and coated with gold. The morphological and structural changes in the cells were observed using FE-SEM (S-4700, Hitachi, Tokyo, Japan).

3. Results and Discussion

3.1. Isolation, Identification, and Characterization of AgNP-Producing Probiotic

Strain MAHUQ-74 was isolated from kimchi, a traditional, fermented food product. Strain MAHUQ-74 was deposited to the Korean Agriculture Culture Collection Center (Deposition number: KACC 22255). The 16S rRNA gene sequence of strain MAHUQ-74 was 1471 bp and the sequence was deposited to NCBI (Accession number: MW488006). Based on the 16S rRNA gene sequence analysis, strain MAHUQ-74 showed the highest sequence similarity with *B. sonorensis* NBRC 101234T (99.93%). To find the phylogenetic location of strain MAHUQ-74, the phylogenetic tree was constructed using the neighbor-joining method in the MEGA6 program package (Figure 1). Cells were Gram-positive, catalase-positive, and oxidase-negative, and were able to grow within 30–50 °C (optimum 37 °C). The strain was positive for hydrolysis of casein, gelatin, starch, and esculin. In the API 20 NE strip, nitrate could be reduced to nitrite; D-glucose was fermented; indole was not produced; urease, arginine dihydrolase, and gelatinase activities were absent; α-glucosidase, α-galactosidase, and β-glucosidase activity were present; and D-glucose, N-acetyl-glucosamine, L-arabinose, D-maltose, D-mannitol, and D-mannose were assimilated. Capric acid, adipic acid, and phenylacetic acid were not assimilated. In the API ZYM strip, alkaline phosphatase, esterase (C4), alkaline phosphatase, leucine arylamidase, naphthol-AS-BI-phosphohydrolase, esterase lipase (C8), acid phosphatase, α-glucosidase, α-galactosidase, and β-glucosidase were present. Lipase (C14), valine arylamidase, cystine arylamidase, trypsin, α-chymotrypsin, β-galactosidase, N-acetyl-β-glycosaminidase, α-mannosidase, and α-fucosidase were absent (Table 1).

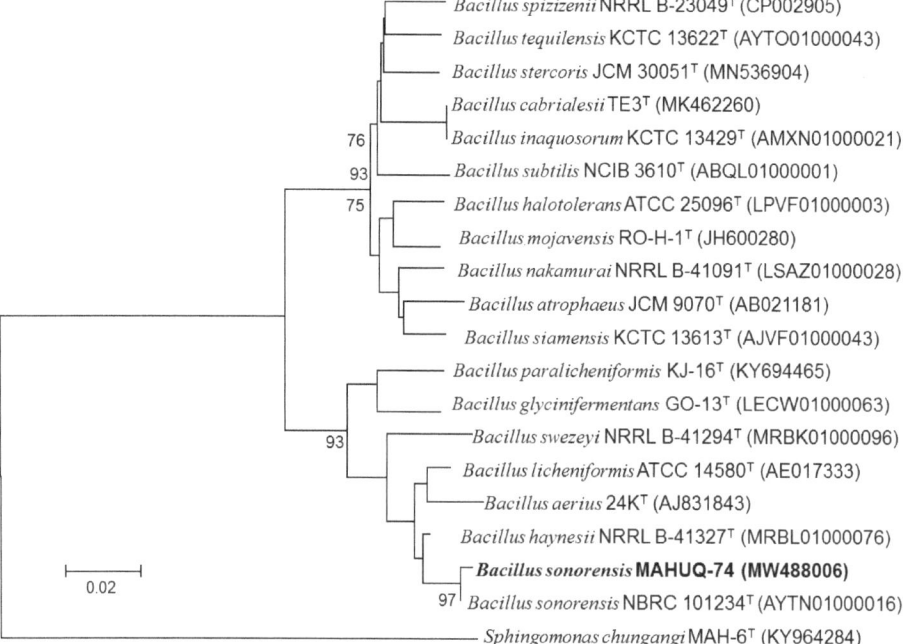

Figure 1. The neighbor-joining (NJ) tree is based on 16S rRNA gene sequence analysis, showing the position of *Bacillus sonorensis* MAHUQ-74 and other *Bacillus* species.

Table 1. Biochemical characterization of *Bacillus sonorensis* MAHUQ-74.

API 20 NE	Result	API ZYM	Result
Reduction of nitrate (API 20 NE)	+	Esterase (C4)	w
		Esterase lipase (C8)	w
		Alkaline phosphatase	+
		Lipase (C-14)	-
Hydrolysis of:		Acid phosphatase	w
Esculine	+	Leucine arylamidase	+
Gelatine	+	Valine arylamidase	-
Urea	-	Cystine arylamidase	-
4-nitrophenyl-BD-galactopyranoside	+	Trypsin	-
Utilization of:		α-chymotrypsin	-
D-glucose	+	β-glucuronidase	-
D-maltose	+	Naphthol-AS-BI-phosphohydrolase	w
D-mannose	+	N-acetyl-β-glucosaminidase	-
D-mannitol	+	α-mannosidase	-
L-arabinose	+	α-glucosidase	+
N-acetyl-glucosamine	+	α-galactosidase	+
Trisodium citrate	+	β-glucosidase	+
Phenylacetic acid	-	β-galactosidase	-
		α-fucosidase	-

+, Positive; -, Negative; w, Weakly positive.

3.2. Biosynthesis of AgNPs Using Strain MAHUQ-74

Biosynthesis of AgNPs using *B. sonorensis* MAHUQ-74 was ensured by monitoring the color of the culture supernatant. The color of the MAHUQ-74 culture supernatant turned to deep brown from watery yellow, which indicated the synthesis of AgNPs. The control sample (without bacterial supernatant) did not show any color change when incubated under the same conditions (Figure 2A,B). Optimum temperature (30 °C) and AgNO$_3$ concentration (final concentration 1 mM) for stable synthesis were determined based on the ultraviolet–visible spectroscopy (UV–Vis) analysis. Two methods are commonly utilized for the biosynthesis of nanoparticles using bacteria—intracellular and extracellular methods. The intracellular method is a more complex and a multi-step process, compared with the extracellular method. In the current study, the extracellular methodology was used to biosynthesize AgNPs, using probiotic bacterial strain MAHUQ-74, which was simple, facile, cost-effective, and ecofriendly.

3.3. Characterization of Biosynthesized AgNPs

Probiotic-mediated synthesized AgNPs showed a peak at 430 nm (Figure 2C), which revealed that AgNPs were fruitfully synthesized. AgNPs are known to exhibit a UV–Visible absorption peak in the range of 400–500 nm [24]. The lower absorption wavelength indicates smaller-sized, spherical NPs [43], which suggested that MAHUQ-74 might synthesize small-sized AgNPs. The results indicated that MAHUQ-74 may be a promising candidate for the biosynthesis of AgNPs. TEM analysis showed that AgNPs were spherical and elliptical, and they dispersed well, without obvious agglomeration. The size of the synthesized AgNPs was 13–50 nm (Figure 2D–F).

The composition and purity of biosynthesized AgNPs were investigated using an EDX spectrometer. The EDX data revealed elemental signals of silver atoms in probiotic-mediated synthesized AgNPs at around 3 keV and indicated the homogenous distribution of AgNPs. Some other peaks of copper were also found in the EDX mode due to the use of a copper grid (Figure 3A). The elemental mapping results showed that the most distributed element in biosynthetic nanoproducts was silver (Figure 3B–D, Table 2).

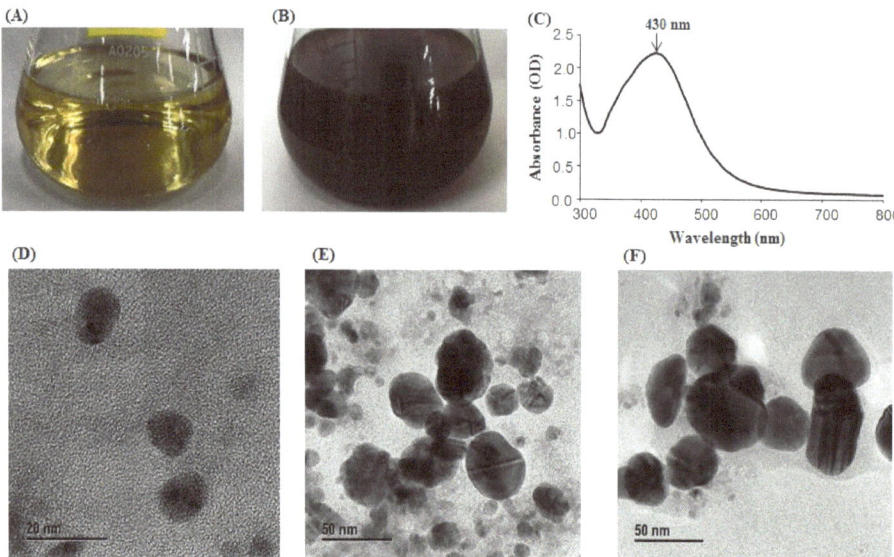

Figure 2. R2A broth as control (**A**), synthesized AgNPs (**B**), UV–Vis spectra (**C**), and FE-TEM images of synthesized AgNPs (**D–F**).

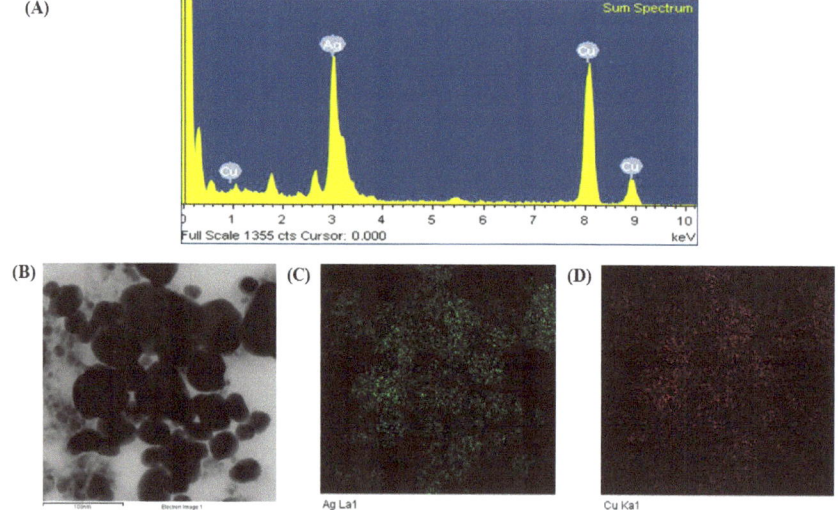

Figure 3. Energy dispersive X-ray (EDX) spectrum of synthesized AgNPs (**A**), FE-TEM image used for elemental mapping (**B**), distribution of silver (**C**), and copper (**D**) in elemental mapping.

Table 2. The number and percentage of chemical elements present in EDX spectrum of probiotic-mediated synthesized AgNPs.

Element	Weight %	Atomic %
Cu K	39.77	52.85
Ag L	60.23	47.15
Totals	100.00	100.00

The XRD data showed diffraction peaks at 2θ values of 38.163°, 44.247°, 64.519°, 77.804°, and 81.474°, which matched with the lattice planes of AgNPs (111, 200, 220, 311, and 222, respectively) (Figure 4A). A recently reported study revealed similar XRD results, in which AgNPs were synthesized using microorganisms [44]. The crystalline structure of probiotic-mediated synthesized AgNPs was confirmed using SAED analysis, which revealed sharp rings corresponding to the reflections of 111, 200, 220, 311, and 222 (Figure 4B). Both the XRD spectrum and SAED pattern confirmed the crystalline structure of AgNPs.

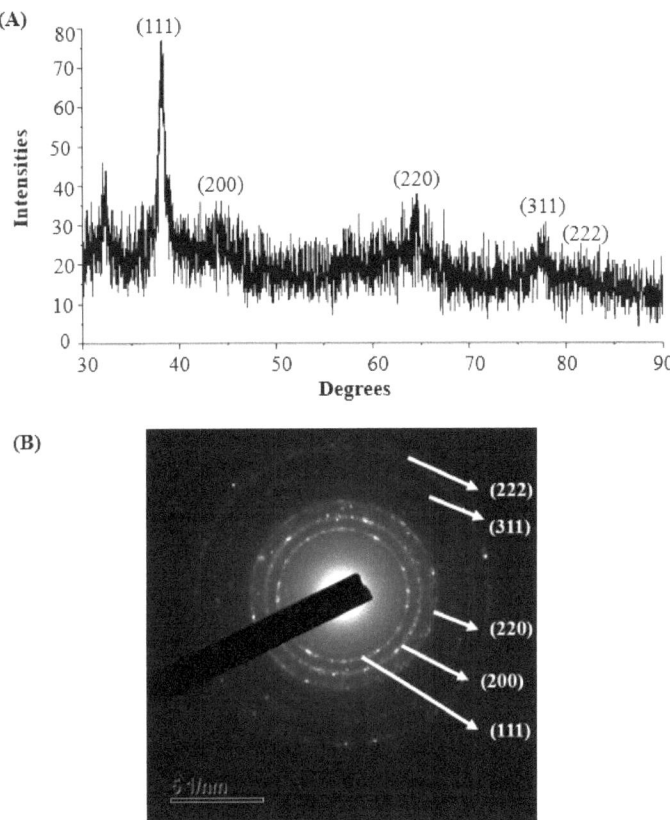

Figure 4. XRD pattern (**A**) and SAED pattern (**B**) of probiotic-mediated synthesized AgNPs.

FTIR analysis revealed that different functional groups surrounded the biosynthesized AgNPs (Figure 5A). The FTIR pattern of AgNPs had several peaks at 3272.36 cm^{-1}, 2917.11 cm^{-1}, 2849.12 cm^{-1}, 2115.72 cm^{-1}, 1737.37 cm^{-1}, 1633.76 cm^{-1}, 1538.72 cm^{-1}, 1454.03 cm^{-1}, 1374.76 cm^{-1}, 1212.81 cm^{-1}, and 1031.17 cm^{-1}. The band found at 3272.36 cm^{-1} was attributed to O–H (alcohol) and/or N–H (amine) stretching. The bands at 2917.11 and 2849.12 cm^{-1} were attributed to C–H (alkane) group. The peaks observed at 2115.72 cm^{-1} were attributed to the N=C=S (isothiocyanate) stretching. The peak found at 1737.37 cm^{-1} represented the C=O (α, β-unsaturated ester) stretching. The peaks at 1633.76, 1538.72, and 1454.03 cm^{-1} were attributed to C=C (olefin) stretching, N-O (nitro compound) stretching, and methyl group C–H (alkane) bending, respectively. The peak observed at 1374.76 cm^{-1} represented the O–H (phenol) bending. The bands at 1212.81 cm^{-1} and 1031.17 cm^{-1} were attributed to C-O (alkyl aryl ether) stretching and S=O (sulfoxide) stretching, respectively. The FTIR spectrum indicated that the functional molecules including biopolymers such as proteins, enzymes, and amino acids secreted by probiotic bacteria could be involved in both the synthesis and stabilization of AgNPs. The particle size of biosynthesized AgNPs was measured with dynamic light scattering (DLS) analysis. Figure 5B shows the particle size distribution of biosynthesized AgNPs based on intensity, volume, and number. The average particle size of probiotic-mediated synthesized AgNPs was 44.6 nm, and the polydispersity value was 0.406.

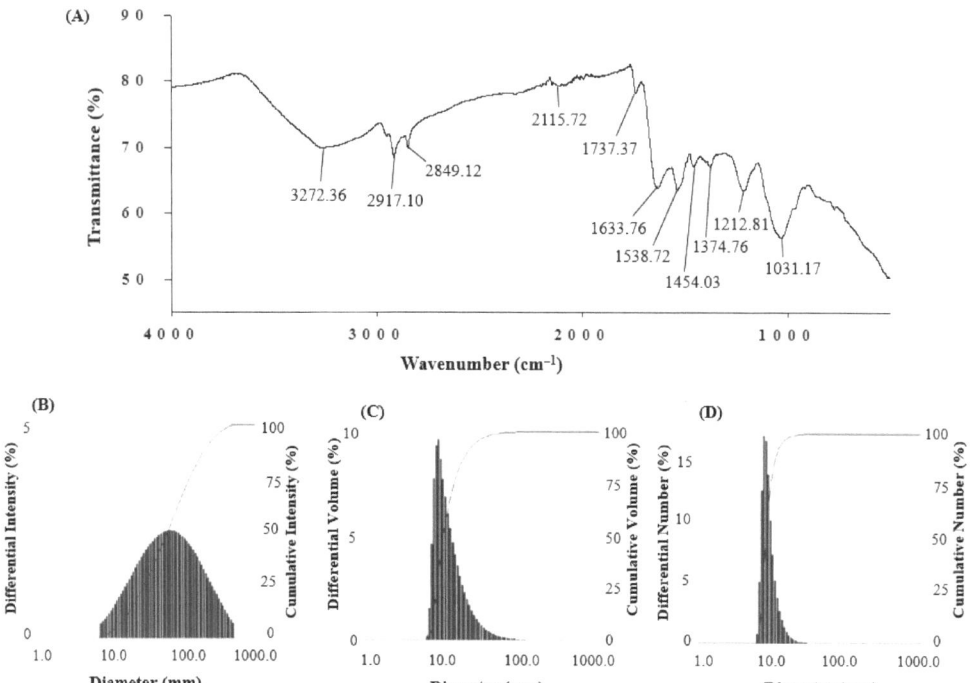

Figure 5. FTIR spectra (**A**) and particle size distribution according to intensity (**B**), volume (**C**), and number (**D**) of synthesized AgNPs.

3.4. Antimicrobial Activity of Probiotic-Mediated Synthesized AgNPs against Different E. coli O157:H7 Strains

The antimicrobial activity of AgNPs is related to their ability to bind microbial DNA, proteins, and enzymes, as well as to alter cell morphology and function [18,45,46]. Small AgNPs have higher antimicrobial activity than large particles, allowing the faster release of Ag+ ions [47]. Many reports have tested the antimicrobial activity of biosynthetic AgNPs against different pathogenic microorganisms, including bacteria, fungi, yeasts, and microbial biofilms. In this study, AgNPs were synthesized using probiotic bacteria *B. sonorensis* MAHUQ-74, and their antibacterial ability was investigated against nine foodborne pathogenic *E. coli* O157:H7 strains (Table 3).

The results showed that the synthesized AgNPs had significant antibacterial activity against all tested foodborne pathogens. The antibacterial efficacy against various pathogenic strains of *E. coli* O157:H7 was determined by calculating the diameter of the zone of inhibition (Figure 6, Table 4). It was also found that AgNPs can inhibit the growth of most *E. coli* O157:H7 strains when treated with a 100 μL AgNP solution. The results of this study revealed that the probiotic bacteria *B. sonorensis* MAHUQ-74-mediated biosynthesized AgNPs are able to control foodborne pathogenic bacteria, especially pathogenic *E. coli* O157:H7 strains.

Table 3. *E. coli* O175:H7 strains used for antibacterial experiments.

No.	Strains	Origin
1	*Escherichia coli* O157:H7 2257	*E. coli* O157:H7 FDIU strain
2	*Escherichia coli* O157:H7 8624	Wild-type *E. coli* O157:H7, a clinical isolate
3	*Escherichia coli* O157: NM 3204-92	International vaccine institute clinical specimen
4	*Escherichia coli* O157:H7 ATCC 43895	About the outbreak of raw hamburger and hemorrhagic colitis
5	*Escherichia coli* O157:H7 ATCC 35150	Feces, human
6	*Escherichia coli* O157:H7 ATCC 43889	Isolated from the stool of a patient with hemolytic uremic syndrome in North Carolina
7	*Escherichia coli* O157:H7 ATCC 43890	Human feces, California
8	*Escherichia coli* O157:H7 ATCC 43894	Human feces from an outbreak of hemorrhagic colitis, Michigan
9	*Escherichia coli* ATCC 25922	Clinical isolate

FDIU, Federal Disease Investigation Unit (Washington State University)

Table 4. Antibacterial efficacy of probiotic-mediated synthesized AgNPs against *E. coli* O157:H7 strains.

No.	Pathogenic Strains	Zone of Inhibition(mm)	
		50 µL	100 µL
1	*E. coli* O157:H7 2257	10.1	24
2	*E. coli* O157:H7 8624	10.1	17
3	*E. coli* O157: NM 3204-92	10.1	16
4	*E. coli* O157:H7 ATCC 43895	-	16
5	*E. coli* O157:H7 ATCC 35150	10	18
6	*E. coli* O157:H7 ATCC 43889	10.1	16
7	*E. coli* O157:H7 ATCC 43890	10	16
8	*E. coli* O157:H7 ATCC 43894	10	19
9	*E. coli* ATCC 25922	10.1	22

-, no inhibition zone

Three antibiotics (erythromycin, vancomycin, and penicillin G) were also tested against the foodborne pathogens *E. coli* O157:H7. Six of the nine foodborne pathogens were found to be resistant to all three antibiotics included in this study (Figure 7, Table 5).

3.5. Determination of MIC and MBC

The MIC and MBC of probiotic-mediated synthesized AgNPs against a total of nine *E. coli* O157:H7 were determined by a standard broth dilution assay. Bacterial growth curves revealed that the MICs of biosynthesized AgNPs were 3.12 µg/mL for eight *E. coli* O157:H7 strains (Figure 8A–H) and 12.5 µg/mL for strain *E. coli* ATCC 25,922 (Figure 8I). This result indicated that the probiotic-mediated synthesized AgNPs extremely suppressed the growth of pathogenic strains of *E. coli* O157:H7.

The MBC of synthesized AgNPs was 25 µg/mL for *E. coli* O157:H7 ATCC 43895, *E. coli* O157:H7 ATCC 35150, *E. coli* O157:H7 ATCC 43889, *E. coli* O157:H7 ATCC 43,890, and *E. coli* ATCC 25,922, and 50 µg/mL for *E. coli* O157:H7 2257, *E. coli* O157: NM 3204-92, *E. coli* O157:H7 8624 and *E. coli* O157:H7 ATCC 43,894 (Figure 9A–I).

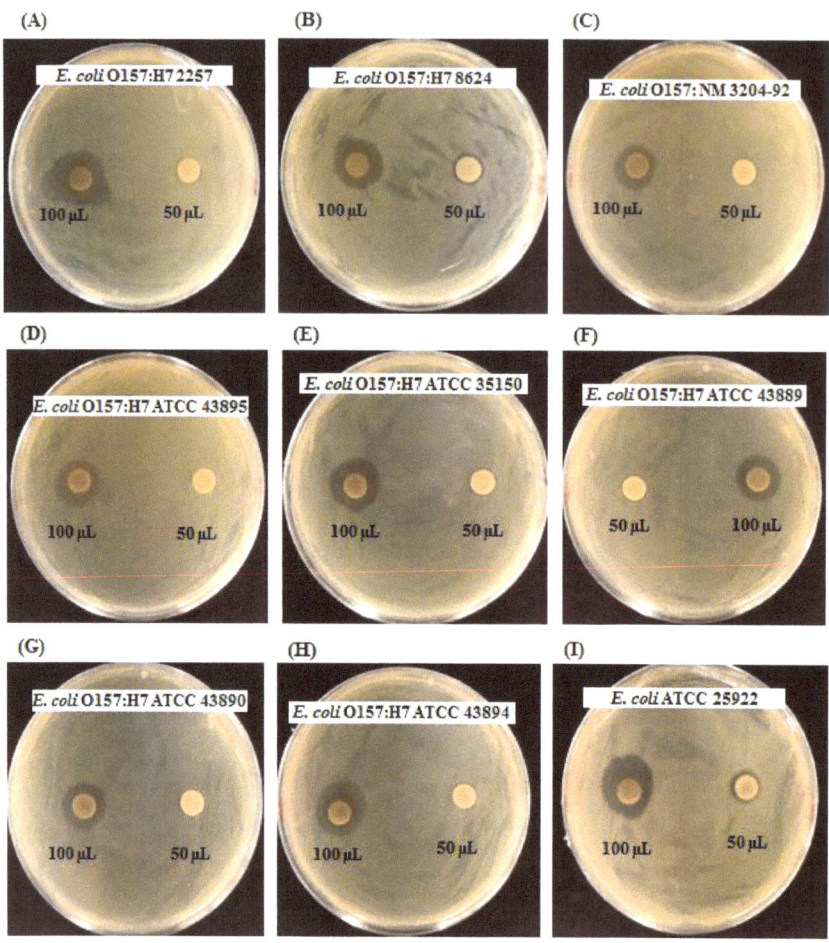

Figure 6. Antibacterial activity of synthesized AgNPs at 1000 µg/mL concentrations in the water against *E. coli* O157:H7 strains (**A–I**).

Table 5. Antibacterial efficacy of some commercial antibiotics against *E. coli* O157:H7 strains—no inhibition zone.

No.	Pathogenic Species	Antibiotic	Zone of Inhibition (mm)
1	*E. coli* O157:H7 2257	Erythromycin	-
		Vancomycin	-
		Penicillin G	-
2	*E. coli* O157:H7 8624	Erythromycin	-
		Vancomycin	-
		Penicillin G	-
3	*E. coli* O157: NM 3204-92	Erythromycin	-
		Vancomycin	-
		Penicillin G	-

Table 5. Cont.

No.	Pathogenic Species	Antibiotic	Zone of Inhibition (mm)
4	E. coli O157:H7 ATCC 43895	Erythromycin	-
		Vancomycin	-
		Penicillin G	-
5	E. coli O157:H7 ATCC 35150	Erythromycin	-
		Vancomycin	-
		Penicillin G	-
6	E. coli O157:H7 ATCC 43889	Erythromycin	15
		Vancomycin	10
		Penicillin G	9
7	E. coli O157:H7 ATCC 43890	Erythromycin	17
		Vancomycin	10
		Penicillin G	9
8	E. coli O157:H7 ATCC 43894	Erythromycin	10
		Vancomycin	8
		Penicillin G	-
9	E. coli ATCC 25922	Erythromycin	-
		Vancomycin	-
		Penicillin G	-

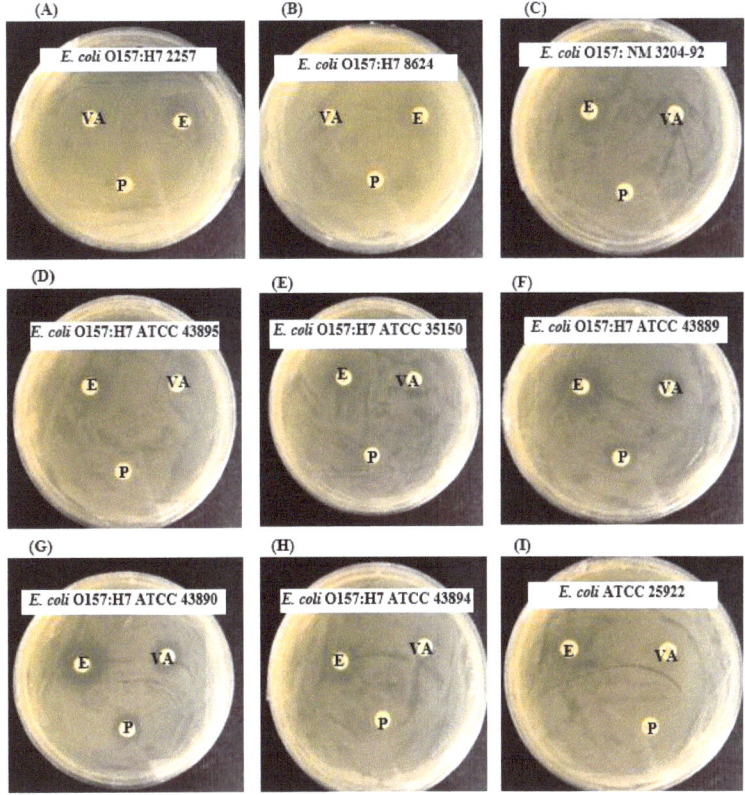

Figure 7. Antibacterial activity of some commercial antibiotics against nine *E. coli* O157:H7 strains (**A–I**). Abbreviations: P (penicillin G, 10 µg/disc), E (erythromycin, 15 µg/disc), and VA (vancomycin, 30 µg/disc).

3.6. Study of Morphogenesis of E. coli O157:H7-Treated Cells Using FE-SEM

The structural and morphological alterations of *E. coli* O157:H7 ATCC 35,150 cells treated with probiotic-mediated synthesized AgNPs were directly observed using FE-SEM (Figure 10). Based on the FE-SEM analysis, it was found that the untreated *E. coli* O157:H7 cells were intact, normal, and rod-shaped, and the structural integrity of the bacterial cells was good without any damage (Figure 10A,B). However, after treatment with $1 \times$ MBC of synthesized AgNPs, the shape of *E. coli* O157:H7 cells became abnormal, irregular, and wrinkled, with the cell membrane entirely collapsed and damaged (Figure 10C,D).

Although the bactericidal mechanism of AgNPs is not yet fully understood, several hypotheses have been proposed in the literature. The attachment of AgNPs to the bacterial cell membrane results in the formation of a "pit" on the bacterial cell wall, thereby allowing the nanoparticles to enter the bacterial cells in the periplasm [48,49]. As a result, subsequent changes in the DNA of the bacterial cells treated with AgNPs lead to the loss of DNA replication ability, which leads to the inactivation of the expression of proteins and enzymes necessary for ATP production. In the present study, the structural and morphological alterations, damage to bacterial cell walls, and cell membrane indicated that the biosynthesized AgNPs might interfere with the metabolic process and normal cell functions, leading to the death of bacterial cells.

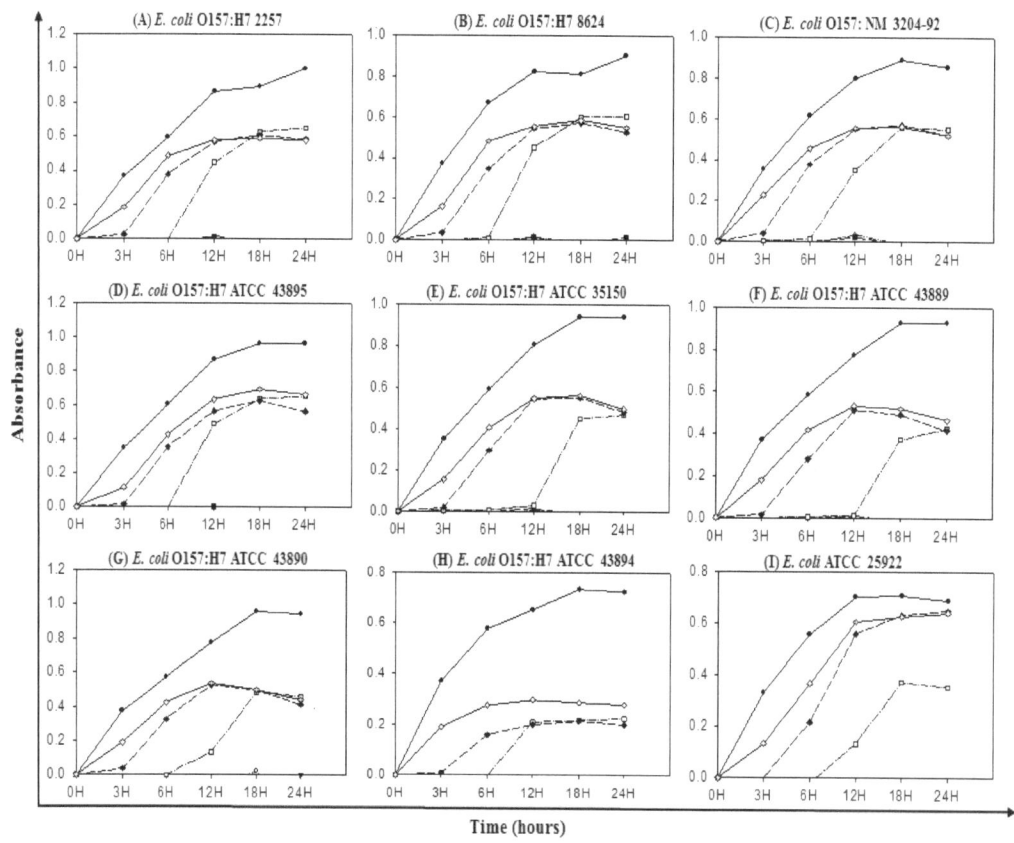

Figure 8. Growth curves of nine *E. coli* O157:H7 strains (**A–I**) cultured in LB broth with various concentrations of synthesized AgNPs to determine MIC. Control (•); 200 µg/mL (○); 100 µg/mL (▼) 50 µg/mL (△); 25 µg/mL (§); 12.5 µg/mL (□); 6.25 µg/mL (w); 3.12 µg/mL (◊).

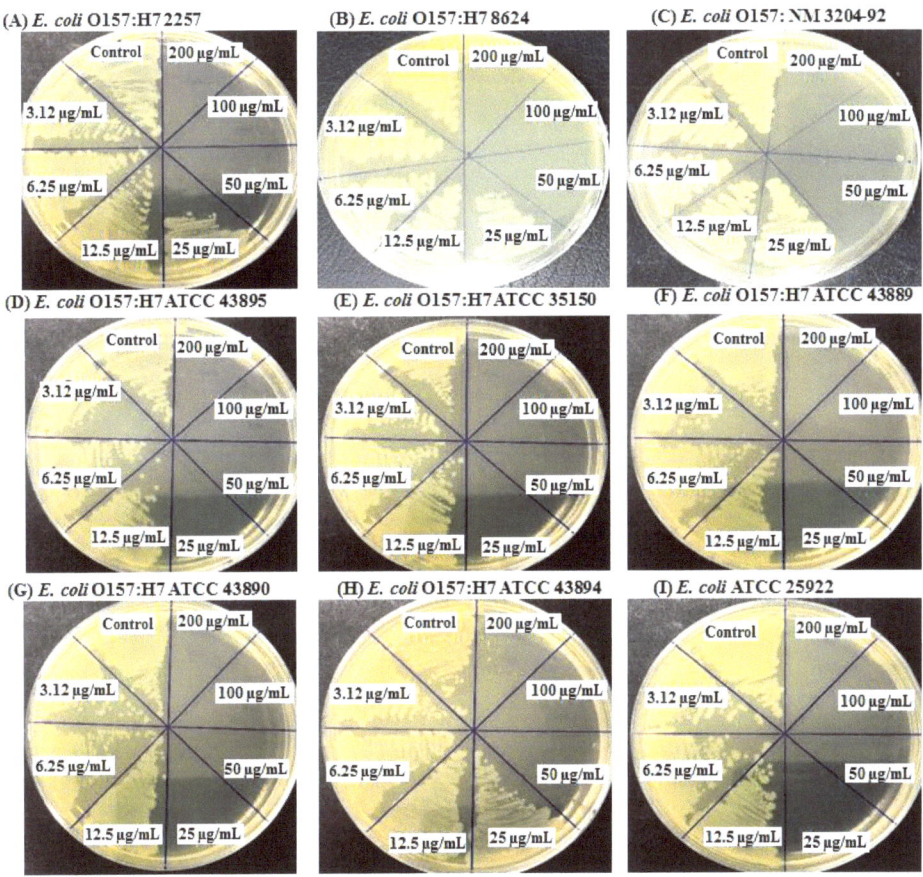

Figure 9. MBC of probiotic-mediated synthesized AgNPs against *E. coli* O157:H7 strains (**A–I**).

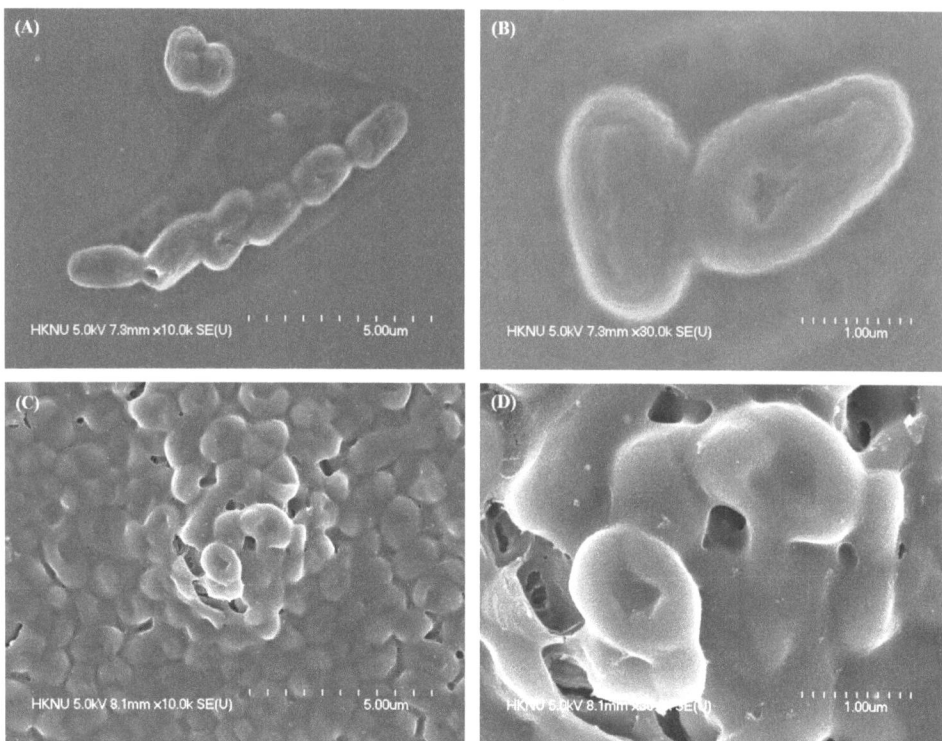

Figure 10. FE-SEM images of normal *E. coli* O157:H7 (ATCC 35150) (**A,B**) and 1 × MBC probiotic-mediated synthesized AgNPs treated *E. coli* O157:H7 (ATCC 35150) (**C,D**).

4. Conclusions

In conclusion, the biological compounds produced by bacteria have high utility value and may play important roles in the production of AgNPs. Biomolecules, including biopolymers such as proteins, enzymes, and amino acids, that are secreted by bacteria are involved in the synthesis process but also may improve the function of synthesized AgNPs. In the present study, we isolated and identified the strain *B. sonorensis* MAHUQ-74 and used their culture supernatant for the biosynthesis of AgNPs. The AgNPs were analyzed using UV–Vis, FE-TEM, XRD, EDX, FTIR, and DLS. FE-TEM images indicated that the AgNPs were mostly circular, and the size ranged from 13 to 50 nm. The FTIR data indicated that various biomolecules may participate in the synthesis and stabilization of AgNPs. The extracellular method was used for the biosynthesis of AgNPs. Moreover, the biosynthesized AgNPs showed potent antibacterial efficacy against nine pathogenic *E. coli* O157:H7 strains. The MICs of the AgNPs synthesized by strain MAHUQ-74 were 3.12 to 12.5 µg/mL for the tested nine *E. coli* O157:H7 strains. The MBCs of the AgNPs synthesized by strain MAHUQ-74 were 25 to 50 µg/mL for the tested *E. coli* O157:H7 strains. FE-SEM analysis showed that the biosynthesized AgNPs can cause changes in the morphology and structure of the foodborne pathogenic *E. coli* O157:H7 strain and destroy the integrity of the membrane, leading to cell death. This is the first study on the use of probiotic *B. sonorensis* MAHUQ-74 isolated from kimchi for the rapid and facile synthesis of bioactive AgNPs. AgNPs manufactured using *B. sonorensis* MAHUQ-74 may be potential antimicrobial agents for controlling antibiotic-resistant microorganisms, especially pathogenic strains of *E. coli* O157:H7.

Author Contributions: Conceptualization, M.A.H.; experiments, X.W., M.A.H. and S.A.; writing—original draft preparation, M.A.H. and X.W.; writing—review and editing, M.A.H., S.A. and S.-Y.L. All authors have read and agreed to the published version of the manuscript.

Funding: This study was performed with the support of the National Research Foundation (NRF) of Korea grant (Project no. NRF-2018R1C1B5041386, Grant Recipient: Md. Amdadul Huq) funded by the Korean government, Korea.

Institutional Review Board Statement: Not applicable.

Informed Consent Statement: Not applicable.

Data Availability Statement: Not applicable.

Conflicts of Interest: The authors declare no conflict of interest.

References

1. Kedi, P.B.E.; Meva, F.E.; Kotsedi, L. Eco-friendly synthesis, characterization, in vitro and in vivo anti-inflammatory activity of silver nanoparticle-mediated *Selaginella myosurus* aqueous extract. *Int. J. Nanomed.* **2018**, *13*, 8537–8548. [CrossRef] [PubMed]
2. El-Naggar, N.E.; Hussein, M.H.; El-Sawah, A.A. Bio-fabrication of silver nanoparticles by phycocyanin, characterization, in vitro anticancer activity against breast cancer cell line and in vivo cytotoxicity. *Sci. Rep.* **2017**, *7*, 10844. [CrossRef] [PubMed]
3. Fouda, A.; Abdel-Maksoud, G.; Abdel-Rahman, M.A.; Salem, S.S.; Hassan, S.E.D.; El-Sadany, M.A.H. Eco-friendly approach utilizing green synthesized nanoparticles for paper conservation against microbes involved in biodeterioration of archaeological manuscript. *Int. Biodeterior. Biodegrad.* **2019**, *142*, 160–169. [CrossRef]
4. Fouda, A.; Abdel-Maksoud, G.; Abdel-Rahman, M.A.; Eid, A.M.; Barghoth, M.G.; El-Sadany, M.A.H. Monitoring the effect of biosynthesized nanoparticles against biodeterioration of cellulose-based materials by *Aspergillus niger*. *Cellulose* **2019**, *26*, 6583–6597. [CrossRef]
5. Cheon, J.Y.; Kim, S.J.; Rhee, Y.H.; Kwon, O.H.; Park, W.H. Shape-dependent antimicrobial activities of silver nanoparticles. *Int. J. Nanomed.* **2019**, *14*, 2773–2780. [CrossRef]
6. Burduşel, A.C.; Gherasim, O.; Grumezescu, A.M.; Mogoanta, L.; Ficai, A.; Andronescu, E. Biomedical applications of silver nanoparticles: An up-to-date overview. *Nanomaterials* **2018**, *8*, 681. [CrossRef]
7. Siddiqi, K.S.; Husen, A.; Rao, R.A.K. A review on biosynthesis of silver nanoparticles and their biocidal properties. *J. Nanobiotechnology* **2018**, *16*, 14. [CrossRef]
8. Kandarp, M.; Mihir, S. Synthesis of silver nanoparticles by using sodium borohydride as a reducing agent. *Int. J. Eng. Res. Technol.* **2013**, *2*, 1–5.
9. Al-Jumaili, A.; Mulvey, P.; Kumar, A. Eco-friendly nanocomposites derived from geranium oil and zinc oxide in one step approach. *Sci. Rep.* **2019**, *9*, 5973. [CrossRef]
10. Wang, L.S.; Wang, C.Y.; Yang, C.H.; Hsieh, C.L.; Chen, S.Y.; Shen, C.Y.; Wang, J.J.; Huang, K.S. Synthesis and anti-fungal effect of silver nanoparticles-chitosan composite particles. *Int. J. Nanomed.* **2015**, *10*, 2685–2696.
11. Raza, S.; Ansari, A.; Siddiqui, N.N.; Ibrahim, F.; Abro, M.I.; Aman, A. Biosynthesis of silver nanoparticles for the fabrication of non-cytotoxic and antibacterial metallic polymer based nanocomposite system. *Sci. Rep.* **2021**, *11*, 10500. [CrossRef] [PubMed]
12. Quintero-Quiroz, C.; Botero, L.E.; Zárate-Triviño, D.; Acevedo-Yepes, N.; Escobar, J.S.; Pérez, V.Z.; Cruz-Riano, L.J. Synthesis and characterization of a silver nanoparticle-containing polymer composite with antimicrobial abilities for application in prosthetic and orthotic devices. *Biomater. Res.* **2020**, *24*, 13. [CrossRef] [PubMed]
13. Huq, M.A. Biogenic Silver Nanoparticles Synthesized by *Lysinibacillus xylanilyticus* MAHUQ-40 to control antibiotic-resistant human pathogens *Vibrio parahaemolyticus* and Salmonella Typhimurium. *Front. Bioeng. Biotechnol.* **2020**, *8*, 1407. [CrossRef] [PubMed]
14. Poulose, S.; Panda, T.; Nair, P.P.; Théodore, T. Biosynthesis of silver nanoparticles. *J. Nanosci. Nanotechnol.* **2014**, *14*, 2038–2049. [CrossRef] [PubMed]
15. Akter, S.; Lee, S.Y.; Siddiqi, M.Z.; Balusamy, S.R.; Ashrafudoulla, M.; Rupa, E.J.; Huq, M.A. Ecofriendly synthesis of silver nanoparticles by *Terrabacter humi* sp. nov. and their antibacterial application against antibiotic-resistant pathogens. *Int. J. Mol. Sci.* **2020**, *21*, 9746. [CrossRef] [PubMed]
16. Abou, E.N.; Kholoud, M.M.; Eftaiha, A.A.; Al-Warthan, A.; Ammar, R.A. Synthesis and applications of silver nanoparticles. *Arab. J. Chem.* **2010**, *3*, 135–140. [CrossRef]
17. David, D. The relationship between biomaterials and nanotechnology. *Biomaterials* **2008**, *29*, 1737–1738.
18. Huq, M.A.; Ashrafudoulla, M.; Rahman, M.M.; Balusamy, S.R.; Akter, S. Green synthesis and potential antibacterial applications of bioactive silver nanoparticles: A review. *Polymers* **2022**, *14*, 742. [CrossRef]
19. Iravani, S.; Korbekandi, H.; Mirmohammadi, S.V.; Zolfaghari, B. Synthesis of silver nanoparticles: Chemical, physical and biological methods. *Res. Pharm. Sci.* **2014**, *9*, 385–406.
20. Kruis, F.E.; Fissan, H.; Rellinghaus, B. Sintering and evaporation characteristics of gas-phase synthesis of size-selected PbS nanoparticles. *Mater. Sci. Eng. B* **2000**, *69*, 329–334. [CrossRef]

21. Tran, Q.H.; Le-Adv, A.T. Silver nanoparticles: Synthesis, properties, toxicology, applications, and perspectives. *Nat. Sci. Nanosci. Nanotechnol.* **2013**, *4*, 033001. [CrossRef]
22. Nasrollahzadeh, M. Green synthesis and catalytic properties of palladium nanoparticles for the direct reductive amination of aldehydes and hydrogenation of unsaturated ketones. *New J. Chem.* **2014**, *38*, 5544–5550. [CrossRef]
23. Firdhouse, M.J.; Lalitha, P.J. Biosynthesis of silver nanoparticles and its applications. *J. Nanotechnol.* **2015**, *1*, 829526. [CrossRef]
24. Huq, M.A. Green synthesis of silver nanoparticles using *Pseudoduganella eburnea* MAHUQ-39 and their antimicrobial mechanisms investigation against drug resistant human pathogens. *Int. J. Mol. Sci.* **2020**, *21*, 1510. [CrossRef] [PubMed]
25. Yurtluk, T.; Akçay, F.A.; Avcı, A. Biosynthesis of silver nanoparticles using novel *Bacillus* sp. SBT8. *Prep. Biochem. Biotechnol.* **2018**, *48*, 151–159. [CrossRef]
26. El-Dein, M.M.N.; Baka, Z.A.; Abou-Dobara, M.I.; El-Sayed, A.K.; El-Zahed, M.M. Extracellular biosynthesis, optimization, characterization, and antimicrobial potential of *Escherichia coli* D8 silver nanoparticles. *Biotech. Food Sci.* **2021**, *10*, 648–656.
27. Markus, J.; Mathiyalagan, R.; Kim, Y.-J.; Abbai, R.; Singh, P.; Ahn, S.; Perez, Z.E.J.; Hurh, J.; Yang, D.C. Intracellular synthesis of gold nanoparticles with antioxidant activity by probiotic *Lactobacillus kimchicus* DCY51T isolated from Korean kimchi. *Enzym. Microb. Technol.* **2016**, *95*, 85–93. [CrossRef]
28. Jo, J.H.; Singh, P.; Kim, Y.J.; Wang, C.; Mathiyalagan, R.; Jin, C.-G.; Yang, D.C. *Pseudomonas deceptionensis* DC5-mediated synthesis of extracellular silver nanoparticles. *Artif. Cells Nanomed. Biotechnol.* **2015**, *44*, 1576–1581. [CrossRef]
29. Singh, P.; Pandit, S.; Mokkapati, S.; Garnæs, J.; Mijakovic, I. A Sustainable approach for the green synthesis of silver nanoparticles from *Solibacillus isronensis* sp. and their application in biofilm inhibition. *Molecules* **2020**, *25*, 2783. [CrossRef]
30. Singh, P.; Pandit, S.; Jers, C.; Abhayraj, S.; Garnæs, J.; Mijakovic, I. Silver nanoparticles produced from *Cedecea sp.* exhibit antibiofilm activity and remarkable stability. *Sci. Rep.* **2021**, *11*, 12619. [CrossRef]
31. Du, J.; Sing, H.; Yi, T.H. Biosynthesis of silver nanoparticles by *Novosphingobium* sp. THG-C3 and their antimicrobial potential. *Artif. Cells. Nanomed. Biotechnol.* **2017**, *45*, 211–217. [CrossRef] [PubMed]
32. Singh, P.; Singh, H.; Kim, Y.J. Extracellular synthesis of silver and gold nanoparticles by *Sporosarcina koreensis* DC4 and their biological applications. *Enzym. Microb. Technol.* **2016**, *86*, 75–83. [CrossRef] [PubMed]
33. Costa, O.; Raaijmakers, J.M.; Kuramae, E.E. Microbial extracellular polymeric substances: Ecological function and impact on soil aggregation. *Front. Microbiol.* **2018**, *9*, 1636. [CrossRef] [PubMed]
34. Turnbull, P.C.B. *Bacillus: Barron's Medical Microbiology*, 4th ed.; University of Texas Medical Branch: Galveston, TX, USA, 1996; ISBN 0-9631172-1-1.
35. Mandell, G.L.; Bennett, J.E.; Dolin, R. *Mandell, Douglas, and Bennett's. Principles and Practice of Infectious Diseases*, 7th ed.; Livingstone, C., Ed.; Elsevier: Amsterdam, The Netherlands, 2010; p. 4028. ISBN 978-0-443-06839-3.
36. Rahal, E.A.; Kazzi, N.; Nassar, F.J.; Matar, G.M. *Escherichia coli* O157:H7-Clinical aspects and novel treatment approaches. *Front. Cell Infect. Microbiol.* **2012**, *2*, 138. [CrossRef] [PubMed]
37. Quan, X.T.; Siddiqi, M.Z.; Liu, Q.Z.; Lee, S.M.; Im, W.T. *Devosia ginsengisoli* sp. nov., isolated from ginseng cultivation soil. *Int. J. Syst. Evol. Microbiol.* **2020**, *70*, 1489–1495. [CrossRef]
38. Weisburg, W.G.; Barns, S.M.; Pelletier, D.A.; Lane, D.J. 16S ribosomal DNA amplification for phylogenetic study. *J. Bacteriol.* **1991**, *173*, 697–703. [CrossRef]
39. Yoon, S.H.; Ha, S.M.; Kwon, S.; Lim, J.; Kim, Y. Introducing EzBioCloud: A taxonomically united database of 16S rRNA gene sequences and whole-genome assemblies. *Int. J. Syst. Evol. Microbiol.* **2017**, *67*, 1613–1617. [CrossRef]
40. Tamura, K.; Stecher, G.; Peterson, D.; Filipski, A.; Kumar, S. MEGA6: Molecular evolutionary genetics analysis version 6.0. *Mol. Biol. Evol.* **2013**, *30*, 2725–2729. [CrossRef]
41. Saitou, N.; Nei, M. The neighbor-joining method: A new method for reconstructing phylogenetic trees. *Mol. Biol. Evol.* **1987**, *4*, 406–425.
42. Siddiqi, M.Z.; Im, W.T. *Hankyongella ginsenosidimutans* gen. nov., sp. nov., isolated from mineral water with ginsenoside coverting activity. *Antonie Van Leeuwenhoek* **2020**, *113*, 719–727. [CrossRef]
43. Akter, S.; Huq, M.A. Biologically rapid synthesis of silver nanoparticles by *Sphingobium* sp. MAH-11 and their antibacterial activity and mechanisms investigation against drug-resistant pathogenic microbes. *Artif. Cells Nanomed. Biotechnol.* **2020**, *48*, 672–682. [CrossRef] [PubMed]
44. Govindaraju, K.; Kiruthiga, V.; Kumar, G.; Singaravelu, G. Extracellular synthesis of silver nanoparticles by a marine alga, *Sargassum Wightii grevilli* and their antibacterial effects. *J. Nanosci. Nanotechnol.* **2009**, *9*, 5497–5501. [CrossRef] [PubMed]
45. Cavaliere, E.; De Cesari, S.; Landini, G.; Riccobono, E.; Pallecchi, L.; Rossolini, G.M.; Gavioli, L. Highly bactericidal Ag nanoparticle films obtained by cluster beam deposition. *Nanomed. Nanotechnol. Biol. Med.* **2015**, *11*, 1417–1423. [CrossRef] [PubMed]
46. Kumar-Krishnan, S.; Prokhorov, E.; Hernández-Iturriaga, M.; Mota-Morales, J.D.; Vázquez-Lepe, M.; Kovalenko, Y.; Luna-Bárcenas, G. Chitosan/silver nanocomposites: Synergistic antibacterial action of silver nanoparticles and silver ions. *Eur. Polym. J.* **2015**, *67*, 242–251. [CrossRef]
47. Sotiriou, G.A.; Pratsinis, S.E. Antibacterial activity of nanosilver ions and particles. *Environ. Sci. Technol.* **2010**, *44*, 5649–5654. [CrossRef]

48. Sondi, I.; Salopek-Sondi, B. Silver nanoparticles as antimicrobial agent: A case study on *E. coli* as a model for Gram-negative bacteria. *J. Colloid Interface Sci.* **2004**, *275*, 177–182. [CrossRef]
49. Yamanaka, M.; Hara, K.; Kudo, J. Bactericidal actions of a silver ion solution on *Escherichia coli*, studied by energy-filtering transmission electron microscopy and proteomic analysis. *Appl. Environ. Microbiol.* **2005**, *71*, 7589–7593. [CrossRef]

Article

Biogenic Synthesis and Characterization of Chitosan-CuO Nanocomposite and Evaluation of Antibacterial Activity against Gram-Positive and -Negative Bacteria

Peace Saviour Umoren [1], Doga Kavaz [1,*], Alexis Nzila [2,3], Saravanan Sankaran Sankaran [2] and Saviour A. Umoren [4,*]

1. Department of Bioengineering, Cyprus International University, via Mersin 10, Nicosia 98258, Turkey; princespeace@yahoo.com
2. Department of Bioengineering, King Fahd University of Petroleum and Minerals (KFUPM), Dhahran 31261, Saudi Arabia; alexisnzila@kfupm.edu.sa (A.N.); saravanan@kfupm.edu.sa (S.S.S.)
3. Interdisciplinary Research Center for Membranes and Water Security, King Fahd University of Petroleum and Minerals (KFUPM), Dhahran 31261, Saudi Arabia
4. Interdisciplinary Research Center for Advanced Materials, King Fahd University of Petroleum and Minerals (KFUPM), Dhahran 31261, Saudi Arabia
* Correspondence: dkavaz@ciu.edu.tr (D.K.); umoren@kfupm.edu.sa (S.A.U.)

Citation: Umoren, P.S.; Kavaz, D.; Nzila, A.; Sankaran, S.S.; Umoren, S.A. Biogenic Synthesis and Characterization of Chitosan-CuO Nanocomposite and Evaluation of Antibacterial Activity against Gram-Positive and -Negative Bacteria. *Polymers* 2022, 14, 1832. https://doi.org/10.3390/polym14091832

Academic Editors: Md. Amdadul Huq and Shahina Akter

Received: 27 March 2022
Accepted: 27 April 2022
Published: 29 April 2022

Publisher's Note: MDPI stays neutral with regard to jurisdictional claims in published maps and institutional affiliations.

Copyright: © 2022 by the authors. Licensee MDPI, Basel, Switzerland. This article is an open access article distributed under the terms and conditions of the Creative Commons Attribution (CC BY) license (https:// creativecommons.org/licenses/by/ 4.0/).

Abstract: Chitosan-copper oxide (CHT-CuO) nanocomposite was synthesized using olive leaf extract (OLE) as reducing agent and $CuSO_4 \cdot 5H_2O$ as precursor. CHT-CuO nanocomposite was prepared using an in situ method in which OLE was added to a solution of chitosan and $CuSO_4 \cdot 5H_2O$ mixture in the ratio of 1:5 (v/v) and heated at a temperature of 90 °C. The obtained CHT-CuO nanocomposite was characterized using field emission scanning electron microscopy (FE-SEM), X-ray diffraction (XRD), ultraviolet-visible (UV-Vis) spectrophotometry, energy-dispersive X-ray spectroscopy (EDAX), Fourier transform infrared spectroscopy (FTIR), and high-resolution transmission electron microscopy (TEM). TEM results indicated that CHT-CuO nanocomposite are spherical in shape with size ranging from 3.5 to 6.0 nm. Antibacterial activity of the synthesized nanocomposites was evaluated against Gram-positive (*Bacillus cereus*, *Staphyloccous haemolytica* and *Micrococcus Luteus*) and Gram-negative (*Escherichia coli*, *Pseudomonas citronellolis*, *Pseudomonas aeruginosa*, *kliebisella* sp., *Bradyrhizobium japonicum* and *Ralstonia pickettii*) species by cup platting or disc diffusion method. Overall, against all tested bacterial strains, the diameters of the inhibition zone of the three nanocomposites fell between 6 and 24 mm, and the order of the antimicrobial activity was as follows: CuO-1.0 > CuO-0.5 > CuO-2.0. The reference antibiotic amoxicillin and ciprofloxacin showed greater activity based on the diameter of zones of inhibition (between 15–32 mm) except for *S. heamolytica* and *P. citronellolis* bacteria strains. The nanocomposites MIC/MBC were between 0.1 and 0.01% against all tested bacteria, except *S. heamolityca* (>0.1%). Based on MIC/MBC values, CuO-0.5 and CuO-1.0 were more active than CuO-2.0, in line with the observations from the disc diffusion experiment. The findings indicate that these nanocomposites are efficacious against bacteria; however, Gram-positive bacteria were less susceptible. The synthesized CHT-CuO nanocomposite shows promising antimicrobial activities and could be utilized as an antibacterial agent in packaging and medical applications.

Keywords: chitosan; copper oxide; olive leaf extract; nanocomposite; antibacterial activity

1. Introduction

Micro-organisms are important in a wide range of life-sustaining functions. On the other hand, some are pathogenic, meaning they can cause illness and even death. Despite the availability of numerous antibiotics for the treatment of bacterial ailments, new infectious illnesses and bacterial resistance continue to emerge [1]. Antibiotic resistance is rising at an unprecedented rate, necessitating the development of novel antimicrobials [2].

Furthermore, recent interest in identifying safe and natural antibiotic substitutes [3,4] has been fuelled by environmental concerns about some antibiotics [1]. Antibiotic misuse has resulted in bacteria developing a biofilm-forming defensive mechanism [3,4], making medications less effective. As a result, several researchers are concentrating on developing further effective antimicrobial drugs to manage multidrug-resistant diseases in order to reduce bacterial growth [2,4].

Nanotechnology has given rise to a slew of novel antibacterial solutions. The nanoparticles' modest size makes them ideal for carrying out antibacterial biological operations [5]. Metal nanoparticles, such as silver, zinc, copper, and iron, have shown great promise as bactericidal and fungicidal components, proving their potential as effective antibiotic therapies in wound care and other medical conditions [6]. Nanomaterials and nanocomposites are a promising platform for alternative strategies to manage bacterial ailments, as they provide long-term antibacterial activity with minimal toxicity, as opposed to tiny-molecule antimicrobial drugs, which have short-term activity and are hazardous to the environment. The antimicrobial nanoparticle physically damages the organism's cell membranes, preventing the formation of drug-resistant bacteria [7]. Cu and CuO nanoparticles (CuO NPs) have long been used in biological systems to prevent harmful bacteria and algae from growing [8–10].

For the synthesis of nanoparticles, diverse approaches have been used, including chemical, physical, and biological synthesis [11]. Nanoparticles are produced by reducing metallic compounds with any biological or microbe, plant, or their extracts [12]. Because it is eco-friendly, economical, simple, and suitable for large-scale manufacturing, biological preparation of metal nanoparticles utilizing plant extracts is recommended [13,14]. Plant extracts are utilized to make metal nanoparticles as an alternative to chemical, physical, and microbiological methods [13]. Metal nanoparticles made with plant extracts have been demonstrated to be very durable and harmless for packing and human medicinal uses [14–16], in addition to the benefits stated above. Furthermore, biologically generated nanoparticles do not require any additional stabilizing agents because the plants used act as capping and stabilizing agents [17]. Biological nanoparticle preparation is a bottom-up strategy in which reductants and stabilizers aid in the creation of nanoparticles. Plant phytochemicals or microbial enzymes that serve as reductants are typically responsible for the reduction of metal compounds [18]. Plant extracts containing bioactive alkaloids, phenolic acids, polyphenols, proteins, and carbohydrates are critical first in the reduction and then in the stabilization of metallic ions [19,20].

Nanocomposites are a type of hybrid material that includes nanoscale reinforcements and a matrix. Nanoscale elements are made in a variety of forms and geometries, with typical dimensions of less than 100 nm. Nanoparticles and nanofibers are two nanoscale constituents commonly employed in nanocomposite systems as reinforcement. There are three types of nanocomposite materials: polymeric matrix nanocomposites (PMNCs), metallic matrix nanocomposites (MMNCs), and ceramic matrix nanocomposites (CMNCs). PMNCs and MMNCs have been extensively researched over the last two decades because of their superior mechanical, electrical, thermal, and chemical characteristics [21–23]. When particles are diminished in size from a micrometre to a nanometre, their characteristics might drastically change. Electrical conductivity, rigidity, active surface area, chemical reactivity, and biological activity, for example, have all been acknowledged to change. The bactericidal activity of metal and metal oxide nanoparticles and nanocomposites has been linked to their size and giant surface-area-to-volume ratio [24–26].

Chitosan is a linear polysaccharide made up of randomly scattered N-acetyl-D-glucosamine and linked D-glucosamine. It is made by treating the chitin shells of shrimp and other crustaceans with an alkaline material, such as sodium hydroxide (NaOH). Due to its bioactive features, such as biodegradability, non-toxicity, biocompatibility, haemostatic action, drug transport, and antibacterial properties, chitosan has a variety of commercial, pharmaceutical, biological, and food applications [27–31]. The availability of protonated groups in the polymer backbone, as well as ionic interactions between the charged groups and bacteria wall components, contribute to its antibacterial activity. Protonation of the $-NH_2$ functional groups in acid solution or structural changes (e.g., methylation, sulfonation, etc.) can produce charges on the chitosan backbone [27,30]. As a result, the peptidoglycans in the microbe wall are hydrolysed, causing intracellular electrolyte leakage and the micro-organism's death [27].

Polymer–metal/metal oxide nanocomposites are new type of hybrid material that have the potential to increase functional qualities dramatically (e.g., biological, electrical conductivity) and enhance antimicrobial properties [32,33]. Metal oxide nanoparticles coupled with chitosan have been demonstrated to be an outstanding biocompatible substance in recent investigations [34,35]. Chitosan–NiO (CS-NiO) and chitosan–MgO (CS-MgO) have been shown to have antibacterial activity against *E. coli* and *S. aureus* bacteria strains [36]. The findings revealed that all of the samples had antibacterial properties against the tested bacteria strains. After 12 h of incubation, the CS-NiO showed greater efficacy as an antibacterial agent, reducing *S. aureus* and *E. coli* viabilities to 2–8%. According to the findings, CS NiO nanocomposites have the potential to be employed as an effective antibacterial agent against dangerous bacterial infections. An environmentally friendly biological synthesis of chitosan/copper oxide (CS-CuO) a nanocomposite utilizing rutin and its antiproliferative efficacy against human lung cancer cell line A549 can be found in the literature [37]. The prepared CS-CuO nanocomposite was reported to have concentration-dependent antiproliferative action against A549 cancer cells, with an IC50 value of 20 0.50 g/mL. Furthermore, using the AO/EtBr fluorescence labelling approach, the produced nanocomposite promotes apoptosis in treated A549 cancer cells. Bharathi et al. [38] also described the simple synthesis of a chitosan-FeO nanocomposite with antibacterial action against Gram-positive and Gram-negative bacterial pathogens.

CuO NPs produced from diverse plant-based materials (extracts) have been shown to have antimicrobial activity in the literature [39–44]. There is no published report on the antibacterial activity of chitosan-CuO nanocomposites produced using olive leaf extract as reducing agents against Gram-negative and Gram-positive bacteria. We adopted a facile technique to synthesize chitosan-CuO nanocomposites utilizing aqueous olive leaf extract as a reducing agent. The advantage is that plant-extract-mediated synthesis is the fastest, cheapest, and most sustainable of the numerous green synthesis processes. Furthermore, using antimicrobial plant extract as an in situ reducing and capping agent/group aids in the synthesis of nanoparticles with enhanced antimicrobial activity.

2. Materials and Methods

2.1. Materials

Chitosan (Merck, Kenilworth, NJ, USA) (MW: 50,000–190,000 Da, degree of deacetylation: 75–85 percent), $CuSO_4 \cdot 5H_2O$ (Merck), and CH_3COOH (Merck) were utilized as received without additional purification. Fresh olive leaves (OLE) were picked near the Haspolat campus of Cyprus International University (CIU). *Bacillus licheniformis* (KF609498), *Staphylococcus haemolytic* (MN388897), *Bacillus cereus* (MN888756), and *Micrococcus luteus* (MN888755) were the Gram-positive bacteria strains used, whereas the Gram-negative bacteria strains were *Pseudomonas aeruginosa* (gene accession number GI482716237), *Pseudomonas citronellolis* (ATCC 25992), *Bacillus japonicum*, *Ralstonia pickettii*, and *Klebisella* sp. (reference on ATCC global resource). The National Center of Biotechnology Institute has these gene accession numbers (NCBI) [45,46].

Different methods were used to prepare the samples for different investigations. For instance, in XRD analysis, the nanoparticles were separated. For TEM, FTIR, and FE-SEM analyses, a simple drying procedure was applied to obtain the solid residue sample used for the analysis, whereas the colloidal solution of the nanocomposite was directly used for UV-vis and Zeta potential analyses.

2.2. Extraction of Plant Leaves

Olive leaves were properly cleaned, then sun-dried for 14 days before being pulverized into powder. To prepare the aqueous olive leaf extract, 5.0 g of the powdered leaf was introduced into 250 mL of distilled H_2O and heated to 100 °C under constant stirring at 200 rpm for 3 h. It was then cooled to room temperature, followed by filtration using Whatman® (US reference) # 1 filter papers. The concentration of the extract in water was determined to be 0.4562 g. The filtrate was kept in a refrigerated environment until it was needed.

2.3. Preparation of Chitosan-CuO Nanocomposite

Chitosan samples, with a mass of 0.5, 1.0 and 2.0 g, respectively, were weighed into a 250 mL capacity conical flask, and 100 mL distilled H_2O containing 1 mL of acetic acid was added. The mixture was agitated with a magnetic stirrer at room temperature until the chitosan was totally dissolved. An appropriate amount of copper sulphate pentahydrate ($CuSO_4 \cdot 5H_2O$) (0.2497 g, 1 mM equivalent) was added to the chitosan solution and agitated. This was followed by the introduction of 5 mL of already made OLE. The mixture was heated using a magnetic hot plate at a constant temperature of 90 °C and continuously stirred for 96 h. The production of chitosan-CuO nanocomposite was detected by a gradual change in colour (Figure 1), which was validated by UV-vis measurement.

Figure 1. Schematic representation of biogenic in situ preparation of CH–CuO nanocomposite.

2.4. Characterization

An ATR-FTIR spectrophotometer (Nicolet iS5, Thermo Scientific, Waltham, MA, USA) covering the 4000 to 400 cm^{-1} range was used to characterize the produced CHT-CuO nanocomposite. The OLE and nanocomposite samples for FTIR analysis were prepared by placing a few drops of liquid samples in a Petri dish, which were evaporated to dryness in an oven at 40 °C and the solid residue was used for FTIR analysis. The UV-vis spectra of the CHT-CuO OLE-mediated nanocomposite was acquired by employing a JASCO770-UV–vis spectrophotometer (Tokyo, Japan). The apparatus was run at a 200 nm min^{-1} scan rate with a 1 nm resolution.

The colloidal solution of chitosan-copper oxide nanocomposite was centrifuged for 25 min at 10,230 rpm. After the treatment, the solid residues obtained from this method were rinsed three times with deionized H_2O. The residues were redissolved in 100% ethanol and oven-dried at 50 °C, and XRD analysis was performed on the powder sample. Using nickel-filtered Cu K radiation = 1.5406 oA at 40 kV and 30 mA, the powder X-ray diffraction pattern was recorded using a Rigaku MiniFlex II (Tokyo, Japan) instrument.

TEM analysis was realized using a JEOL JEM-2100F instrument (Tokyo, Japan) to determine the size and shape of CuO nanoparticles in the composite. A small amount of CHT-CuO nanocomposite solution was placed on a Cu sample holder (carbon-coated), dried at ambient temperature, and imaged at 200 kV accelerating voltage.

The prepared CHT-CuO nanocomposite's dynamic light-scattering analysis and zeta potential was measured utilizing a Malvern zetasizer ver. 7.12 (UK). A 1 M HCl or 1 M NaOH solution was used to alter the pH.

A LYRA3 TESCAN apparatus was used to conduct an FE-SEM investigation. The sample was made by placing a small amount of the nanocomposite solution on a sample holder made of Al, which was dried at room temperature to form thin films of the sample on the sample holder's surface. The sample was subjected to FE-SEM analysis utilizing 20 kV accelerating voltage.

An energy-dispersive spectroscope (EDS), which was connected to the FE-SEM apparatus, was utilized to elucidate the elemental composition of the sample.

2.5. Bacteria Cultures

Cryopreserved bacterial cultures were resuscitated by 3 days of incubation in rich Luria broth medium (LB) at 37 degrees Celsius and 120 revolutions per minute. For determination of growth retardation in liquid culture (Section 2.7), bacteria were grown in LB medium for one day at 37 °C, 120 rpm, in the presence of the produced nanocomposite as antibacterial agents, and their growth was visually and quantitatively evaluated [47]. Solid agar plate cultures for "cup plating" were incubated for 12 h (Section 2.6), whereas bacterial enumeration was incubated for a day (Section 2.7). All of the tests were performed twice, and the average value was calculated.

2.6. Antibacterial Activity by 'Cup Plating' or "Disc Diffusion" Technique

The gross inhibiting activity of the chitosan-CuO nanocomposite was determined using the 'cup and plating' protocol published elsewhere [47,48]. Briefly, the approach is based on the use of a solid agar plate made in rich medium (Luria Broth, LB), on which is streaked around 100 microlitres of a fresh bacterial culture of approximately 10^6 CFU/mL. Then, a 5 mm diameter by 2 mm height hole was made and filled with 100 µL of the antimicrobial chitosan-CuO nanocomposite. The chitosan-CuO nanocomposite then diffused around the hole, creating a zone of inhibition of bacterial growth (disc diffusion), the diameter of which was measured 12 h after incubation at 37 °C. The greater the diameter of this zone, the higher the inhibition effect of the prepared chitosan-CuO nanocomposite [48]. The antibiotics amoxicillin and ciprofloxacin were employed as known inhibitors at 10 and 50 mg/mL, respectively.

2.7. Determination of the Minimum Inhibitory Concentration (MIC) and Minimum Bactericidal Concentration (MBC)

Bacterial growth was monitored visually in LB liquid culture at 37 °C for 24 h in the presence of varied nanocomposite concentrations (10^{-1}–10^{-5}%), and growth was assessed visually by observing the culture turbidity, which indicates the presence of both active and lifeless bacteria [49]. The bacteriostatic effect or the lowest antimicrobial agent concentration in which turbidity is not detected was determined from the culture. According to this method, both viable and non-viable bacterial cells were used to calculate the MIC.

In the assessment of only viable bacterial cells, 100 microlitres of culture was placed on an agar LB plate and incubated at 37 °C for 12 h (after visual monitoring, as indicated in the preceding section). Only live bacteria can grow and produce colony-forming units (CFUs) on an agar plate. If the bacterial concentration in the original liquid media was high, the culture was diluted (by a factor of 10 for five dilutions) before being transferred to the solid plate. The lowest concentration of nanocomposite that limited bacterial growth was designated as MBC, and viable bacteria were counted as CFUs. The negative control in these trials was culture medium that was free of micro-organisms and antimicrobial agents. The goal of these checks was to make sure the culture medium was not contaminated by micro-organisms. Contrarily, the positive control, comprised of bacterial cultures devoid of nanocomposite, was used to determine the bacteria's maximal growth rate. In this investigation, the following bacteria were examined: *Bacillus licheniformis*, *Staphylococcus haemolyticus*, *Bacillus cereus*, and *Micrococcus luteus* (Gram-positive), as well as *Pseudomonas aeruginosa*, *Pseudomonas citronellolis*, *E. coli*, *Klebsiella* sp., *Bradyrhizobium japonicum*, and *Ralstonia pickettii* (Gram-negative).

3. Results and discussion

3.1. Synthesis of Chitosan-CuO Nanocomposite

The bio-inspired technique was used to synthesize CHT-CuO nanocomposite utilizing OLE. The synthesis of the CHT-CuO nanocomposite was initially confirmed by the change of colour from colourless (chitosan-CuO solution) (due to low concentration of the precursor used) to yellowish colour with the addition of OLE and eventually to dark-brown colour, signalling the formation of CHT-CuO nanocomposite, as illustrated in Figure 1. For the creation of CHT-CuO nanocomposite, OLE worked as a reducing agent, and chitosan molecules may have played a major role as stabilizing and capping agents [50]. Olive leaf water extract is expected to contain secoiridoids, such as oleuropein, ligstroside, 1-methyloleuropein, and oleoside; flavanoids, such as apigenin, kaempferol, luteolin, and chrysoeriol; and phenolic compounds, such as caffeic acid, tyrosol, and hydroxytyrosol, in agreement with reports in the literature [51]. These phytoconstituents could have played an important role in the reduction process.

3.2. Analysis of UV–Visible Spectroscopy

UV-vis is a great tool for recognizing, characterizing, and researching nanomaterials because nanoparticles have distinctive optical characteristics that are sensitive to size, shape, concentration, aggregation state, and refractive index near the nanoparticle surface, generally attributed to the so-called surface plasmon resonance (SPR) effect. UV-visible spectroscopy (UV-vis) analyses the attenuation of light (scattered or absorbed) passing through a substance. UV-visible spectroscopy was used to monitor the synthesis of the CHT-CuO nanocomposite utilizing OLE. CuO nanoparticles have a pronounced blue shift in their UV absorption spectra in comparison to bulk CuO. CuO nanoparticles show a high density of surface defects, interstitials, and oxygen vacancies due to their high surface-area-to-volume ratio. In the current study, the spectral absorbance peak of the prepared CHT-CuO nanocomposite shifted around 285 nm (Figure 2). This wavelength is lower than the value of 316 nm reported by Bharathi et al. [37]. The difference in the peak position could be attributed to many factors, such as the influence of particle size and shape, the dosage of chitosan, the local refractive index, and methods of preparation, amongst others. Figure 2 also depicts the UV-vis spectrum of the extract (OLE) alone for comparison purposes. The spectrum of OLE exhibited two peaks at 231 and 281 nm. Neither of these peaks is observed in the spectrum of the chitosan-CuO nanocomposite, indicating the successful conversion of copper sulphate to copper oxide utilizing some of the phytoconstituents present in OLE.

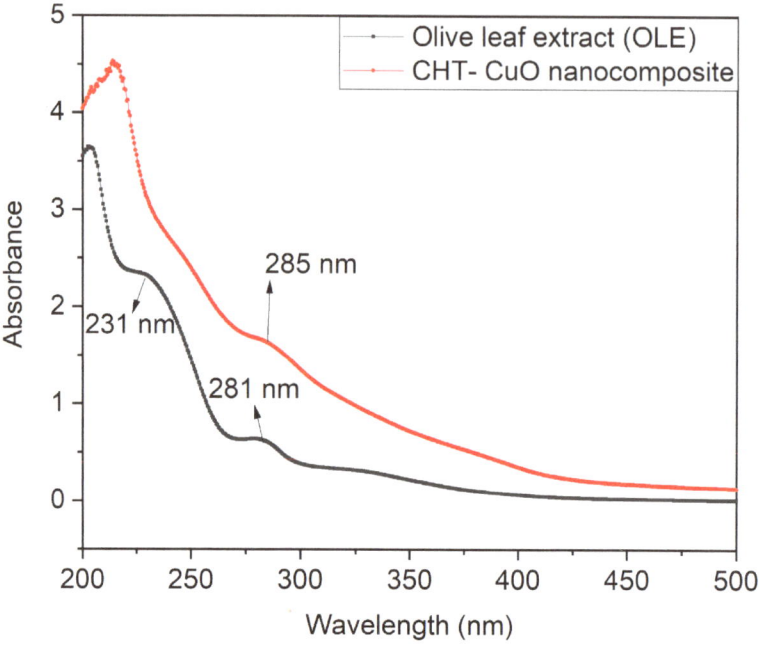

Figure 2. UV-vis spectra of OLE and CHT−CuO nanocomposite.

3.3. FTIR Analysis

FTIR spectra of OLE, chitosan, and chitosan-CuO nanocomposite samples are depicted in Figure 3. FTIR spectra of the chitosan-CuO nanocomposite sample reveal important peaks at 3286, 2971, 2879, 2358, 1557, 1378, 1152, 1022, 895, and 577 cm^{-1}. The bands at 3286 cm^{-1} could be attributed to O-H stretching, indicating the presence of hydroxyl groups, whereas the peak at 2879 cm^{-1} is assigned to C-H stretching. Peaks at 1634 cm^{-1} and 1022 cm^{-1} show N-H bending and C-N stretching of amine groups, respectively. The bands at 1557 and 1153 cm^{-1} may be attributed to C=C stretching and strong C-O stretching, respectively. Peaks observed in the range of 593–637 cm^{-1} are attributed to the metal–oxygen bond, which could be due to CuO in the nanocomposite [52]. A similar observation was previously reported [53]. The FTIR spectrum of chitosan is presented in Figure 3, showing characteristic peaks at 3358, 2873, 1589, 1375, 1149, 1026, 893, and 592 cm^{-1}. The absorption bands in the vicinity of 1149 and 1028 cm^{-1} correspond to typical saccharide moiety characteristic peaks [36], but the peak at 2877 cm^{-1} corresponds to the –CH stretching vibration. The band at 3358 cm^{-1} is assigned to –OH and –NH$_2$ stretching vibrations, whereas the separate peaks at 1674 and 1589 cm^{-1} are related to –NH$_2$ bending vibrations, and the peak at 1375 cm^{-1} is attributed to amide III: C-N stretching vibrations.

The OLE FTIR spectrum (Figure 3) exhibited distinguished peaks at 3285, 2927, 1602, 1391, 1261, 1070, 1030, 692, and 549 cm^{-1}. The intense peak in the vicinity of 3285 cm^{-1} in the spectrum is typical of the O–H stretching of polyphenols [54]. The intense bands at 2927 cm^{-1} correspond to –CH stretching, which may come from the phytoconstituent of bioactive compounds present in OLE [55]. The distinct band at 1602 cm^{-1} is attributed to -NH bending, indicating the presence of -NH bending of amine. The characteristic peak at 1261 cm^{-1} is also attributed to strong C-O stretching, indicating the presence of a carboxylate group. The presence of bioactive phytochemicals, such as triterpenes, proteins, steroids, carbohydrates, alkaloids, and other substances, in the OLE may cause unsaturated C-H bending and C-N stretching between 1152 and 1000 cm^{-1}. These chemicals may have a capping effect, which contributes to the stability of the chitosan-CuO nanocomposite.

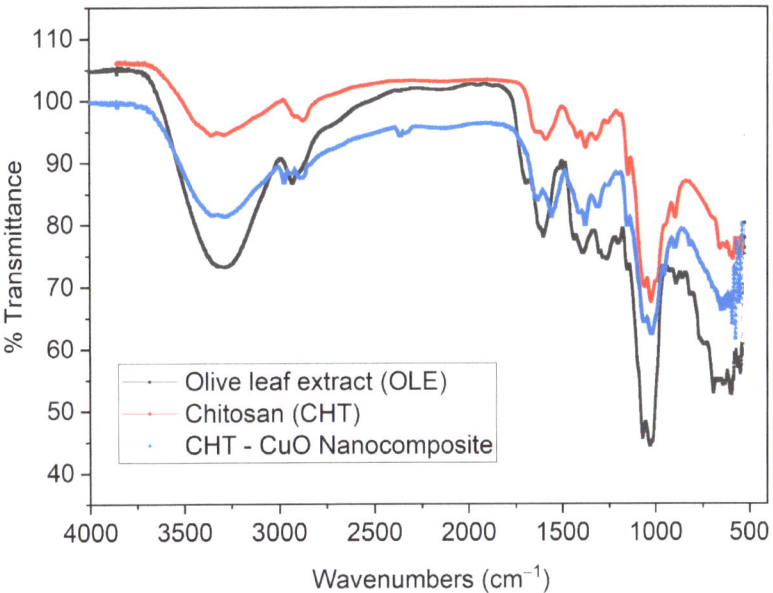

Figure 3. FTIR spectra of OLE, chitosan, and CHT− CuO nanocomposite.

3.4. Studies on the Zeta Potential (ZP)/Dynamic Light Scattering (DLS) Analysis

The ZP value was utilized to determine nanoparticle colloidal stability [56–58]. The following categorization was adopted for ZP: A colloidal solution with a ZP value in the range of 0–10 mV was considered unstable. A ZP value of 10–20 mV or ±20–30 mV indicates a relatively or moderately stable colloid, respectively. The ZP value for a very stable nanoparticle colloid is predicted to be higher than 30 mV [56,57]. The ZP value for the chitosan-CuO nanocomposite in the present work (Figure 4) is +28.2 mV under an acidic pH of 2. These results imply that the tested nanocomposite is moderately stable. The greater ZP value of the OLE-mediated chitosan-CuO nanocomposite demonstrates the influence of the reducing agent on nanoparticle formation and stability.

The pH of a sample in aqueous conditions is one of the most critical elements affecting its ZP. Consider a particle suspended in a liquid with a negative zeta potential. When a large quantity of alkali is introduced to the suspension, the particles become more negatively charged. If acid is introduced to the suspension, it reaches a point where the charge is neutralized. If the ions are specifically adsorbed, adding more acid may induce a build-up of positive charge. A zeta potential versus pH curve in this situation would be positive under lower pH and lower or negative under greater pH. The isoelectric point is when the graph passes through zero ZP and is particularly essential from a practical standpoint. It is usually where aggregation is most frequent and the colloidal system is the least stable. Figure 4 depicts a typical zeta potential vs. pH curve. The isoelectric point of the sample can be seen on the plot to be around pH 7.5. The figure can also be utilized to estimate that the sample will be stable at pH values lower than 4 (more positive charge) and greater than pH 7.5. (more negative charge). Because the ZP values are between +30 and −30 mV [58–60], problems with dispersion stability are likely for pH levels between 6 and 12.

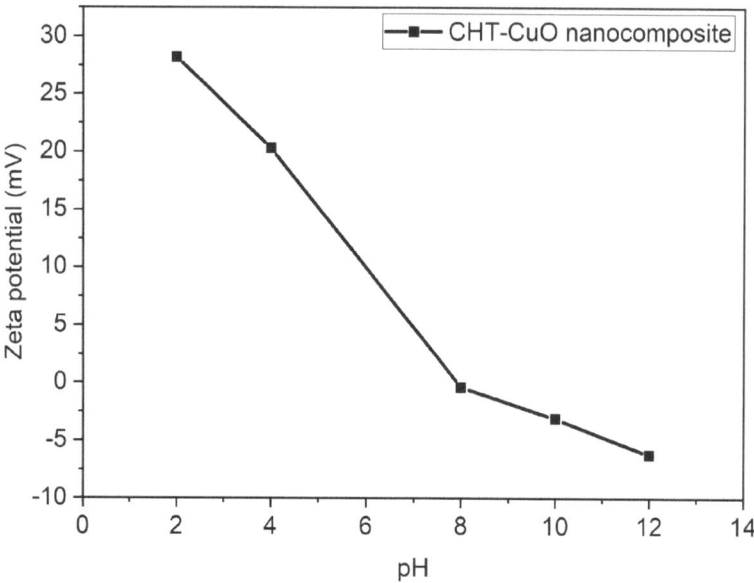

Figure 4. Zeta potential plot for CHT− CuO nanocomposite at different pH values.

For particle size analysis in the nanoscale range, dynamic light scattering is a well-established, standardized approach. DLS gives information on particle size distribution, as well as mean particle size. Figure S1 (Supplementary Information) shows the particle size distribution of CuO produced using the DLS approach. The average particle size is roughly 50 nm, as shown in the diagram. This value is higher than what TEM measurements revealed. There have been some proposed explanations for such disparities. TEM measurement is performed in the dry state, whereas DLS measurement is performed in the solvated condition. Furthermore, TEM is a number-based particle size measurement that places a greater emphasis on the smallest components of the size distribution, whereas DLS is an intensity-based measurement that places a greater emphasis on larger particle sizes.

3.5. XRD Studies

Figure 5a shows the CuO NP XRD pattern in the composite formed by treating 5 mL of OLE with 1.0 g chitosan + 1 mM aqueous $CuSO_4 \cdot 5H_2O$ solution, and Figure 5b shows the obtained diffraction peaks (b). The crystalline structure, crystalline grain size, phase nature, lattice parameter, and amorphous nature of the nanocomposite were all investigated using XRD [61]. Near 2-Theta = 19.98°, 25.43°, 28.98°, and 34.38°, prominent peaks can be seen. These peaks correspond to orientation planes (220), (−311), (400), and (−113), respectively, which are indexed to a typical monoclinic structure and benchmarked with the Joint Committee on Powder Diffraction Standards (JCPDS Card No. 01-085-1693). Some of the peaks, particularly those at 25.43° and 34.38°, which correspond to (−311) and (−113) orientations, respectively, are outside of known CuO nanoparticle diffraction peaks. The existence of contaminants in the produced nanocomposite could explain these peaks [47]. Mineral components in plant-based materials that covered the surface of the nanoparticles [47] or unused copper sulphate could be the source of the contaminants.

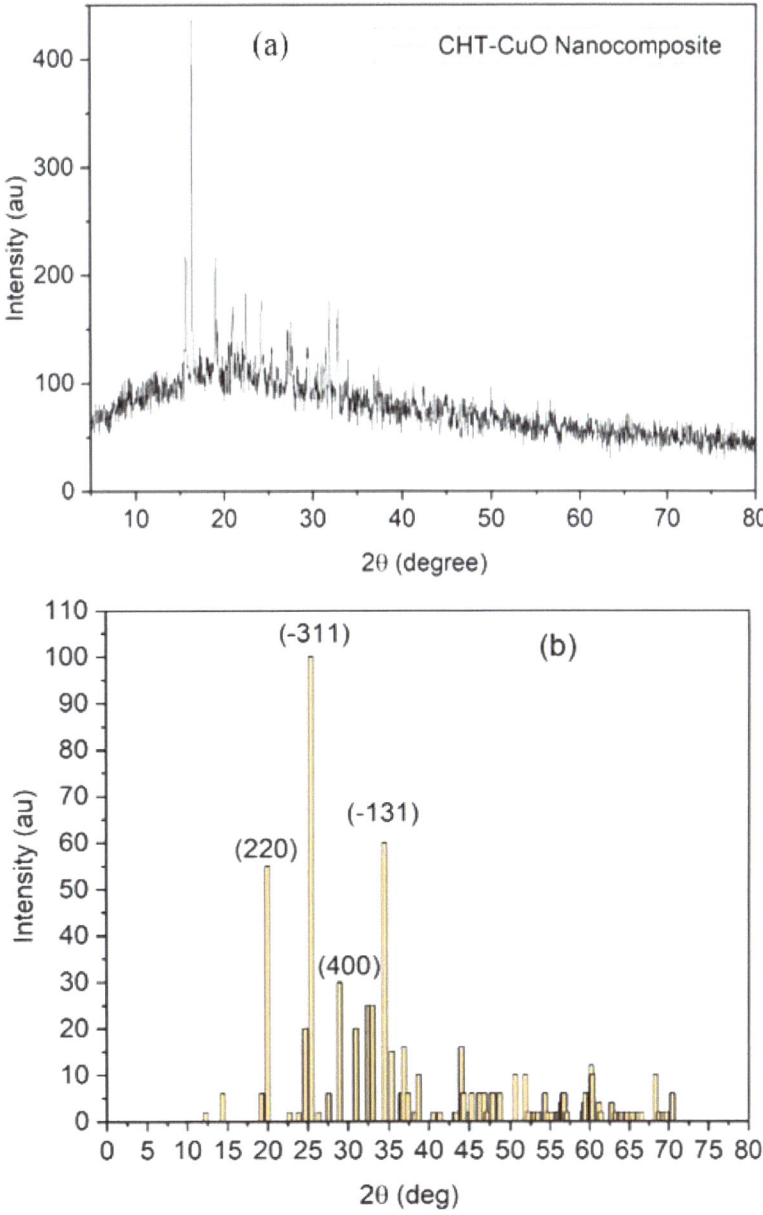

Figure 5. (**a**) XRD pattern and (**b**) XRD peak plots for CHT−CuO nanocomposites.

3.6. FESEM/EDS Analysis

FE-SEM images of the prepared chitosan-CuO nanocomposite are presented in Figure 6, corresponding EDS images are depicted in Figure 7a, and elemental mapping is presented in Figure 7b. As can be clearly seen in Figure 6, OLE successfully reduced $CuSO_4 \cdot 5H_2O$ to CuO nanoparticles, which are spherical in shape and polydispersed. In the corresponding EDS images (Figure 7a,b), typical optical absorption peaks of metallic Cu nanocrystals at 1 and 8 keV can be clearly observed (Figure 7a), thus validating the formation of CuO NPs. Figure 7b shows the elemental mapping of the chitosan-CuO nanocomposite. EDS revealed a high-intensity metallic peak of elements such as aluminium (Al), copper (Cu), oxygen (O) and low-intensity peaks of carbon (C) [62]; their elemental mapping is clearly shown (Figure 7b). The high-intensity peak of Al emanates from the sample holder. For CuO nanoparticles (NPs), EDS analysis revealed a fairly homogeneous copper-rich composition (Figure 7b). The chitosan that surrounds the CuO NPs may be responsible for the carbon and oxygen signals [63,64].

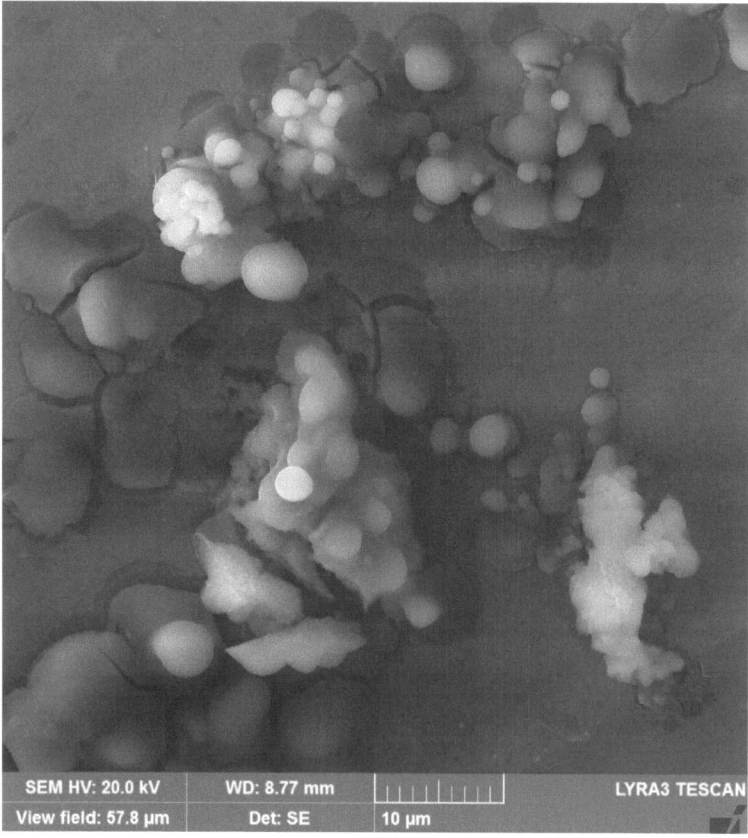

Figure 6. FE-SEM image of CHT−CuO nanocomposite.

The XRD, SEM, TEM, and DLS analysis results were similar for all the nanocomposites prepared in the presence of varying amounts of chitosan. This suggests that the concentration of chitosan did not influence the quality of nanoparticles formed.

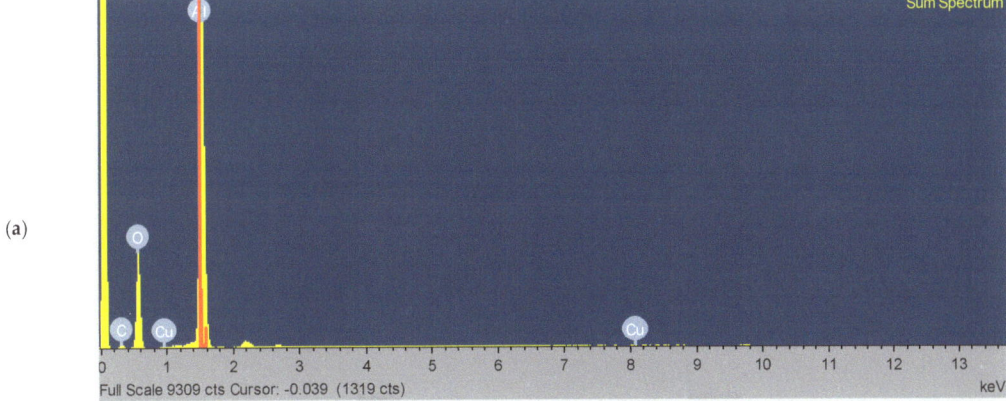

Figure 7. (**a**) EDS spectrum and (**b**) elemental mapping of CHT−CuO nanocomposite.

3.7. TEM Analysis

TEM examination was used to identify the shapes and sizes of the biosynthesized chitosan-CuO nanocomposite. TEM pictures at 500 nm and 20 nm magnifications are shown in Figure 8a,b. Figure 8c depicts the SAED (selected area electron diffraction) patterns of the chitosan-CuO nanocomposite at a magnification of 101 nm. The tested chitosan-CuO nanocomposite is polydispersed, spherical, and of various sizes. In an OLE-mediated nanocomposite, the size of the CuO NPs is in the range of 3.2–6.0 nm (Figure 8b). Previously, several researchers [62] related size diversity to formation time variations. According to the SAED pattern, the CuO NPs are embedded in the chitosan matrix. The SAED pattern further reveals that the CuO nanoparticles are polycrystalline [62], as the diffraction point (Figure 8c) is dispersed on concentric rings. The brilliant circular spots found in the SAED patterns in Figure 8c support the crystalline character of the nanoparticles indicated by the XRD results (Figure 5a).

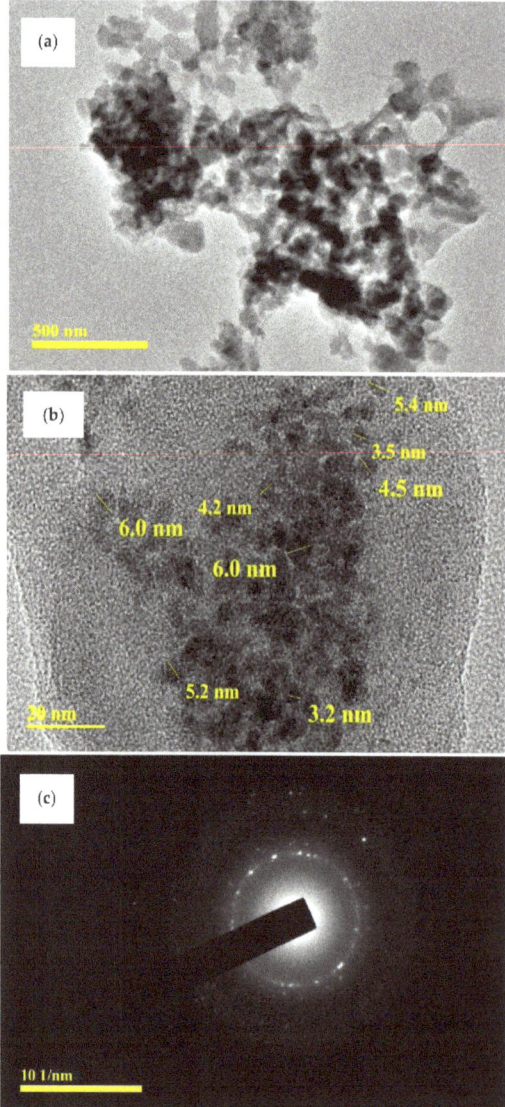

Figure 8. (a,b) TEM micrographs of CHT-CuO nanocomposite at different magnifications and (c) SAED pattern.

3.8. Antimicrobial Studies

The antimicrobial activities of the synthesized nanocomposites, using different amounts of chitosan (0.5 [CuO-0.5], 1.0 [CuO-1] and 2.0 g [CuO-2]) were studied using the cup-platting (or disc diffusion) technique against Gram-positive bacteria (*B. licheniformis*, *S. haemolyticus*, *B. cereus*, and *M. luteus*) and Gram-negative bacteria (*P. aeruginosa*, *P. citronellolis*, *E. coli*, *Klebisiella* sp., *B. japonicum*, and *R. pickettii*) (Figure 9, Figures S2 and S3) [46]. The known antibiotics ciprofloxacin and amoxicillin were employed for comparison.

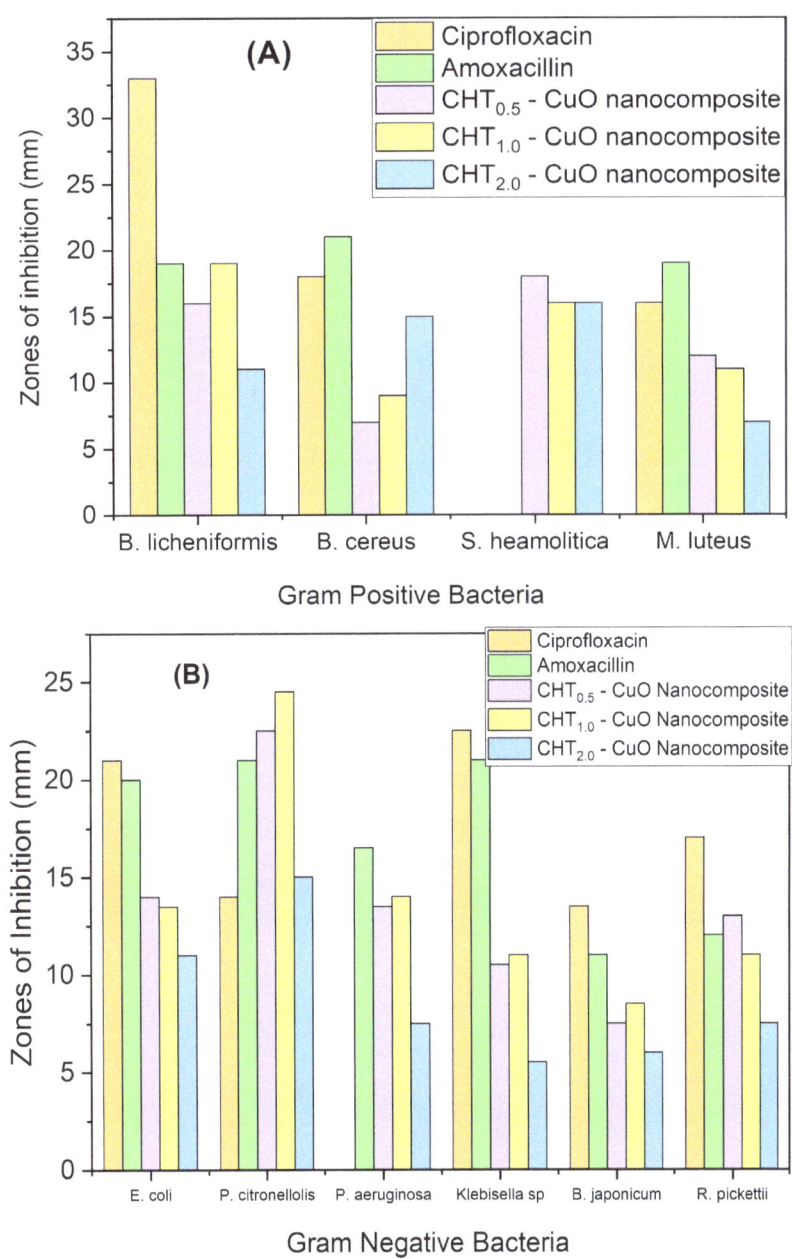

Figure 9. Zones of inhibition showing antibacterial activity of the chitosan-CuO nanocomposite in comparison with antibiotics (ciprofloxacin and amoxicillin) against (**A**) Gram-positive and (**B**) Gram-negative bacteria.

Overall, against all tested bacterial strains, the diameters of the inhibition of the three nanocomposites fell between 6 and 24 mm, and no noticeable difference was observed between Gram-negative and Gram-positive bacteria (Figure 9a,b). Figures S2 and S3 provide pictures of the diameter of inhibition of each tested bacteria. Interestingly, nanocomposites made with 2.0 g of chitosan (CuO-2) were associated with the smallest diameters of the inhibition zone against the tested bacteria (except in *B. cereus*), reflecting their lower antimicrobial activities. The order of the antimicrobial activity was CuO-1 > CuO-0.5 > CuO-2. An inhibition zone greater than 1 mm indicates satisfactory antibacterial potential on the basis of the SNV 195920-1992 Standard Antibacterial Test [65–67].

The reference antibiotics, amoxicillin and ciprofloxacin, showed higher antimicrobial activity, with a diameter of inhibition zone falling between 15 and 32 cm. However, it is important to note that these reference antibiotics were less active than nanocomposites in three bacterial strains. For instance, the two tested antibiotics did not show an inhibition zone against *S. Heamolityca*; likewise, no inhibition zone was observed with the use of ciprofloxacin against *P. aeruginosa*. Interestingly, against *P. citronellolis*, the two tested antibiotics were less active than the three nanocomposites, illustrating the potential of this new nanocomposite material against some bacterial strains.

Further assessment of the nanocomposite's antibacterial effect was performed in a liquid medium. As shown in Table S1, complete inhibition of bacterial growth was observed at a concentration of 0.1–0.01% of the nanocomposite (Supplementary Information). Furthermore, as expected, as the dilution of the nanocomposite increases, the growth inhibition decreases (Table S1). The use of nanocomposites at 0.1% (irrespective of chitosan concentration) inhibited all tested bacteria, except *S. heamoylitica*. At 0.01%, the nanocomposite CuO-0.5 inhibited more bacterial strains than CuO-2 (mainly Gram-negative bacteria, Table S1). Overall, the nanocomposites CuO-0.5 and CuO-1 were more active than CuO-2.

Table 1 summarizes MIC and MBC of the nanocomposites. MIC is the lowest concentration at which turbidity is visible, and this turbidity is associated with both with viable and dead bacteria, whereas MBC is associated with viable bacteria only. Overall, against each bacterial strain, MIC and MBC values of the three nanocomposites were identical, except in *M. luteus*. In this strain, the three nanocomposites have MIC and MBC values of 0.01 and 0.1%, respectively (Table 1). The fact that MIC and MBC are the same suggests that the turbidity of the cultures was mostly caused by live bacteria. Thus, in the case of *M. luteus*, viable and non-viable cells account for this turbidity.

Table 1. Minimum inhibitory concentration that inhibits bacteria growth based on visual turbidity or bacteriostatic effect (MIC, in %) and minimum inhibitory concentration that inhibit growth in solid plate culture or bactericidal effect (MBC, in %). For CHT-CuO nanocomposites.

Type	Bacteria	$CHT_{0.5}$-CuO Nanocomposite (%)		$CHT_{1.0}$-CuO Nanocomposite (%)		$CHT_{2.0}$-CuO Nanocomposite (%)	
		MIC	MBC	MIC	MBC	MIC	MBC
Gram-positive	B. licheniformis	0.1	0.1	0.1	0.1	0.1	0.1
	B. cereus	0.01	0.01	0.01	0.01	0.01	0.01
	S. heamolytica	>0.1	>0.1	>0.1	>0.1	>0.1	>0.1
	M. luteus	0.01	0.1	0.01	0.1	0.01	0.1
Gram-negative	E. coli	0.1	0.1	0.1	0.1	0.01	0.01
	P. citronellolis	0.01	0.01	0.01	0.01	0.01	0.01
	P. aeruginosa	0.01	0.01	0.01	0.01	0.01	0.01
	Klebsiella sp.	0.01	0.01	0.01	0.01	0.1	0.1
	B. japonicum	0.01	0.01	0.01	0.01	0.1	0.1
	R. pickettii	0.01	0.01	0.01	0.01	0.1	0.1

Interestingly, the nanocomposite MIC/MBC values were > 0.1% against *S. heamolitica*, indicating this strain is less susceptible to these antibacterial agents. Based on the disc diffusion experiment (Figures S2 and S3), this strain was not inhibited by the reference antibiotics ciprofloxacin and amoxicillin (at 10 mg/L), whereas the inhibition zone was observed with the nanocomposites (Figures S2 and S3). This indicates that nanocomposites could inhibit these strains at concentrations higher than 0.1%. Another interesting information is that against Gram-negative bacteria, the nanocomposites CuO-0.5 and CuO-1.0 have MIC/MBC values of 0.01%, whereas values for CuO-2 were 0.1% against *Klebisella* sp., *B. japonicum*, and *R. pickettii* and 0.01% against the two tested *Pseudomonas* species (*P. citronellolis* and *P. aeruginosa*). The only exception was *E. coli*, against which CuO-2 was more active than the other two tested nanocomposites. Thus, overall, CuO-0.5 and CuO-1.0 were more active than CuO-2, which is in line with the observations from the disc diffusion experiment (Figures S2, S3 and 9). Furthermore, these findings revealed that Gram-positive bacteria were less vulnerable to nanocomposites than Gram-negative bacteria. The strong antibacterial activity demonstrated by OLE-mediated chitosan-CuO nanocomposites can be attributed to the release of CuO NPs, which kill bacteria by any or all of the methods previously mentioned. Low-molecular-weight chitosan has also been shown to have antibacterial properties. Costa et al. [68] speculate that the action emanates from chitosan's capacity to commune with and destroy micro-organism cell walls via hole formation or membrane breakdown.

In our previous studies, we reported that chitosan exhibited antibacterial activity against the tested bacteria strains [46]. In order to assess the origin of the observed antibacterial activity, olive leaf extract (OLE) alone was also tested. The obtained results are depicted in Table 2 and Figure S4. Plant extracts (OLE) had no inhibitory effect against eight (8) of the studied bacteria, with little inhibition of growth of *P. aeruginosa* and *B. japonicum*, with inhibition zone diameters of 3.5 ± 0.0 mm and 2.0 ± 0.0 mm, respectively. Many reports in the literature have also shown that CuO nanoparticles exhibit excellent antibacterial activity [39–42]. It is therefore pertinent that the combined action of chitosan and CuO nanoparticles could be responsible for the excellent antibacterial action of OLE-mediated chitosan-CuO nanocomposites.

Table 2. Antimicrobial effect of olive leaf extract (OLE) using cup plate experiments. Values stand for the diameter (in mm) of zones of inhibition.

Type	Bacteria	Diameter of Zones of Inhibition (mm)		
		1st Run	2nd Run	Average Values
Gram-positive	*B. licheniformis*	0.0 ± 0.0	0.0 ± 0.0	0.0 ± 0.00
	B. cereus	0.0 ± 0.0	0.0 ± 0.0	0.0 ± 0.0
	S. heamolitica	0.0 ± 0.0	0.0 ± 0.0	0.0 ± 0.0
	M. luteus	0.0 ± 0.0	0.0 ± 0.0	0.0 ± 0.00
Gram-negative	*E. coli*	0.0 ± 0.0	0.0 ± 0.0	0.0 ± 0.0
	P. citronellolis	0.0 ± 0.0	0.0 ± 0.0	0.0 ± 0.0
	P. aeruginosa	3.0 ± 0.0	4.0 ± 0.0	3.5 ± 0.0
	Klebsiella sp.	0.0 ± 0.00	0.0 ± 0.0	0.0 ± 0.0
	B. japonicum	2.0 ± 0.0	2.0 ± 0.0	2.0 ± 0.0
	R. pickettii	0.0 ± 0.0	0.0 ± 0.0	0.0 ± 0.0

Antibacterial activity increases when the concentration of chitosan increases from 0.5 g to 1.0 g; however, increasing the concentration to 2.0 g resulted in a decline in antibacterial activity for the majority of the bacteria strains examined. Similar findings have been reported in the literature, where cotton fabrics treated with chitosan concentrations of 0.5–0.75% showed the highest antibacterial activity and increasing the chitosan content to 1% resulted in a decrease in antibacterial activity [69]. The possible reasons for this, according to the literature report, are as follows: when the concentration of chitosan is lower, chitosan combines with the surface of bacteria cells with a negative charge, disrupts the membrane of bacteria cells, and induces component leakage in bacteria cells, eventually leading to the death of bacteria cells. When the concentration of chitosan is higher, however, protonated chitosan can be wrapped around the surface of bacterial cells to prevent component leakage, and positively charged bacterial cells repel each other to avoid agglutination [70].

The findings reveal that the OLE-mediated chitosan-CuO nanocomposite is particularly effective against the bacteria tested, which is consistent with previous studies on the antimicrobial activity of natural polymers/CuO nanocomposites [66,67]. The nanocomposite's activity could be attributed to its small size and capacity to increase bacteria's surface contact, perhaps inflicting injury. The bactericidal mechanism of metal and metal oxide nanocomposites is based on the formation of reactive oxygen species, such as superoxide radical anions, hydrogen peroxide anions, and hydrogen peroxide, which commune with bacterial cell walls, giving rise to the destruction of the cell membrane, inhibiting further cell growth, and causing leakage of internal cellular components, ultimately leading to bacterial death [43]. Reactive oxygen species created by metal oxide nanocomposites obstruct bacterial cell processes, such as glutathione depletion, DNA rupture, protein denaturation, and enzyme damage, preventing bacterial cell reproduction.

4. Conclusions

The green synthesis, characterisation, and antibacterial efficacy of chitosan-CuO nanocomposites made with olive leaves extract (OLE) as the reducing agent were described in this paper. FTIR, UV-vis, XRD, FE-SEM, and TEM were used to confirm the effective synthesis of an OLE-mediated chitosan-CuO nanocomposite. XRD, SEM, TEM, and DLS analysis results were similar for all the nanocomposites prepared in the presence of varying amounts of chitosan. This suggests that the concentration of chitosan did not influence the quality of nanoparticles formed. The CuO nanoparticles in the composite are very stable and polydispersed. The CHT-CuO nanocomposite's zeta potential varies with pH. The largest zeta potential (+28 mV) was found at pH 2, whereas the lowest (−7.5 mV) was found at pH 12. According to the FTIR data, the loose carboxylate functional groups in the plant-based material (extract) stabilize the nanoparticles in the polymer matrix. OLE works well as a reducing agent, resulting in smaller nanoparticles with sizes in the range of 3.2–6.0 nm based on TEM analysis. These values are lower than those obtained from DLS analysis, which were found to be around 50 nm. Gram-positive bacteria, including *B. licheniformis*, *B. cereus*, and *M. luteus*, as well as Gram-negative bacteria, such as *E. coli*, *P. citronellolis*, *P. aeruginosa*, *kliebisella* sp., *Bradyrhizobium japonicum*, and *Ralstonia pickettii*, are successfully inhibited by OLE-mediated chitosan-CuO nanocomposites. Commercially available antibiotics (ciprofloxacin and amoxicillin) had little effect on *Staphylococcus haemolytica*, but an OLE-mediated chitosan-CuO nanocomposite did. The plant extracts (OLE) had no inhibitory effect against eight (8) of the studied bacteria, with little inhibition of growth of *P. aeruginosa* and *B. japonicum* with inhibition zone diameters of 3.5 0.0 mm, 2.0 and 0.0 mm, respectively. The smaller nanoparticles and capacity to overcome biological barriers may explain the greater antibacterial activity of OLE-mediated chitosan-CuO nanocomposites. It is pertinent that the antibacterial activity of the prepared nanocomposites could be largely attributed to the combined effect of chitosan and CuO nanoparticles. This environmentally friendly chitosan-CuO nanocomposite is a cost-effective, biogenic molecule capable of acting as an antibacterial agent against Gram-positive and Gram-negative bacteria.

Supplementary Materials: The following supporting information can be downloaded at: https://www.mdpi.com/article/10.3390/polym14091832/s1, Figure S1: Particle size distribution of CuO nanoparticles obtained by DLS method; Figure S2: Inhibition of bacterial growth by cup plate experiment. A, B, C and D show inhibition of *Bacillus licheniformies*, *Staphylococcus heamolitica*, *Bacillus cereus* and *Micrococcus luteus* respectively in the presence of 1- Ciprofloxacin, 2-Amoxacillin, 3-CHT$_{0.5}$-CuO nanocomposite, 4- CHT$_{1.0}$-CuO nanocomposite and 5- CHT$_{2.0}$-CuO nanocomposite; Figure S3: Inhibition of bacterial growth by cup plate experiment. A, B, C, D, E and F show inhibition of *Escherichia coli*, *Pseudomonas citronellolis*, *Pseudomonas aeruginosa*, *Klebisella* sp., *Bradyrhizobium* *japonicum* and *Ralstonia pickettii* respectively in the presence of 1- Ciprofloxacin, 2-Amoxacillin, 3- $CHT_{0.5}$-CuO nanocomposite, 4- $CHT_{1.0}$-CuO nanocomposite and 5- $CHT_{2.0}$-CuO nanocomposite; Figure S4: Inhibition of bacterial growth by cup plate experiment in the presence of OLE for all bacteria strains; Table S1: Assessment of the antimicrobial effect of CHT -CuO nanocomposites on gram positive and gram negative bacteria based on turbidity visualization of the culture and bacterial count on solid Agar plate. Values in brackets represent bacteria counts or colony forming units (CFU $\times$ 10^6/mL) after 24 h culture. Positive control presents culture liquid without bacterial inhibitor substance, and negative control is the culture without bacteria. '–' infers complete growth inhibition where '+' indicate bacteria growth.

Author Contributions: Conceptualization, D.K. and S.A.U.; methodology, P.S.U., S.S.S. and A.N.; validation, D.K., S.A.U., A.N. and S.S.S.; formal analysis, P.S.U., S.S.S. and A.N.; investigation, P.S.U. and S.S.S.; resources, D.K., S.A.U. and A.N.; data curation, P.S.U. and S.S.S.; writing—original draft preparation, P.S.U.; writing—review and editing, S.A.U., D.K. and A.N.; supervision, D.K. and S.A.U.; project administration, D.K. and S.A.U. All authors have read and agreed to the published version of the manuscript.

Funding: This research received no external funding.

Institutional Review Board Statement: Not applicable.

Informed Consent Statement: Not applicable.

Data Availability Statement: The data presented in this study are available on request from the corresponding authors.

Conflicts of Interest: The authors declare no conflict of interest.

References

1. Liu, Y.; Jiang, Y.; Zhu, J.; Huang, J.; Zhang, H. Inhibition of bacterial adhesion and biofilm formation of sulfonated chitosan against *Pseudomonas aeruginosa*. *Carbohydr. Polym.* **2019**, *206*, 12–419. [CrossRef] [PubMed]
2. Bassetti, M.; Vena, A.; Croxatto, A.; Righi, E.; Guery, B. How to manage *Pseudomonas aeruginosa* infections. *Drugs Context* **2018**, *27*, 212527. [CrossRef] [PubMed]
3. Hiramatsu, K. Vancomycin-resistant *Staphylococcus aureus*: A new model of antibiotic resistance. *Lancet Infect. Dis.* **2001**, *1*, 147–155. [CrossRef]
4. Das, B.; Moumita, S.; Ghosh, S.; Khan, M.I.; Indira, D.; Jayabalan, R.; Tripathy, S.K.; Mishra, A.; Balasubramanian, P. Biosynthesis of magnesium oxide (MgO) nanoflakes by using leaf extract of *Bauhinia purpurea* and evaluation of its antibacterial property against *Staphylococcus aureus*. *Mater. Sci. Eng. C* **2018**, *91*, 436–444. [CrossRef]
5. Azam, A.; Ahmed, A.S.; Oves, M.; Khan, M.S.; Memic, A. Size-dependent antimicrobial properties of CuO nanoparticles against Gram-positive and -negative bacterial strains. *Int. J. Nanomed.* **2021**, *7*, 3527–3535. [CrossRef]
6. Dizaj, S.M.; Lotfipour, F.; Barzegar-Jalali, M.; Zarrintan, M.H.; Adibkia, K. Antimicrobial activity of the metals and metal oxide nanoparticles. *Mater. Sci. Eng. C Mater. Biol. Appl.* **2014**, *44*, 278–284. [CrossRef]
7. Nas, F.S.; Ali, M.; Muhammad, A. Application of Nanomaterials as antimicrobial agents. *Arch. Nanomed.* **2018**, *1*, 59–63.
8. Srivastava, A. Antiviral activity of copper complexes of isoniazid against RNA tumor viruses. *Resonance* **2009**, *14*, 60–754. [CrossRef]
9. Santo, C.E.; Quaranta, D.; Grass, G. Antimicrobial metallic copper surfaces kill *Staphylococcus haemolyticus* via membrane damage. *Microbiol. Open* **2012**, *1*, 46–52. [CrossRef]
10. Grass, G.; Rensing, C.; Solioz, M. Metallic copper as an antimicrobial surface. *Appl. Environ. Microbiol.* **2011**, *77*, 7–1541. [CrossRef]
11. Gamboa, S.M.; Rojas, E.R.; Martínez, V.V.; Vega-Baudrit, J. Synthesis and characterization of silver nanoparticles and their application as an anti-bacterial agent. *Int. J. Biosens. Bioelectron.* **2019**, *5*, 166–173.
12. Zhang, D.; Ma, X.L.; Gu, Y.; Huang, H.; Zhang, G.W. Green synthesis of metallic nanoparticles and their potential applications to treat cancer. *Front. Chem.* **2020**, *8*, 799. [CrossRef]
13. Elahi, N.; Kamali, M.; Baghersad, M.H. Recent biomedical applications of gold nanoparticles: A review. *Talanta* **2018**, *184*, 537–556. [CrossRef]
14. Nakkala, J.R.; Bhagat, E.; Suchiang, K.; Sadras, S.R. Comparative study of antioxidant and catalytic activity of silver and gold nanoparticles synthesized from *Costus pictus* leaf extract. *J. Mater. Sci. Technol.* **2015**, *13*, 986–994. [CrossRef]
15. Bindhu, M.R.; Umadevi, M. Synthesis of monodispersed silver nanoparticles using *Hibiscus cannabinus* leaf extract and its antimicrobial activity. *Spectrochim. Acta Part A Mol. Biomol. Spectrosc.* **2013**, *101*, 184–190. [CrossRef]
16. Chowdhury, I.H.; Ghosh, S.; Roy, M.; Naskar, M.K. Green synthesis of water-dispersible silver nanoparticles at room temperature using green carambola (star fruit) extract. *J. Sol-Gel Sci. Technol.* **2014**, *73*, 199–207. [CrossRef]

17. Makarov, V.V.; Love, A.J.; Sinitsyna, O.V.; Makarova, S.S.; Yaminsky, I.V.; Taliansky, M.E.; Kalinina, N.O. Green nanotechnologies: Synthesis of metal nanoparticles using plants. *Acta Nat.* **2014**, *6*, 35–44. [CrossRef]
18. Condorelli, G.G.; Costanzo, I.L.; Fragala, I.L.; Giuffrida, S.; Ventimiglia, G. A single photochemical route for the formation of both copper nanoparticles and patterned nanostructured films. *J. Mater. Chem.* **2003**, *13*, 2409–2411. [CrossRef]
19. Breakspear, I.; Guillaume, C. A quantitative phytochemical comparison of olive leaf extracts on the Australian market. *Molecules* **2020**, *25*, 4099. [CrossRef]
20. Kabbash, E.; Ayoub, I.; Abdel-shakour, Z.; El-Ahmady, S. A phytochemical study on *Olea europaeal*. olive leaf extract (cv. Koroneiki) growing in Egypt. *Arch. Pharm. Sci. Ain Shams Univ.* **2019**, *3*, 99–105.
21. Rahimi, M.; Noruzi, E.B.; Sheykhsaran, E.; Ebadi, B.; Kariminezhad, Z.; Molaparast, M.; Mehrabani, M.G.; Mehramouz, B.; Yousefi, M.; Ahmadi, R.; et al. Carbohydrate polymer-based silver nanocomposites: Recent progress in the antimicrobial wound dressings. *Carbohydr. Polym.* **2019**, *231*, 115696. [CrossRef] [PubMed]
22. Yang, Y.; Lan, J.; Li, X. Study on bulk aluminum matrix nano-composite fabricated by ultrasonic dispersion of nano-sized SiC particles in molten aluminum alloy. *Mater. Sci. Eng. A* **2004**, *380*, 378–383. [CrossRef]
23. Cao, G.; Konishi, H.; Li, X. Mechanical properties and microstructure of Mg/SiC nanocomposites fabricated by ultrasonic cavitation based nanomanufacturing. *J. Manuf. Sci. Eng.* **2008**, *130*, 031105. [CrossRef]
24. Qu, N.; Chan, K.; Zhu, D. Pulse co-electrodeposition of nano Al_2O_3 whiskers nickel composite coating. *Scr. Mater.* **2004**, *50*, 1131–1134. [CrossRef]
25. Chen, L.Y.; Xu, J.Q.; Choi, H.; Pozuelo, M.; Ma, X.; Bhowmick, S.; Yang, J.M.; Mathaudhu, S.; Li, X.C. Processing and properties of magnesium containing a dense uniform dispersion of nanoparticles. *Nature* **2015**, *528*, 539–543. [CrossRef]
26. Ribeiro, A.I.; Dias, A.M.; Zille, A. Synergistic effects between metal nanoparticles and commercial antimicrobial agents: A Review. *ACS Appl. Nano Mater.* **2022**, *5*, 3030–3064. [CrossRef]
27. Li, J.; Zhuang, S. Antibacterial activity of chitosan and its derivatives and their interaction mechanism with bacteria: Current state and perspectives. *Eur. Polym. J.* **2020**, *138*, 109984. [CrossRef]
28. Balagangadharan, K.; Dhivya, S.; Selvamurugan, N. Chitosan based nanofibers in bone tissue engineering. *Int. J. Biol. Macromol.* **2017**, *104*, 1372–1382. [CrossRef]
29. Smagghe, G.; Steurbaut, W. Chitosan as antimicrobial agent: Applications and mode of action. *Biomacromolecules* **2003**, *4*, 1457–1465.
30. Dodane, V.; Vilivalam, V.D. Pharmaceutical applications of chitosan. *Pharm. Sci. Technol. Today* **1998**, *1*, 246–253. [CrossRef]
31. Ravi Kumar, M.N. A review of chitin and chitosan applications. *React. Funct. Polym.* **2000**, *46*, 1–27. [CrossRef]
32. Arjunan, N.; Singaravelu, C.M.; Kulanthaivel, J.; Kandasamy, J.A. Potential photocatalytic, antimicrobial and anticancer activity of chitosan-copper nanocomposite. *Int. J. Biol. Macromol.* **2017**, *104*, 1774–1782. [CrossRef]
33. Syame, S.M.; Mohamed, R.K.; Omara, T. Synthesis of copper-chitosan Nanocomposites and its Application in treatment of local pathogenic Isolates Bacteria. *Orient. J. Chem.* **2017**, *33*, 2959–2969. [CrossRef]
34. Yusof, N.A.A.; Zain, N.M.; Pauzi, N. Synthesis of ZnO nanoparticles with chitosan as stabilizing agent and their antibacterial properties against Gram-positive and Gram-negative bacteria. *Int. J. Biol. Macromol.* **2019**, *124*, 1132–1136. [CrossRef]
35. Perelshtein, I.; Ruderman, E.; Perkas, N.; Tzanov, T.; Beddow, J.; Joyce, E.; Mason, T.J.; Blanes, M.; Mollá, K.; Patolla, A. Chitosan and chitosan-ZnO-based complex nanoparticles: Formation, characterization, and antibacterial activity. *J. Mater. Chem. B* **2013**, *1*, 1968–1976. [CrossRef]
36. Mizwari, Z.M.; Oladipo, A.A.; Yilmaz, E. Chitosan/metal oxide nanocomposites: Synthesis, characterization, and antibacterial activity. *Int. J. Polym. Mater. Polym. Biomater.* **2021**, *70*, 83–391. [CrossRef]
37. Bharathi, D.; Ranjithkumar, R.; Chandarshekar, B.; Bhuvaneshwari, V. Bio-inspired synthesis of chitosan/copper oxide nanocomposite using rutin and their anti-proliferative activity in human lung cancer cells. *Int. J. Biol. Macromol.* **2019**, *141*, 476–483. [CrossRef]
38. Bharathi, D.; Ranjithkumar, R.; Vasantharaj, S.; Chandarshekar, B.; Bhuvaneshwari, V. Synthesis and characterization of chitosan/iron oxide nanocomposite for biomedical applications. *Int. J. Biol. Macromol.* **2019**, *132*, 880–887. [CrossRef]
39. Shameem, P.P.N.U.; Pratap Kollu, R.L.; Kalyani, S.; Pammi, V.N. Green synthesis of copper oxide nanoparticles using *Aloe vera* leaf extract and its antibacterial activity against fish bacterial pathogens. *BioNanoScience* **2015**, *5*, 135–139.
40. Qamar, H.; Rehman, S.; Chauhan, D.K.; Tiwari, A.K.; Upmanyu, V. Green Synthesis, Characterization and antimicrobial activity of copper oxide nanomaterial derived from *Momordica charantia*. *Int. J. Nanomed.* **2020**, *15*, 2541–2553. [CrossRef]
41. Naikaa, H.R.; Lingarajua, K.; Manjunathb, K.; Kumar, D.; Nagarajuc, G.; Sureshd, D.; Nagabhushanae, H. Green synthesis of CuO nanoparticles using *Gloriosa superba* L. extract and their antibacterial activity. *J. Taibah Univ. Sci.* **2015**, *9*, 7–12. [CrossRef]
42. Chandraker, S.K.; Lal, M.; Ghosh, M.K.; Tiwari, V.; Ghorai, T.K.; Shukla, R. Green synthesis of copper nanoparticles using leaf extract of *Ageratum houstonianum* Mill. and study of their photocatalytic and antibacterial activities. *Nano Express* **2020**, *1*, 010033. [CrossRef]
43. Letchumanan, D.; Sok, S.P.; Ibrahim, M.S.; Nagoor, N.H.; Arshad, N.M. Plant-based biosynthesis of copper/copper oxide nanoparticles: An update on their applications in biomedicine, mechanisms, and toxicity. *Biomolecules* **2021**, *11*, 564. [CrossRef]
44. Pachaiappan, R.; Rajendran, S.; Show, P.L.; Manavalan, K.; Naushad, M. Metal/metal oxide nanocomposites for bactericidal effect: A review. *Chemosphere* **2021**, *272*, 128607. [CrossRef]

45. National Center for Biotechnology Information (NCBI). Genomic Sequence. Available online: https://www.ncbi.nlm.nih.gov/ (accessed on 3 December 2019).
46. Umoren, S.A.; Solomon, M.M.; Nzila, A.; Obot, I.B. Preparation of silver/chitosan nanofluids using selected plant extracts: Characterization and antimicrobial studies against Gram-positive and Gram-negative bacteria. *Materials* **2020**, *13*, 1627. [CrossRef]
47. Umoren, S.A.; Nzila, A.M.; Sankaran, S.; Solomon, M.M.; Umoren, P.S. Green synthesis, characterization and antibacterial activities of silver nanoparticles from strawberry fruit extract. *Pol. J. Chem. Technol.* **2017**, *19*, 128–136. [CrossRef]
48. Rios, J.L.; Recio, M.C.; Villar, A. Screening methods for natural products with antimicrobial activity: A review of the literature. *J. Ethnopharmacol.* **1988**, *23*, 127–149. [CrossRef]
49. Luksiene, Z. Nanoparticles and their potential application as antimicrobials in the food industry. In *Nanotechnology in the Agri-Food Industry, Food Preservation*; Grumezescu, A.M., Ed.; Academic Press: Cambridge, UK, 2017; pp. 567–601.
50. Marslin, G.; Siram, K.; Maqbool, Q.; Selvakesavan, R.; Kruszka, D.; Kachlicki, P.; Franklin, G. Secondary metabolites in the green synthesis of metallic nanoparticles. *Materials* **2018**, *11*, 940. [CrossRef]
51. Acar-Tek, N.; Ağagündüz, D. Olive leaf (*Olea europaea* L. folium): Potential effects on glycemia and lipidemia. *Ann. Nutr. Metab.* **2020**, *76*, 10–15. [CrossRef]
52. Ren, G.; Hu, D.; Cheng, E.W.; Vargas-Reus, M.A.; Reip, P.; Allaker, R.P. Characterization of copper oxide nano particles for antimicrobial applications. *Int. J. Antimicrob. Agents* **2009**, *33*, 587–590. [CrossRef]
53. Kumari, P.; Panda, P.K.; Jha, E.; Kumari, K.; Nisha, K.; Mallick, M.A.; Verma, S.K. Mechanistic insight to ROS and apoptosis regulated cytotoxicity inferred by green synthesized CuO nanoparticles from *Calotropis gigantea* to embryonic zebrafish. *Sci. Rep.* **2017**, *7*, 16284. [CrossRef] [PubMed]
54. Al Aboody, M.S. Silver/silver chloride (Ag/AgCl) nanoparticles synthesized from *Azadirachta indica* lalex and its antibiofilm activity against fluconazole resistant *Candida tropicalis*. *Artif. Cells Nanomed. Biotechnol.* **2019**, *47*, 2107–2113. [CrossRef] [PubMed]
55. Rashid, M.M.; Akhter, K.N.; Chowdhury, J.A.; Hossen, F.; Hussain, M.S.; Hossain, M.T. Characterization of phytoconstituents and evaluation of antimicrobial activity of silver-extract nanoparticles synthesized from *Momordica charantia* fruit extract. *BMC Complement. Altern. Med.* **2017**, *17*, 336. [CrossRef] [PubMed]
56. Melendrez, M.F.; Ardenas, G.C.; Arbiol, J. Synthesis and characterization of gallium colloidal nanoparticles. *J. Colloid Interface Sci.* **2010**, *346*, 279–287. [CrossRef]
57. Saeb, A.T.M.; Alshammari, A.S.; Al-Brahim, H.; Al-Rubeaan, K.A. Production of silver nanoparticles with strong and stable antimicrobial activity against highly pathogenic and multidrug resistant bacteria. *Sci. World J.* **2014**, *2014*, 704708. [CrossRef]
58. Essien, E.A.; Kavaz, D.; Solomon, M.M. Olive leaves extract mediated zero-valent iron nanoparticles: Synthesis, characterization, and assessment as adsorbent for nickel (II) ions in aqueous medium. *Chem. Eng. Commun.* **2018**, *205*, 1568–1582. [CrossRef]
59. Ardani, H.K.; Imawan, C.; Handayani, W.; Djuhana, D.; Harmoko, A.; Fauzia, V. Enhancement of the stability of silver nanoparticles synthesized using aqueous extract of Diospyros discolor Wild. leaves using polyvinyl alcohol. In *IOP Conference Series: Materials Science and Engineering*; Institute of Physics Publishing: Bristol, UK, 2017; Volume 188, p. 012056.
60. Bhattacharjee, S. DLS and zeta potential—What they are and what they are not? *J. Control. Release* **2016**, *235*, 337–351.
61. Lateef, M.; Azeez, A.; Asafa, T.B.; Yekeen, T.A.; Akinboro, A.; Oladipo, I.C.; Azeez, L.; Ajibade, S.E.; Ojo, S.A.; Gueguim-Kana, E.B.; et al. Biogenic synthesis of silver nanoparticles using a pod extract of *Cola nitida*: Antibacterial and antioxidant activities and application as a paint additive. *J. Taibah Univ. Sci.* **2016**, *10*, 551–562. [CrossRef]
62. Manikandan, A.; Sathiyabama, M. Green synthesis of copper-chitosan Nanoparticles and study of its Antibacterial activity. *J. Nanomed. Nanotechnol.* **2015**, *6*, 251.
63. Tokarek, K.; Hueso, J.L.; Kustrowski, P.; Stochel, G. Green synthesis of chitosan-stabilized copper nanoparticles. *Eur. J. Inorg. Chem.* **2013**, *28*, 940–4947. [CrossRef]
64. Jayaseelan, C.; Ramkumar, R.; Rahuman, A.A.; Perumal, P. Green synthesis of gold nanoparticles using seed aqueous extract of *Abelmoschus esculentus* and its antifungal activity. *Ind. Crops Prod.* **2013**, *45*, 423–429. [CrossRef]
65. Sarac, T.N.; Ugur, A.; Karaca, I.R. Antimicrobial characteristics and biocompatibility of the surgical sutures coated with biosynthesized silver nanoparticles. *Bioorg. Chem.* **2019**, *86*, 254–258.
66. Ravindra, S.; Mohan, Y.M.; Reddy, N.N.; Raju, K.M. Fabrication of antibacterial cotton fibres loaded with silver nanoparticles via Green Approach. *Colloids Surf. A Physicochem. Eng. Asp.* **2010**, *367*, 31–40. [CrossRef]
67. Padil, V.V.T.; Cernik, M. Green synthesis of copper oxide nanoparticles using gum karaya as a biotemplate and their antibacterial application. *Int. J. Nanomed.* **2013**, *8*, 889–898.
68. Costa, E.M.; Silva, S.; Pina, C.; Tavaria, F.K.; Pintado, M.M. Evaluation and insights into chitosan antimicrobial activity against anaerobic oral pathogens. *Anaerobe* **2012**, *18*, 305–309. [CrossRef]
69. El-Tahlawy, K.F.; El-Bendary, M.A.; Elhendawy, A.G.; Hudson, S.M. The antimicrobial activity of cotton fabrics treated with different crosslinking agents and chitosan. *Carbohydr. Polym.* **2005**, *60*, 421–430. [CrossRef]
70. Lim, S.H.; Hudson, S.M. Synthesis and antimicrobial activity of a water-soluble chitosan derivative with a fiber-reactive group. *Carbohydr. Res.* **2004**, *339*, 313–319. [CrossRef]

Article

Gelatin/Chitosan Films Incorporated with Curcumin Based on Photodynamic Inactivation Technology for Antibacterial Food Packaging

Fan Wang [1], Ronghan Wang [1], Yingjie Pan [1,2,3], Ming Du [4], Yong Zhao [1,2,3,*] and Haiquan Liu [1,2,3,5,*]

[1] College of Food Science and Technology, Shanghai Ocean University, Shanghai 201306, China; wangfan970622@163.com (F.W.); wrh13775771822@163.com (R.W.); yjpan@shou.edu.cn (Y.P.)
[2] Shanghai Engineering Research Center of Aquatic-Product Processing & Preservation, Shanghai 201306, China
[3] Laboratory of Quality & Safety Risk Assessment for Aquatic Products on Storage and Preservation (Shanghai), Ministry of Agriculture and Rural Affairs, Shanghai 201306, China
[4] Collaborative Innovation Center of Seafood Deep Processing, Dalian Polytechnic University, Dalian 116034, China; duming121@163.com
[5] Engineering Research Center of Food Thermal-Processing Technology, Shanghai Ocean University, Shanghai 201306, China
* Correspondence: yzhao@shou.edu.cn (Y.Z.); hqliu@shou.edu.cn (H.L.)

Citation: Wang, F.; Wang, R.; Pan, Y.; Du, M.; Zhao, Y.; Liu, H. Gelatin/Chitosan Films Incorporated with Curcumin Based on Photodynamic Inactivation Technology for Antibacterial Food Packaging. *Polymers* 2022, *14*, 1600. https://doi.org/10.3390/polym14081600

Academic Editors: Md. Amdadul Huq and Shahina Akter

Received: 5 March 2022
Accepted: 11 April 2022
Published: 14 April 2022

Publisher's Note: MDPI stays neutral with regard to jurisdictional claims in published maps and institutional affiliations.

Copyright: © 2022 by the authors. Licensee MDPI, Basel, Switzerland. This article is an open access article distributed under the terms and conditions of the Creative Commons Attribution (CC BY) license (https://creativecommons.org/licenses/by/4.0/).

Abstract: Photodynamic inactivation (PDI) is a new type of non-thermal sterilization technology that combines visible light with photosensitizers to generate a bioactive effect against foodborne pathogenic bacteria. In the present investigation, gelatin (GEL)/chitosan (CS)-based functional films with PDI potency were prepared by incorporating curcumin (Cur) as a photosensitizer. The properties of GEL/CS/Cur (0.025, 0.05, 0.1, 0.2 mmol/L) films were investigated by evaluating the surface morphology, chemical structure, light transmittance, and mechanical properties, as well as the photochemical and thermal stability. The results showed a strong interaction and good compatibility between the molecules present in the GEL/CS/Cur films. The addition of Cur improved different film characteristics, including thickness, mechanical properties, and solubility. More importantly, when Cur was present at a concentration of 0.1 mM, the curcumin-mediated PDI inactivated >4.5 Log CFU/mL (>99.99%) of *Listeria monocytogenes*, *Escherichia coli*, and *Shewanella putrefaciens* after 70 min (15.96 J/cm^2) of irradiation with blue LED (455 ± 5) nm. Moreover, *Listeria monocytogenes* and *Shewanella putrefaciens* were completely inactivated after 70 min of light exposure when the Cur concentration was 0.2 mM. In contrast, the highest inactivation effect was observed in *Vibrio parahaemolyticus*. This study showed that the inclusion of Cur in the biopolymer-based film transport system in combination with photodynamic activation represents a promising option for the preparation of food packaging films.

Keywords: gelatin; chitosan; photodynamic inactivation; curcumin; antimicrobial activity

1. Introduction

In recent years, the increase in the level of consumption escalated the demand for goods and the development of the packaging industry. The world production of packaging materials has increased at an alarming rate of 8% per year [1,2]. However, more than 90% of these materials are discarded. Since most of them are petroleum-based plastic, they are difficult to degrade [3], resulting in serious environmental pollution and destruction of biodiversity. In addition, they may affect human health through water, soil, and air pollution. Therefore, in order to find the perfect substitute, natural polymers, such as proteins, carbohydrates, and lipids, as well as their derivatives, which are biodegradable and edible substances, have been used as packaging materials, attracting substantial interest from researchers in recent decades.

GEL is a promising bio-based natural polymer material. It is made of animal protein derived from the partial denaturation of collagen that widely exists in nature (animal skin, connective tissues or organs, and bones). GEL is a water-soluble molecule with rich biological functional groups, such as hydroxyl, carboxyl, and amino groups, that provide this material with excellent biocompatibility and degradability properties [4]. Therefore, it is widely used in the preparation of bio-based food packaging films. However, the pure GEL film often exhibits poor thermal stability and mechanical properties, due to its high hydrophilicity, which limits the application of the pure GEL film [5]. In order to solve these problems, many studies have improved the functional and physicochemical properties of GEL films by adding polysaccharides, such as chitosan (CS) [6], tara gum [7], carboxymethylcellulose [8], and starch [9]. Among them, GEL/CS is the most widely studied edible composite film that presents good film formation, biocompatibility, biodegradability, and safety [10–13]. CS, a deacylated derivative of chitin, is the second most abundant alkaline biological polysaccharide, consisting of β-(1–4)-2-acetamido-D-glucose and β-(1–4)-2-amino-D-glucose units. Many studies have reported that CS displays excellent selective gas permeability and good mechanical properties [14,15]. In addition, CS has been widely used as a bio-based food packaging material due to its abundant availability, low cost, and non-toxicity, among other characteristics.

Photodynamic therapy (PDT) presents antitumor, antibacterial, and antivirus effects and is extensively used in clinical treatments in the medical field [16–18]. However, Photodynamic inactivation (PDI) has rarely been reported in the food sector. Sterilization methods in the food industry can be divided into thermal sterilization and non-thermal sterilization techniques. Thermal sterilization technology is traditional, widely used, and has a wide sterilization spectrum, but its energy consumption is high, and it may cause food to lose its original nutrition and flavor under high temperature. Non-thermal sterilization technologies, such as irradiation, pulsed electric field, and ultrasound, have limited their application in food, due to expensive equipment and high energy consumption [19,20]. PDI is considered to be a promising microbial control strategy due to its environmental protection, low energy consumption, and low cost [21]. It is expected that this new non-thermal sterilization technology will serve as an auxiliary methodology to solve the new challenges faced by the food industry. The mechanism of PDI is based on three indispensable components, including a light source, photosensitizer, and oxygen. In the presence of blue light (450–460 nm), the photosensitizer is excited to produce reactive oxygen species (ROS), including hydroxyl radical ($\cdot$OH), and excited state singlet oxygen ($^1O_2$) through electron transfer or energy transfer. These species are able to destroy almost all types of biomolecules (proteins, lipids, and nucleic acids) and, in consequence, may cause bacteria to die [22]. In the present study, Cur was selected as the active photosensitizer. Cur is a natural plant extract containing polyphenolic active substances isolated from the rhizome of the herb turmeric. At present, it is one of the main natural edible pigments worldwide. A large number of clinical trials have evaluated the safety and efficacy of Cur in humans [23–26], and it has been proven that this pigment presents different functional properties, including anticancer, anti-inflammatory, antioxidant, and antibacterial [27–30]. Therefore, it is often added to biomedical composites as a functional active substance [31,32].

The purpose of this study was to develop bioactive packaging films with antibacterial properties combined with PDI technology. Moreover, the characterizations of GEL/CS/Cur films with concentrations of Cur (0.025, 0.05, 0.1, and 0.2 mM) were performed by scanning electron microscopy (SEM), thermogravimetric analysis (TGA), moisture content (MC), water solubility (WS) and water vapor permeability (WVP), UV–visible absorption, Fourier-transform infrared (FTIR), and X-ray diffraction (XRD). Furthermore, the effects of the GEL/CS/Cur films against pathogenic bacteria (*E. coli, L. monocytogenes, V. parahaemolyticus*) and spoilage bacteria (*S. putrefaciens*) were investigated in pure culture, using a blue LED system. This novel non-thermal technology has great potential in food packaging and can potentially be used to prevent microbial contamination, extend the shelf-life of products, and ensure food safety.

2. Materials and Methods

2.1. Materials and Bacterial Strains

GEL (purity $\geq$ 99%), CS (degree of deacetylation $\geq$95%; viscosity, 100–200 mpa.s) and Cur (purity >98%) were purchased from Shanghai Macklin Biochemical Co., Ltd. (Shanghai, China). Glycerol was obtained from Shanghai Yuanye Biotechnology Co., Ltd.(Shanghai, China). All other chemicals and solvents were of analytical grade. *Listeria monocytogenes* (ATCC19115 and ATCC7644), *Vibrio parahaemolyticus* (ATCC17802), *Escherichia coli* (ATCC 43895), and *Shewanella putrefaciens* (SP 05 and SP 08) were isolated from salmon in our laboratory and stored at -80 °C, with a glycerol concentration of 50% (v/v). Cells were activated, isolated, and cultured to obtain a suspension of ~8.0 Log_{10} CFU/mL.

2.2. Films Preparation

In the present investigation, GEL/CS/Cur films were prepared according to the procedures published by Reference [33], but with slight modifications. For this purpose, GEL/CS/Cur film-forming solutions (FFSs) were obtained. In the first step, the GEL FFS (1.5%, m/v) was prepared by dissolving GEL in deionized water under continuous stirring for 1 h at 55 °C. The CS FFS (1%, m/v) was obtained by dissolving CS in an acetic acid solution (1% v/v), under continuous stirring at 55 °C for 2 h. Later, a 0.01 mmol/mL Cur solution was prepared by adding the proper amount of Cur in 95% ethanol. The mixture was stirred at 300 rpm until Cur was dissolved. Subsequently, the GEL FFS and the CS FFS were mixed according to Table 1, and glycerol (0.3%, m/v) was added. Magnetic stirring was applied for 3 h to achieve homogenization and remove bubbles. Different concentrations of FFSs are shown in Figure 1. Finally, all films were obtained by adding 30 mL of the FFSs to polystyrene Petri dishes (10 cm × 10 cm × 1 cm) and dried in an oven with the air-flow circulation at 40 °C for 24–36 h. Samples were stored at 25 °C and 55% RH in a brown desiccator before further analyses.

Table 1. Film nomenclature and final formulation of film-forming dispersions.

Film Nomenclature	Gelatin (%, *w/v*)	Chitosan (%, *w/v*)	Ratio (%, *w/v*)	Glycerol (%, *w/v*)	Curcumin (mmol/L)
GEL/CS	1.5	1	4:6	0.3	-
GEL/CS /Cur0.025	1.5	1	4:6	0.3	0.025
GEL/CS /Cur0.05	1.5	1	4:6	0.3	0.05
GEL/CS /Cur0.1	1.5	1	4:6	0.3	0.1
GEL/CS /Cur 0.2	1.5	1	4:6	0.3	0.2

Figure 1. Film-forming solutions (FFSs) with different concentrations of Cur (prepared with a 4:6 mixture of GEL and CS).

2.3. Characterizations of the Composite Films

2.3.1. Scanning Electron Microscopy (SEM)

The microstructure of the surface and cross-section of the films was acquired by using SEM (Quanta FEG 250, Hillsboro, OR, USA). The composite films were fixed on a stainless-steel support with a double-sided adhesive, and the analysis was conducted in low vacuum (0.6 Torr), at an acceleration voltage of 20 and 10 kV respectively.

Four bacterial suspensions (1 mL) were centrifuged for 10 min at 4000× g. The supernatants were discarded, and the pellets were mixed with 500 µL of glutaraldehyde (2.5%) and formaldehyde (4%) in 0.1 M cacodylate buffer for 8 h at 4 °C. Subsequently the samples were dehydrated in serial dilutions of ethanol solutions (30%, 50%, 70%, 90%, and 100%) for 10 min. The samples were separately placed on the support, with a double-sided adhesive, and coated with gold. The microstructures of cells were observed by using SEM.

2.3.2. Color

Color parameters were measured by using a portable Minolta colorimeter (JZ-300, Osaka, Japan) with a standard white color plate (L_0 = 99.44, a_0 = −0.28, b_0 = 0.54) as the background reference. Results of L* (lightness), a* (red to green), and b* (yellow to blue) were directly read from the colorimeter. The total color difference (ΔE) of the films was calculated according to Equation (1) [34]:

$$\Delta E = \sqrt{(\Delta L^*)^2 + (\Delta a^*)^2 + (\Delta b^*)^2} \tag{1}$$

where ΔL*, Δa*, and Δb* are the differences between each color value of the standard color plate and film specimen, respectively. Values were expressed as the means of ten measurements on different areas of each film.

2.3.3. UV–Visible Spectra

The UV–Vis transmission spectra of the composite films (1 cm × 4 cm) were obtained in order to evaluate the effect of the addition of Cur on the barrier properties of films to ultraviolet (UV) and visible (Vis) light. Spectra were recorded by using a UV spectrophotometer (UV-3600, Shimadzu, Tokyo, Japan). The film opacity was calculated by using Equation (2) [35]:

$$\text{Opacity value} = \frac{-\log T_{600}}{x} \tag{2}$$

where T_{600} is the transmittance at 600 nm, and x is the film thickness (mm).

2.3.4. Thickness and Mechanical Properties

The thickness of film samples was measured by randomly taking the average of 10 points on the film, using a micrometer caliper (Mitutoyo, Japan) with a precision of 0.001 mm. The tensile strength (TS) and elongation at break (EB) of the films were obtained by using an Auto Tensile Tester (XLW-EC, PARAM, Jinan, China). Before testing, film samples were cut into strips (15 mm × 100 mm) and mounted in the tensile grip at an initial distance of 65 mm. Later, samples were stretched at a cross-head speed of 50 mm/min until breaking occurred. At least five replicates were tested for each film.

2.3.5. Moisture Content (MC), Water Solubility (WS), and Water Vapor Permeability (WVP)

The film pieces (2 cm × 2 cm, n = 3, M_1) were dried in an oven at 105 °C for 24 h to reach a constant weight (M_2). Samples were then completely immersed in centrifuge tube with 30 mL distilled water. Tubes were shaken at 180 r/min and 26 °C for 24 h. Later, samples were filtered to remove excess water and dried at 105 °C for 24 h until constant weight (M_3). The MC and WS (%) of film samples were calculated by using Equations (3) and (4) [36].

$$\text{MC } (\%) = \frac{M_1 - M_2}{M_1} \times 100 \tag{3}$$

$$WS\% = \frac{M_2 - M_3}{M_2} \times 100 \quad (4)$$

The water vapor permeability (WVP) of the films was measured gravimetrically according to E96-05 (ASTM, 2005), but with some modifications. In this assay, the glass weighing bottle was filled with 20 mL of distilled water (100% RH). Later, the film sample was placed over the circular opening and sealed tightly with parafilm to prevent the leakage of water vapor. The glass weighing bottles were maintained at a constant temperature of 25 °C. Weight changes of the glass weighing bottles were monitored at intervals of 2 h for a total of 12 h. The slope of weight changes versus time plot was obtained by using linear regression ($r^2 > 0.99$). WVTR and WVP were calculated according to Equations (5) and (6) [37].

$$WVTP = \frac{\Delta w}{A \times \Delta t} \quad (5)$$

$$WVP = \frac{WVTR \times L}{\Delta P \times \Delta RH} \quad (6)$$

where $\Delta W/\Delta t$ indicated the weight change as a function of time (g/h), A is the area of the exposed film surface (m^2), L corresponds to the mean film thickness (m), Δp is the water vapor pressure difference (kPa) between two sides of the film, and ΔRH is the relative humidity gradient across the film (%).

2.3.6. Fourier-Transform Infrared (FTIR) Spectroscopy

Fourier-transform infrared (FTIR) spectroscopy analysis was performed to obtain chemical information of the films' surface. The spectra of Cur and composite films were obtained by using a spectrophotometer (Thermo IS10, Thermo Fisher, MA, USA) system from 400 to 4000 cm^{-1}, with a resolution of 4 cm^{-1} and 32 scans.

2.3.7. X-Ray Diffraction (XRD)

In order to examine the crystalline structure of the films, X-ray diffraction patterns (XRD) were recorded by using a Rigaku Ultima IV X-ray diffractometer (RINT2000, Tokyo, Japan), equipped with a Cu-Kα radiation, at 40 kV voltage and 30 mA current. Samples were scanned over the 2θ range of 5–50°, at a speed of 2°/min (RINT2000, Tokyo, Japan).

2.3.8. Thermogravimetric Analysis (TGA)

The thermal properties of the composite films were determined by using a thermal analyzer (NETZSCH STA 449C, Selb, German). The measured samples were kept in the range of 30–800 °C, and the heating rate was 10 °C/min. Nitrogen was used as the protective gas [38].

2.4. Photodynamic Inactivation of the Composite Films

2.4.1. Light-Emitting Diodes (LEDs) System

The blue LEDs (10 W, 450–460 nm, 30 cm) were used as the light source for the photodynamic treatment [39,40]. These LEDs were surrounded by deep photo accessories to prevent the interference of external light sources. The film samples were cut into squares (2 cm × 2 cm) and placed directly on 6-well plates. The distance was adjusted to 5 cm between the light source and film samples. The blue light intensity was 3.8 mW/cm^2, which was determined by using a PM100D energy meter console (Newton, MA, USA). The obtained energy dosage of each composite film sample was calculated by using Equation (7) [41].

$$E = Pt \quad (7)$$

where E = dose (energy density) in J/cm^2, P = irradiance (power density) in W/cm^2, and t = time in s.

2.4.2. Antimicrobial Activity

The bacterial suspension (100 μL) was evenly distributed on the surface of the GEL/CS/Cur films containing different concentrations of Cur (0, 0.025, 0.05, 0.1, and 0.2 mM). Later, films were exposed to light for 70 min (15.96 J/cm^2) and then maintained in the dark for another 10 min to ensure that the bacteria were able to attach to the film before irradiation. In addition, GEL/CS/Cur films containing 0.1 mM Cur were irradiated for 30 min (6.84 J/cm^2), 50 min (11.4 J/cm^2), 70 min (15.96 J/cm^2), and 90 min (20.52 J/cm^2) in order to explore their potential use in PDI of food-borne pathogenic bacteria. After treatment, the films containing bacteria were homogenized with sodium chloride (0.85%, w/v) for 5 min. The antibacterial effect of the GEL/CS/Cur films was investigated by spreading 100 μL of the suspension onto agar plates and incubated for 12–48 h. Viable cells were quantified as Log CFU/mL. Herein, samples without light treatment and Cur were labeled as negative control (L−C−). In addition, samples with light treatment but without Cur were labeled as illumination control (L+C−). Moreover, samples with Cur but without light treatment were identified as Cur control (L−C+). All the experiments were performed in triplicate.

2.5. Statistical Analysis

The experimental data were analyzed by using SPSS (SPSS 17.0 Software, Inc., Chicago, IL, USA). One-way analysis of variance (ANOVA) was used to compare differences between pairs of means ($p < 0.05$).

3. Results and Discussion

3.1. Optical Properties of Films

Transparency and color of edible food packaging materials are critical properties that influence consumer acceptance, as they directly affect the appearance of the product. The color values (L*, a*, and b*), total color difference (ΔE*), opacity, and picture of the GEL/CS film and GEL/CS /Cur films are shown in Table 2. The L* values of the composite film were in the range of 86.09–93.38, and the brightness of the composite films were not affected by the presence of Cur. However, the a* and b* values of the GEL/CS/Cur composite films were higher than those of the GEL/CS films, which indicated that the yellowness and redness of the composite films significantly improved with the addition of Cur. As a result, the total color difference (ΔE*) of the curcumin-added composite films increased as compared to GEL/CS. Data in Table 2 indicated that the increase in Cur concentration resulted in an increase in film opacity, for which the values were lower than 5 at 600 nm in all cases. Thus, we can conclude that all the films prepared in the present study were transparent. The higher the opacity value, the lower the transparency of the film.

Table 2. Color parameter (L*, a*, b*, and ΔE*; n = 10), opacity (n = 3) values and digital images of the composite films.

Film Samples	L*	a*	b*	ΔE	Opacity	Image
GEL/CS	93.38 ± 0.07 [a]	0.04 ± 0.03 [d]	3.23 ± 0.82 [c]	4.13 ± 0.46 [c]	1.22 ± 0.04 [e]	CS/GEL
GEL/CS /Cur0.025	90.73 ± 0.47 [b]	1.23 ± 0.26 [c]	15.00 ± 2.12 [b]	15.64 ± 1.95 [b]	1.73 ± 0.06 [d]	CS/GEL/Cur
GEL/CS /Cur0.05	89.91 ± 0.20 [bc]	1.69 ± 0.10 [bc]	16.63 ± 0.39 [b]	17.47 ± 0.43 [b]	2.09 ± 0.07 [c]	CS/GEL/Cur
GEL/CS /Cur0.1	89.45 ± 0.41 [c]	2.22 ± 0.43 [b]	17.80 ± 2.03 [b]	18.79 ± 2.07 [b]	2.83 ± 0.07 [b]	CS/GEL/Cur
GEL/CS /Cur 0.2	86.09 ± 0.85 [d]	6.47 ± 0.63 [a]	30.76 ± 3.03 [a]	32.65 ± 3.17 [a]	4.69 ± 0.06 [a]	CS/GEL/Cur

[a,b,c,d,e] Different superscript letters between columns indicate significant difference between the results ($p < 0.05$) (ANOVA).

Figure 2 presented the UV–visible light transmission spectra (200–800 nm) of the composite films with and without Cur. The packaging material with good light-barrier properties effectively prevents light transmission and reduces light-induced oxidation of packaged foods, consequently inhibiting the lipid oxidation, nutrient loss, and degradation of active compounds [42]. All composite films exhibited high protection against UV light, which was probably the result of the excellent UV-absorbing properties of aromatic amino acids found in the GEL [43,44]. It was also observed that light absorption by the GEL/CS/Cur composite films decreased in the visible light wavelength region of 390–450 nm. This probably occurred because Cur absorbs light in the range of 400–500 nm, which is similar to the tested wavelength range. Since the phenolic compounds in Cur display excellent light absorption properties, the addition of Cur improved the light-barrier characteristics of the films. With the increase of the Cur concentration, the light transmittance of the composite films decreased. However, all the analyzed films presented good transparency.

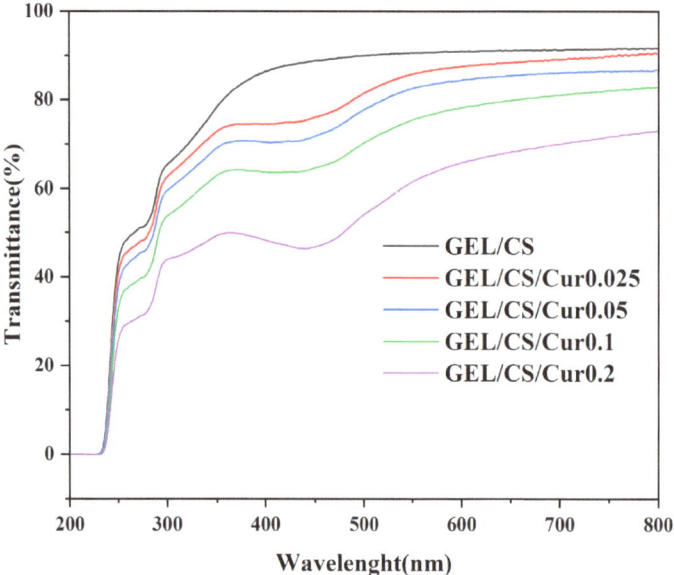

Figure 2. UV–Vis transmittance spectra of GEL/CS and GEL/CS/Cur composite films.

3.2. Microstructure of Films

The surface topography and cross-section are used to characterize the microstructure of packaging materials. This information is helpful in determining different properties, including sealing and flexibility. The surface and cross-section of the GEL/CS film and GEL/CS/Cur composite films were observed by using SEM (Figure 3). GEL/CS, GEL/CS/Cur 0.025 and GEL/CS/Cur 0.05 films presented homogeneous and smooth surfaces. However, with increasing Cur concentrations, slight protuberances appeared on the surface of GEL/CS/Cur films. In addition, when films contained Cur, the cross-section of the composite films appeared slightly rough in contrast to the control films. However, the films also displayed a uniform thickness and regular texture. Moreover, no macroscopic phase separation was observed, indicating that Cur was properly dispersed in GEL/CS FFSs. This mainly occurred because of the good materials compatibility. Roy et al. (2017) also found that Cur was well distributed in GEL/Cur composite films [30,45].

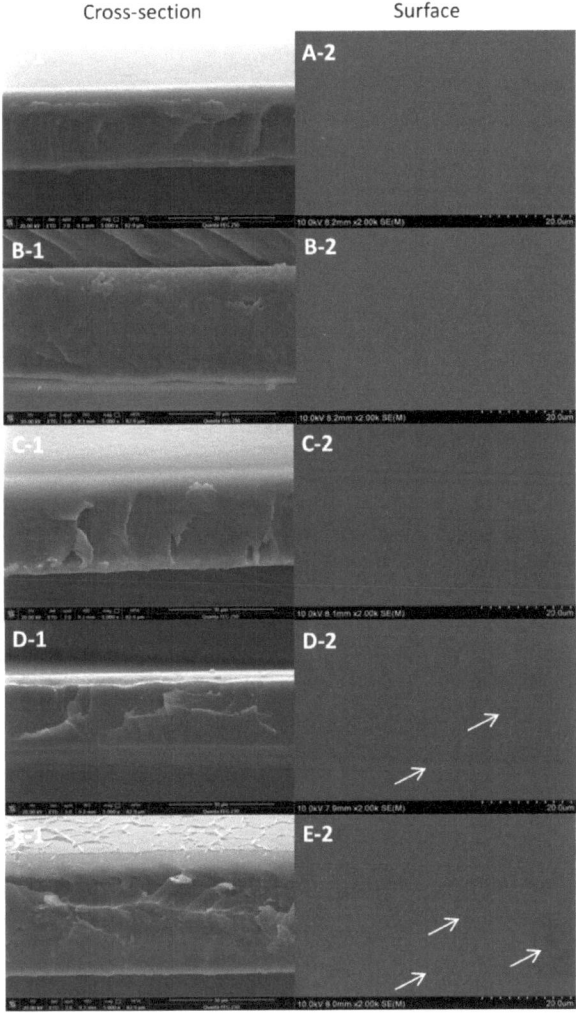

Figure 3. Scanning electron microscopy (SEM) images of the surface (**A-1–E-1**) and cross section (**A-2–E-2**) of composite films. (**A-1,A-2**) GEL/CS films; (**B-1–E-1**) and (**B-2–E-2**) GEL/CS/Cur films with Cur contents of 0.025, 0.05, 0.1, and 0.2, respectively.

3.3. Mechanical Properties

Because of the low content of Cur, the thickness of the films varied in the range of 0.33–0.38, with little difference. Data in Table 3 indicate that the tensile strength (TS) and elongation at break (EB) of the composite films increased with respect to controls. The mechanical properties of the fabricated GEL/CS/Cur films were affected by the addition of Cur. It was believed that the hydrogen bond interaction between Cur and GEL/CS was responsible for the improved TS of the films [30]. These results agree with the XRD data. Other studies have reported that EB increases because Cur improves the adhesion between the filler and the polymer [46,47].

Table 3. Thickness, tensile strength (TS), elongation at break (EB), moisture content (MC), water solubility (WS), and water vapor permeability (WVP) of the GEL/CS films and those GEL/CS/Cur films with different concentrations of Cur.

	Thickness (μm)	TS (MPa)	EB (%)	MC (%)	WS (%)	WVP (g·mm/m²·h·kPa)
GEL/CS	0.33 ± 0.06 [b]	14.12 ± 0.57 [e]	53.19 ± 1.27 [e]	21.96 ± 0.42 [a]	20.46 ± 1.53 [d]	0.304 ± 0.029 [ab]
GEL/CS/Cur0.025	0.34 ± 0.05 [b]	14.74 ± 0.44 [d]	56.01 ± 3.15 [d]	21.29 ± 0.56 [b]	22.06 ± 0.57 [c]	0.296 ± 0.013 [ab]
GEL/CS/Cur0.05	0.36 ± 0.01 [a]	15.43 ± 0.66 [c]	59.08 ± 1.40 [c]	20.04 ± 0.11 [c]	22.33 ± 1.16 [c]	0.289 ± 0.008 [b]
GEL/CS/Cur0.1	0.37 ± 0.02 [a]	16.85 ± 0.45 [b]	60.72 ± 1.59 [b]	19.38 ± 0.73 [d]	23.42 ± 0.63 [b]	0.319 ± 0.026 [ab]
GEL/CS/Cur0.2	0.38 ± 0.02 [a]	18.12 ± 0.31 [a]	65.26 ± 0.62 [a]	18.54 ± 1.12 [e]	24.75 ± 0.73 [a]	0.325 ± 0.014 [a]

Reported values for each film are means ± standard deviation (n = 10 for thickness; n = 3 for TS, EB, MC, WS, and WVP). [a,b,c,d,e] Different superscript letters in the same column indicate significant differences between samples ($p < 0.05$), according to ANOVA.

3.4. Moisture Content (MC), Water Solubility (WS), and Water Vapor Permeability (WVP)

Water sensitivity plays an important role in the wide applications of biodegradable films. The results of the moisture content (MC), water solubility (WS), and water vapor permeability (WVP) of GEL/CS and GEL/CS/Cur films were presented in Table 3.

According to our data, GEL/CS films displayed an MC of 21.96% higher as that quantified in GEL/CS/Cur films. Cur is a hydrophobic molecule with a very small capacity for water retention. This resulted in a continuous MC decrease from 21.29 to 18.54%. The UV–Vis spectra indicated that Cur-containing films absorbed UV–Vis irradiation between 390 and 450 nm, which is in agreement with the maximum absorption of Cur that occurs between 400 and 500 nm. As the Cur concentration gradually increased, the maximum absorption shifted to blue. This explains why the WS of GEL/CS/Cur films increased by increasing the concentration of Cur. Gómez-Estaca et al. [48] also observed a blue shift in the absorption spectrum of Cur, which occurred because of the formation of a Cur–GEL complex. This complex is responsible for the increase in Cur solubility in water. In addition, the increase in WS could be explained by the fact that the GEL swell in water and was partially soluble at 25 °C. As a result, with the increase in Cur content, the WS value of the GEL/CS/Cur films exhibited a continuous increase, reaching values between 20.46 and 24.75%.

The water barrier properties meet the requirements for food preservation. Therefore, a lower WVP should be considered in the application of biodegradable films. In the present study, by increasing the concentration of Cur, the overall value of WVP increased, with small differences. The complex composition of polymers results in weak cohesion between their components, leading to a less disordered crystal structure [49]. This phenomenon was the possible cause of the WVP increase.

3.5. Physicochemical and Structural Properties of Films

The chemical nature of the interactions in the GEL/CS/Cur films were investigated by using FTIR (Figure 4). The Cur spectra showed a peak at 3510 cm^{-1}, which was attributed to the phenol O-H stretching vibrations. Additional peaks at 1630 and 1505 cm^{-1}, corresponding to C=O and C=C stretching vibrations of the Cur structures, were also observed. Furthermore, peaks at 1275 cm^{-1} referred to ether C-O stretching vibration, and those at 810 and 964 cm^{-1} represented the C-H bending vibration, which was consistent with the alkene structure in Cur [30]. In the GEL/CS films, a broad band at 3287 cm^{-1} was attributed to the overlapped −OH stretching vibration of CS and GEL. The peak at 1536 cm^{-1} corresponded to the amide II C=O stretching vibration of CS, which overlapped with that of GEL [6]. Moreover, the spectra of GEL/CS showed bands corresponding to different amide types present in GEL (A, B, I, II, and III). These peaks were observed at 3285, 2936, 1637, 1541, and 1242 cm^{-1}, respectively, which indicated the existence of N-H, −CH$_2$, and C=O stretching vibration, with N-H bending coupled to C-N stretching, and C-N stretching coupled to N-H bending [50]. Interestingly, after adding different Cur concentrations during the preparation of the GEL/CS/Cur composite films, the spectra of these materials displayed similar main peaks as those of the GEL/CS films. This indicated that the addition of Cur preserved the chemical structures originally present in GEL/CS films. Thus, the addition of Cur did not produce new chemical structures [51]. Moreover, changes in peak intensities were the result of physical interactions, non-covalent interactions, and hydrogen bonding [52,53].

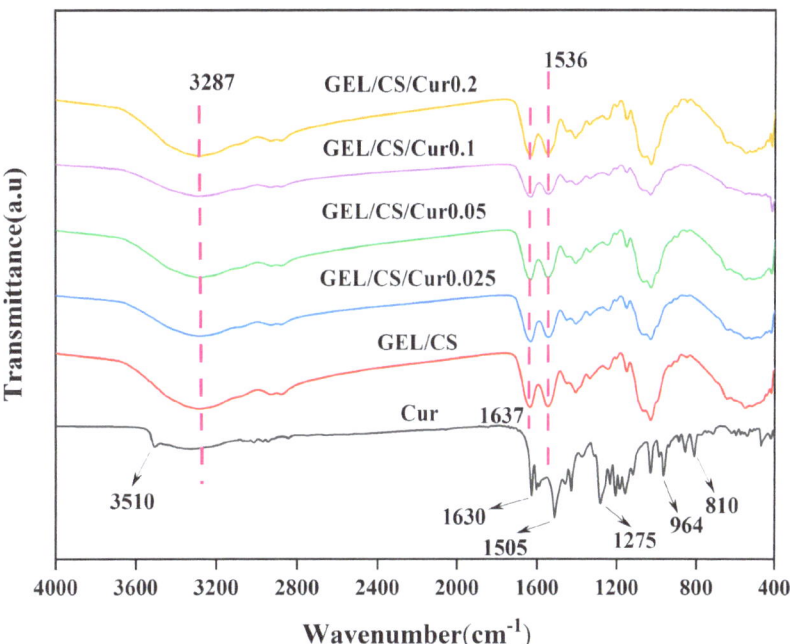

Figure 4. Fourier–transform infrared (FTIR) spectra of GEL/CS and GEL/CS/Cur composite films.

The crystalline structures of the composite films were discussed by XRD analysis (Figure 5). All composite films presented a similar wide diffraction peak at $2\theta = 20°$, which was considered to be the random coiled conformation of GEL, and this meant that the film with Cur added was amorphous [54]. The XRD data of the GEL/CS/Cur composite films showed that the Cur addition resulted in an increased crystallinity. For this reason, the diffraction peak intensity increased [55]. However, according to the literature [56], pure Cur was a crystalline material that presented a series of diffraction peaks between 7° and 30°. The disappearance of Cur peaks in the GEL/CS/Cur XRD pattern probably occurred because Cur was present at low concentrations. Thus, the characteristic peaks of GEL/CS overlapped with that of Cur [46].

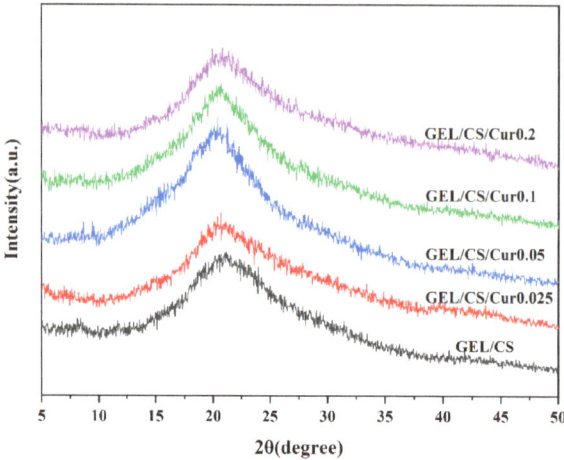

Figure 5. X-ray diffraction patterns of GEL/CS and GEL/CS/Cur composite films.

3.6. Thermal Properties

The thermal stability of the composite films was investigated by using thermogravimetric analysis (TGA) and derivative thermogravimetric analysis (DTG), and the results are shown in Figure 6. Cur is a hydrophobic molecule with a very low capacity for water retention and is stable until 200 °C. After this point, Cur is continuously degraded before 450 °C. However, it reaches the maximum degradation rate at 375 °C, and the residue rate was 35% at 800 °C. These results were consistent with the reported literature [48].

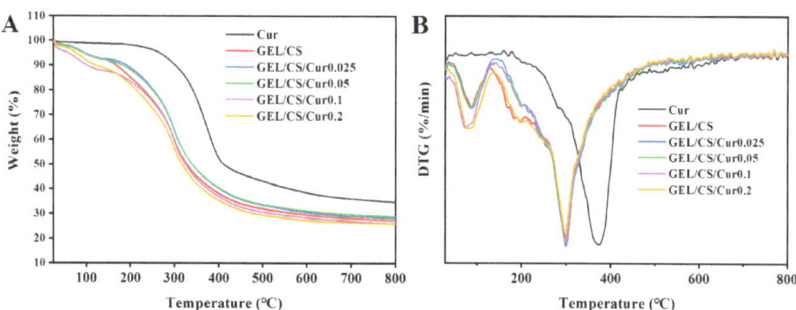

Figure 6. Thermogravimetric (TG) (**A**) and derivative thermogravimetric (DTG) (**B**) thermograms of Cur powder, GEL/CS, and GEL/CS/Cur composite films.

The weight loss rates of the GEL/CS/Cur films were faster than those of Cur and showed the similar weight-loss characteristics of the GEL/CS films. The weight-loss process was divided into two main stages: (1) water loss in the range between 50 and 150 °C, and (2) degradation of composites at 200–800 °C [57]. The maximum degradation rate occurred at about 300 °C, and the rapid degradation of Cur disappeared at 375 °C. In addition, the decomposition temperature for Cur was higher than that for composite films. This proved that Cur had good compatibility with GEL and CS, and that Cur was well embedded in GEL/CS films, forming a uniform system [58]. When the Cur was added at a 0.05 mM concentration, the residual rate of the film was the highest one (32%), which also indicated that the thermal stability of GEL/CS/Cur film was improved [57].

3.7. In Vitro Antimicrobial Properties

In the present research, we evaluated the antibacterial potency of the composite films against *E. coli*, *L. monocytogenes*, *V. parahaemolyticus*, and *S. putrefaciens* (Figures 7 and 8). For this purpose, we selected blue LED illumination and Cur concentration as study variables. As it was shown in Figure 7, individual LED illumination (L+C−) and Cur treatment (L−C+) did not cause significant changes in the antimicrobial activity of the films as compared to the negative control (L−C−).

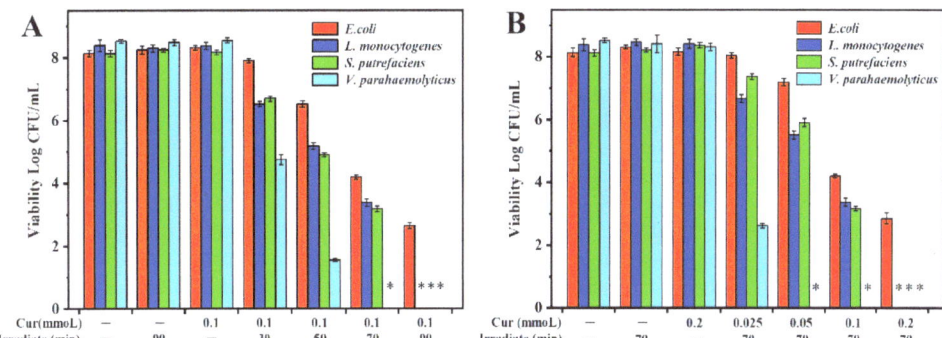

Figure 7. Antibacterial effect of the PDI-treated composite films against *E. coli*, *L. monocytogenes*, *S. putrefaciens*, and *V. parahaemolyticus* at different illumination times (**A**) and Cur concentrations (**B**). * Indicates absence of growth.

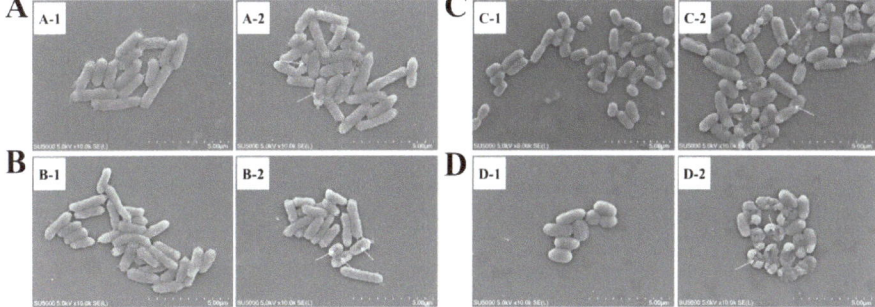

Figure 8. Effects of the curcumin-medicated PDI on the outer membranes of the four bacterial cells. (**A–D**) *E. coli*, *L. monocytogenes*, *S. putrefaciens*, and *V. parahaemolyticus*, respectively. (**A-1–D-1**) L−C−; (**A-2–D-2**) 15.96 J/cm^2 and 0.1 mM Cur.

When the illumination time was increased, the irradiation dose increased and the antibacterial activity improved. These results were shown in Figure 7A. Obviously, the fabricated GEL/CS/Cur films exhibited good antibacterial activity when the Cur concentration was treated by 0.1 mM after 30 min of irradiation (6.84 J/cm^2). In the case of *V. parahaemolyticus* cells, a decrease from 8.55 to 4.76 Log CFU/mL was observed. Moreover, after the GEL/CS/Cur film with 0.1 mM Cur was irradiated for 70 min (15.96 J/cm^2), no bacterial cells were detected. Furthermore, *L. monocytogenes* and *S. putrefaciens* cells were killed and could not be detected after 90 min of irradiation (20.52 J/cm^2). It was also observed that, when the PDI illumination time increased from 30 min (6.84 J/cm^2) to 90 min (20.52 J/cm^2), *E. coli* cells presented a continuous decrease from 8.15 to 2.64 Log CFU/mL.

The effects of the Cur concentration on the inactivation of the four bacterial cells were evaluated in Figure 7B. In all cases, an increase in Cur concentration from 0.025 to

0.2 mM led to a significant decrease in the number of bacterial cells. Among the four species, *V. parahaemolyticus* was the most affected, as the number of cells decreased to 2.63 Log CFU/mL after 70 min of irradiation (15.96 J/cm^2). When Cur concentration in the GEL/CS/Cur films increased to 0.05 mM, none of *V. parahaemolyticus* cells was detected. After 70 min of irradiation with 0.1 mM Cur, *L. monocytogenes* and *S. putrefaciens* cells decreased to 3.38 and 3.17 Log CFU/mL, respectively. Likewise, none of the bacterial cells was detectable when the Cur concentration increased to 0.2 mM. In the case of *E. coli*, when the Cur concentration augmented from 0.025 to 0.2 mM, the number of cells decreased from 8.15 to 3.02 Log CFU/mL.

In the negative control (L−C−), four bacterial cells were plump and rod-shaped (Figure 8). Under the PDI treatment with the illumination time of 70 min (15.96 J/cm^2) and the Cur concentration of 0.1 mM, the morphological deformation and groove appeared in the *E. coli*, *L. monocytogenes*, and *S. putrefaciens* cells, and the cell surfaces of *S. putrefaciens* and *V. parahaemolyticus* appeared atrophied and ruptured. Obviously, after PDI treatment, *V. parahaemolyticus* cells were most seriously damaged.

Based on the obtained results, it was concluded that the PDI-induced antimicrobial films displayed a typical Cur concentration and irradiation dosage-dependent feature. The photosensitizer Cur produced singlet oxygen and some reactive oxygen species when exposed to blue light (450–460 nm) irradiation, which presents strong oxidative effects and is able to destroy different macromolecular structures, such as proteins, DNA, and lipids. This process may result in the destruction and even death of cells and, in consequence, can be used to eliminate pathogenic bacteria. Previous studies have reported that PDI-induced antimicrobial films exhibited a broad-spectrum antibacterial activity [27,59].

4. Conclusions

Degradable and environmentally friendly GEL/CS/Cur composite films were prepared via solution casting. As the photosensitizer of PDI technology, Cur was uniformly dispersed in the fabricated films. Cur presented a high compatibility with GEL and CS. For this reason, this compound improved the mechanical properties and thermal stability of the films. Moreover, low Cur concentrations contributed to smooth the continuous surfaces. However, the MC value decreased and WVP slightly changed, suggesting that the water barrier property of the film was enhanced. The resulting films were resistant to UV and visible light. We would also like to emphasize the fact that the PDI-mediated GEL/CS/Cur films presented a great antibacterial effect in regard to *E. coli*, *L. monocytogenes*, *V. parahaemolyticus*, and *S. putrefaciens* in pure cultures. Furthermore, with the increase in Cur concentration and illumination time, the inactivation effect was enhanced. According to our results, we believe that edible composite films combined with Cur have a great potential application in PDI technology for microbial control in the food packaging industry.

Author Contributions: Conceptualization, Y.P., M.D., Y.Z. and H.L.; methodology, F.W., Y.Z. and H.L.; software, F.W. and R.W.; validation, F.W., Y.Z. and M.D.; investigation, F.W. and R.W.; resources, F.W., M.D., Y.Z. and H.L.; data curation, F.W. and R.W.; writing—original draft preparation, F.W.; writing—review and editing, M.D., Y.Z. and H.L.; supervision, Y.Z. and H.L.; funding acquisition, Y.P., M.D., Y.Z. and H.L. All authors have read and agreed to the published version of the manuscript.

Funding: This research was supported by the National Natural Science Foundation of China (31671779), "Plan of Action for Scientific and Technological innovation" of Science and Technology Commission of Shanghai Municipality (19391901600, 22N31900600), Capacity building program of local colleges and universities in Shanghai of Science and Technology Commission of Shanghai Municipality (22010502300), and Shanghai Municipal Science and Technology Committee of Shanghai outstanding academic leaders plan (21XD14012009).

Institutional Review Board Statement: Not applicable.

Informed Consent Statement: Not applicable.

Data Availability Statement: Not applicable.

Conflicts of Interest: The authors declare no conflict of interest.

References

1. Sutherland, W.J.; Aveling, R.; Bennun, L.; Chapman, E.; Clout, M.; Côté, I.M.; Depledge, M.H.; Dicks, L.V.; Dobson, A.P.; Fellman, L.; et al. A horizon scan of global conservation issues for 2012. *Trends Ecol. Evol.* **2012**, *27*, 12–18. [CrossRef] [PubMed]
2. Beikzadeh, S.; Khezerlou, A.; Jafari, S.M.; Pilevar, Z.; Mortazavian, A.M. Seed mucilages as the functional ingredients for biodegradable films and edible coatings in the food industry. *Adv. Colloid Interface Sci.* **2020**, *280*, 102164. [CrossRef] [PubMed]
3. Espitia, P.J.P.; Du, W.-X.; Avena-Bustillos, R.D.J.; Soares, N.D.F.F.; McHugh, T.H. Edible films from pectin: Physical-mechanical and antimicrobial properties—A review. *Food Hydrocoll.* **2014**, *35*, 287–296. [CrossRef]
4. Huang, Z.-M.; Zhang, Y.Z.; Ramakrishna, S.; Lim, C.T. Electrospinning and mechanical characterization of gelatin nanofibers. *Polymer* **2004**, *45*, 5361–5368. [CrossRef]
5. Ma, L.; Zhang, M.; Bhandari, B.; Gao, Z. Recent developments in novel shelf life extension technologies of fresh-cut fruits and vegetables. *Trends Food Sci. Technol.* **2017**, *64*, 23–38. [CrossRef]
6. Rodrigues, M.Á.V.; Marangon, C.A.; Martins, V.d.C.A.; Plepis, A.M.D.G. Chitosan/gelatin films with jatobá resin: Control of properties by vegetal resin inclusion and degree of acetylation modification. *Int. J. Biol. Macromol.* **2021**, *182*, 1737–1745. [CrossRef]
7. Nuvoli, L.; Conte, P.; Fadda, C.; Reglero Ruiz, J.A.; García, J.M.; Baldino, S.; Mannu, A. Structural, thermal, and mechanical properties of gelatin-based films integrated with tara gum. *Polymer* **2021**, *214*, 123244. [CrossRef]
8. Łupina, K.; Kowalczyk, D.; Lis, M.; Raszkowska-Kaczor, A.; Drożłowska, E. Controlled release of water-soluble astaxanthin from carboxymethyl cellulose/gelatin and octenyl succinic anhydride starch/gelatin blend films. *Food Hydrocoll.* **2022**, *123*, 107179. [CrossRef]
9. Cheng, Y.; Gao, S.; Wang, W.; Hou, H.; Lim, L.-T. Low temperature extrusion blown ε-polylysine hydrochloride-loaded starch/gelatin edible antimicrobial films. *Carbohydr. Polym.* **2022**, *278*, 118990. [CrossRef]
10. Ahmadi, S.; Hivechi, A.; Bahrami, S.H.; Milan, P.B.; Ashraf, S.S. Cinnamon extract loaded electrospun chitosan/gelatin membrane with antibacterial activity. *Int. J. Biol. Macromol.* **2021**, *173*, 580–590. [CrossRef]
11. Ahmad, A.A.; Sarbon, N.M. A comparative study: Physical, mechanical and antibacterial properties of bio-composite gelatin films as influenced by chitosan and zinc oxide nanoparticles incorporation. *Food Biosci.* **2021**, *43*, 101250. [CrossRef]
12. Zhao, J.; Wei, F.; Xu, W.; Han, X. Enhanced antibacterial performance of gelatin/chitosan film containing capsaicin loaded MOFs for food packaging. *Appl. Surf. Sci.* **2020**, *510*, 145418. [CrossRef]
13. Zhang, C.; Yang, Z.; Shi, J.; Zou, X.; Zhai, X.; Huang, X.; Li, Z.; Holmes, M.; Daglia, M.; Xiao, J. Physical properties and bioactivities of chitosan/gelatin-based films loaded with tannic acid and its application on the preservation of fresh-cut apples. *LWT* **2021**, *144*, 111223. [CrossRef]
14. Miao, Q.; Mi, Y.; Cui, J.; Zhang, J.; Tan, W.; Li, Q.; Guo, Z. Determination of chitosan content with Schiff base method and HPLC. *Int. J. Biol. Macromol.* **2021**, *182*, 1537–1542. [CrossRef]
15. Kumar, D.; Kumar, P.; Pandey, J. Binary grafted chitosan film: Synthesis, characterization, antibacterial activity and prospects for food packaging. *Int. J. Biol. Macromol.* **2018**, *115*, 341–348. [CrossRef]
16. Yi, E.; Yang, C.K.; Leem, C.; Park, Y.; Chang, J.-E.; Cho, S.; Jheon, S. Clinical outcome of photodynamic therapy in esophageal squamous cell carcinoma. *J. Photochem. Photobiol. B Biol.* **2014**, *141*, 20–25. [CrossRef]
17. Machado, F.C.; de Matos, R.P.A.; Primo, F.L.; Tedesco, A.; Rahal, P.; Calmon, M.F. Effect of curcumin-nanoemulsion associated with photodynamic therapy in breast adenocarcinoma cell line. *Bioorg. Med. Chem.* **2019**, *27*, 1882–1890. [CrossRef]
18. Paiva, A.d.C.M.d.; Ferreira, M.D.C.; Fonseca, A.D.S.D. Photodynamic therapy for treatment of bacterial keratitis. *Photodiagn. Photodyn. Ther.* **2022**, *37*, 102717. [CrossRef]
19. Li, X.; Farid, M. A review on recent development in non-conventional food sterilization technologies. *J. Food Eng.* **2016**, *182*, 33–45. [CrossRef]
20. Su, J.; Cavaco-Paulo, A. Effect of ultrasound on protein functionality. *Ultrason. Sonochem.* **2021**, *76*, 105653. [CrossRef]
21. Luksiene, Z.; Zukauskas, A. Prospects of photosensitization in control of pathogenic and harmful micro-organisms. *J. Appl. Microbiol.* **2009**, *107*, 1415–1424. [CrossRef] [PubMed]
22. Broekgaarden, M.; Weijer, R.; Van Gulik, T.M.; Hamblin, M.R.; Heger, M. Tumor cell survival pathways activated by photodynamic therapy: A molecular basis for pharmacological inhibition strategies. *Cancer Metastasis Rev.* **2015**, *34*, 643–690. [CrossRef] [PubMed]
23. Prasad, S.; Gupta, S.C.; Tyagi, A.K.; Aggarwal, B.B. Curcumin, a component of golden spice: From bedside to bench and back. *Biotechnol. Adv.* **2014**, *32*, 1053–1064. [CrossRef] [PubMed]
24. Yadav, A.K.; Srikrishna, S.; Gupta, S.C. Cancer drug development using drosophila as an in vivo tool: From bedside to bench and back. *Trends Pharmacol. Sci.* **2016**, *37*, 789–806. [CrossRef]
25. Khamrai, M.; Banerjee, S.L.; Paul, S.; Samanta, S.; Kundu, P.P. Curcumin entrapped gelatin/ionically modified bacterial cellulose based self-healable hydrogel film: An eco-friendly sustainable synthesis method of wound healing patch. *Int. J. Biol. Macromol.* **2019**, *122*, 940–953. [CrossRef]
26. Vaughn, A.R.; Haas, K.N.; Burney, W.; Andersen, E.; Clark, A.K.; Crawford, R.; Sivamani, R.K. Potential role of curcumin against biofilm-producing organisms on the skin: A review. *Phytother. Res.* **2017**, *31*, 1807–1816. [CrossRef]
27. Li, T.; Zhao, Y.; Matthews, K.; Gao, J.; Hao, J.; Wang, S.; Han, J.; Jia, Y. Antibacterial activity against *Staphylococcus aureus* of curcumin-loaded chitosan spray coupled with photodynamic treatment. *LWT* **2020**, *134*, 110073. [CrossRef]

28. Lin, Y.-L.; Liu, Y.-K.; Tsai, N.-M.; Hsieh, J.-H.; Chen, C.-H.; Lin, C.-M.; Liao, K.-W. A Lipo-PEG-PEI complex for encapsulating curcumin that enhances its antitumor effects on curcumin-sensitive and curcumin-resistance cells. *Nanomed. Nanotechnol. Biol. Med.* 2012, *8*, 318–327. [CrossRef]
29. Yu, T.; Ji, J.; Guo, Y.-L. MST1 activation by curcumin mediates JNK activation, Foxo3a nuclear translocation and apoptosis in melanoma cells. *Biochem. Biophys. Res. Commun.* 2013, *441*, 53–58. [CrossRef]
30. Roy, S.; Rhim, J.-W. Preparation of antimicrobial and antioxidant gelatin/curcumin composite films for active food packaging application. *Colloids Surf. B Biointerfaces* 2020, *188*, 110761. [CrossRef]
31. Ma, T.; Chen, Y.; Zhi, X.; Du, B. Cellulose laurate films containing curcumin as photoinduced antibacterial agent for meat preservation. *Int. J. Biol. Macromol.* 2021, *193*, 1986–1995. [CrossRef]
32. Filho, J.G.D.O.; Bertolo, M.R.V.; Rodrigues, M.Á.V.; Marangon, C.A.; Silva, G.D.C.; Odoni, F.C.A.; Egea, M.B. Curcumin: A multifunctional molecule for the development of smart and active biodegradable polymer-based films. *Trends Food Sci. Technol.* 2021, *118*, 840–849. [CrossRef]
33. Benbettaïeb, N.; Chambin, O.; Karbowiak, T.; Debeaufort, F. Release behavior of quercetin from chitosan-fish gelatin edible films influenced by electron beam irradiation. *Food Control* 2016, *66*, 315–319. [CrossRef]
34. Da Silva, A.O.; Cortez-Vega, W.R.; Prentice, C.; Fonseca, G.G. Development and characterization of biopolymer films based on bocaiuva (*Acromonia aculeata*) flour. *Int. J. Biol. Macromol.* 2020, *155*, 1157–1168. [CrossRef]
35. Haghighi, H.; Leugoue, S.K.; Pfeifer, F.; Siesler, H.W.; Licciardello, F.; Fava, P.; Pulvirenti, A. Development of antimicrobial films based on chitosan-polyvinyl alcohol blend enriched with ethyl lauroyl arginate (LAE) for food packaging applications. *Food Hydrocoll.* 2020, *100*, 105419. [CrossRef]
36. Sun, L.; Sun, J.; Chen, L.; Niu, P.; Yang, X.; Guo, Y. Preparation and characterization of chitosan film incorporated with thinned young apple polyphenols as an active packaging material. *Carbohydr. Polym.* 2017, *163*, 81–91. [CrossRef]
37. Li, Y.; Tang, Z.; Lu, J.; Cheng, Y.; Qian, F.; Zhai, S.; An, Q.; Wang, H. The fabrication of a degradable film with high antimicrobial and antioxidant activities. *Ind. Crop. Prod.* 2019, *140*, 111692. [CrossRef]
38. Liu, X.; You, L.; Tarafder, S.; Zou, L.; Fang, Z.; Chen, J.; Lee, C.H.; Zhang, Q. Curcumin-releasing chitosan/aloe membrane for skin regeneration. *Chem. Eng. J.* 2019, *359*, 1111–1119. [CrossRef]
39. Chen, B.; Huang, J.; Li, H.; Zeng, Q.-H.; Wang, J.J.; Liu, H.; Pan, Y.; Zhao, Y. Eradication of planktonic *Vibrio parahaemolyticus* and its sessile biofilm by curcumin-mediated photodynamic inactivation. *Food Control* 2020, *113*, 107181. [CrossRef]
40. Huang, J.; Chen, B.; Zeng, Q.-H.; Liu, Y.; Liu, H.; Zhao, Y.; Wang, J.J. Application of the curcumin-mediated photodynamic inactivation for preserving the storage quality of salmon contaminated with L. monocytogenes. *Food Chem.* 2021, *359*, 129974. [CrossRef]
41. Li, H.; Tan, L.; Chen, B.; Huang, J.; Zeng, Q.; Liu, H.; Zhao, Y.; Wang, J.J. Antibacterial potency of riboflavin-mediated photodynamic inactivation against *Salmonella* and its influences on tuna quality. *LWT—Food Sci. Technol.* 2021, *146*, 111462.
42. Bing, S.; Zang, Y.; Li, Y.; Zhang, B.; Mo, Q.; Zhao, X.; Yang, C. A combined approach using slightly acidic electrolyzed water and tea polyphenols to inhibit lipid oxidation and ensure microbiological safety during beef preservation. *Meat Sci.* 2022, *183*, 108643. [CrossRef] [PubMed]
43. Bonilla, J.; Sobral, P.J.A. Investigation of the physicochemical, antimicrobial and antioxidant properties of gelatin-chitosan edible film mixed with plant ethanolic extracts. *Food Biosci.* 2016, *16*, 17–25. [CrossRef]
44. Mu, C.; Guo, J.; Li, X.; Lin, W.; Li, D. Preparation and properties of dialdehyde carboxymethyl cellulose crosslinked gelatin edible films. *Food Hydrocoll.* 2012, *27*, 22–29. [CrossRef]
45. Musso, Y.S.; Salgado, P.R.; Mauri, A.N. Smart edible films based on gelatin and curcumin. *Food Hydrocoll.* 2017, *66*, 8–15. [CrossRef]
46. Wu, C.; Sun, J.; Chen, M.; Ge, Y.; Ma, J.; Hu, Y.; Pang, J.; Yan, Z. Effect of oxidized chitin nanocrystals and curcumin into chitosan films for seafood freshness monitoring. *Food Hydrocoll.* 2019, *95*, 308–317. [CrossRef]
47. Liu, J.; Wang, H.; Wang, P.; Guo, M.; Jiang, S.; Li, X.; Jiang, S. Films based on κ-carrageenan incorporated with curcumin for freshness monitoring. *Food Hydrocoll.* 2018, *83*, 134–142. [CrossRef]
48. Gómez-Estaca, J.; Balaguer, M.; López-Carballo, G.; Gavara, R.; Hernández-Muñoz, P. Improving antioxidant and antimicrobial properties of curcumin by means of encapsulation in gelatin through electrohydrodynamic atomization. *Food Hydrocoll.* 2017, *70*, 313–320. [CrossRef]
49. Łupina, K.; Kowalczyk, D.; Zięba, E.; Kazimierczak, W.; Mężyńska, M.; Basiura-Cembala, M.; Wiącek, A.E. Edible films made from blends of gelatin and polysaccharide-based emulsifiers—A comparative study. *Food Hydrocoll.* 2019, *96*, 555–567. [CrossRef]
50. Xu, T.; Gao, C.; Feng, X.; Huang, M.; Yang, Y.; Shen, X.; Tang, X. Cinnamon and clove essential oils to improve physical, thermal and antimicrobial properties of chitosan-gum Arabic polyelectrolyte complexed films. *Carbohydr. Polym.* 2019, *217*, 116–125. [CrossRef]
51. Wang, K.; Lim, P.N.; Tong, S.Y.; Thian, E.S. Development of grapefruit seed extract-loaded poly(ε-caprolactone)/chitosan films for antimicrobial food packaging. *Food Packag. Shelf Life* 2019, *22*, 100396. [CrossRef]
52. Abral, H.; Ariksa, J.; Mahardika, M.; Handayani, D.; Aminah, I.; Sandrawati, N.; Sugiarti, E.; Muslimin, A.N.; Rosanti, S.D. Effect of heat treatment on thermal resistance, transparency and antimicrobial activity of sonicated ginger cellulose film. *Carbohydr. Polym.* 2020, *240*, 116287. [CrossRef]

53. Su, L.; Huang, J.; Li, H.; Pan, Y.; Zhu, B.; Zhao, Y.; Liu, H. Chitosan-riboflavin composite film based on photodynamic inactivation technology for antibacterial food packaging. *Int. J. Biol. Macromol.* **2021**, *172*, 231–240. [CrossRef]
54. Liu, J.; Zhang, L.; Liu, C.; Zheng, X.; Tang, K. Tuning structure and properties of gelatin edible films through pullulan dialdehyde crosslinking. *LWT* **2021**, *138*, 110607. [CrossRef]
55. Cai, L.; Shi, H.; Cao, A.; Jia, J. Characterization of gelatin/chitosan ploymer films integrated with docosahexaenoic acids fabricated by different methods. *Sci. Rep.* **2019**, *9*, 8375. [CrossRef]
56. Wang, H.; Hao, L.; Wang, P.; Chen, M.; Jiang, S.; Jiang, S. Release kinetics and antibacterial activity of curcumin loaded zein fibers. *Food Hydrocoll.* **2017**, *63*, 437–446. [CrossRef]
57. Herniou-Julien, C.; Mendieta, J.R.; Gutiérrez, T.J. Characterization of biodegradable/non-compostable films made from cellulose acetate/corn starch blends processed under reactive extrusion conditions. *Food Hydrocoll.* **2019**, *89*, 67–79. [CrossRef]
58. Wang, T.; Ke, H.; Chen, S.; Wang, J.; Yang, W.; Cao, X.; Liu, J.; Wei, Q.; Ghiladi, R.A.; Wang, Q. Porous protoporphyrin IX-embedded cellulose diacetate electrospun microfibers in antimicrobial photodynamic inactivation. *Mater. Sci. Eng. C* **2021**, *118*, 111502. [CrossRef]
59. Chen, L.; Dong, Q.; Shi, Q.; Du, Y.; Zeng, Q.; Zhao, Y.; Wang, J.J. Novel 2,3-dialdehyde cellulose-based films with photodynamic inactivation potency by incorporating the β-cyclodextrin/curcumin inclusion complex. *Biomacromolecules* **2021**, *22*, 2790–2801. [CrossRef]

Review

Green Synthesis and Potential Antibacterial Applications of Bioactive Silver Nanoparticles: A Review

Md. Amdadul Huq [1,*,†], Md. Ashrafudoulla [2,†], M. Mizanur Rahman [3], Sri Renukadevi Balusamy [4,*] and Shahina Akter [5,*]

1. Department of Food and Nutrition, College of Biotechnology and Natural Resource, Chung-Ang University, Anseong 17546, Korea
2. Department of Food Science and Technology, Chung-Ang University, Anseong 17546, Korea; ashrafmiu584@gmail.com
3. Department of Biotechnology and Genetic Engineering, Faculty of Biological Science, Islamic University, Kushtia 7003, Bangladesh; rahmanmm@btge.iu.ac.bd
4. Department of Food Science and Biotechnology, Sejong University, Gwangjin-gu, Seoul 05006, Korea
5. Department of Food Science and Biotechnology, Gachon University, Seongnam 461701, Korea
* Correspondence: amdadbge@gmail.com or amdadbge100@cau.ac.kr (M.A.H.); renucoimbatore@gmail.com (S.R.B.); shahinabristy16@gmail.com (S.A.)
† These authors contributed equally to this work.

Citation: Huq, M.A.; Ashrafudoulla, M.; Rahman, M.M.; Balusamy, S.R.; Akter, S. Green Synthesis and Potential Antibacterial Applications of Bioactive Silver Nanoparticles: A Review. *Polymers* 2022, 14, 742. https://doi.org/10.3390/polym14040742

Academic Editor: Zhen Zhang

Received: 20 January 2022
Accepted: 12 February 2022
Published: 15 February 2022

Publisher's Note: MDPI stays neutral with regard to jurisdictional claims in published maps and institutional affiliations.

Copyright: © 2022 by the authors. Licensee MDPI, Basel, Switzerland. This article is an open access article distributed under the terms and conditions of the Creative Commons Attribution (CC BY) license (https://creativecommons.org/licenses/by/4.0/).

Abstract: Green synthesis of silver nanoparticles (AgNPs) using biological resources is the most facile, economical, rapid, and environmentally friendly method that mitigates the drawbacks of chemical and physical methods. Various biological resources such as plants and their different parts, bacteria, fungi, algae, etc. could be utilized for the green synthesis of bioactive AgNPs. In recent years, several green approaches for non-toxic, rapid, and facile synthesis of AgNPs using biological resources have been reported. Plant extract contains various biomolecules, including flavonoids, terpenoids, alkaloids, phenolic compounds, and vitamins that act as reducing and capping agents during the biosynthesis process. Similarly, microorganisms produce different primary and secondary metabolites that play a crucial role as reducing and capping agents during synthesis. Biosynthesized AgNPs have gained significant attention from the researchers because of their potential applications in different fields of biomedical science. The widest application of AgNPs is their bactericidal activity. Due to the emergence of multidrug-resistant microorganisms, researchers are exploring the therapeutic abilities of AgNPs as potential antibacterial agents. Already, various reports have suggested that biosynthesized AgNPs have exhibited significant antibacterial action against numerous human pathogens. Because of their small size and large surface area, AgNPs have the ability to easily penetrate bacterial cell walls, damage cell membranes, produce reactive oxygen species, and interfere with DNA replication as well as protein synthesis, and result in cell death. This paper provides an overview of the green, facile, and rapid synthesis of AgNPs using biological resources and antibacterial use of biosynthesized AgNPs, highlighting their antibacterial mechanisms.

Keywords: green synthesis; silver nanoparticles; antibacterial application; antibacterial mechanisms

1. Introduction

Nanotechnology is an emerging field of research, with numerous applications in science and technology, especially in the development of different nanomaterials and nanoparticles. Nanoparticles (NPs) are small particles of size from 1 nm to 100 nm and have gained significant interest from scientists due to their multiple applications in diverse fields of science such as biomedicine, agriculture, pharmaceutics, textile, food technology, catalysis, sensors, mechanics, electronics, and optics [1,2]. There are different varieties of nanoparticles, including silver, gold, zinc, cadmium sulfide, copper, iron, titanium dioxide, etc., with unique characteristics [2–5]. Among different nanoparticles, silver nanoparticles

(AgNPs) have been one of the most popular subjects of study in recent decades due to their wide scope of application in different branches of biomedical science as antibacterial, antifungal, antioxidant, anti-cancer, anti-inflammatory, drug delivery, wound dressings, biosensors, and biocatalysis, etc. [6–12]. Some recent studies have shown the strong antimicrobial, antioxidant, and anti-cancer activities of green synthesized AgNPs [6–8]. The biosynthesized AgNPs were also effectively used to degrade various toxic chemicals [9]. Moreover, green synthesized AgNPs have many other applications in different branches of biotechnology such as water filtration, sanitization, food preservation, production of cosmetics, nano-insecticides, and nanopesticides, etc. [10,11,13]. Green synthesized AgNPs have been reported as potential antibacterial agents against various Gram-positive and Gram-negative pathogenic bacteria, including *Salmonella epidermidis, Salmonella Typhimurium, Pseudomonas aeruginosa, Staphylococcus aureus, Streptococcus pyogenes, Escherichia coli, Bacillus subtilis, Vibrio parahaemolyticus, Streptococcus pneumoniae, Enterobacter hormaechei, Salmonella paratyphi, Klebsiella pneumoniae, Aeromonas hydrophila, Pseudomonas fluorescens, Flavobacterium branchiophilum, Enterobacter aerogenes, Shigella flexneri, Xanthomonas axonopodis, Salmonella enterica*, etc. [2,3,6,11–13].

Various physical and chemical methods such as physiochemical [14], electrochemical [15], photochemical [16], chemical reduction [17], and microwave irradiation [18] are commonly used for the synthesis of these nanoparticles. The main drawbacks of these methods are that they are expensive and hazardous because of the usage of toxic ingredients, costly, demand labor-intensive equipment and the generation of hazardous byproducts [2,5,19]. Due to the various drawbacks of physicochemical methods, researchers are currently focusing more on biological approaches for eco-friendly, non-toxic, inexpensive, and facile synthesis of nanoparticles (Figure 1). Green synthesis is an efficient process that uses natural compounds as reducing, capping, and stabilizing agents instead of expensive toxic chemicals. Various biological resources such as plants and their different parts (roots, leaves and fruit, etc.), bacteria, fungi, algae, etc. could be utilized for the green synthesis of bioactive nanoparticles [20–23]. Recently, green synthesis of AgNPs using plant extracts or microbes and their antimicrobial activity were widely investigated.

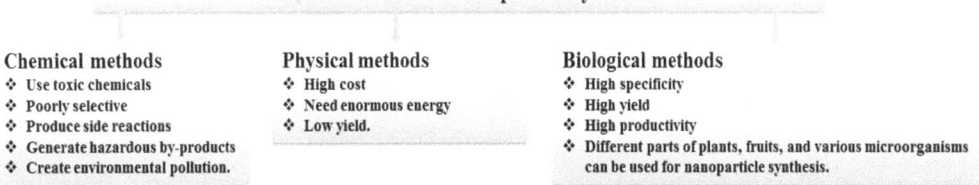

Figure 1. Different methods of nanoparticle synthesis.

Multidrug-resistant microorganisms are a serious threat to public health worldwide as different life-threatening infectious diseases are caused by these pathogens. There is a continuous increase in the number of multidrug-resistant bacterial strains due to mutation, pollution, changing environmental conditions and excessive use of drugs. To overcome this problem, scientists are trying to develop new drugs for the treatment of such microbial infections. Green synthesized AgNPs have been found to be effective for controlling these multidrug-resistant bacterial strains. This review provides an overview of green synthesis of AgNPs using different biological resources, various parameters essential for stable, easy and high yields, antibacterial applications and mechanisms of biosynthesized AgNPs as well as describing the prospect for their future development and potential antibacterial applications.

2. Green Synthesis of AgNPs

Green synthesis of AgNPs using different biological agents such as plants, bacteria, fungi, algae and yeast is an economical, facile, and eco-friendly approach without generating any toxic byproducts. In recent years, both microbes and plants were extensively investigated for the green synthesis of AgNPs. Figure 2 illustrates the various steps of green synthesis of AgNPs using plants and microbes.

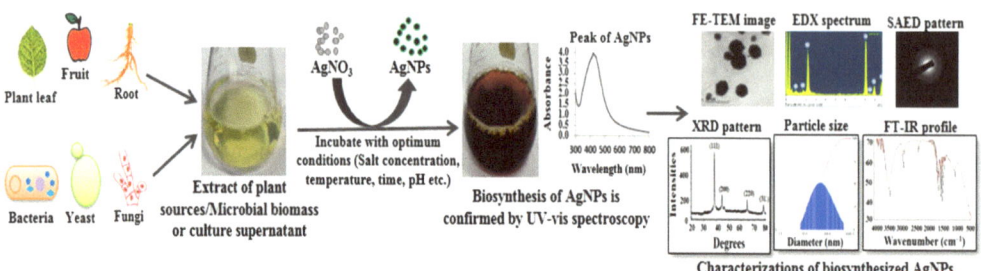

Figure 2. Schematic illustration of green synthesis and characterizations of AgNPs.

3. Plant Mediated Synthesis of AgNPs

Plant-mediated synthesis of AgNPs is a widely adopted technique due to the availability of various plants and their easy and safe utilization. Different parts of the plant including fruits, roots, flowers, leaves, peels, etc., have been successfully utilized for the green synthesis of bioactive AgNPs (Table 1). Plant extracts contain numerous bioactive compounds such as alkaloids, flavonoids, terpenoids, tannins, saccharides, phenols, vitamins, as well as various enzymes, amino acids, and proteins [21,24,25]. Due to the presence of these active biomolecules in plant extracts, synthesis of bioactive AgNPs using plants is more stable and easier. In the last few years, many studies have been conducted for the green synthesis of bioactive AgNPs using different parts of plants such as fruits, seeds, roots, flowers, stems, leaves, peels, etc. For instance, the leaf extract of *Clerodendrum viscosum* was used for facile, rapid, and eco-friendly synthesis of bioactive AgNPs [26]. They also investigated the antimicrobial efficacy of biosynthesized AgNPs against various pathogenic bacteria. Pawar and Patil [27] synthesized AgNPs using tuber extract of *Eulophia herbacea*. Fruit extract of *Amomum villosum* was used by Soshnikova et al. [28] for the facile synthesis of AgNPs. The seeds and roots of *Durio zibethinus* and *Rheum palmatum*, respectively, were used for green synthesis of AgNPs [29,30]. Peel extracts of different vegetables such as *Lagenaria siceraria*, *Luffa cylindrica*, *Solanum lycopersicum*, *Solanum melongena* and *Cucumis sativus* were investigated for synthesis of bioactive AgNPs [31]. Synthesis time, size and shape of synthesized AgNPs and their bioactivity varies greatly depending on the plant or part of the plant which was used for synthesis. For example, AgNPs of 10 to 30 nm in size were synthesized using root extract of *Panax ginseng* by two hours' reaction [32]. On the other hand, AgNPs of 5 to 15 nm were synthesized using leaf extract of *Panax ginseng* within 45 mins of reaction [33]. According to Adeyemi et al. [34], the leaf extract of *Spondias mombin* produced rod- or triangular-shaped AgNPs. However, the plant extract of *Prunus africana*, and *Camellia sinensis* produced spherical-shaped AgNPs [35]. Various parameters such as the extract salt ratio, incubation time, incubation temperature, pH, etc. also greatly affected the easy, rapid, high, and stable synthesis of AgNPs using plant extracts [3,6]. The probable mechanism of plant-mediated synthesis of AgNPs is the chemistry of reduction and oxidation. It has been proposed that the plant extract contains vitamins, amino acids, proteins, enzymes, organic acid, flavonoids, terpenoids, alkaloids, polyphenols, and polysaccharides, which have significant roles for the reduction of silver salts as well as serve as capping and stabilizing agents [21,24,25].

Table 1. Green synthesis of AgNPs using plants and their antibacterial applications.

Plants	Used Parts	Size (nm)	Shape	Optimum Synthesis Parameters	Target Pathogens	References
Plantago major	Leaf extract	10–20	Spherical	1 mM, 70 °C, 60 min	*S. aureus*, *E. coli*, *P. aeruginosa*	[25]
Prunus africana, Camellia sinensis	Plant extract	10–19	Spherical	0.5 mM, 25 °C, 24 h	*E. coli, K. pneumoniae*	[35]
Tasmanian flax-lily	Dried leaves extract	Av. 70	Spherical	0.1 mM, 60 °C, 25 min	*S. aureus, S. epidermidis, P. aeruginosa, C. albican*	[36]
Carduus crispus	Plant extract	33–131	NA	1 mM, room temperature, 24 h	*E. coli, M. luteus*	[37]
Anastatica hierochuntica, Artemisia absinthium	Plant and seed extracts	Av. 114, 125.5	Spherical	1 mM, room temperature, 48 h	*P. aeruginosa, E. coli, S. aureus, C. albicans*	[38]
Lantana trifolia	Leaf extract	5–70	Spherical	1.5 M, 35 °C, 2 h	*S. aureus, C. albicans, E. coli, P. aeruginosa, B. subtilis*	[39]
Blumea eriantha	Plant extract	10–60	Spherical	1%, ambient temperature, 24 h	*S. aureus, B. subtilis, B. cereus, E. coli*	[40]
Cucumis prophetarum	Leaf extract	30–50	Polymorphic	1 mM, 80 °C, 3 h	*S. typhi, S. aureus*	[41]
Clerodendrum viscosum	Leaf extract	36–74	Spherical	1 mM, 60 °C, 60 min	*E. coli, P. aeruginosa, B. subtilis, S. aureus*	[26]
Grape	Proanthocyanidin from seed	100–120	Aggregated	Ambient temperature, 2–3 h	*S. aureus, P. aeruginosa, E. coli*	[42]
Spondias mombin	Leaf extract		Rod or triangular	1 mM, room temperature	*S. aureus, P. aeruginosa, E. coli*	[34]
Eulophia herbacea	Tuber extract	Av. 11.7	NA	1 mM, room temperature, 5 h	*E. coli, S. aureus, P. aruginosa, B. subtilis*	[27]
Torreya nucifera	Leaf extract	10–125	Spherical	1 M, 20 °C, 24 h	*S. typhimurium*	[43]
Chlorophytum borivilianum	Callus extracts	35–168	Spherical	1 mM, room temperature, 5 h	*B. subtilis, S. aureus, P. aeruginosa, E. coli*	[44]
Purple heart plant	Leaves extract	Av. 104.6	NA	50 mM, 65 °C	*E. coli, S. aureus*	[45]
Phoenix dactylifera	Root hair extract	21–41	Spherical	0.1 mM, 50 °C, 48 h	*C. albicans, E. coli*	[46]
Taraxacum officinale	Leaf extract	5–30	Spherical	1 mM, room temperature, 15 min	*X. axonopodis, P. syringae*	[47]
Chicory	Seed exudates	≤25	Spherical	5 mM, 30 °C	*P. aeruginosa, K. pneumoniae, A. baumannii, F. solani*	[48]
Punica granatum	Peel extract	20–40	Spherical	0.1 mM, room temperature, 72 h	*E. coli, S. epidermidis, P. aeruginosa, S. typhi, P. vulgaris, S. aureus, K. pneumonia*	[49]
Durio Zibethinus	Seed extract	20–75	Spherical, rod	1.5 mM, in sunlight, 30 min	*E. coli, B. subtilis, S. typhimurium, S. typhi*	[29]
Market vegetable	Vegetable waste extract	10–90	Spherical	1 mM, 37 °C, 5 h	*Klebsiella* sp., *Staphylococcus* sp.	[50]
Rheum palmatum	Root extract	44–113	Hexagonal, spherical	2 mM, room temperature, 24 h	*S. aureus, P. aeruginosa*	[30]
Angelica pubescens	Root extract	20–50	Quasi-spherical	5 mM, 80 °C, 50 min	*S. aureus, P. aeruginosa, E. coli, S. enterica*	[51]
Protium serratum	Leaf extract	Av. 74.5	Spherical	1 mM, 25 °C, 4 h	*P. aeruginosa, E. coli, B. subtilis*	[52]
Amomum villosum	Dried fruit extract	5–15	Spherical	1 mM, room temperature, 3 s	*S. aureus, E. coli*	[28]
Glycyrrhiza uralensis	Root extract	5–15	Spherical	1 mM, 80 °C, 40 min	*E. coli, S. aureus, P. aeruginosa, S. enterica*	[53]
Ficus palmata	Leaf extract	28–33	Spherical	2 mM, room temperature, 6 h	*S. pneumonia, E. coli, P. aeruginosa, K. pneumonia, P. vulgaris*	[54]
Euphorbia antiquorum	Latex extract	10–50	Spherical	1 mM, room temperature, 24 h	*K. Pneumoniae, P. mirabilis, V. cholerae, E. faecalis*	[55]
Ocimum Sanctum	Leaf extract	Av. 14.6	Spherical	2 mM, 35 °C, 4 h	*E. coli*	[56]
Moringa stenopetala	Leaf extract	Av. 11.4	NA	1 mM, 60 °C, 15 min	*S. aureus, E. coli*	[57]
Euphrasia officinalis	Leaf extract	Av. 40.3	Quasi-spherical	1 mM, 65 °C, 19 min	*P. aeruginosa, E. coli, S. aureus, V. parahaemolyticus.*	[58]
Siberian ginseng	Dried stem	Av. 14.6	Spherical	1 mM, 80 °C, 1.5 h	*S. aureus, B. anthracis, V. parahaemolyticus. E. coli*	[59]
Borago officinalis	Leaf extract	30–80	Spherical, hexagonal, irregular	1 mM, 65 °C, 68 s	*P. aeruginosa, E. coli, V. parahaemolyticus, S. aureus*	[60]
Cocoa pod	Husk extract	4–32	Spherical	1 mM, 30 °C, few minutes	*E. coli, K. pneumoniae, S. pyogenes, S. aureus, P. aeruginosa*	[61]
Lagenaria siceraria, Luffa cylindrica, Solanum lycopersicum, Solanum melongena, Cucumis sativus	Vegetable peel extract	up to 20	Spherical	2 mM, 80 °C, 10 min	*E. coli, K. pneumoniae*	[31]
Azadirachta indica	Leaf extract	Av. 34	Spherical	1 mM, room temperature, 24 h	*S. aureus, E. coli*	[62]

Table 1. Cont.

Plants	Used Parts	Size (nm)	Shape	Optimum Synthesis Parameters	Target Pathogens	References
Pedalium murex	Leaf extract	10–50	Spherical	10 mM, 20 min	B. subtilis, S. aureus, E. coli, M. flavus, P. aeruginosa, B. pumilus, K. pneumoniae	[63]
Cassia fistula	Leaf extract	40–50	Spherical	1 mM, room temperature, overnight	B. subtilis, S. aureus, C. kruseii, T. mentagrophytes	[64]
Psidium guajava	Leaf extract	10–90	Spherical	1 mM, 30 °C, 10 min	P. aeruginosa	[65]
Coffea arabica	Seed extract	10–150	Spherical, ellipsoidal	20 mM, room temperature, 2 h	E. coli, S. aureus	[66]
Styrax benzoin	Benzoin gum extract	12–38	Spherical	1 mM, 60 °C, 5 h	E. coli, P. aeruginosa, S. aureus, C. tropicalis	[21]
Cardiospermum halicacabum	Leaf extract	Av. 23	Cubic	1 mM, room temperature, 16 h	P. vulgaris, P. aeruginosa, S. aureus, B. subtilis, S. paratyphi, A. solani, F. oxysporum	[67]
Atrocarpus altilis	Leaf extract	20–50	Spherical	1 mM, 25 °C, 24 h	S. aureus, P. aeruginosa, E. coli, A. vesicolor	[68]
Ficus benghalensis, Azadirachta indica	Bark extracts	Av. 60	Spherical	1 M, 80 °C, 30 min	E. coli, P. aeruginosa, V. cholera, B. subtilis	[69]
Thevetia peruviana	Leaf extract	Av. 18.1	Spherical	1 mM, 30 °C, 4 h	E. coli, K. pneumonia, P. aeruginosa, S. aureus, B. subtilis, S. typhi	[70]
Capparis spinosa	Leaf extract	5–30	Spherical	10 mM, room temperature, 15 min	E. coli, S. typhimurium, S. aureus, B. cereus	[71]
Potentilla fulgens	Root extract	10–15	Spherical	1 mM, 35 °C, 18 h	E. coli, B. subtilis	[72]
Petroselinum crispum	Leaf extract	30–32	Spherical	10 mM, room temperature, 24 h	K. pneumoniae, E. coli, S. aureus	[73]
Eucalyptus globulus	Leaf extract	5–25	Spherical, oval	1 mM, 37 °C, 60 min	P. aeruginosa, E. coli, S. aureus	[74]
Banana plant	Banana peel extract	23.7	Spherical	1.75 mM, 30 °C, 72 h	E. coli, P. aeruginosa, B. subtilis, S. aureus	[75]
Zingiber officinale	Rhizome	1.4–5.7	Spherical	1 mM, room temperature, 1 h	S. aureus, E. coli	[76]
Erythrina indica	Root extract	20–118	Spherical	1 mM, room temperature, overnight	S. aureus, M. luteus, E. coli, B. subtilis, S. typhi, S. paratyphi	[77]
Prosopis farcta	Plant extract	Av. 10.8	Spherical	1 mM, room temperature, 1 h	S. aureus, B. subtilis, E. coli, P. aeruginosa	[78]
Cassia roxburghii	Aqueous extract	10–30	Spherical	1 mM, room temperature, overnight	B. subtilis, S. aureus, M. luteus, P. aeruginosa, E. coli, E. aerogenes	[79]
Garcinia mangostana	Fruit extract	30–50	Various	1 mM, 80 °C, 15 min	E. coli, P. aeruginosa, S. aureus	[80]
Panax ginseng	Root extract	10–30	Spherical	1 mM, 80 °C, 2 h	B. anthracis, E. coli, V. parahaemolyticus, S. aureus, B. cereus	[32]
Panax ginseng	Leaf extract	5–15	Spherical	1 mM, 80 °C, 45 min	E. coli, S. enterica, V. parahaemolyticus, S. aureus, B. anthracis, B. cereus	[33]
Clitoria ternatea, Solanum nigrum	Leaf extract	20–28	Spherical	100 mM, room temperature, 60 min	B. subtilis, S. aureus, S. pyogenes, E. coli, P. aeruginosa, K. aerogenes	[81]
Mukia maderaspatana	Leaf extract	Av. 158	Spherical	1 mM, room temperature, 15–20 min	B. subtilis, K. pneumoniae, S. aureus, S. typhi	[82]
Terminalia arjuna	Plant extract	8–16	Spherical	1 mM, room temperature, 15 min	S. aureus, E. coli	[83]
Eclipta alba	Leaf extract	310–400	Cubic	1 mM, 32 °C, 24 h	E. coli, S. aureus, P. aeruginosa	[84]
Alternanthera dentata	Leaf extract	50–100	Spherical	1 mM, 60 °C, 45 min	E. coli, P. aeruginosa, K. pneumonia, E. faecalis	[85]
Dalbergia spinosa	Leaf extract	Av. 18	Spherical	100 mM, room temperature, 30 min	B. subtilis, P. aeruginosa, S. aureus, E. coli	[86]
Pulicaria glutinosa	Plant extract	40–60	Spherical	1 mM, 90 °C, 2 h	E. coli, P. aeruginosa, S. aureus, M. luteus	[87]
Phyllanthus amarus	Aqueous extract	15.7–29.9	Spherical	1 mM, 70 °C, 20 min	P. aeruginosa	[88]
Withania somnifera	Leaf powder	5–30	Spherical	1 mM, room temperature, 12 h	S. aureus, E. coli	[89]
Acorous calamus	Rhizome extract	Av. 31.8	Spherical	1 mM, room temperature, 12 h	B. subtilis, B. cereus, S. aureus	[90]
Cocos nucifera	Plant extract	Av. 22	Spherical	0.9 mM, 36 °C, 24 h	K. pneumoniae, B. subtilis, P. aeruginosa, S. paratyphi	[91]
Boerhaavia diffusa	Plant extract	Av. 25	Spherical	100 mM, 24 h	A. hydrophila, P. fluorescens, F. branchiophilum	[92]
Azadirachta indica	Leaf extract	4.7–18.9	Spherical	0.1 N, room temperature, 2 h	B. subtilis, S. typhimorium	[93]

Table 1. Cont.

Plants	Used Parts	Size (nm)	Shape	Optimum Synthesis Parameters	Target Pathogens	References
Coriandrum sativum	Seed extract	9.9–12.6	Spherical	0.1 N, room temperature, 2 h	B. subtilis	[94]
Hibiscus cannabinus	Leaf extract	7–25	Spherical	5 mM, 30 °C, 40 min	E. coli, P. mirabilis, S. flexneri.	[95]
Ocimum tenuiflorum	Leaf extract	25–40	NA	1 mM, room temperature, 10 min	E. coli, Corney bacterium, B. substilus	[96]
Tribulus terrestris	Fruit bodies	16–28	Spherical	1 mM, room temperature, 36 h	S. pyogens, P. aeruginosa, E. coli, B. subtilis, S. aureus	[97]
Lantana camara	Fruit extract	12.5–13.0	Spherical	1 mM, room temperature, 1 h	M. luteus, B. subtilis, S. aureus, V. cholerae, K. pneumoniae, S. typhi	[98]
Morinda citrifolia	Leaf extract	10–60	Spherical	1 mM, 90 °C, 60 min	E. coli, P. aeroginosa, K. pneumoniae, E. aerogenes, B. cereus, Enterococci sp.	[99]
Terminalia chebula	Plant extract	less than 100	Pentagons, spherical, triangular	2 mM, room temperature, 15–20 min	S. aureus, E. coli	[100]
Solanum xanthocarpum	Berry extract	4–18	Spherical	1 mM, 45 °C, 25 min	H. pylori	[101]
Dioscorea bulbifera	Tuber extract	8–20	Nanorods, triangles	0.7 mM, 50 °C, 5 h	E. coli, P. aeruginosa, S. typhi, B. subtilis	[102]
Garcinia mangostana	Leaf extract	Av. 35	Spherical	1 mM, 75 °C, 60 min	E. coli, S. aureus	[103]
Cymbopogan citratus	Leaf extract	Av. 32	Spherical	1 mM, 37 °C, 24 h	E.coli, S. aureus, P. mirabilis, S. typhi, K. pnuemoniae	[104]
Sesuvium portulacastrum L.	Callus and leaf extracts	5–20	Spherical	1 mM, room temperature, 24 h	P. aeruginosa, S. aureus, L. monocytogenes, M. luteu, K. pneumoniae, A. alternata, P. italicum, F. equisetii, C. albicans	[105]

Av., average; NA, not available; s, second; min, minute; h, hour.

4. Microbe Mediated Synthesis of AgNPs

In the last few years, the potential of green synthesis of AgNPs using microorganisms has been realized (Table 2). Microorganisms have been shown to be excellent biological agents for the facile, cost effective, and ecofriendly synthesis of AgNPs, avoiding toxic and expensive chemicals and the high energy demands required for physiochemical approaches. Various microorganisms such as bacteria, yeast, fungi, and algae are often favored for the green synthesis of AgNP because of their rapid growth, simpler cultivation and ease of handling. There are two methods for the green synthesis of AgNP using microorganisms, such as the extracellular and intracellular methods [12,24]. Microorganisms synthesize various extracellular and intracellular biomolecules, including amino acid, enzymes, proteins, sugar molecules, organic materials, and many other primary and secondary metabolites [12,24]. The exact mechanism of biosynthesis of AgNP using microorganisms is still not fully known. The widely accepted mechanism of microbe-mediated synthesis of AgNPs is the chemistry of reduction and oxidation, similar to plant-mediated synthesis. First, the metal ions are reduced to NPs with the presence of microbial enzymes including reductase enzyme. Then, various extracellular and intracellular biomolecules of microorganisms serve as the capping and stabilizing agents [2,24]. Huq and Akter [106] have reported the extracellular synthesis of AgNPs from Massilia sp. MAHUQ-52. The interaction of 1 mM AgNO$_3$ with the bacterial culture supernatant at 30 °C temperature yielded nanoparticles within 48 h of reaction. The size of synthesized AgNPs from FE-TEM analysis was found to range between 15 and 55 nm.

Table 2. Green synthesis of AgNPs using microorganisms and their antibacterial applications.

Microorganisms	Method	Size (nm)	Shape	Optimum Synthesis Parameters	Target Pathogens	References
Massilia sp. MAHUQ-52	Extracellular	15–55	Spherical	1 mM, 30 °C, 48 h	K. pneumoniae, S. Enteritidis	[106]
Streptomyces strains	Intracellular	1.17–13.3	Spherical	5 mM, 30 °C, 120 h	B. cereus, E. faecalis, S. aureus, E. coli, S. typhi, P. aeruginosa, K. pneumoniae, P. vulgaris	[107]
Cedecea sp.	Extracellular	10–40	Spherical	2 mM, 37 °C, 48 h	E. coli, P. aeruginosa, S. epidermis, S. aureus	[108]
Arthrobacter bangladeshi	Extracellular	12–50	Spherical	1 mM, 30 °C, 24 h	S. typhimurium, Y. enterocolitica	[109]
Aspergillus terreus	Extracellular	60–100	Spherical	100 mM, 27 °C, 48 h	K. pneumoniae, S. aureus, S. typhi, P. aeruginosa, E. coli, S. epidermidis, E. faecalis, P. mirabilis, B. subtilis	[110]
Penicillium chrysogenum	Extracellular	18–60	Spherical	1 mM, 28 °C, 24 h	C. albicans, C. krusei, C. tropicalis, C. parapsilosis, C. glabrata	[111]
Paenarthrobacter nicotinovorans	Extracellular	13–27	Spherical	1 mM, 30 °C, 24 h	B. cereus, P. aeruginosa	[12]
Aspergillus fumigatus	Intracellular	<100	Spherical	3.5 mM, 25 °C, 72 h	11 different pathogenic bacteria	[112]
Paenibacillus sp.	Extracellular	17.4–52.8	Spherical	0.1 mM, room temperature, 120 h	S. aureus, E. faecalis, S. pneumoniae, E. coli	[113]
Lysinibacillus xylanilyticus	Extracellular	8–30	Spherical	1 mM, 30 °C, 48 h	V. parahaemolyticus, S. Typhimurium	[114]
Cyanobacteria Desertifilum sp.	Intracellular	4.5–26	Spherical	1 mM, room temperature, 24 h	B. cereus, P. aeruginosa, B. cercus, B. subtilis, S. flexneri, S enterica	[115]
Chlorella ellipsoidea	Intracellular	Av. 220	Spherical, cubic, rod, triangular	1 mM, room temperature, 24 h	S. aureus, E. coli, K. pneumoniae, P. aeruginosa	[116]
Citrobacter spp. MS5	Extracellular	5–15	Spherical	1 mM, 40 °C, 180 min	E. hormaechei, K. pneumoniae	[117]
Sphingobium sp. MAH-11	Extracellular	7–22	Spherical	1 mM, 30 °C, 48 h	E. coli, S. aureus	[2]
Padina sp.	Intracellular	25–60	Spherical	10 mM, 60 °C, 48 h	S. aureus, B. subtilis, P. aeruginosa, S. typhi, E. coli	[118]
Chaetoceros sp., Skeletonema sp., Thalassiosira sp.	Biomass	149–239	Rectangular, square, regular	room temperature, 48 h	E. coli, B. subtilis, S. pneumonia, Aeromonas sp., S. aureus	[119]
Penicillium oxalicum	Extracellular	60–80	Spherical	1 mM, 37 °C, 72 h	S. aureus, S. dysenteriae, S. typhi	[120]
Lactobacillus plantarum	Intracellular	Av. 14.0	Spherical	2 mM, 37 °C, 24 h	S. aureus, E. coli, S. epidermidis, Salmonella sp.	[121]
Escherichia coli	Extracellular	5–50	Spherical	1 mM, 37 °C, 72 h	B. subtilis, S. aureus, B. cereus, P. aeruginosa, K. pneumoniae, E. coli, S. typhi, E. vermicularis	[122]
Terrabacter humi	Extracellular	6–24	Spherical	1 mM, 30 °C, 48 h	E. coli, P. aeruginosa	[20]
Bacillus subtilis	Intracellular	3–20	Spherical	1 mM, 30 °C, 24 h	S. aureus, E. coli, S. epidermidis, K. pneumoniae, C. albicans	[123]
Pseuduganella eburnea MAHUQ-39	Extracellular	8–24	Spherical	1 mM, 30 °C, 24 h	S. aureus, P. aeruginosa	[6]
Oscillatoria limnetica	Extracellular	3.3–17.9	quasi-spherical	10 mM, room temperature, 48 h	E. coli, B. cereus	[3]
Acinetobacter baumannii	Extracellular	37–168	Spherical	1 mM, 37 °C	E. coli, P. aeruginosa, K. pneumoniae	[124]
Pseudomonas sp. THG-LS1.4	Extracellular	10–40	Irregular	1 mM, 28 °C, 48 h	B. cereus, S. aureus, C. tropicalis, V. parahaemolyticus, E. coli, P. aeruginosa	[125]
Novosphingobium sp. THG-C3	Extracellular	8–25	Spherical	1 mM, 25 °C, 48 h	S. aureus, C. tropicalis, P. aeruginosa, E. coli, V. parahaemolyticus, C. albicans, S. enterica	[126]
Sporosarcina koreensis DC4	Extracellular	30–50	Spherical	1 mM, 25 °C, 48 h	B. subtilis, B. cereus, V. parahaemolyticus, E. coli, S. enterica, B. anthracis	[127]
Bacillus sp. AZ1	Extracellular	7–31	Spherical	1 mM, 40 °C, 24 h	B. cereus, S. aureus, S. typhi, E. coli, S. epidermis, S. aureus	[128]
Aeromonas sp. THG-FG1.2	Extracellular	8–16	Spherical	1 mM, 28 °C, 48 h	S. enterica, E. coli, P. aeruginosa, V. parahaemolyticus, B. cereus, B. subtilis, S. aureus, C. albicans	[129]

Table 2. Cont.

Microorganisms	Method	Size (nm)	Shape	Optimum Synthesis Parameters	Target Pathogens	References
Kinneretia THG-SQI4	Extracellular	15–20	Spherical	1 mM, 28 °C, 48 h	C. albicans, E. coli, C. tropicalis, B. cereus, B. subtilis, S. aureus, S. enterica, P. aeruginosa, V. parahaemolyticus	[130]
Bacillus safensis	Extracellular	5–30	Spherical	1 mM, 30 °C, 2 h	E. coli	[131]
Aspergillus niger	Intracellular	43–63	Spherical	1 mM, 35 °C, 48 h	K. planticola, E. coli, Pseudomonas sp., B. subtilis, B. cereus	[132]
Weissella oryzae	Extracellular	10–30	Spherical	1 mM, 25 °C, 48 h	V. parahaemolyticus, B. cereus, B. anthracis, S. aureus, E. coli, C. albicans	[133]
Microbacterium resistens	Extracellular	10–20	Spherical	1 mM, 30 °C, 48 h	S. enterica, S. aureus, B. anthracis, B. cereus, E. coli, C. albicans	[134]
Bacillus methylotrophicus	Extracellular	10–30	Spherical	1 mM, 28 °C, 48 h	V. parahaemolyticus, S. enterica, E. coli, C. albicans	[135]
Pseudomonas deceptionensis	Extracellular	10–30	Spherical	1 mM, 25 °C, 48 h	S. aureus, S. enterica, V. parahaemolyticus, B. anthracis, C. albicans	[136]
Bhargavaea indica	Extracellular	30–100	Pentagon, spherical, hexagonal, triangle, nanobar	1 mM, 25 °C, 48 h	V. parahaemolyticus, S. enterica, S. aureus, B. anthracis, B. cereus, E. coli, C. albicans	[137]
Actinomycetes	Extracellular, Intracellular	65–80	Spherical	1 mM, 37 °C, 72 h	S. aureus, E. coli, K. pneumoniae, P. vulgaris, P. aeruginosa	[138]
Bacillus flexus	Extracellular	12–61	Spherical, triangular	1 mM, room temperature, 8 h	S. pyogenes, B. subtilis, P. aeruginosa, E. coli	[139]

Singh et al. [108] have demonstrated an extracellular synthesis of AgNPs using the culture supernatant of a bacterial strain *Cedecea* sp. within 48 h of reaction and found spherical-shaped nanoparticles of 10–40 nm in size. Mondal et al. [117] have also reported the rapid synthesis of AgNPs (within 180 min) using the culture supernatants of *Citrobacter* spp. MS5. Another report showed that AgNPs were synthesized through bioreduction of AgNO$_3$ by the culture supernatant of *Penicillium chrysogenum* [111]. *Sphingobium* sp. MAH-11 and *Pseudoduganella eburnea* MAHUQ-39 have the ability to produce AgNPs with higher antibacterial activities against pathogenic microbes [2,6]. Eltarahony et al. [107] have reported the intracellular synthesis of AgNPs (within 5 min) using *Streptomyces* strains. Hamida et al. [115] have also reported intracellular synthesis of AgNPs using *Cyanobacteria Desertifilum* sp. They found spherical-shaped nanoparticles of a small size, in the range of 4.5–26 nm. Various fungi and algae were also used for facile, rapid, and ecofriendly synthesis of AgNPs. For instance, the culture supernatant of *Aspergillus terreus* was used to produce AgNPs with a size of 60–100 nm [110]. Raza et al. [112] have reported the intracellular synthesis of AgNPs using a fungus strain *Aspergillus fumigatus* KIBGE-IB33.

5. Critical Parameters for Rapid, Facile, and Stable Synthesis of AgNPs

Several factors play a key role for rapid, stable, and mass production of AgNPs such as the concentration of plant extracts and metal salts, incubation time, temperature, pH, etc. (Figure 3). The shape and size of synthesized nanoparticles also depend on these factors. Extracts of the medicinal plant *Potentilla fulgens* was used by Mittal et al. [72] for the green synthesis of AgNPs and they found that the various physico-chemical parameters including concentrations of plant extract and metal ions, incubation time and temperature, and the pH of the reaction time greatly affected the rate of synthesis as well as their shape, size, and yield. They used different concentrations of plant extract (1 to 200 mg in 50 mL water) and found that 4 mg extract in 50 mL water was able to produce the highest concentration of AgNPs. They also used different concentrations of AgNO$_3$ from 0.5 to 5 mM and revealed that the yield of AgNPs increased with the increase of AgNO$_3$ concentration from 0.5 to 1 mM, beyond which, there was again a fall in the absorbance. They found that 45 °C is the best temperature for maximum yield and concluded that at a higher temperature, the rate of synthesis of smaller size nanoparticles increased. The synthesis was also influenced by the pH of reaction mixture. They revealed that at an alkaline pH, smaller size nanoparticles were formed, whereas at an acidic pH, larger size nanoparticles were observed. Moreover, incubation time had a great effect on the synthesis process as well as the particle size distribution [72]. Nayak et al. [69] have reported the effect of temperature, pH, and incubation time for the green synthesis of AgNPs using bark extracts of *A. indica* and *F. benghalensis* and concluded that 80°C temperature, a pH of 10 and 30 min incubation are the optimum conditions for rapid and stable synthesis. Similarly, Hamouda et al. [3] have shown the effect of plant extracts and $AgNO_3$ concentrations for biosynthesis of AgNPs using an aqueous extract of *Oscillatoria limnetica* and reported that concentrations of the aqueous extract of *Oscillatoria limnetica* and $AgNO_3$ affected the characteristics of synthesized AgNPs through controlling its size and shape. As with plant-mediated synthesis, microbe-mediated synthesis is also significantly influenced by these parameters. According to Huq [6], extracellular synthesis of AgNPs using culture supernatant of *Pseudoduganella eburnea* MAHUQ-39 was affected by temperature and metal salt ($AgNO_3$) concentration. It was found that 30 °C temperature, 1 mM $AgNO_3$ (final concentration) and 24 h incubation time are the best conditions for the rapid and stable synthesis of AgNPs using *P. eburnea*. Many other recent studies also showed the effect of concentration of plant extract and metal salt, incubation time, temperature, and pH for the rapid and stable synthesis of homogenous AgNPs with a high yield using both plants and microbes [25,26,48,108].

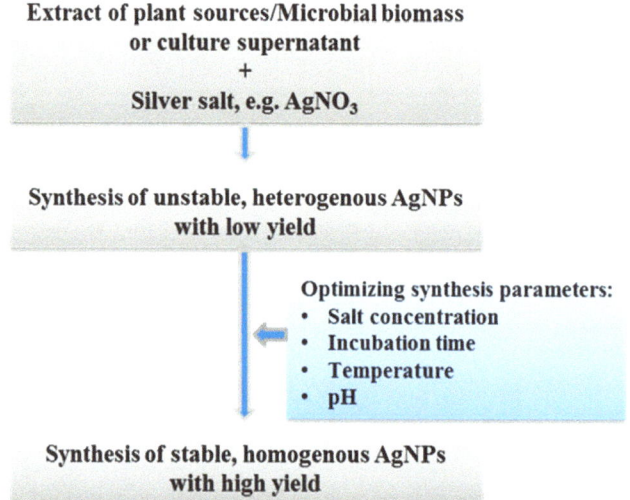

Figure 3. Optimization of parameters for stable, monodispersed, rapid and high-yield of AgNPs.

6. Characterization of Green Synthesized AgNPs

Characterization of AgNPs is an important step of green synthesis to check their morphology, size, shape, purity, surface chemistry, etc. Several instruments have been utilized for characterizations of green synthesized AgNPs such as UV-visible spectrophotometry, X-ray diffraction (XRD), Scanning electron microscope (SEM), Transmission electron microscope (TEM), Fourier Transform Infrared Spectroscopy (FTIR), Dynamic light scattering (DLS), and Zeta potential analyzer, etc. Synthesis of AgNPs is initially observed by the naked eye due to the change of color. Generally, the dark brown color of the reaction mixture indicates the synthesis of AgNPs. Then, the formation of AgNPs is confirmed by UV-visible spectrophotometry. Synthesized AgNPs showed a strong peak at around 400–470 nm in UV-visible spectrophotometry. The absorption spectra depended on the morphology, size and shape of biosynthesized of AgNPs [12,140]. SEM and TEM are the powerful tools to characterize the nanoparticles. Both SEM and TEM are used to observe the morphology, shape, size, and the degree of particle aggregation and purity of synthesized nanoparticles [21,108,114]. XRD is an analytical technique which has been utilized to evaluate the structural features of nanoparticles such as the degree of crystallinity, particle sizes, etc. [20]. Dynamic light scattering (DLS) is used to investigate the hydrodynamic size and polydispersity index of synthesized nanoparticles. Measurement of Zeta potential is very important to check the stability of AgNPs in aqueous suspensions. AgNPs with a Zeta potential less than −25 mV or greater than +25 mV typically have high stability [108,141].

FTIR spectroscopy is a very important tool to investigate the biomolecules responsible for the capping and stabilizing of nanoparticles [2]. Biosynthesis of AgNPs using culture supernatant of *Sphingobium* sp. MAH-11 and their characterization by UV–vis, TEM, XRD, DLS, and FTIR has been reported by Akter and Huq [2]. Synthesis of AgNPs was initially observed by changing of color into dark brown and finally the synthesis was confirmed on the basis of the appearance of a sharp peak at 423 nm in the UV–vis region of the spectrum. The TEM analysis revealed the spherical shape and the size was 7–22 nm. The SAED pattern revealed sharp rings which indicated the crystalline nature of synthesized AgNPs. The XRD pattern also showed the crystalline structure of AgNPs. The FTIR spectrum showed that various biomolecules acted as reducing agents as well as capping and stabilizing agents during the synthesis process (Figure 4), [2]. Sukweenadhi et al. [25] have reported the green synthesis of AgNPs from leaf extract of *Plantago major* and the synthesized AgNPs were characterized by UV–vis, TEM, SEM, XRD, DLS and FTIR.

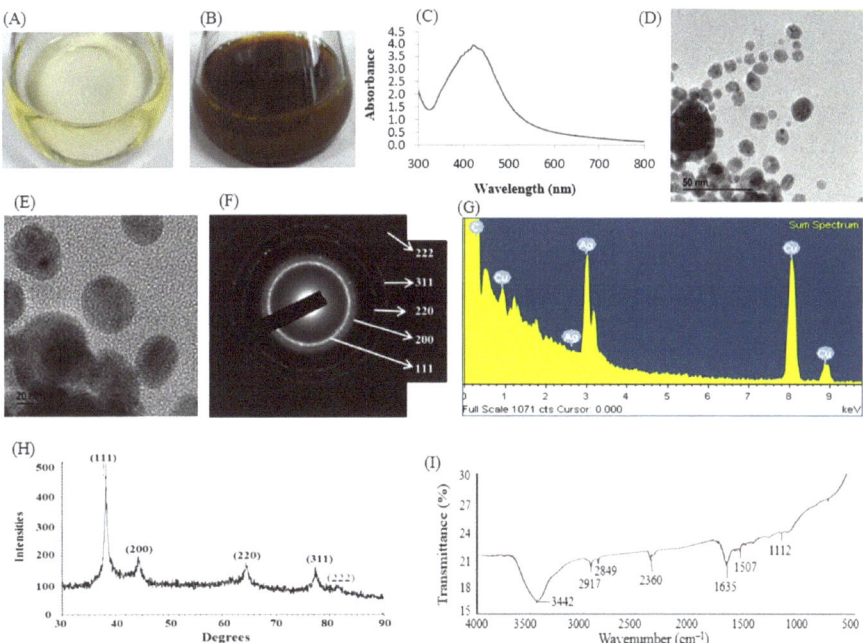

Figure 4. R2A broth with AgNO$_3$ as control (**A**); biosynthesized AgNPs (**B**); UV–vis spectra (**C**); FE-TEM images (**D,E**); SAED pattern (**F**); EDX spectrum (**G**); X-ray diffraction pattern (**H**); and FT-IR spectra of biosynthesized AgNPs (**I**). This figure has been reprinted with permission from Ref. [2], copyright 2020, Informa UK Limited.

7. Antibacterial Application of Green Synthesized AgNPs

At the present time, nanoparticles have gained lots of attention by reason of the continuous improvement in treatment of bacterial infections and diseases, as well as inefficient treatment [142]. Among many applied nanoparticles, AgNPs have shown significant application in the reduction of pathogenic microbes and also in the treatment of microbial infections. Due to the rapid increase of antibiotic resistance in this period, this has revived the attention of the researchers investigating the therapeutic abilities of AgNPs systems as potential antimicrobial agents [142]. The published articles proposed the antibacterial activities of AgNPs, and explored them as a promising strategy which could be used as effective growth inhibitors in various microorganisms, antimicrobial control systems and for developing silver-coated medicinal devices, and silver-based dressings, such as nanogels, nanolotions, etc. [12,143,144]. The smaller particle size and greater surface volume of AgNPs holds an extensive contact area with the microbes. These features of AgNPs strongly increases their biological and chemical properties, which greatly helps them to show a robust bactericidal material [145]. This study also showed that AgNPs inhibited the growth of *E. coli* ATCC-15224 on both liquid as well as solid growth media. AgNPs with the concentration of 60 µg/mL have shown a complete cytoxicity to the *E. coli* bacterial strain, and the lower concentration of 60 µg/mL inhibited bacterial cell growth and multiplication [145].

Recently, AgNPs were synthesized using different plants and their various parts as well as bacteria, and the AgNPs produced were tested against various pathogenic microbes including multidrug-resistant bacteria (Tables 1 and 2). Huq and Akter [106] reported bacterial-mediated synthesis of AgNPs and their antibacterial activity against pathogenic strains of *K. pneumoniae* and *S.* Enteritidis. The synthesized AgNPs showed a 17.6 and a 16.8 mm zone of inhibition (ZOI) against *K. pneumoniae* and *S.* Enteritidis, respectively, whereas some commercial antibiotics such as erythromycin, penicillin, vancomycin, oleandomycin, novobiocin, and lincomycin were resistant or displayed very weak activity against these pathogens. The minimum inhibitory concentration (MIC)/minimum bactericidal concentration (MBC) values of synthesized AgNPs against *K. pneumoniae* and *S.* Enteritidis were 12.5/50.0 and 25.0/50.0 µg/mL, respectively. These MIC/MBC values were well below other antimicrobial agents including zinc oxide and gold nanoparticles against *K. pneumoniae* and *S.* Enteritidis. Another study reported *Arthrobacter bangladeshi* mediated green synthesis of AgNPs and investigated their antibacterial activity against pathogenic strains of *S. typhimurium* and *Y. enterocolitica*. The green synthesized AgNPs showed a 18.3 and a 20.4 mm ZOI against *S. typhimurium* and *Y. enterocolitica*, respectively. The MIC/ MBC values of synthesized AgNPs against *S. typhimurium* and *Y. enterocolitica* were 6.2/12.5 and 3.1/12.5 µg/mL, respectively. These MIC/MBC values were significantly lower than some other antimicrobial agents against *S. typhimurium* and *Y. enterocolitica* [109].

Ahmed et al. [36] reported the green synthesis of AgNPs using dried leaf extract of Tasmanian flax-lily and evaluated their antibacterial activity against several microbes including *S. aureus*, *S. epidermidis*, *P. aeruginosa* and *C. albicans*. *Chlorophytum borivilianum* callus extract was utilized for the green synthesis of AgNPs and the synthesized nanoparticle was used to investigate the antimicrobial activity towards the human pathogens such as *B. subtilis*, *S. aureus*, *P. aeruginosa* and *E. coli*. This result revealed that the synthesized AgNPs showed strong inhibitory activity against tested pathogens [44]. *Plantago major*, *Prunus africana* and *Camellia sinensis* were also reported to synthesize small-size AgNPs and evaluated against *S. aureus*, *E. coli*, *P. aeruginosa* and *K. pneumoniae* [25,35]. It has been reported that smaller size NPs showed higher antibacterial activities due to the larger surface area [146]. It was reported that AgNPs has shown remarkable antibacterial efficacy against antibiotic-resistant human pathogenic strains *S. aureus*, *E. coli*, and *P. aeruginosa* [2]. Hasnain et al. [45] reported on the purple heart plant leaves extract -mediated synthesis of AgNPs and evaluated their antibacterial activity against *E. coli*, and *S. aureus*. They found that the purple heart plant leaves extract-mediated synthesized AgNPs showed significantly strong antibacterial activity against both *E. coli*, and *S. aureus* compared to the purple heart plant leaves extract. Another report also proposed the excellent antimicrobial activity of biosynthesized AgNPs against various Gram-negative and Gram-positive pathogenic microorganisms which showed the way to use it as a potential application of antibacterial agent against multidrug-resistant bacteria [126].

Recently a few studies have stated that the conjugation of AgNPs with bactericidal agents may reduce the toxic effect towards the mammalian cells whilst increasing the bactericidal activity. This conjugation helps to increase the amount of antibacterial agent in the specific bacterial site and thus the therapeutic activity of the antibiotic agents could be enhanced against the bacterial infection [147,148]. It was also reported that AgNPs can be applied on a clinical platform against human pathogenic strains *C. albicans*, *S. enterica*, *E. coli*, and *V. parahemolyticus* [134,135]. It was demonstrated that green synthesis AgNPs has shown antimicrobial activity against multidrug-resistant pathogenic microbes. They also mentioned that it was ecofriendly, safe, facile, effective, and economical, which could be applied in both medical and non-medical sectors, especially as an antimicrobial agent to control drug-resistant pathogens [20]. The biosynthesized AgNPs presented great antimicrobial effect against multidrug-resistant pathogens such as *S. aureus* and *P. aeruginosa*. The MBCs to inhibit *S. aureus* and *P. aeruginosa* were 200 and 50 µg/mL, respectively [6]. In another study this author proposed that the AgNPs synthesized by strain MAHUQ-40 showed significant antibacterial activity against *V. parahaemolyticus* and *S. Typhimurium* with MICs 3.12 and 6.25 µg/mL, respectively [114], whereas some commercial antibiotics such as penicillin G, erythromycin, oleandomycin, lincomycin, and vancomycin were resistant or displayed very weak activity against these pathogens. Another study investigated antimicrobial activity against both Gram-positive *B. cereus* and Gram-negative bacteria *P. aeruginosa*. The bacterial-mediated synthesized AgNPs inhibited the growth of pathogenic strains *B. cereus* and *P. aeruginosa* through developing a clear zone of inhibition [12]. Due to this killing ability, AgNPs are recognized for their remarkable antibacterial activity. Moreover, the modification in AgNPs surface developed the interactions of the constituents and this surface modification of AgNPs through chemical functionalization has gained much consideration which could be useful in numerous areas such as medical, engineering, and biological uses [149,150].

8. Antibacterial Mechanisms of AgNPs

The most important thing about nanoparticles is their mechanism of action and this mechanism mostly depends upon the size, pH, and ionic strength of the medium, and also on the type of capping agent. However, the complete antibacterial mechanism of AgNPs is still not fully known and has not been completely explained. According to the previous studies, it could be considered that AgNPs may frequently release the silver ions (Ag+), which might be considered as one of the mechanisms behind the bactericidal activity of AgNPs [142,151]. It has been demonstrated that the Ag+ ion forms complexes with the nucleic acids and interacts with the nucleosides of nucleic acids to show antibacterial activities. Nanoparticles altered the membrane permeability as evident from the release of sugars, proteins, and nuclear material through the damaged membrane [152]. The electrostatic attractions as well as an affinity towards the sulfur proteins enhanced the adhesion of Ag+ ion to the cytoplasm and cell membrane and lead to the disruptions of bacterial casings with enhancing the permeability of bacterial cell membrane [153].

The production of reactive oxygen species (ROS) is increased due to the production of free Ag+ ions by the cells, which may interrupt adenosine triphosphate (ATP) release [154]. This ROS may play an important role to disrupt the cellular membrane and the alteration in the deoxyribonucleic acid, which could cause different issues related to DNA, including DNA replication and cell propagation. On the other hand, free Ag+ ions may efficiently interfere the protein synthesis by denaturing cytoplasmic ribosomal components [155]. The release percentage of Ag+ ions can inhibit the growth of bacteria because the nanoscale size of AgNPs has the ability to penetrate the bacterial cell wall as well as denaturation of the cell membranes [156]. Due to denaturation of the cell membrane the intracellular and extracellular components of bacterial cell membrane may be ruptured which also causes cell lysis [157]. The antibacterial mechanisms of the AgNPs are mainly influenced by their dissolution profile in the reaction media and dissolution efficacy also depend on the synthesis and processing parameters [158]. Although the exact antibacterial mechanism of AgNPs has not been entirely clarified, different antibacterial actions of AgNPs have been proposed in Figure 5.

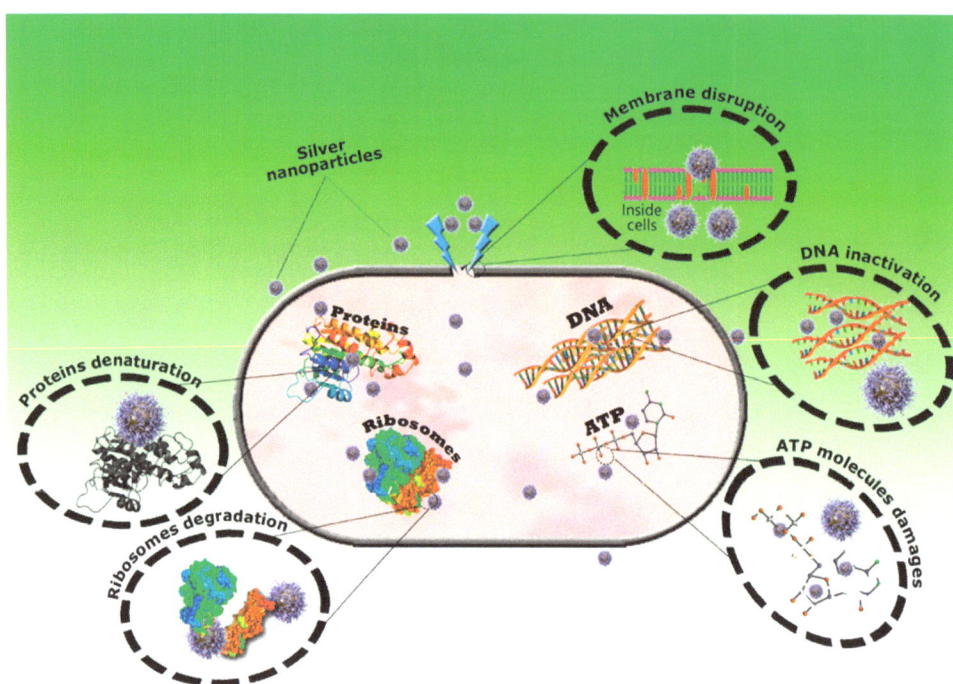

Figure 5. Possible antibacterial mechanisms of AgNPs. Disruption of cell wall and cell membrane, damage of ATP molecules due to the production of reactive oxygen species, DNA inactivation, protein denaturation and ribosome degradation.

The researchers proposed that the biosynthesized AgNPs may affect the bacterial cell morphology and penetrate the cell membrane by damaging the of cell wall of *E. coli* and *S. aureus*, which may decrease the reproduction of cell and ultimately lead to cell death. The FE-SEM images proved the strong antibacterial mechanism of AgNPs against pathogenic bacteria and promoted the application of AgNPs as an antimicrobial agent [2]. Another study demonstrated that synthesized AgNPs changed the structural function of bacterial cells like *S. aureus* and *P. aeruginosa*. These mechanical activities of the proposed AgNPs create a promising hope to recognize it as an effective antimicrobial agent for various therapeutic applications against *S. aureus* and *P. aeruginosa* infections [6]. It was stated that AgNPs show the efficacy to alter the cell morphology as well as damage the cell membrane of tested pathogens (Figure 6). The main mechanism of AgNPs is that these nanoparticles strongly attach to the bacterial cell membrane surface and disturb its proper function, because of the enhancement of DNA damage [126]. The AgNPs also have the capability to penetrate the cell membrane and when it penetrates the cell membrane it potentially disrupts the cellular components by reacting with the sulphur-mediated proteins and phosphorus-mediated complexes like deoxyribonucleic acid [159]. Scanning electron microscopy (SEM) and TEM studies demonstrated that AgNPs shown the ability to adhere and interact with *E. coli* and penetrate into the bacterial cells. This adhesion and interaction ability increases the antibacterial activity of AgNPs, which are attributed with total surface area of nanoparticles [145]. Thus, the ecofriendly synthesis of AgNPs could be useful in various applications in both pharmaceutical and non- pharmaceutical sectors to eradicate drug-resistant pathogens [20].

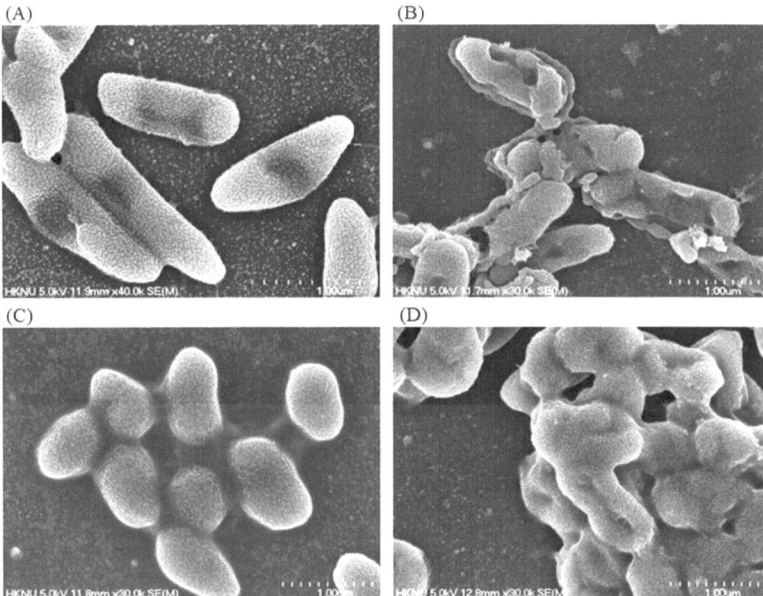

Figure 6. FE–SEM images of normal *P. aeruginosa* cells (**A**); 1 × MBC AgNPs treated *P. aeruginosa* cells (**B**); normal *S. aureus* cells (**C**); 1 × MBC AgNPs treated *S. aureus* cells (**D**). This figure has been reprinted with permission from Ref. [6], copyright 2020, MDPI.

AgNPs have changed and damaged *V. parahaemolyticus* and *S. typhimurium* membrane integrity, which reduced the metabolic activity and normal cell function caused bacterial cells' death [114]. The field emission scanning electron microscopy analysis demonstrated that AgNPs were responsible for damaging the cell wall and altering the cell morphology of treated Gram-positive and Gram-negative pathogenic bacteria, leading to the death of cells [12]. The literature demonstrated that AgNPs trigger the inhibition of protein synthesis as well as cell wall synthesis, which provides strong evidence about the protein disruption of the outer cellular membrane and increasing ATP leakage, resulting in cell death [160].

Apart from these, the size and shape of the AgNPs increase the release of Ag+ ions owing to their greater surface area which influence potential activity against bacterial disease. The dissolution rate of AgNPs also interferes with its antimicrobial level. If the dissolution rate is high, then the potential activity could be increased [161]. It is generally proposed that AgNPs smaller than 10 nm may directly penetrate cell membranes, enter into the bacterial cells, and initiate cell lysis [162]. Therefore, the finding may provide a meaningful statement about AgNPs to use as an alternative antibacterial agent to protect pathogenic bacteria as well as to treat bacterial infectious diseases.

9. Conclusions and Future Prospects

Green synthesis of AgNPs is preferred due to its eco-friendly nature. The utilization of various parts of plant, bacteria, fungi, algae is an efficient, facile and environmentally friendly way to synthesize AgNPs. Plant extracts contain different biomolecules such as amino acids, proteins, enzymes, terpene, alkaloids, flavonoids, phenols, tannins, vitamins, etc., which act as reducing, capping, and stabilizing agents. Similarly, microorganisms synthesize various extracellular and intracellular biomolecules such as enzymes, amino acid, proteins, and many other primary and secondary metabolites that act as reducing agents as well as capping and stabilizing agents during the synthesis process. Biosynthesized AgNPs have great bactericidal potential against various Gram-positive and Gram-negative bacteria. In this review, green synthesis of AgNPs using plants and microbes has been comprehensively reviewed. The antibacterial applications and mechanisms of the biosynthesized AgNPs against pathogenic microbes have also been highlighted. Although the rapid, facile and eco-friendly synthetic methods using plants and microbes have shown great potential in AgNPs, the exact mechanism of synthesis and the mode of antimicrobial action are still not fully understood. Hence, several points might be considered for the future synthesis of AgNPs from plants or microbes. First, the selection of plant or microbes for easy, rapid and eco-friendly synthesis. For plant selection, researchers should consider the availability of plants and their extraction process. Plants should be available and the extraction process should be simple for facile and mass production of AgNPs. Similarly, researchers should focus on non-pathogenic and rapid growth microbes for safe and easy handling during the synthesis process. In this case, probiotic microbes could be the great synthetic agent. Second, investigation of the biomolecules present in plant extracts or in microbial biomass or culture supernatant. It is believed that different biomolecules present in plant extracts or in microbial culture supernatant are mainly responsible for the synthesis and stabilization of AgNPs. The role of various enzymes for biosynthesis needs to be studied in detail. Additionally, these biomolecules are also responsible to enhance the antibacterial efficacy of synthesized AgNPs. Therefore, it is important to investigate the biomolecules present in the plant extract or in microbial culture supernatant for successful synthesis of AgNPs. Third, optimization of parameters for rapid, stable and mass production of AgNPs. Several studies reported that various parameters such as concentration of the plant extract and $AgNO_3$, incubation time and temperature, pH of reaction, etc. have great effect on synthesis process. Hence, mass production on an industrial scale can be achieved by optimizing these reaction conditions. Fourth, investigation of the antibacterial mechanisms. Most of the studies reported the efficacy of AgNPs in the screening level without investigating the exact mechanisms. It is very important to find out the mode of action of AgNPs against pathogens. Fifth, investigation of cytotoxic effect of biosynthesized AgNPs on human cells. Some studies reported that AgNPs have cytotoxic effects on human cells. Hence, it is essential to investigate the potential toxicity of biosynthesized AgNPs on healthy human cells to ensure their safe use for human and the environment.

Author Contributions: Conceptualization, M.A.H.; writing—original draft preparation, M.A.H., M.A. and S.A.; writing—review and editing, M.A.H., M.M.R. and S.R.B. All authors have read and agreed to the published version of the manuscript.

Funding: This research received no external funding.

Institutional Review Board Statement: Not applicable.

Informed Consent Statement: Not applicable.

Data Availability Statement: Not applicable.

Conflicts of Interest: The authors declare no conflict of interest.

References

1. Bachheti, R.K.; Fikadu, A.; Bachheti, A.; Husen, A. Biogenic fabrication of nanomaterials from flower-based chemical compounds, characterization and their various applications: A review. *Saudi J. Biol. Sci.* **2020**, *27*, 2551–2562. [CrossRef] [PubMed]
2. Akter, S.; Huq, M.A. Biologically rapid synthesis of silver nanoparticles by *Sphingobium* sp. MAH-11 T and their antibacterial activity and mechanisms investigation against drug-resistant pathogenic microbes. *Artif. Cells Nanomed. Biotechnol.* **2020**, *48*, 672–682. [CrossRef] [PubMed]
3. Hamouda, R.A.; Hussein, M.H.; Abo-Elmagd, R.A.; Bawazir, S.S. Synthesis and biological characterization of silver nanoparticles derived from the cyanobacterium *Oscillatoria limnetica*. *Sci. Rep.* **2019**, *9*, 13071. [CrossRef]
4. Kulkarni, N.; Muddapur, U. Biosynthesis of metal nanoparticles: A review. *J. Nanotechnol.* **2014**, *2014*, 510246. [CrossRef]
5. Jamkhande, P.G.; Ghule, N.W.; Bamer, A.H.; Kalaskar, M.G. Metal nanoparticles synthesis: An overview on methods of preparation, advantages and disadvantages, and applications. *J. Drug Deliv. Sci. Technol.* **2019**, *53*, 101174. [CrossRef]

6. Huq, M.A. Green synthesis of silver nanoparticles using *Pseudoduganella eburnea* MAHUQ-39 and their antimicrobial mechanisms investigation against drug resistant human pathogens. *Int. J. Mol. Sci.* **2020**, *21*, 1510. [CrossRef]
7. Kedi, P.B.E.; Meva, F.E.; Kotsedi, L. Eco-friendly synthesis, characterization, in vitro and in vivo anti-inflammatory activity of silver nanoparticle-mediated *Selaginella myosurus* aqueous extract. *Int. J. Nanomed.* **2018**, *13*, 8537–8548. [CrossRef] [PubMed]
8. El-Naggar, N.E.; Hussein, M.H.; El-Sawah, A.A. Bio-fabrication of silver nanoparticles by phycocyanin, characterization, in vitro anticancer activity against breast cancer cell line and in vivo cytotoxicity. *Sci. Rep.* **2017**, *7*, 10844. [CrossRef]
9. Fouda, A.; Abdel-Maksoud, G.; Abdel-Rahman, M.A.; Eid, A.M.; Barghoth, M.G.; El-Sadany, M.A.H. Monitoring the effect of biosynthesized nanoparticles against biodeterioration of cellulose-based materials by *Aspergillus niger*. *Cellulose* **2019**, *26*, 6583–6597. [CrossRef]
10. Burdusel, A.C.; Gherasim, O.; Grumezescu, A.M.; Mogoanta, L.; Ficai, A.; Andronescu, E. Biomedical applications of silver nanoparticles: An up-to-date overview. *Nanomaterials* **2018**, *8*, 681. [CrossRef]
11. Rafique, M.; Sadaf, I.; Rafique, M.S.; Tahir, M.B. A review on green synthesis of silver nanoparticles and their applications. *Artif. Cells Nanomed. Biotechnol.* **2017**, *45*, 1272–1291. [CrossRef] [PubMed]
12. Huq, M.A.; Akter, S. Bacterial mediated rapid and facile synthesis of silver nanoparticles and their antimicrobial efficacy against pathogenic microorganisms. *Materials* **2021**, *14*, 2615. [CrossRef] [PubMed]
13. Salem, S.S.; Fouda, A. Green synthesis of metallic nanoparticles and their prospective biotechnological applications: An overview. *Biol. Trace Elem. Res.* **2021**, *199*, 344–370. [CrossRef] [PubMed]
14. Sharma, A.K.; Kaith, B.S.; Shanker, U.; Gupta, B. γ-radiation induced synthesis of antibacterial silver nanocomposite scaffolds derived from natural gum *Boswellia serrata*. *J. Drug Deliv. Sci. Technol.* **2020**, *56*, 101550. [CrossRef]
15. Elemike, E.E.; Onwudiwe, D.C.; Fayemi, O.E.; Botha, T.L. Green synthesis and electrochemistry of Ag, Au, and Ag–Au bimetallic nanoparticles using golden rod (*Solidago canadensis*) leaf extract. *Appl. Phys. A Mater. Sci. Process.* **2019**, *125*, 42. [CrossRef]
16. Dos Santos, M.A.; Paterno, L.G.; Moreira, S.G.C.; Sales, M.J.A. Original photochemical synthesis of Ag nanoparticles mediated by potato starch. *SN Appl. Sci.* **2019**, *1*, 554. [CrossRef]
17. Guzman, M.G.; Dille, J.; Godet, S. Synthesis of silver nanoparticles by chemical reduction method and their antibacterial activity. *Int. J. Chem. Biomol. Eng.* **2009**, *2*, 104–111.
18. Pauzi, N.; Zain, N.M.; Yusof, N.A.A. Microwave-assisted synthesis of ZnO nanoparticles stabilized with Gum Arabic: Effect of microwave irradiation time on ZnO nanoparticles size and morphology. *Bull. Chem. React. Eng. Catal.* **2019**, *14*, 182. [CrossRef]
19. Iravani, S.; Korbekandi, H.; Mirmohammadi, S.V.; Zolfaghari, B. Synthesis of silver nanoparticles: Chemical, physical and biological methods. *Res. Pharm. Sci.* **2014**, *9*, 385–406.
20. Akter, S.; Lee, S.-Y.; Siddiqi, M.Z.; Balusamy, S.R.; Ashrafudoulla, M.; Rupa, E.J.; Huq, M.A. Ecofriendly synthesis of silver nanoparticles by *Terrabacter humi* sp. nov. and their antibacterial application against antibiotic-resistant pathogens. *Int. J. Mol. Sci.* **2020**, *21*, 9746. [CrossRef]
21. Du, J.; Sing, H.; Yi, T.H. Antibacterial, anti-biofilm and anticancer potentials of green synthesized silver nanoparticles using benzoin gum (*Styrax benzoin*) extract. *Bioprocess Biosyst. Eng.* **2016**, *39*, 1923–1931. [CrossRef] [PubMed]
22. Vigneshwaran, N.; Kathe, A.A.; Varadarajan, P.V. Silve-protein (core-shell) nanoparticle production using spent mushroom substrate. *Langmuir* **2007**, *23*, 7113–7117. [CrossRef] [PubMed]
23. Huq, M.A. *Paenibacillus anseongense* sp. nov. a silver nanoparticle producing bacterium isolated from rhizospheric soil. *Curr. Microbiol.* **2020**, *77*, 2023–2030. [CrossRef] [PubMed]
24. Singh, P.; Kim, Y.J.; Zhang, D.; Yang, D.C. Biological synthesis of nanoparticles from plants and microorganisms. *Trends Biotechnol.* **2016**, *34*, 588–599. [CrossRef] [PubMed]
25. Sukweenadhi, J.; Setiawan, K.I.; Avanti, C.; Kartini, K.; Rupa, E.J.; Yang, D.C. Scale-up of green synthesis and characterization of silver nanoparticles using ethanol extract of *Plantago major* L. leaf and its antibacterial potential. *S. Afr. J. Chem. Eng.* **2021**, *38*, 1–8. [CrossRef]
26. Nahar, K.; Yang, D.C.; Rupa, E.J.; Khatun, M.K.; Al-Reza, S.M. Eco-friendly synthesis of silver nanoparticles from *Clerodendrum viscosum* leaf extract and its antibacterial potential. *Nanomed. Res. J.* **2020**, *5*, 276–287.
27. Pawar, J.S.; Patil, R.H. Green synthesis of silver nanoparticles using *Eulophia herbacea* (Lindl.) tuber extract and evaluation of its biological and catalytic activity. *SN Appl. Sci.* **2020**, *2*, 52. [CrossRef]
28. Soshnikova, V.; Kim, Y.J.; Singh, P.; Huo, Y.; Markus, J.; Ahn, S.; Castro-Aceituno, V.; Kang, J.; Chokkalingam, M.; Mathiyalagan, R.; et al. Cardamom fruits as a green resource for facile synthesis of gold and silver nanoparticles and their biological applications. *Artif. Cells Nanomed. Biotechnol.* **2018**, *46*, 108–117. [CrossRef]
29. Sumitha, S.; Vasanthi, S.; Shalini, S.; Chinni, S.V.; Gopinath, S.C.B.; Anbu, P.; Bahari, M.B.B.; Harish, R.; Kathiresan, S.; Ravichandran, V. Phyto-mediated photo catalysed green synthesis of silver nanoparticles using *Durio Zibethinus* seed extract: Antimicrobial and cytotoxic activity and photocatalytic applications. *Molecules* **2018**, *23*, 3311. [CrossRef]
30. Arokiyaraj, S.; Vincent, S.; Saravanan, M.; Lee, Y.; Oh, Y.K.; Kim, K.H. Green synthesis of silver nanoparticles using *Rheum palmatum* root extract and their antibacterial activity against *Staphylococcus aureus* and *Pseudomonas aeruginosa*. *Artif. Cells Nanomed. Biotechnol.* **2017**, *45*, 372–379. [CrossRef]
31. Sharma, K.; Kaushik, S.; Jyoti, A. Green Synthesis of Silver Nanoparticles by Using Waste Vegetable Peel and its Antibacterial Activities. *J. Pharm. Sci. Res.* **2016**, *8*, 313–316.

32. Singh, P.; Kim, Y.J.; Wang, C.; Mathiyalagan, R.; Yang, D.C. The development of a green approach for the biosynthesis of silver and gold nanoparticles by using *Panax ginseng* root extract, and their biological applications. *Artif. Cells Nanomed. Biotechnol.* **2015**, *44*, 1150–1157. [PubMed]
33. Singh, P.; Kim, Y.J.; Yang, D.C. A strategic approach for rapid synthesis of gold and silver nanoparticles by *Panax ginseng* leaves. *Artif. Cells Nanomed. Biotechnol.* **2016**, *44*, 1949–1957. [CrossRef] [PubMed]
34. Adeyemi, D.K.; Adeluola, A.O.; Akinbile, M.J.; Johnson, O.O.; Ayoola, G.A. Green synthesis of Ag, Zn and Cu nanoparticles from aqueous extract of *Spondias mombin* leaves and evaluation of their antibacterial activity. *Afr. J. Clin. Exper. Microbiol.* **2020**, *21*, 106–113. [CrossRef]
35. Ssekatawa, K.; Byarugaba, D.K.; Kato, C.D.; Wampande, E.M.; Ejobi, F.; Nakavuma, J.L.; Maaza, M.; Sackey, J.; Nxumalo, E.; Kirabira, J.B. Green strategy-based synthesis of silver nanoparticles for antibacterial applications. *Front. Nanotechnol.* **2021**, *3*, 697303. [CrossRef]
36. Ahmed, S.R.; Anwar, H.; Ahmed, S.W.; Shah, M.R.; Ahmed, A.; Abid Ali, S.A. Green synthesis of silver nanoparticles: Antimicrobial potential and chemosensing of a mutagenic drug nitrofurazone in real samples. *Measurement* **2021**, *180*, 109489. [CrossRef]
37. Urnukhsaikhan, E.; Bold, B.E.; Gunbileg, A.; Sukhbaatar, N.; Mishig-Ochir, T. Antibacterial activity and characteristics of silver nanoparticles biosynthesized from *Carduus crispus*. *Sci. Rep.* **2021**, *26*, 21047. [CrossRef]
38. Aabed, K.; Mohammed, A.E. Synergistic and antagonistic effects of biogenic silver nanoparticles in combination with antibiotics against some pathogenic microbes. *Front. Bioeng. Biotechnol.* **2021**, *9*, 249. [CrossRef]
39. Madivoli, E.S.; Kareru, P.G.; Gachanja, A.N.; Mugo, S.M.; Makhanu, D.S.; Wanakai, S.I.; Gavamukulya, Y. Facile synthesis of silver nanoparticles using *Lantana trifolia* aqueous extracts and their antibacterial activity. *J. Inorg. Organomet. Polym.* **2020**, *30*, 2842–2850. [CrossRef]
40. Chavan, R.R.; Bhinge, S.D.; Bhutkar, M.A.; Randive, D.S.; Wadkar, G.H.; Todkar, S.S. Characterization, antioxidant, antimicrobial and cytotoxic activities of green synthesized silver and iron nanoparticles using alcoholic Blumea eriantha DC plant extract. *Urade. Mater. Today Commun.* **2020**, *24*, 101320. [CrossRef]
41. Hemlata; Meena, P.R.; Singh, A.P.; Tejavath, K.K. Biosynthesis of silver nanoparticles using *Cucumis prophetarum* aqueous leaf extract and their antibacterial and antiproliferative activity against cancer cell lines. *ACS Omega* **2020**, *5*, 5520–5528. [CrossRef]
42. Shejawal, K.P.; Randive, D.S.; Bhinge, S.D.; Bhutkar, M.A.; Wadkar, G.H.; Jadhav, N.R. Green synthesis of silver and iron nanoparticles of isolated proanthocyanidin: Its characterization, antioxidant, antimicrobial, and cytotoxic activities against COLO320DM and HT. *J. Genet. Eng. Biotechnol.* **2020**, *18*, 43. [CrossRef] [PubMed]
43. Kalpana, D.; Han, J.H.; Park, W.S.; Lee, S.M.; Wahab, R.; Lee, Y.S. Green biosynthesis of silver nanoparticles using *Torreya nucifera* and their antibacterial activity. *Arab. J. Chem.* **2019**, *12*, 1722–1732. [CrossRef]
44. Huang, F.; Long, Y.; Liang, Q.; Purushotham, B.; Swamy, M.K.; Duan, Y. Safed musli (*Chlorophytum borivilianum* L.) callus-mediated biosynthesis of silver nanoparticles and evaluation of their antimicrobial activity and cytotoxicity against human colon cancer cells. *J. Nanomater.* **2019**, *2019*, 2418785. [CrossRef]
45. Hasnain, M.S.; Javed, N.; Alam, S.; Rishishwar, P.; Rishishwar, S.; Ali, S.; Nayak, A.K.; Beg, S. Purple heart plant leaves extract-mediated silver nanoparticle synthesis: Optimization by Box-Behnken design. *Mater. Sci. Eng. C* **2019**, *99*, 1105–1114. [CrossRef]
46. Oves, M.; Aslam, M.; Rauf, M.A.; Qayyum, S.; Qari, H.A.; Khan, M.S. Antimicrobial and anticancer activities of silver nanoparticles synthesized from the root hair extract of *Phoenix dactylifera*. *Mater. Sci. Eng. C* **2018**, *89*, 429–443. [CrossRef] [PubMed]
47. Saratale, R.G.; Benelli, G.; Kumar, G.; Kim, D.S.; Saratale, G.D. Bio-Fabrication of silver nanoparticles using the leaf extract of an ancient herbal medicine, dandelion (*Taraxacum officinale*), evaluation of their antioxidant, anticancer potential, and antimicrobial activity against phytopathogens. *Environ. Sci. Pollut. Res.* **2017**, *25*, 10392–10406. [CrossRef] [PubMed]
48. Khatami, M.; Zafarnia, N.; Heydarpoor Bami, M.; Sharifi, I.; Singh, H. Antifungal and antibacterial activity of densely dispersed silver nanospheres with homogeneity size which synthesized using chicory: An in vitro study. *J. Mycol. Med.* **2018**, *28*, 637–644. [CrossRef]
49. Devanesan, S.; AlSalhi, M.S.; Balaji, R.V.; Ranjitsingh, A.J.A.; Ahamed, A.; Alfuraydi, A.A.; AlQahtani, F.Y.; Aleanizy, F.S.; Othman, A.H. Antimicrobial and cytotoxicity effects of synthesized silver nanoparticles from *Punica granatum* peel extract. *Nanoscale Res. Lett.* **2018**, *13*, 315. [CrossRef]
50. Mythili, R.; Selvankumar, T.; Kamala-Kannan, S.; Sudhakar, C.; Ameen, F.; Al-Sabri, A.; Selvam, K.; Govarthanan, M.; Kim, H. Utilization of market vegetable waste for silver nanoparticle synthesis and its antibacterial activity. *Mater. Lett.* **2018**, *225*, 101–104. [CrossRef]
51. Markus, J.; Wang, D.; Kim, Y.J.; Ahn, S.; Mathiyalagan, R.; Wang, C.; Yang, D.C. Biosynthesis, characterization, and bioactivities evaluation of silver and gold nanoparticles mediated by the roots of Chinese Herbal *Angelica pubescens* maxim. *Nanoscale Res. Lett.* **2017**, *12*, 46. [CrossRef] [PubMed]
52. Mohanta, Y.K.; Panda, S.K.; Bastia, A.K.; Mohanta, T.K. Biosynthesis of silver nanoparticles from *Protium serratum* and investigation of their potential impacts on food safety and control. *Front. Microbiol.* **2017**, *8*, 626. [CrossRef]
53. Huo, Y.; Singh, P.; Kim, Y.J.; Soshnikova, V.; Kang, J.; Markus, J.; Ahn, S.; Castro-Aceituno, V.; Mathiyalagan, R.; Chokkalingam, M.; et al. Biological synthesis of gold and silver chloride nanoparticles by *Glycyrrhiza uralensis* and in vitro applications. *Artif. Cells Nanomed. Biotechnol.* **2018**, *46*, 303–312. [CrossRef] [PubMed]

54. Nasar, S.; Murtaza, G.; Mehmood, A.; Bhatti, T.M. Green approach to synthesis of silver nanoparticles using *Ficus Palmata* leaf extract and their antibacterial profile. *Pharm. Chem. J.* **2017**, *51*, 811–817. [CrossRef]
55. Rajkuberan, C.; Prabukumar, S.; Sathishkumar, G.; Wilson, A.; Ravindran, K.; Sivaramakrishnan, S. Facile synthesis of silver nanoparticles using *Euphorbia antiquorum* L. latex extract and evaluation of their biomedical perspectives as anticancer agents. *J. Saudi Chem. Soc.* **2017**, *21*, 911–919. [CrossRef]
56. Jain, S.; Mehata, M.S. Medicinal plant leaf extract and pure flavonoid mediated green synthesis of silver nanoparticles and their enhanced antibacterial property. *Sci. Rep.* **2017**, *7*, 15867. [CrossRef]
57. Mitiku, A.A.; Yilma, B. Antibacterial and antioxidant activity of silver nanoparticles synthesized using aqueous extract of *Moringa stenopetala* leaves. *Afr. J. Biotechnol.* **2017**, *16*, 1705–1716.
58. Singh, H.; Du, J.; Singh, P.; Yi, T.H. Ecofriendly synthesis of silver and gold nanoparticles by *Euphrasia officinalis* leaf extract and its biomedical applications. *Artif. Cells Nanomed. Biotechnol.* **2017**, *46*, 1163–1170. [CrossRef]
59. Abbai, R.; Ramya Mathiyalagan, J.M.; Kim, Y.J.; Wang, C.; Singh, P.; Ahn, S.; Farh, M.E.A.; Yang, D.C. Green synthesis of multifunctional silver and gold nanoparticles from the oriental herbal adaptogen: *Siberian ginseng. Int. J. Nanomed.* **2016**, *11*, 3131–3143.
60. Singh, H.; Du, J.; Yi, T.H. Green and rapid synthesis of silver nanoparticles using *Borago officinalis* leaf extract: Anticancer and antibacterial activities. *Artif. Cells Nanomed. Biotechnol.* **2017**, *45*, 1310–1316. [CrossRef]
61. Lateef, A.; Azeez, M.A.; Asafa, T.B.; Yekeen, T.A.; Akinboro, A.; Oladipo, I.C.; Azeez, L.; Ojo, S.A.; Gueguim-Kana, E.B.; Beukes, L.S. Cocoa pod husk extract-mediated biosynthesis of silver nanoparticles: Its antimicrobial, antioxidant and larvicidal activities. *J. Nanostruct. Chem.* **2016**, *6*, 159–169. [CrossRef]
62. Ahmed, S.; Saifullah; Ahmad, M.; Swami, B.L.; Ikram, S. Green synthesis of silver nanoparticles using *Azadirachta indica* aqueous leaf extract. *J. Radiat. Res. Appl. Sci.* **2016**, *9*, 1–7. [CrossRef]
63. Anandalakshmi, K.; Venugobal, J.; Ramasamy, V. Characterization of silver nanoparticles by green synthesis method using *Pedalium murex* leaf extract and their antibacterial activity. *Appl. Nanosci.* **2016**, *6*, 399–408. [CrossRef]
64. Mohanta, Y.K.; Panda, S.K.; Biswas, K.; Tamang, A.; Bandyopadhyay, J.; Bastia, A.K. Biogenic synthesis of silver nanoparticles from *Cassia fistula* (Linn.): In vitro assessment of their antioxidant, antimicrobial and cytotoxic activities. *IET Nanobiotechnol.* **2016**, *10*, 438–444. [CrossRef] [PubMed]
65. Bose, D.; Chatterjee, S. Biogenic synthesis of silver nanoparticles using guava (*Psidium guajava*) leaf extract and its antibacterial activity against *Pseudomonas aeruginosa*. *Appl. Nanosci.* **2016**, *6*, 895–901. [CrossRef]
66. Dhand, V.; Soumya, L.; Bharadwaj, S.; Chakra, S.; Bhatt, D.; Sreedhar, B. Green synthesis of silver nanoparticles using *Coffea arabica* seed extract and its antibacterial activity. *Mater. Sci. Eng. C* **2016**, *58*, 36–43. [CrossRef]
67. Sundararajan, B.; Mahendran, G.; Thamaraiselvi, R.; Kumari, B.D. Biological activities of synthesized silver nanoparticles from *Cardiospermum halicacabum* L. Bull. *Mater. Sci.* **2016**, *39*, 423–431.
68. Ravichandran, V.; Vasanthi, S.; Shalini, S.; Ali-Shah, S.A.; Harish, R. Green synthesis of silver nanoparticles using *Atrocarpus altilis* leaf extract and the study of their antimicrobial and antioxidant activity. *Mater. Lett.* **2016**, *180*, 264–267. [CrossRef]
69. Nayak, D.; Ashe, S.; Rauta, P.R.; Kumari, M.; Nayak, B. Bark extract mediated green synthesis of silver nanoparticles: Evaluation of antimicrobial activity and antiproliferative response against osteosarcoma. *Mater. Sci. Eng. C* **2016**, *58*, 44–52. [CrossRef]
70. Oluwaniyi, O.O.; Adegoke, H.I.; Adesuji, E.T.; Alabi, A.B.; Bodede, S.O.; Labulo, A.H.; Oseghale, C.O. Biosynthesis of silver nanoparticles using aqueous leaf extract of *Thevetia peruviana* Juss and its antimicrobial activities. *Appl. Nanosci.* **2016**, *6*, 903–912. [CrossRef]
71. Benakashania, F.; Allafchian, A.R.; Jalali, S.A.H. Biosynthesis of silver nanoparticles using *Capparis spinosa* L. leaf extract and their antibacterial activity. *Karbala Int. J. Mod. Sci.* **2016**, *2*, 251–258. [CrossRef]
72. Mittal, A.K.; Tripathy, D.; Choudhary, A.; Aili, P.K.; Chatterjee, A.; Singh, I.P.; Banerjee, U.C. Bio-synthesis of silver nanoparticles using potentilla fulgens wall. ex Hook. and its therapeutic evaluation as anticancer and antimicrobial agent. *Mater. Sci. Eng. C* **2015**, *53*, 120–127. [CrossRef] [PubMed]
73. Roy, K.; Sarkar, C.; Ghosh, C. Plant-mediated synthesis of silver nanoparticles using parsley (*Petroselinum crispum*) leaf extract: Spectral analysis of the particles and antibacterial study. *Appl. Nanosci.* **2015**, *5*, 945–951. [CrossRef]
74. Ali, K.; Ahmed, B.; Dwivedi, S.; Saquib, Q.; Al-Khedhairy, A.A.; Musarrat, J. Microwave accelerated green synthesis of stable silver nanoparticles with *Eucalyptus globulus* leaf extract and their antibacterial and antibiofilm activity on clinical isolates. *PLoS ONE* **2015**, *10*, e0131178. [CrossRef]
75. Ibrahim, H.M. Green synthesis and characterization of silver nanoparticles using banana peel extract and their antimicrobial activity against representative microorganisms. *J. Radiat. Res. Appl. Sc.* **2015**, *8*, 265–275. [CrossRef]
76. Shalaby, T.I.; Mahmoud, O.A.; Batouti, G.A.E.; Ibrahim, E.E. Green synthesis of silver nanoparticles: Synthesis, characterization and antibacterial activity. *Nanosci. Nanotechnol.* **2015**, *5*, 23–29.
77. Sre, P.R.; Reka, M.; Poovazhagi, R.; Kumar, M.A.; Murugesan, K. Antibacterial and cytotoxic effect of biologically synthesized silver nanoparticles using aqueous root extract of *Erythrina indica* lam. *Spectrochim. Acta Part A Mol. Biomol. Spectrosc.* **2015**, *135*, 1137–1144.
78. Miri, A.; Sarani, M.; Bazaz, M.R.; Darroudi, M. Plant-mediated biosynthesis of silver nanoparticles using *Prosopis farcta* extract and its antibacterial properties. *Spectrochim. Acta A Mol. Biomol. Spectrosc.* **2015**, *141*, 287–291. [CrossRef]

79. Balashanmugam, P.; Kalaichelvan, P.T. Biosynthesis characterization of silver nanoparticles using *Cassia roxburghii* DC. aqueous extract, and coated on cotton cloth for effective antibacterial activity. *Int. J. Nanomed.* **2015**, *10*, 87. [CrossRef]
80. Rajakannu, S.; Shankar, S.; Perumal, S.; Subramanian, S.; Dhakshinamoorthy, G.P. Biosynthesis of silver nanoparticles using *Garcinia mangostana* fruit extract and their antibacterial, antioxidant activity. *Int. J. Curr. Microbiol. Appl. Sci.* **2015**, *4*, 944–952.
81. Krithiga, N.; Rajalakshmi, A.; Jayachitra, A. Green synthesis of silver nanoparticles using leaf extracts of *Clitoria ternatea* and *Solanum nigrum* and study of its antibacterial effect against common nosocomial pathogens. *J. Nanosci.* **2015**, *8*, 928204. [CrossRef]
82. Harshiny, M.; Matheswaran, M.; Arthanareeswaran, G.; Kumaran, S.; Rajasree, S. Enhancement of antibacterial properties of silver nanoparticles–ceftriaxone conjugate through *Mukia maderaspatana* leaf extract mediated synthesis. *Ecotoxicol. Environ. Saf.* **2015**, *121*, 135–141. [CrossRef] [PubMed]
83. Ahmed, S.; Ikram, S. Silver nanoparticles: One pot green synthesis using *Terminalia arjuna* extract for biological application. *Nano Res. Appl.* **2015**, *1*, 309.
84. Premasudha, P.; Venkataramana, M.; Abirami, M.; Vanathi, P.; Krishna, K.; Rajendran, R. Biological synthesis and characterization of silver nanoparticles using *Eclipta alba* leaf extract and evaluation of its cytotoxic and antimicrobial Potential. *Bull. Mater. Sci.* **2015**, *38*, 965–973. [CrossRef]
85. Kumar, D.A.; Palanichamy, V.; Roopan, S.M. Green synthesis of silver nanoparticles using *Alternanthera dentata* leaf extract at room temperature and their antimicrobial activity. *Spectrochim. Acta A Mol. Biomol. Spectrosc.* **2014**, *127*, 168–171. [CrossRef]
86. Muniyappan, N.; Nagarajan, N.S. Green synthesis of silver nanoparticles with *Dalbergia spinosa* leaves and their applications in biological and catalytic activities. *Process Biochem.* **2014**, *49*, 1054–1061. [CrossRef]
87. Khan, M.; Khan, S.T.; Khan, M.; Adil, S.F.; Musarrat, J.; Al-Khedhairy, A.A.; Al-Warthan, A.; Siddiqui, M.; Alkhathlan, H.Z. Antibacterial properties of silver nanoparticles synthesized using *Pulicaria glutinosa* plant extract as a green bioreductant. *Int. J. Nanomed.* **2014**, *9*, 3551–3565.
88. Singh, K.; Panghal, M.; Kadyan, S.; Chaudhary, U.; Yadav, J.P. Green silver nanoparticles of *Phyllanthus amarus*: As an antibacterial agent against multi drug resistant clinical isolates of *Pseudomonas aeruginosa*. *J. Nanobiotechnol.* **2014**, *12*, 40. [CrossRef]
89. Raut, R.W.; Mendhulkar, V.D.; Kashid, S.B. Photosensitized synthesis of silver nanoparticles using *Withania somnifera* leaf powder and silver nitrate, *J. Photochem. Photobiol. B* **2014**, *132*, 45–55. [CrossRef]
90. Nakkala, J.R.; Mata, R.; Gupta, A.K.; Sadras, S.R. Biological activities of green silver nanoparticles synthesized with *Acorous calamus* rhizome extract, *Eur. J. Med. Chem.* **2014**, *85*, 784–794. [CrossRef]
91. Mariselvam, R.; Ranjitsingh, A.; Nanthini, A.U.R.; Kalirajan, K.; Padmalatha, C.; Selvakumar, P.M. Green synthesis of silver nanoparticles from the extract of the inflorescence of *Cocos nucifera* (Family: *Arecaceae*) for enhanced antibacterial activity. *Spectrochim. Acta Part A Mol. Biomol. Spectrosc.* **2014**, *129*, 537–541. [CrossRef] [PubMed]
92. Kumar, P.V.; Pammi, S.; Kollu, P.; Satyanarayana, K.; Shameem, U. Green synthesis and characterization of silver nanoparticles using *Boerhaavia diffusa* plant extract and their anti bacterial activity. *Ind. Crops Prod.* **2014**, *52*, 562–566. [CrossRef]
93. Nazeruddin, G.M.; Prasad, N.R.; Waghmare, S.R.; Garadkar, K.M.; Mulla, I.S. Extracellular biosynthesis of silver nanoparticle using *Azadirachta indica* leaf extract and its anti-microbial activity. *J. Alloy. Compd.* **2014**, *583*, 272–277. [CrossRef]
94. Nazeruddin, G.M.; Prasad, N.R.; Prasadd, S.R.; Shaikha, Y.I.; Waghmare, S.R.; Adhyapak, P. *Coriandrum sativum* seed extract assisted in situ green synthesis ofsilver nanoparticle and its anti-microbial activity. *Ind. Crops Prod.* **2014**, *60*, 212–216. [CrossRef]
95. Bindhu, M.R.; Umadevi, M. Synthesis of monodispersed silver nanoparticles using *Hibiscus cannabinus* leaf extract and its antimicrobial activity. *Spectrochim. Acta A Mol. Biomol. Spectrosc.* **2013**, *101*, 184–190. [CrossRef]
96. Patil, R.S.; Kokate, M.R.; Kolekar, S.S. Bioinspired synthesis of highly stabilized silver nanoparticles using *Ocimum tenuiflorum* leaf extract and their antibacterial activity. *Spectrochim. Acta A* **2012**, *91*, 234–238. [CrossRef]
97. Gopinath, V.; MubarakAli, D.; Priyadarshini, S.; Priyadharsshini, N.M.; Thajuddin, N.; Velusamy, P. Biosynthesis of silver nanoparticles from *Tribulus terrestris* and its antimicrobial activity: A novel biological approach. *Colloids Surf. B Biointerfaces* **2012**, *96*, 69–74. [CrossRef]
98. Sivakumar, P.; Nethradevi, C.; Renganathan, S. Synthesis of silver nanoparticles using *Lantana camara* fruit extract and its effect on pathogens, *Asian J. Pharm. Clin. Res.* **2012**, *5*, 97–101.
99. Sathishkumar, G.; Gobinatha, C.; Karpagama, K.; Hemamalini, V.; Premkumar, K.; Sivaramakrishnan, S. Phyto-synthesis of silver nanoscale particles using *Morinda citrifolia* L. and its inhibitory activity against human pathogens. *Colloids Surf. B Biointerfaces* **2012**, *95*, 235–240. [CrossRef]
100. Kumar, K.M.; Sinha, M.; Mandal, B.K.; Ghosh, A.R.; Siva Kumar, K.; Sreedhara Reddy, P. Green synthesis of silver nanoparticles using *Terminalia chebula* extract at room temperature and their antimicrobial studies. *Spectrochim. Acta A Mol. Biomol. Spectrosc.* **2012**, *91*, 228–233. [CrossRef]
101. Amin, M.; Anwar, F.; Janjua, M.R.S.A.; Iqbal, M.A.; Rashid, U. Green Synthesis of Silver Nanoparticles through Reduction with *Solanum xanthocarpum* L. Berry Extract: Characterization, Antimicrobial and Urease Inhibitory Activities against *Helicobacter pylori*. *Int. J. Mol. Sci.* **2012**, *13*, 9923–9941. [CrossRef] [PubMed]
102. Ghosh, S.; Patil, S.; Ahire, M.; Kitture, R.; Jabgunde, A.; Kale, S.; Pardesi, K.; Cameotra, S.S.; Bellare, J.; Dhavale, D.D.; et al. Synthesis of silver nanoparticles using *Dioscorea bulbifera* tuber extract and evaluation of its synergistic potential in combination with antimicrobial agents. *Int. J. Nanomed.* **2012**, *7*, 483–496.

103. Veerasamy, R.; Xin, T.Z.; Gunasagaran, S.; Xiang, T.F.W.; Yang, E.F.C.; Jeyakumar, N.; Dhanaraj, S.A. Biosynthesis of silver nanoparticles using mangosteen leaf extract and evaluation of their antimicrobial activities. *J. Saudi Chem. Society.* **2011**, *15*, 113–120. [CrossRef]
104. Masurkar, S.A.; Chaudhari, P.R.; Shidore, V.B.; Kamble, S.P. Rapid biosynthesis of silver nanoparticles using *Cymbopogan Citratus* (Lemongrass) and its antimicrobial activity. *Nano-Micro Lett.* **2011**, *3*, 189–194. [CrossRef]
105. Nabikhan, A.; Kandasamy, K.; Raj, A.; Alikunhi, N.M. Synthesis of antimicrobial silver nanoparticles by callus and leaf extracts from saltmarsh plant, *Sesuvium portulacastrum* L. *Colloids Surf. B* **2010**, *79*, 488–493. [CrossRef]
106. Huq, M.A.; Akter, S. Biosynthesis, characterization and antibacterial application of novel silver nanoparticles against drug resistant pathogenic *Klebsiella pneumoniae* and *Salmonella enteritidis*. *Molecules* **2021**, *26*, 5996. [CrossRef] [PubMed]
107. Eltarahony, M.; Ibrahim, A.; El-shall, H.; Ibrahim, E.; Althobaiti, F.; Fayad, E. Antibacterial, antifungal and antibiofilm activities of silver nanoparticles supported by crude bioactive metabolites of bionanofactories isolated from lake mariout. *Molecules* **2021**, *26*, 3027. [CrossRef]
108. Singh, P.; Pandit, S.; Jers, C.; Abhayraj, S.; Garnæs, J.; Mijakovic, I. Silver nanoparticles produced from *Cedecea* sp. exhibit antibiofilm activity and remarkable stability. *Sci. Rep.* **2021**, *11*, 12619. [CrossRef]
109. Huq, M.A.; Akter, S. Characterization and genome analysis of *Arthrobacter bangladeshi* sp. nov., applied for the green synthesis of silver nanoparticles and their antibacterial efficacy against drug-resistant human pathogens. *Pharmaceutics* **2021**, *13*, 1691. [CrossRef]
110. Vellingiri, M.M.; Ashwin, J.K.M.; Soundari, A.J.P.G.; Sathiskumar, S.; Priyadharshini, U.; Paramasivam, D.; Liu, W.C.; Balasubramanian, B. Mycofabrication of AgONPs derived from *Aspergillus terreus* FC36AY1 and its potent antimicrobial, antioxidant, and anti-angiogenesis activities. *Mol. Biol. Rep.* **2021**, *48*, 7933–7946. [CrossRef]
111. Soliman, A.M.; Abdel-Latif, W.; Shehata, I.H.; Fouda, A.; Abdo, A.M.; Ahmed, Y.M. Green approach to overcome the resistance pattern of *Candida* spp. using biosynthesized silver nanoparticles fabricated by *Penicillium chrysogenum* F9. *Biol. Trace Elem. Res.* **2021**, *199*, 800–811. [CrossRef] [PubMed]
112. Raza, S.; Ansari, A.; Siddiqui, N.N.; Ibrahim, F.; Abro, M.I.; Aman, A. Biosynthesis of silver nanoparticles for the fabrication of non cytotoxic and antibacterial metallic polymer based nanocomposite system. *Sci. Rep.* **2021**, *11*, 10500. [CrossRef] [PubMed]
113. Sreenivasa, N.; Meghashyama, B.P.; Pallavi, S.S.; Bidhayak, C.; Dattatraya, A.; Muthuraj, R.; Shashiraj, K.N.; Halaswamy, H.; Dhanyakumara, S.B.; Vaishnavi, M.D. Biogenic synthesis of silver nanoparticles using *Paenibacillus* sp. in-vitro and their antibacterial, anticancer activity assessment against human colon tumour cell line. *J. Environ. Biol.* **2021**, *42*, 118–127. [CrossRef]
114. Huq, M.A. Biogenic silver nanoparticles synthesized by *Lysinibacillus xylanilyticus* MAHUQ-40 to control antibiotic-resistant human pathogens vibrio parahaemolyticus and salmonella typhimurium. *Front. Bioeng. Biotechnol.* **2020**, *8*, 1407. [CrossRef] [PubMed]
115. Hamida, R.S.; Abdelmeguid, N.E.; Ali, M.A.; Bin-Meferij, M.M.; Khalil, M.I. Synthesis of silver nanoparticles using a novel cyanobacteria *Desertifilum* sp. extract: Their antibacterial and cytotoxicity effects. *Int. J. Nanomed.* **2020**, *15*, 49–63. [CrossRef]
116. Borah, D.; Das, N.; Das, N.; Bhattacharjee, A.; Sarmah, P.; Ghosh, K.; Chandel, M.; Rout, J.; Pandey, P.; Ghosh, N.N.; et al. Alga-mediated facile green synthesis of silver nanoparticles: Photophysical, catalytic and antibacterial activity. *Appl. Organ. Chem.* **2020**, *34*, e5597. [CrossRef]
117. Mondal, A.H.; Yadav, D.; Ali, A.; Khan, N.; Jin, J.O.; Haq, Q.M.R. Anti-bacterial and anti-candidal activity of silver nanoparticles biosynthesized using *Citrobacter* spp. Ms5 culture supernatant. *Biomolecules* **2020**, *10*, 944. [CrossRef]
118. Bhuyar, P.; Rahim, M.H.A.; Sundararaju, S.; Ramaraj, R.; Maniam, G.P.; Govindan, N. Synthesis of silver nanoparticles using marine macroalgae Padina sp. and its antibacterial activity towards pathogenic bacteria. *Beni-Suef Univ. J. Basic Appl. Sci.* **2020**, *9*, 3. [CrossRef]
119. Mishra, B.; Saxena, A.; Tiwari, A. Biosynthesis of silver nanoparticles from marine diatoms *Chaetoceros* sp., *Skeletonema* sp., *Thalassiosira* sp., and their antibacterial study. *Biotechnol. Rep.* **2020**, *28*, e00571. [CrossRef] [PubMed]
120. Feroze, N.; Arshad, B.; Younas, M.; Afridi, M.I.; Saqib, S.; Ayaz, A. Fungal mediated synthesis of silver nanoparticles and evaluation of antibacterial activity. *Microsc. Res. Tech.* **2020**, *83*, 72–80. [CrossRef]
121. Mohd-Yusof, H.; Rahman, A.; Mohamad, R.; Zaidan, U.H. Microbial mediated synthesis of silver nanoparticles by *Lactobacillus Plantarum* TA4 and its antibacterial and antioxidant activity. *Appl. Sci.* **2020**, *10*, 6973. [CrossRef]
122. Saeed, S.; Iqbal, A.; Ashraf, M.A. Bacterial-mediated synthesis of silver nanoparticles and their significant effect against pathogens. *Environ. Sci. Pollut. Res.* **2020**, *27*, 37347–37356. [CrossRef] [PubMed]
123. Alsamhary, K.I. Eco-friendly synthesis of silver nanoparticles by *Bacillus subtilis* and their antibacterial activity. *Saudi J. Biol. Sci.* **2020**, *27*, 2185–2191. [CrossRef] [PubMed]
124. Shaker, M.A.; Shaaban, M.I. Synthesis of silver nanoparticles with antimicrobial and anti-adherence activities against multidrug-resistant isolates from *Acinetobacter Baumannii*. *J. Taibah Univ. Med. Sci.* **2017**, *12*, 291–297. [CrossRef]
125. Singh, H.; Du, J.; Singh, P.; Yi, T.H. Extracellular synthesis of silver nanoparticles by *Pseudomonas* sp. THG-LS1.4 and their antimicrobial application. *J. Pharm. Anal.* **2018**, *8*, 258–264. [CrossRef]
126. Du, J.; Sing, H.; Yi, T.H. Biosynthesis of silver nanoparticles by *Novosphingobium* sp. THG-C3 and their antimicrobial potential. *Artif. Cells Nanomed. Biotechnol.* **2017**, *45*, 211–217. [CrossRef]
127. Singh, P.; Singh, H.; Kim, Y.J. Extracellular synthesis of silver and gold nanoparticles by *Sporosarcina koreensis* DC4 and their biological applications. *Enzym. Microb. Technol.* **2016**, *86*, 75–83. [CrossRef]

128. Deljou, A.; Goudarzi, S. Green extracellular synthesis of the silver nanoparticles using thermophilic *Bacillus* Sp. AZ1 and its antimicrobial activity against several human pathogenetic bacteria. *Iran. J. Biotechnol.* **2016**, *14*, 25–32. [CrossRef]
129. Singh, H.; Du, J.; Yi, T.H. Biosynthesis of silver nanoparticles using *Aeromonas* sp. THG-FG1.2 and its antibacterial activity against pathogenic microbes. *Artif. Cells Nanomed. Biotechnol.* **2017**, *45*, 584–590. [CrossRef]
130. Singh, H.; Du, J.; Yi, T.H. *Kinneretia* THG-SQI4 mediated biosynthesis of silver nanoparticles and its antimicrobial efficacy. *Artif. Cells Nanomed. Biotechnol.* **2017**, *45*, 602–608. [CrossRef]
131. Lateef, A.; Adelere, I.A.; Gueguim-Kana, E.B.; Asafa, T.B.; Beukes, L.S. Green synthesis of silver nanoparticles using keratinase obtained from a strain of *Bacillus safensis* LAU 13. *Int. Nano Lett.* **2015**, *5*, 29–35. [CrossRef]
132. Rajeshkumar, S.; Paulkumar, K.; Gnanajobitha, G.; Chitra, K.; Malarkodi, C.; Annadurai, G. Fungal assisted intracellular and enzyme based synthesis of silver nanoparticles and its bactericidal efficiency. *Int. Res. J. Pharm. Biosci.* **2015**, *2*, 8–19.
133. Singh, P.; Kim, Y.J.; Wang, C.; Mathiyalagan, R.; Yang, D.C. *Weissella oryzae* DC6-facilitated green synthesis of silver nanoparticles and their antimicrobial potential. *Artif. Cells Nanomed. Biotechnol.* **2015**, *44*, 1569–1575. [CrossRef] [PubMed]
134. Wang, C.; Singh, P.; Kim, Y.J.; Mathiyalagan, R.; Myagmarjav, D.; Wang, D.; Jin, C.-G.; Yang, D.C. Characterization and antimicrobial application of biosynthesized gold and silver nanoparticles by using *Microbacterium resistens*. *Artif. Cells Nanomed. Biotechnol.* **2016**, *44*, 1714–1721. [CrossRef] [PubMed]
135. Wang, C.; Kim, Y.J.; Singh, P.; Mathiyalagan, R.; Jin, Y.; Yang, D.C. Green synthesis of silver nanoparticles by *Bacillus methylotrophicus*, and their antimicrobial activity. *Artif. Cells Nanomed. Biotechnol.* **2015**, *44*, 1127–1132. [PubMed]
136. Jo, J.H.; Singh, P.; Kim, Y.J.; Wang, C.; Mathiyalagan, R.; Jin, C.-G.; Yang, D.C. *Pseudomonas deceptionensis* DC5-mediated synthesis of extracellular silver nanoparticles. *Artif. Cells Nanomed. Biotechnol.* **2015**, *44*, 1576–1581. [CrossRef]
137. Singh, P.; Kim, Y.J.; Singh, H.; Mathiyalagan, R.; Wang, C.; Yang, D.C. Biosynthesis of anisotropic silver nanoparticles by *Bhargavaea indica* and their synergistic effect with antibiotics against pathogenic microorganisms. *J. Nanomater.* **2015**, *2*, 4.
138. Abdeen, S.; Geo, S.; Praseetha, P.K.; Dhanya, R.P. Biosynthesis of silver nanoparticles from Actinomycetes for therapeutic applications. *Int. J. Nano Dimen.* **2014**, *5*, 155–162.
139. Priyadarshini, S.; Gopinath, V.; Priyadharsshini, N.M.; Mubarak, A.D.; Velusamy, P. Synthesis of anisotropic silver nanoparticles using novel strain, *Bacillus flexus* and its biomedical application. *Coll. Surf. B Biointerface* **2013**, *102*, 232–237. [CrossRef]
140. Tomaszewska, E.; Soliwoda, K.; Kadziola, K.; Tkacz-Szczesna, B.; Celichowski, G.; Cichomski, M.; Szmaja, W.; Grobelny, J. Detection limits of DLS and UV-Vis spectroscopy in characterization of polydisperse nanoparticles colloids. *J. Nanomater.* **2013**, *2013*, 313081. [CrossRef]
141. Rajeshkumar, S.; Bharath, L.V. Mechanism of plant-mediated synthesis of silver nanoparticles e A review on biomolecules involved, characterisation and antibacterial activity. *Chem.-Biol. Interact.* **2017**, *273*, 219–227. [CrossRef] [PubMed]
142. Anees-Ahmad, S.; Sachi-Das, S.; Khatoon, A.; Ansari, M.T.; Afzal, M.; Saquib-Hasnain, M.; Kumar-Nayak, A. Bactericidal activity of silver nanoparticles: A mechanistic review. *Mater. Sci. Energy Technol.* **2020**, *3*, 756–769. [CrossRef]
143. Mahendra, R.; Yadav, A.; Gade, A. Silver nanoparticles as a new generation of antimicrobials. *Biotechnol. Adv.* **2009**, *27*, 76–83.
144. Kim, J.S.; Kuk, E.; Yu, K.N.; Kim, J.-H.; Park, S.J.; Lee, H.J.; Kim, S.H.; Park, Y.K.; Park, Y.H.; Hwang, C.-Y.; et al. Antimicrobial effects of silver nanoparticles. *Nanomed. Nanotechnol. Biol. Med.* **2007**, *3*, 95–101. [CrossRef] [PubMed]
145. Raffi, M.; Hussain, F.; Bhatti, T.M.; Akhter, J.I.; Hameed, A.; Hasan, M.M. Antibacterial characterization of silver nanoparticles against *E. coli* ATCC-15224. *J. Mater. Sci. Technol.* **2008**, *24*, 192–196.
146. Goswami, S.R.; Sahareen, T.; Singh, M.; Kumar, S. Role of biogenic silver nanoparticles in disruption of cell-cell adhesion in *Staphylococcus aureus* and *Escherichia coli* biofilm. *J. Ind. Eng. Chem.* **2015**, *26*, 73–80. [CrossRef]
147. Allahverdiyev, A.M.; Kon, K.V.; Abamor, E.S.; Bagirova, M.; Rafailovich, M. Coping with antibiotic resistance: Combining nanoparticles with antibiotics and other antimicrobial agents. Expert Rev. Anti. Infect. Ther. **2011**, *9*, 1035–1052. [CrossRef]
148. Sekhon, B.S. Metalloantibiotics and antibiotic mimics-an overview. *J. Pharm. Educ. Res.* **2010**, *1*, 1.
149. Das, S.S.; Neelam; Hussain, K.S.; Singh, S.; Hussain, A.; Faruk, A.; Tebyetekerwa, M. Laponite-based nanomaterials for biomedical applications: A review. *Curr. Pharm. Des.* **2019**, *25*, 424–443. [CrossRef]
150. Shetti, N.P.; Nayak, D.S.; Malode, S.J.; Reddy, K.R.; Shukla, S.S.; Aminabhavi, T.M. Electrochemical behavior of flufenamic acid at amberlite XAD-4 resin and silver-doped titanium dioxide/amberlite XAD-4 resin modified carbon electrodes. *Colloids Surf. B Biointerfaces* **2019**, *177*, 407–415. [CrossRef]
151. Bapat, R.A.; Chaubal, T.V.; Joshi, C.P.; Bapat, P.R.; Choudhury, H.; Pandey, M.; Gorain, B.; Kesharwani, P. An overview of application of silver nanoparticles for biomaterials in dentistry. *Mater. Sci. Eng. C* **2018**, *91*, 881–898. [CrossRef] [PubMed]
152. Chouhan, S.; Guleria, S. Green synthesis of AgNPs using *Cannabis sativa* leaf extract: Characterization, antibacterial, anti-yeast and α-amylase inhibitory activity. *Mater. Sci. Energy Technol.* **2020**, *3*, 536–544. [CrossRef]
153. Khorrami, S.; Zarrabi, A.; Khaleghi, M.; Danaei, M.; Mozafari, M. Selective cytotoxicity of green synthesized silver nanoparticles against the MCF-7 tumor cell line and their enhanced antioxidant and antimicrobial properties. *Int. J. Nanomed.* **2018**, *13*, 8013–8024. [CrossRef] [PubMed]
154. Das, S.S.; Alkahtani, S.; Bharadwaj, P.; Ansari, M.T.; ALKahtani, M.D.F.; Pang, Z.; Hasnain, M.S.; Nayak, A.K.; Aminabhavi, T.M. Molecular insights and novel approaches for targeting tumor metastasis. *Int. J. Pharm.* **2020**, *585*, 119556. [CrossRef]
155. Pareek, V.; Gupta, R.; Panwar, J. Do physico-chemical properties of silver nanoparticles decide their interaction with biological media and bactericidal action? A Review. *Mater. Sci. Eng. C Mater. Biol. Appl.* **2018**, *90*, 739–749. [CrossRef] [PubMed]

156. Liao, C.; Li, Y.; Tjong, S.C. Bactericidal and cytotoxic properties of silver nanoparticles. *Int. J. Mol. Sci.* **2019**, *20*, 449. [CrossRef] [PubMed]
157. Ahmed, K.B.R.; Nagy, A.M.; Brown, R.P.; Zhang, Q.; Malghan, S.G.; Goering, P.L. Silver nanoparticles: Significance of physico-chemical properties and assay interference on the interpretation of in vitro cytotoxicity studies. *Toxicol. Vitr.* **2017**, *38*, 179–192. [CrossRef] [PubMed]
158. Noronha, V.T.; Paula, A.J.; Duran, G.; Galembeck, A.; Cogo-Mueller, K.; Franz-Montan, M.; Duran, N. Silver nanoparticles in dentistry. *Dent. Mater.* **2017**, *33*, 1110–1126. [CrossRef]
159. Soumya, E.A.; Saad, I.K.; Hassan, L.; Ghizlane, Z.; Hind, M.; Adnane, R. Carvacrol and thymol components inhibiting *Pseudomonas aeruginosa* adherence and biofilm formation. *Afr. J. Microbiol. Res.* **2011**, *5*, 3229–3232.
160. Park, J.; Lim, D.H.; Lim, H.J.; Kwon, T.; Choi, J.S.; Jeong, S.; Choi, I.H.; Cheon, J. Size dependent macrophage responses and toxicological effects of Ag nanoparticles. *Chem. Commun.* **2011**, *47*, 4382–4384. [CrossRef]
161. Jacob, J.M.; John, M.S.; Jacob, A.; Abitha, P.; Kumar, S.S.; Rajan, R.; Natarajan, S.; Pugazhendhi, A. Bactericidal coating of paper towels via sustainable biosynthesis of silver nanoparticles using *Ocimum sanctum* leaf extract. *Mater. Res. Express* **2019**, *6*, 45401. [CrossRef]
162. Saravanan, M.; Arokiyaraj, S.; Lakshmi, T.; Pugazhendhi, A. Synthesis of silver nanoparticles from *Phenerochaete chrysosporium* (MTCC-787) and their antibacterial activity Antibacterial Activity against Human Pathogenic Bacteria. *Microb. Pathog.* **2018**, *117*, 68–72. [CrossRef] [PubMed]

MDPI AG
Grosspeteranlage 5
4052 Basel
Switzerland
Tel.: +41 61 683 77 34

Polymers Editorial Office
E-mail: polymers@mdpi.com
www.mdpi.com/journal/polymers

Disclaimer/Publisher's Note: The title and front matter of this reprint are at the discretion of the Guest Editors. The publisher is not responsible for their content or any associated concerns. The statements, opinions and data contained in all individual articles are solely those of the individual Editors and contributors and not of MDPI. MDPI disclaims responsibility for any injury to people or property resulting from any ideas, methods, instructions or products referred to in the content.

www.ingramcontent.com/pod-product-compliance
Lightning Source LLC
LaVergne TN
LVHW072319090526
838202LV00019B/2308